Lecture Notes on Data Engineering and Communications Technologies

Volume 138

T0200208

Series Editor

Fatos Xhafa, Technical University of Catalonia, Barcelona, Spain

The aim of the book series is to present cutting edge engineering approaches to data technologies and communications. It will publish latest advances on the engineering task of building and deploying distributed, scalable and reliable data infrastructures and communication systems.

The series will have a prominent applied focus on data technologies and communications with aim to promote the bridging from fundamental research on data science and networking to data engineering and communications that lead to industry products, business knowledge and standardisation.

Indexed by SCOPUS, INSPEC, EI Compendex.

All books published in the series are submitted for consideration in Web of Science.

More information about this series at https://link.springer.com/bookseries/15362

Vijayan Sugumaran · A. G. Sreedevi ·
Zheng Xu
Editors

Application of Intelligent Systems in Multi-modal Information Analytics

The 4th International Conference
on Multi-modal Information Analytics
(ICMMIA 2022), Volume 2

Set 1

Editors
Vijayan Sugumaran
School of Business Administration
Oakland University
Rochester, MI, USA

A. G. Sreedevi
Amrita School of Engineering
Chennai, Tamil Nadu, India

Zheng Xu
Shanghai Polytechnic University
Shanghai, China

ISSN 2367-4512 ISSN 2367-4520 (electronic)
Lecture Notes on Data Engineering and Communications Technologies
ISBN 978-3-031-05483-9 ISBN 978-3-031-05484-6 (eBook)
https://doi.org/10.1007/978-3-031-05484-6

This Springer imprint is published by the registered company Springer Nature Switzerland AG
The registered company address is: Gewerbestrasse 11, 6330 Cham, Switzerland

Foreword

We are living in the era of data deluge. The world of big data exhibits a rich and complex set of cross-media contents, such as text, image, video, audio, and graphics. So far, great research efforts have been dedicated separately to big data processing and cross-media mining, with well-grounded theoretical underpinnings and great practical success. However, studies jointly considering cross-media big data analytics are relatively sparse. This research gap needs our further attention, since it will benefit a variety of real-world applications. Despite its significance and value, it is non-trivial to analyze cross-media big data due to their heterogeneity, large-scale volume, increasing size, unstructured nature, correlations, and noise. Multi-modal Information systems, which can be treated as the most significant breakthrough in the past 10 years, have greatly impacted the methodologies utilized in computer vision and achieved terrific progress in both academia and industry.

Building on the previous years' successes of online meeting (2021 and 2020 due to COVID-19), and in-person meeting in Shenyang, China (2019), the 4th International Conference on Multi-modal Information Analytics (ICMMIA 2022) is proud to be in the fourth consecutive conference year.

We would like to express our thanks to Professor Vijayan Sugumaran, Oakland University and Dr. A. G. Sreedevi, Amrita School of Engineering, for being the keynote speakers at the conference. We thank the General Chairs, Program Committee Chairs, Organizing Chairs, and Workshop Chairs for their hard work. The local organizers' and the students' help are also highly appreciated.

Our special thanks are also due to editors Dr. Thomas Ditzinger and Suresh Dharmalingam for publishing the proceedings in LNDECT book series of Springer.

Vijayan Sugumaran
A. G. Sreedevi
Zheng Xu

Organization

General Chair

Tharam Dillon   La Trobe University, Australia

Program Chairs

Vijayan Sugumaran   Oakland University, USA
A. G. Sreedevi   Amrita School of Engineering, Chennai Campus, India
Zheng Xu   Shanghai University, China

Publication Chairs

Juan Du   Shanghai University, China
Junyu Xuan   University of Technology Sydney, Australia

Publicity Chairs

Shunxiang Zhang   Anhui University of Science and Technology, China
Neil. Y. Yen   University of Aizu, Japan

Program Committee Members

William Bradley Glisson   University of South Alabama, USA
George Grispos   University of Limerick, Ireland
Abdullah Azfar   KPMG Sydney, Australia
Aniello Castiglione   Università di Salerno, Italy
Florin Pop   University Politehnica of Bucharest, Romania
Ben Martini   University of South Australia, Australia

Wei Wang	The University of Texas at San Antonio, USA
Neil Yen	University of Aizu, Japan
Meng Yu	The University of Texas at San Antonio, USA
Shunxiang Zhang	Anhui Univ. of Sci. & Tech., China
Guangli Zhu	Anhui Univ. of Sci. & Tech., China
Tao Liao	Anhui Univ. of Sci. & Tech., China
Xiaobo Yin	Anhui Univ. of Sci. & Tech., China
Xiangfeng Luo	Shanghai Univ., China
Xiao Wei	Shanghai Univ., China
Huan Du	Shanghai Univ., China
Zhiguo Yan	Fudan University, China
Rick Church	UC Santa Barbara, USA
Tom Cova	University of Utah, USA
Susan Cutter	University of South Carolina, USA
Zhiming Ding	Beijing University of Technology, China
Yong Ge	University of North Carolina at Charlotte, USA
T. V. Geetha	Anna University, India
Danhuai Guo	Computer Network Information Center, Chinese Academy of Sciences, China
Jeng-Neng Hwang	University of Washington, USA
Jianping Fang	University of North Carolina at Charlotte, USA
Jianhui Li	Computer Network Information Center, Chinese Academy of Sciences, China
Yi Liu	Tsinghua University, China
Foluso Ladeinde	SUNU, South Korea
Kuien Liu	Pivotal Inc., USA
Feng Lu	Institute of Geographic Science and Natural Resources Research, Chinese Academy of Sciences, China
Ricardo J. Soares Magalhaes	University of Queensland, Australia
D. Manjula	Anna University, India
Alan Murray	Drexel University, USA
S. Murugan	Sathyabama Institute of Science and Technology, India
Yasuhide Okuyama	University of Kitakyushu, Japan
S. Padmavathi	Amrita University, India
Latha Parameswaran	Amrita University, India
S. Suresh	SRM University, India
Wei Xu	Renmin University of China, China
Chaowei Phil Yang	George Mason University, USA
Enwu Yin	China CDC, USA
Hengshu Zhu	Baidu Inc., China
Morshed Chowdhury	Deakin University, Australia

Min Hu   Shanghai University, China
Gang Luo   Shanghai University, China
Juan Chen   Shanghai University, China
Qigang Liu   Shanghai University, China

ICMMIA 2022 Keynotes

Vijayan Sugumaran is Professor of Management Information Systems and Chair of the Department of Decision and Information Sciences at Oakland University, Rochester, Michigan, USA. He is also Co-Director of the Center for Data Science and Big Data Analytics at Oakland University. He received his Ph.D. in Information Technology from George Mason University, Fairfax, Virginia, USA. His research interests are in the areas of big data management and analytics, ontologies and semantic web, intelligent agent and multi-agent systems. He has published over 270 peer-reviewed articles in journals, conferences, and books. He has edited twenty books and serves on Editorial Board of eight journals. He has published in top-tier journals such as Information Systems Research, ACM Transactions on Database Systems, Communications of the ACM, IEEE Transactions on Big Data, IEEE Transactions on Engineering Management, IEEE Transactions on Education, and IEEE Software. He is the editor-in-chief of the International Journal of Intelligent Information Technologies. He is Chair of the Intelligent Agent and Multi-Agent Systems mini-track for Americas Conference on Information Systems (AMCIS 1999–2022). He has served as Program Chair for the 14th Workshop on E-Business (WeB2015), the International Conference on Applications of Natural Language to Information Systems (NLDB 2008, NLDB 2013, NLDB 2016, and NLDB 2019), the 29th Australasian Conference on Information Systems

(ACIS 2018), the 14th Annual Conference of Midwest Association for Information Systems, 2019, the 5th IEEE International Conference on Big Data Service and Applications, 2019, and the Midwest Decision Sciences Institute Annual Conference (MWDSI 2022). He also regularly serves as a program committee member for numerous national and international conferences.

Dr. A. G. Sreedevi is currently serving as Program Coordinator for Cyber Security and Assistant Professor in the Department of Computer Science and Engineering, Amrita Vishwa Vidyapeetham, Chennai, India. She received her Ph.D. from SRM Institute of Science and Technology, Chennai, India, in the area of Telecommunication Engineering, specializing in reinforcement learning. She has seven years of research and three years of academic experience. She was invited as an academic guest by ETH Switzerland where she worked with their communication research group in simulating UAVs as part of the SWARMIX Project. She is currently working on designing Multi-Agent Systems for Weather Forecasting and malware analyses. She has published 16 research papers and one patent. Her other research interests include reinforcement learning, cyber security, and device-to-device communications.

Contents

Multi-modal Informatics in Industrial, Robot, and Smart City

Design and Application of Grid Management Database Based on Data Sharing Technology

Li Tang[✉]

Basic Teaching Department of Lanzhou Resources and Environment Vocational and Technical University, Lanzhou 730021, Gansu, China
lc860513@163.com

Abstract. The grid management system borrows the concept of the grid management system in the computer, divides the management object into a series of grid units according to certain specifications, and uses the latest computer technology and the coordination mechanism between different grid units. Each grid is just a kind of unit. Manage the resources of the organization in a transparent manner, and finally achieve the latest management concept of collecting and managing the resources of the organization. This paper aims to study the design and application of grid management database based on data sharing technology. On the basis of analyzing the characteristics of grid management, the requirements of grid management and the method of business submission and sorting, combined with data sharing technology, the network is designed. The database is managed in a grid and its performance is tested. The test results show that the login page, log information page, information query page, and map browsing page tested in this article perform well in terms of response time, CPU usage, and memory usage, and basically meet the needs of this article.

Keywords: Data sharing technology · Grid management · Database design · Design and implementation

1 Introduction

The level of informatization of a city has long formed one of the main indicators for judging a city's national comprehensive national strength, modernization degree, international competitiveness, and even the overall national economic development capability [1, 2]. Since entering the 21st century, the vigorous development of the electronic information industry has also promoted the vigorous development of urban communication technology, but the management methods of various communication industries cannot keep up with the development of the city [3, 4].

With the progress of information system realization technology, many fields are studying in depth how to apply system technology development. As a new method of applying management concepts, gridding is actually an important result of research and application of information technology. In recent years, many developed countries have begun to scramble to develop and use this technology. Some researchers pointed out that grid computing will eventually achieve general computing functions that are completely unaffected by conditions and geography, thereby reducing resource islands

© The Author(s), under exclusive license to Springer Nature Switzerland AG 2022
V. Sugumaran et al. (Eds.): ICMMIA 2022, LNDECT 138, pp. 3–10, 2022.
https://doi.org/10.1007/978-3-031-05484-6_1

and achieving a high degree of integration and sharing [5, 6]. In EU countries, grid-related data has been investigated for two years. The vigorous promotion of the British National Grid aims to use the grid to connect the high-performance server resources of many key domestic research and development institutions. Market-oriented partnerships are more open to user capabilities. In addition, the nomadic project in the UK is mainly to improve the service platform and the effective management of the municipal authorities, and through the integration of the management experience of different cities and governments on mobile government services, the overall theoretical and technological changes have been realized [7, 8]. Some researchers have provided a grid method to evaluate and analyze the public logistics management related policies of various local cities in France, including the determinants of urban public logistics policies, and provide some guidance for local municipalities and enterprises in urban logistics management [9, 10]. Some researchers have put forward an application platform for urban forestry management applications developed by combining geographic information grids and MVC models to effectively manage urban forestry projects and reduce management costs. Spatial data connectivity can further contribute to urban forest management for researchers and community management services, and provide better living conditions for the urban population [11, 12]. However, in response to the above development, foreign grid technology is not very common in the comprehensive management of civil society due to different national conditions.

Based on a large number of "grid management" and "data sharing technology", this paper designs a grid management database based on the related concepts of grid management. The database is mainly composed of four parts, which are user permissions. Database, log information database, basic information database and map information database, and finally tested the performance of the database.

2 Design and Application of Grid Management Database Based on Data Sharing Technology

2.1 Features of Grid Management

(1) Grid management is to achieve a hierarchical management mode by dividing geographical features. For example, divide the area into four levels, clarify the level of management responsibilities, monitor all the doors in real time, solve the regional management problems in a timely manner, and realize true, active, and prudent regional management.

(2) Grid management adopts geographic classification and coding rules to realize the classified management of regional resource sub-elements. After the surrounding objects are classified and subdivided through inventory, the geocoding of each regional resource object is recorded on the map, the data is structured, and everything is imported into the computer environment for management. After the classification and sub-processing of the geographic classification coding technology are adopted in this way, the content management of the local natural resources can be fully refined, and the responsible unit can be added according to the classification technology and local management, and it can be extended to accurate type conversion.

(3) Grid management ideas can be used in the context of digital city technology to create new ways of sending information in real time. Use the most advanced wireless technology and computer information management technology for integrated management of peripheral resource objects. Integrating information resources to realize the rapid transmission and exchange of spatial information is the modernization of regional positioning resource management. The organic combination of mobile communication services and wireless technology can be used to display a full range of information resource collection to ensure real-time management and monitoring of natural objects in the area.

(4) The application characteristic of grid management is that according to the rules of e-government, the integration of government functions is based on the principle of separation of management, evaluation and monitoring. It can effectively solve the problems of multiple management, interrelated responsibilities, and incomplete responsibilities of specialized management departments in local resource management. According to the changes in new models and the modern evolution of land resource management based on geographic grids, the settings and functional planning of professional departments can be adjusted in time.

3 Requirements for Grid Management

(1) A three-level management system of reasonable division of labor, division of responsibilities, township, community, and grid has been formed. Grid management is the main part of grid partitioning. According to the national geographic distribution and demographic statistics, it is responsible for a reasonable range and divides the area into several grids. The general principle of the grid is "full coverage, clearly defined, no gaps, no overlaps". Therefore, grid partitioning is necessary to facilitate work, accountability, site surveys, and provide direct communication services.

(2) Investigating baseline conditions and establishing a comprehensive system covering business planning and comprehensive management are the core tasks of grid management. Make all tasks clearer, smarter and more convenient. It is necessary to establish a scientific and effective information collection and supervision mechanism, feedback and monitoring systems, real-world accurate information, channels and timely feedback to save costs, centralize control and better serve the public.

(3) Vigorously develop and continuously improve the level of national informatization construction, and establish an effective network platform to operate and share grid management information resources. The grid management platform creates a set of services for the smallest administrative unit, community or village. Integrating marital status, house, family, education, occupation, etc. information on a computer grid platform will help people understand the time and proficiency of the project, and dynamically, conveniently, and accurately classify and summarize. At the same time, we attach importance to the daily maintenance and update of information to enhance the effect of information application. Focus on intelligent application management to facilitate effective notification of personnel and improve work efficiency.

4   Business Submission Ranking Method

Multifunctional analysis refers to the evaluation of ranking results in the evaluations generated by multiple decision-making choices and multiple factors. It is best to choose something that can determine the level of available benefits based on the settings of these factors, and compare the range of selected results. Design an effective test sequence to determine the target to be selected. The key to this analysis is the calculation of substitution factors, the calculation of factor distribution weights, and the determination of the classification of related factors.

$$\sum_{i=1}^{n} w_i = 1 \tag{1}$$

The sorting order obtained after analysis obtains the required selection result. Definition a_{ij} is the j-th attribute of the i-th option, and the obtained attribute judgment matrix is expressed as formula (2):

$$A = (a_{ij})_{m \times n} = \begin{pmatrix} a_{11} & a_{12} & \cdots & a_{1n} \\ a_{21} & a_{22} & \cdots & a_{2n} \\ \vdots & \vdots & \ddots & \vdots \\ a_{m1} & a_{m2} & \cdots & a_{mn} \end{pmatrix} \tag{2}$$

5   Experiment

The overall structure of the database designed in this article is shown in Fig. 1:

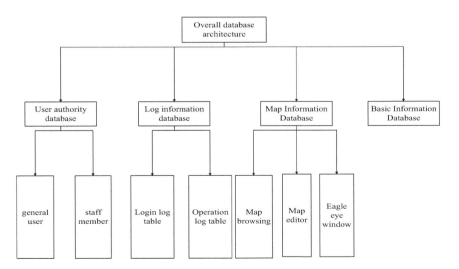

Fig. 1. Overall database design

5.1 User Authority Database

The application maintenance module in the system manages user rights, allows different users to process system data at different levels, and provides a certain degree of data security. After entering the URL, enter the homepage of the website, match the user name and password with the user information in the database, verify the legitimacy and authority of the user, and whether the visitor is an administrator, a staff member, or a normal user is a user.

Ordinary users can browse the workflow, view notifications from the grid control center and command center, and click Web. Click the report button to enter the report interface, enter the relevant description, report the problem in the surrounding city/event column, and enter other sites through the link.

After successfully completing the input, the staff will enter the interface of the central office. This interface mainly includes functional modules such as electronic map, monitoring and approval, comprehensive evaluation, online reporting, and application maintenance. To perform basic map functions, fill in city management business forms such as approval forms, inquiry forms, and view and manage historical documents. The above applicable functional interfaces are mainly functional interfaces operated by employees.

5.2 Log Information Database

Login logs record the IP address, physical address, and transaction name of the current account that you log in to, and run the server for a period of time, facilitating periodic information security checks. We determined who made the transaction. The fields drawn in the log table include account D, IP address, MAC address, and mode name.

The function log records the behaviors of users querying, adding, modifying, deleting data after logging in to the system, preventing data from being accidentally deleted or modified, and is based on data recovery. Data integrity can be protected programmatically. The field information displayed in the function log table mainly includes function name, operation type, user login IP and MAC address, and currently logged-in account, etc.

5.3 Basic Information Database

The basic information database mainly contains a large amount of data such as building information data, housing information data, housing information data, population information data, and business information data. There is a one-to-one relationship between these data. There are many relationships between police station information and building and community information. Community information has a one-to-many relationship with community information and building information. There is a one-to-one correspondence between community information and building information, and building information and family information are interrelated. It can be associated with all other information data through any of the above basic information data, which can facilitate the query correlation work of developers, significantly reduce the redundancy of the back-end interface SQL code, reduce the query time waiting for users, and the system response time.

5.4   Map Information Database

(1) Map browsing
    Map navigation functions include full screen, fast zoom, fast zoom, pan, zoom in, zoom out, previous view and next view.
(2) Map editing
    Only administrator users can use this function to add, delete, modify and edit various geographic objects on the map. First, you need to set the level of the editing object in the layer list box to be editable, otherwise you cannot operate on any object of the layer.
(3) Hawkeye window
    The specific meaning of the eagle eye function mainly includes two kinds: one is through the eagle eye window, you can determine the place or area of the map currently displayed in the map window, and the area in the red rectangle box of the eagle eye map window is in the map window. The map display area; the second is to quickly modify the map information in the map window by moving the red rectangular frame in the eagle eye window. When the mouse clicks anywhere in the eagle eye window, the map window is quickly positioned at any place clicked by the mouse.

6   Discussion

This database mainly provides services to the outside world in the form of Web, and users access it through a browser. By using the test tool, after recording the automated test script, stress test the login page, log information page, information query page, and map browsing page. The list page only displays a fixed number of records on each page, which is displayed in pages. The test results are shown in Table 1 and Fig. 2:

Table 1. Performance test results

Test item	Concurrency	Response time/s	Business success rate	CPU usage	Memory usage
Log in page	300	0.43	100%	22%	24%
Log information page	300	0.52	100%	41%	25%
Information query page	300	1.14	100%	54%	43%
Map browse page	300	1.07	100%	53%	47%

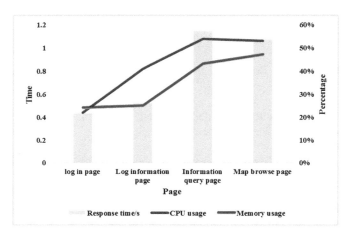

Fig. 2. Performance test results

As can be seen from Table 1 and Fig. 2, in the case of 300 concurrent users, the login page, log information page, information query page, and map browsing page tested in this article are in response time, CPU usage, and memory usage. The performance on the rate is better, and basically meets the needs of this article.

7   Conclusions

Grid management is an inevitable product of rapid social progress and an inevitable requirement for progress. However, as people's ideological awareness in the process of socialist modernization is improved, organizational management methods also need to keep pace with the times. How to adapt to the different needs of the masses is an important issue that society faces and solves.

Acknowledgments. Fund Project: Provincial Education Working Committee (gxdj-2020).

References

1. Eissa, M.M.: First time real time incentive demand response program in smart grid with "i-Energy" management system with different resources. Appl. Energy **212**, 607–621 (2018)
2. Brdjanin, D., Banjac, G., Banjac, D., Maric, S.: An experiment in model-driven conceptual database design. Softw. Syst. Model. **18**(3), 1859–1883 (2019)
3. Yallamilli, R.S., Mishra, M.K.: Instantaneous symmetrical component theory based parallel grid side converter control strategy for microgrid power management. IEEE Trans. Sustain. Energy **10**(2), 682–692 (2019)
4. Gomes, L., Vale, Z.A., Corchado, J.M.: Multi-agent microgrid management system for single-board computers: a case study on peer-to-peer energy trading. IEEE Access **8**, 64169–64183 (2020)

5. Virji, M., Randolf, G., Ewan, M., Rocheleau, R.: Analyses of hydrogen energy system as a grid management tool for the Hawaiian Isles. Int. J. Hydrog. Energy **45**(15), 8052–8066 (2020). https://doi.org/10.1016/j.ijhydene.2020.01.070
6. Anderson, J.: Americas energy ceo series: nyiso's dewey eyes high renewable energy grid management. Platts Megawatt Daily **24**(109), 2–3 (2019)
7. Watson, M.: Distributed energy resources adding to grid management complexity: panelists. Platts Megawatt Daily **24**(76), 6–8 (2019)
8. Nizami, M., Hossain, M.J., Mahmud, K.: A coordinated electric vehicle management system for grid-support services in residential networks. IEEE Syst. J. **15**(2), 2066–2077 (2020)
9. Latifi, M., Khalili, A., Rastegarnia, A., Bazzi, W.M., Sanei, S.: Demand-side management for smart grid via diffusion adaptation. IET Smart Grid **3**(1), 69–82 (2020)
10. Prudhviraj, D., Kiran, P.B.S., Pindoriya, N.M.: Stochastic energy management of microgrid with nodal pricing. J. Mod. Power Syst. Clean Energy **8**(1), 102–110 (2020)
11. Morsali, R., Kowalczyk, R.: Demand response based day-ahead scheduling and battery sizing in microgrid management in rural areas. IET Renew. Power Gener. **12**(14), 1651–1658 (2018)
12. Ito, H., Hanai, K., Saito, N., Kojima, T., Yoshiyama, S., Fukui, S.: Electricity system reform requirements: a novel implementation to grid management and control. IEEE Power Energy Mag. **16**(2), 46–56 (2018)

Route Tracking Model of Rock Climbing Activity Based on Probabilistic Hough Transform Algorithm

Jing Huang[1], Dan Zhang[2(✉)], Luyuan Cai[3], Fangyao Xie[1],
and Huajie Huang[1]

[1] School of Physical Education, China University of Geosciences,
Wuhan, Hubei, China
[2] School of Sports and Leisure, Sichuan Tourism University, Chengdu,
Sichuan, China
wanxmm66@163.com
[3] Mountaineering Management Center, General Administration
of Sport of China, Beijing, China

Abstract. In order to study the cognitive characteristics of the decision-making process of athletes, the expert model analyzes the visual search mode with theoretical and practical value, which helps athletes improve their decision-making ability and exercise level. This article focuses on the research of rock climbing activity route tracking based on probabilistic Hough transform algorithm. After understanding the theory of probabilistic Hough transform algorithm and rock climbing activity route tracking based on literature data, the rock climbing activity route tracking model based on probabilistic Hough transform algorithm is studied. It is constructed, and the constructed model is tested through experiments. The test results show that the model constructed in this paper has a high recognition accuracy and a certain degree of robustness.

Keywords: Hough algorithm · Probability model · Target tracking · Rock climbing

1 Introduction

Target tracking is an important link in the field of computer vision [1, 2]. Its main purpose is to find and identify moving objects for each frame of the video sequence, and get the next step: position, direction of motion, speed, size and other information, which is the basis for higher-level analysis of the target object [3, 4]. The continuous improvement of computer performance and extensive in-depth research provide the hardware and software requirements needed to track the target [5, 6]. A large amount of experimental data also provides sufficient support for the realization of the target tracking algorithm. Target tracking is now widely used in intelligent transportation systems, intelligent transportation, human-computer interaction, military and other fields [7, 8].

For target tracking research, some researchers have proposed a DLT algorithm that applies a deep model to the task of tracking a target, and applies a powerful object representation trained with detailed image classification data to the tracking problem, which solves the problem of insufficient target tracking samples problem. This problem is also very helpful to improve the tracking ability [9]. Some researchers have pointed out that there are still many challenges in the current tracking field, especially in the actual tracking process, the target object is affected by various factors such as illumination changes, distortion, background clutter, scale changes, obstacles, etc.[10]. Some researchers use related HOG filter tracking system and color histogram color probability locator to detect targets, and merge the response maps of the two trackers to improve tracker performance. However, most current fusion strategies do not make full use of different features by directly linking different features to each other or by linearly inserting different feature tracking results using predefined parameters. At the same time, the model needs to be updated in real time during the monitoring process to adapt to the changes in the target environment during the monitoring process. The traditional method of updating fixed frame weights involves inputting noisy data, which will lead to the risk of monitoring distortion [11]. In summary, because target tracking is in the computer field, the academic community pays more attention to target tracking, but there are still some problems in the target tracking process that need to be resolved, such as the impact of the target in the external environment and the tracking of moving targets.

This paper studies the rock climbing activity route tracking based on the probabilistic Hough transform algorithm, and analyzes the difficulties of rock climbing activity route tracking and the application of the probabilistic Hough transform algorithm on the basis of literature data, and then analyzes the rock climbing activity route based on the probabilistic Hough transform algorithm. The tracking model is constructed, and the constructed model is tested, and relevant conclusions can be drawn through the test results.

2   Research on Probabilistic Hough Transform Algorithm and Route Tracking of Rock Climbing Activities

2.1   Difficulties in the Realization of Route Tracking for Rock Climbing Activities

Regarding active target tracking, due to the uncertainty of tracking, this is the main problem that must be overcome in target tracking. Uncertainty roughly has the following aspects: First, the movement state of the target is ambiguous, because the movement state of the target is unknown. Therefore, the trajectory of rock climbing will cause factors such as the climber's subjective will. Second, since the target monitoring data obtained is taken from the observation source, the sensor will be affected by noise. In addition, the influence of the current surrounding environment will also cause uncertainty in the data received from the sensor. Therefore, the key problem in the target tracking process is to estimate and predict the state of the target flow. By establishing the target flow pattern, filtering the received ambiguous data, the uncertainty of the target flow is minimized. Environmental factors include:

(1) No prior knowledge

Target detection cannot be used with other prior information target detection (footprint, face recognition, etc.), because it lacks prior information about the shape and quantity of the target—use the previous information to train the target model, and then find the sorted target.

(2) Shadow

Shadows increase the difficulty of tracking the target and cause the algorithm to detect foreground errors. Therefore, the target detection algorithm should reduce the detection of error areas.

(3) Light changes

According to the speed of light change, it can be divided into two types: gradual light change and instantaneous light change. Light gradients mainly appear in outdoor scenes and are caused by changes in sunlight. Temporary lighting conditions mainly occur in indoor scenes and are caused by artificial light sources (such as lights). Changing the light will cause strong interference to the moving target detection system, which will cause the system to produce incorrect results. Therefore, the moving target detection algorithm must have strong resistance to changes in light.

(4) Dynamic background

Some backgrounds (leaves, snowflakes, etc.) may be moving, but these moving parts also need to be correctly identified as backgrounds by the target detection algorithm.

2.2 Application of Probabilistic Hough Transform Algorithm

Probabilistic random graph modeling fully considers the dynamic information in the sequence, and is suitable for modeling and cognition of the uncertainty of complex human motion processing. Although the current research on the prospect of independent human motion perception based on random graph modeling has made preliminary progress, the research in this field is still in its infancy, and many studies have yet to be resolved, and there are still problems.

1) Feature robustness is not strong: Existing graphical model probabilistic algorithms use most common features based on a single frame at the feature level, and the local blocking problem that is resistant to viewing angle changes is ignored. This is because most studies mainly display local attributes in the bag of words, it lacks the internal temporal characteristics of actions, and it is difficult to combine with probabilistic graphical models with powerful modeling capabilities.

2) The complexity of the algorithm is very high. The random graph model is used in the process of modeling actions that are not related to the viewpoint. Different behaviors, different appearances and behavior habits of different people, and different observations from different camera perspectives need to be considered at the same time. The parameter space is huge and requires a lot of specific training data and processes, which results in slow algorithm training.

3) The graphic model has no perspective transformation constraint: Because the action recognition method based on the probabilistic graphic model deals with perspective changes, most algorithms do not consider the perspective

transformation constraint and are directly used for human body modeling, and build a multi-view gallery for characterizing human posture models or characterizing human behavior models. This makes it difficult to obtain human body characteristic parameters. For convenience, a two-dimensional image sequence with a limited viewing angle can be used to directly construct a probabilistic graph model of human behavior that has nothing to do with the viewing angle.

2.3   Probabilistic Hough Transform Algorithm

After observing the particle image block set $\{p_i\}$ randomly sampled by the observation model, the weight of each particle is combined with its detection result in the HF classifier to weighted voting in the Hough space. The particle weight is.

$$\omega_i = G_s(d_i) \cdot G_p(d_p) \tag{1}$$

$G_s()$ and $G_p()$, respectively, the center position and the Gaussian kernel corresponding to the HF sparse detection sampling, d_i is the displacement from the particle position to the corresponding center during random sampling, and d_p is the displacement from the corresponding center to the prior target state center during random sampling. According to the voting method of the HF classifier, the voting coefficient of the image block corresponding to particle i is:

$$v_i = \frac{p_k}{|\{d_i\}_k|} \tag{2}$$

p_k is the particle image block.

3   Construction of Rock Climbing Activity Route Tracking Model Based on Probabilistic Hough Transform Algorithm

3.1   Selection of Characteristics of Rock Climbing Activities

There are two main observation indicators for eye movement research in the field of sports psychology. One is the time of the gaze, the number of gazes, the distance, the degree of the field of view, the position of attachment, the trajectory of eye movement, etc. By recording and analyzing these eye movement characteristics, it is possible to summarize the athlete's visual search pattern (problems close to the actual game state), reaction time (keyboard reaction), and action response (ball return, control), etc. The first category is a direct measurement of eye movement response, and the second category is a behavioral indicator of cognitive activity. There are three main types of eye movements: fixation, saccades and smooth monitoring.

Main parameters of line-of-sight monitoring technology.

(1)   Total number of gazes: an important indicator to measure the effectiveness of search. The more the total number of fixations, the more absurd the layout of the display area.

(2) Duration: Reflects how easy it is for the participants to extract information. A longer time usually means that it is more difficult for the subject to obtain information from the display area.

(3) Gaze display: the total time in the area during the gaze.

3.2 Feature Extraction

For the first feature selection, this paper selects the flow-specific field of view in the typical motion mode and the standard spatiotemporal point features in the local mode, and the typical contour information of the shape attributes of the static features. The above feature extraction methods are relatively mature. These features can be extracted based on the prior art first, and then the robustness of various features and the impact of the feature on the change of environmental factors such as noise and brightness and the dimensional influence of the viewing angle change can be determined to determine the impact of the feature. Based on the analysis of the experimental results, this paper combines the spatiotemporal interest points with the characteristics of the visual flow, weighs the dimension and resolution of the features, selects the appropriate feature parameters, and finally uses the probability-based Hough transform algorithm to perform human actions recognition.

3.3 Based on Probability Hough Transform Algorithm

(1) Algorithm framework

Figure 1 shows the overall framework of the algorithm in this paper. It is divided into two stages, offline training and online monitoring. In the offline training phase, the commonly used Hough Forests algorithm is used, and the general human target detection operator is used in the training sample set, including the ontology model (displayed as FDC) and the model of each part under each view (displayed as SPC).

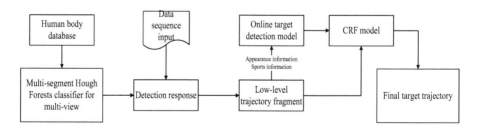

Fig. 1. The overall framework of the algorithm in this paper

(2) Division of human body models

The online target model of the algorithm in this paper is used to create a CRF model with real-time monitoring, isolate various targets, detect blockages, and finally establish a suitable target path. The appearance model representing the human target is M, including appearance models of various perspectives, segments, and overall models. m_1 and m_2 are the overall model in front/back and side perspective, respectively. And m_1 to m_{18} and m_{18} to m_{23} are two fragment models respectively. Each m contains the detection response r (that is, the confidence probability of the direction histogram), the direction histogram e and the color histogram c, namely M = {r,e,c}.

(3) Target association method

In recent years, some researchers have used the CRF model to simulate the correlation between detection responses [12]. The CRF model is an undirected graph, and its biggest feature is that the internal relationship between nodes and the relationship between nodes can be simultaneously modeled by the energy function of the node and the energy function of the edge. For the tracking problem, it is modeled as the correlation problem between the input tracking segments. This process allows the use of nodes to represent the correlation between two trajectory segments, and the node energy function of the CRF model to represent the relationship between the trajectory segments correlation, at the same time, the edge between two nodes is used to represent the consistency between the two correlations. That is, the edge energy function is used to express the coherence between the two sets of orbital correlations. So this article uses the CRF model, but for the energy function of the CRF model, this article learns the energy function online in the time window.

4 Model Testing

In order to verify the effectiveness of the algorithm in this document, this paper verifies the algorithm skill-based experiment. The test method uses the remove actor (LOAO) method. Each time the behavior data of one person is selected from the sample area as the test sample, and the behavior data of the other person is used as the training sample for the experiment. Repeat this process until all human behavior data samples have been tested, while ensuring that the tester's behavior data samples will not appear in the training samples to ensure the validity of the experimental data test. The experimental results are shown in Table 1:

Table 1. Algorithm test results

	Camera1	Camera 2	Camera 3	Camera 4
Perspective 1	77.21	80.23	84.76	77.21
Perspective 2	90.89	90.89	90.89	89.02
Perspective 3	90.89	88.87	88.78	83.56
Average	86.23	86.56	88.01	83.34

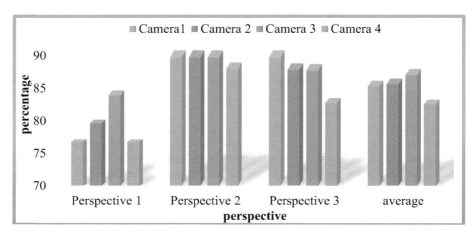

Fig. 2. Algorithm test results

It can be seen from Fig. 2 that the recognition rate of this algorithm is high. Because the algorithm in this article is a local attribute, it only needs to use information from some important parts of the human body to characterize human behavior. Even if the angle of view changes, as long as these important parts can be accurately positioned, the action can be accurately positioned, so it is also robust to the angle of view.

5   Conclusions

This paper focuses on the research of the rock climbing activity route tracking based on the probabilistic Hough transform algorithm. After understanding the relevant theory, the probabilistic Hough transform algorithm is used to construct the rock climbing activity route tracking model, and the constructed model is tested by experiments. The test results are the model constructed in this paper has a high recognition rate and a certain degree of robustness.

References

1. Mazurik, N.A.: The twenty third international conference LIBCOM–2019 – information technologies, computer systems and publications for libraries. (Review of events). Sci. Tech. Libraries (3), 93–112 (2020)
2. Lersteau, C., Rossi, A., Sevaux, M.: Robust scheduling of wireless sensor networks for target tracking under uncertainty. Eur. J. Oper. Res. **252**(2), 407–417 (2016)
3. Zhang, Z., Salous, S., Li, H., Tian, Y.: Optimal coordination method of opportunistic array radars for multi-target-tracking-based radio frequency stealth in clutter. Radio Sci. **50**(11), 1187–1196 (2016)
4. Hogan, B.G., Cuthill, I.C., Scott-Samuel, N.E.: Dazzle camouflage, target tracking, and the confusion effect. Behav. Ecol. **27**(5), 1547–1551 (2016)

5. Yang, X., Zhang, W.-A., Li, Y., Xing, K.: Multi-rate distributed fusion estimation for sensor network-based target tracking. IEEE Sens. J. **16**(5), 1233–1242 (2016)
6. Bjelić, S.N., Marković, N.A., Bogićević, Z.S., Bjelić, I.S.: Information Technology and Computer Science. Int. J. Inform. Technol. Comput. Sci. **11**(9), 20–30 (2019)
7. Atiquzzaman, M., Yen, N., Xu, Z.: Big Data Analytics for Cyber-Physical System in Smart City BDCPS 2019, 28–29 December 2019. Advances in Intelligent Systems and Computing, Shenyang, China (2020)
8. Meyer, F., et al.: Message passing algorithms for scalable multitarget tracking. Proc. IEEE **106**(2), 221–259 (2018)
9. Kumar, C.A.: Attitude and opinion towards computer technology. i-manager's J. Educ. Technol. **12**(1), 36–51 (2021)
10. Chen, Y., Zhao, Q., An, Z., Lv, P., Zhao, L.: Distributed multi-target tracking based on the K-MTSCF algorithm in camera networks. IEEE Sensors Journal **16**(13), 5481–5490 (2016)
11. Sindhu, M.: Recent trends in exports of computer software and information technology enabled services from India. Int. J. Sci. Res. Manag. **8**(2), 1584–1589 (2020)
12. Zhang, H., Ge, H., Yang, J., Yuan, Y.: A GM-PHD algorithm for multiple target tracking based on false alarm detection with irregular window. Signal Processing **120**, 537–552 (2016)

Application of Ant Colony Optimization Algorithm in the Design of Laser Methane Telemetry System

Zhichen Zhang[1,2(✉)]

[1] China Coal Technology and Engineering Group Shenyang Research Institute,
Fushun, Liaoning 113000, China
zzc9900@126.com
[2] State Key Laboratory of Coal Mine Safety Technology,
Fushun, Liaoning 113122, China

Abstract. With the advancement of science and technology and the transformation of development concepts, green economy has gradually become a major direction. Methane leakage will not only pollute the environment, but even the slightest negligence will lead to more harmful industrial safety accidents, which is very detrimental to the development of green economy. Therefore, appropriate measures must be taken to solve the problems it causes. This shows that the real-time monitoring of methane gas is of great significance. This article aims to study the application of ant colony optimization algorithm in the design of laser methane telemetry system. Based on the analysis of the systematic characteristics of ant colony algorithm, the mathematical model of ant colony algorithm and the method of measuring methane gas concentration, a laser methane telemetry system is designed. And finally tested the repeatability and stability of the system. The test results show that the system basically handles a straight line when the methane concentration is 5.05%, with small fluctuations, and good stability, which can meet the actual measurement requirements in the detection.

Keywords: Ant colony optimization algorithm · Laser methane detection · Gas telemetry · System design

1 Introduction

The speed of industrial development is also getting faster and faster, causing great pressure and damage to the environment, leading to environmental degradation, especially the increase in air pollution and smog caused by various toxic and harmful greenhouse gas emissions, which have seriously affected people's health and quality of life [1, 2]. In recent years, explosions caused by gas leaks and other reasons have occurred from time to time, causing immeasurable casualties and huge economic losses [3, 4].

Overseas research and development of TDLAS technology started very early. With the advent of TDLAS and wavelength modulation technology (WMS) in the 1970s and 1980s, as well as the development and application of gas detection, the accuracy of gas detection has been significantly improved [5, 6]. However, lead salt semiconductor

© The Author(s), under exclusive license to Springer Nature Switzerland AG 2022
V. Sugumaran et al. (Eds.): ICMMIA 2022, LNDECT 138, pp. 19–27, 2022.
https://doi.org/10.1007/978-3-031-05484-6_3

lasers had very strict requirements on the working environment at that time, which severely restricted the further development of TDLAS in this direction. After continuous development and improvement in the 1990s, semiconductor lasers have realized the characteristics of normal temperature operation, narrow bandwidth, strong suppression, and single mode, and have been widely used in the field of electronic communications [7, 8]. Since then, semiconductor lasers have gradually entered the market. Due to the continuous improvement and mass production of laser devices, their costs continue to decrease, and tunable semiconductor lasers have become a high-quality light source for gas composition and concentration. After that, the research and large-scale application of TDLAS technology began [9, 10]. In terms of gas detection, many foreign research and development institutions have conducted research and development and have achieved many results. Some gas analyzer manufacturers in Europe, the United States, Japan and other countries are also developing gas monitoring devices based on TDLAS technology [11, 12].

On the basis of consulting a large number of domestic and foreign references related to methane detection, this paper combines the systematic characteristics of ant colony algorithm, the mathematical model of ant colony algorithm and the method of measuring methane gas concentration, and designs a laser methane telemetry system, which mainly includes three modules are the laser methane detection system module, the remote monitoring center module and the authority management module. At the end, the repeatability and stability of the system were tested.

2   Application of Ant Colony Optimization Algorithm in the Design of Laser Methane Telemetry System

2.1   Systematic Characteristics of Ant Colony Algorithm

(1) Distributed features
     Ant colony algorithm embodies the distributed characteristics of group behavior. Each artificial ant starts to construct a problem-solving strategy autonomously from each part of the problem area at the same time, and the whole problem-solving is not affected by the failure of a certain artificial ant to find a solution. However, in specific computing optimization problems, this distributed feature not only improves the security of computing, but also enables computing to have a powerful overall search capability.

(2) Self-organization
     The characteristics of self-organization and no control center make the ant colony algorithm quite powerful. Self-organization refers to the process of obtaining space, time, or functional structure without external special intervention. It is not difficult to see that the most typical self-organizing system is the biota. The role of individual ants is very simple, but the synergy between individuals is particularly obvious. The current process of the algorithm is the process of individual artificial ant colonies irregularly seeking solutions. In this process, individual artificial ant colonies are more inclined to seek the best solution in the world, thus presenting a self-organizing evolution from disorder to order.

(3) Indirect connection between distributed individuals

The ant colony communicates through the pheromone sensation in the pathway. Individual ants do not exchange information directly, but interact by changing the coexistence environment. Changing the environment will affect the behavior of other individuals.

(4) Positive and negative feedback

In the ant colony algorithm, ants are more likely to choose the path with the highest pheromone concentration, and these ants will further increase the pheromone concentration in the path to the maximum pheromone concentration. Therefore, the intelligent ant colony system is a positive feedback system. At the same time, the ant colony algorithm also has a negative feedback mechanism. The negative feedback mechanism is provided by the probabilistic search technology used in the algorithm construction and solution process. This technique improves the randomness of the generated solution. On the one hand, the effect of randomness is to accept some degradation of the solution while keeping the search range large enough for a period of time.

2.2 Mathematical Model of Ant Colony Algorithm

(1) Setting of initial parameters

The basic parameters of the ant colony algorithm are as follows: n is the number of cities; m is the number of ants; d_{ij} is the distance between the two cities i and j; $\tau_{ij}(t)$ is the trajectory strength of the edge arc (i, j) at time t, that is, i and j Concentration of pheromone on the line; suppose the initial pheromone concentration of the line between cities is a constant, namely $\tau_{ij}(0) = C$, where C is a constant, i, j = 1,2 ,3 .n, and i $\neq$ j.

(2) The movement of ants

The pheromone concentration and heuristic information on each path jointly affect the choice of the next city for the kth ant (k = 1, 2...m) during the movement. At time t, the probability of the k-th ant moving from city i to city j can be expressed as:

$$p_{ij}^k(t) = \left\{ \begin{array}{c} \dfrac{\tau_{ij}^\alpha \cdot \eta_{ij}^\beta}{\sum_{j=\text{allowed}} (\tau_{ij}^\alpha - \eta_{ij}^\beta)} \\ 0 \end{array} \right. \tag{1}$$

In the formula, α is the information heuristic factor, which represents the coefficient of the importance of pheromone concentration; β is the expected heuristic factor, which represents the coefficient of the importance of heuristic information.

$$\eta_{ij}(t) = \frac{1}{d_{ij}} \tag{2}$$

From Eq. (2), we can see that for the kth ant, the larger the d_{ij} is, the smaller the $\eta_{ij}(t)$ is.

(3) Update of pheromone
In order to avoid the excessive concentration of residual pheromone over-whelming the enlightening information, each ant updates the pheromone after taking a step or completing a cycle (that is, the end of the tour to all cities).

2.3   Methane Gas Concentration Measurement Method

(1) Ultrasonic technology
Ultrasonic technology mainly uses the correlation between the propagation velocity of the ultrasonic technology in the gas and the current characteristics of the gas to predict the gas content. Because the ultrasonic frequency is high, it can reduce the impact on low-frequency noise, thereby greatly improving the detection accuracy. In addition, although ultrasonic gas content measurement equipment has the advantages of small size and low price. However, the measurement results are easily affected by various factors such as air pressure, air temperature, and relative humidity in the measurement environment, so compensation measures are needed to reduce errors.

(2) Differential absorption spectroscopy technology
Differential absorption spectroscopy (DOAS) technology uses narrow-band absorption characteristics to determine the gas composition and inverts the concentration of gas traces by adjusting the narrow-band absorption intensity. DOAS technology can not only meet the requirements of on-site testing, but also minimize the connection between the equipment and analysis results due to its non-invasive characteristics. Therefore, the accuracy and sensitivity are very high. However, because the DOAS technology will decompose the spectrum due to the change of the absorption gas spectrum, the gas receiving device has great requirements on the detection system and the test environment.

(3) Fourier transform infrared spectroscopy technology
The Fourier Transform Infrared Spectroscopy (FTIR) technology uses a near-infrared light source, which is aimed at the target through an adjustment device, and is generated in the form of parallel light, and is received by the telescope system using an effective light path of hundreds of meters. The optical signal is first obtained from the interferometer, and then the infrared detector converts the optical signal into an electrical signal, and finally uses the fast Fourier transform to receive the spectral information of each gas component. Although the measurement system using FTIR technology has a simple structure and is easy to use, however, a large amount of random noise is generated, and the accuracy of the measurement is reduced, so it is necessary to perform radiation calibration frequently.

(4) Tunable semiconductor laser absorption spectroscopy technology
Tunable Semiconductor Laser Absorption Spectroscopy (TDLAS) technology uses the fine line and current adjustment characteristics of a tuned semiconductor laser to scan the laser output wavelength of the gas to be measured and measure the inversion of gas-based gas parameters. TDLAS technology has the advantages of high spectral resolution, non-invasiveness, and fast response. As the laser is used as the carrier, it has a strong ability to resist electromagnetic interference. In addition,

the weighing system based on this technology is easy to operate, easy to maintain, and can be put into use at any time. Online measurement of gas state and gas concentration. Especially in the near-infrared region, the maturity of semiconductor laser technology has greatly promoted the wide application of TDLAS technology in the detection of trace pollutants in the atmospheric environment.

3 Experiment

3.1 Laser Methane Detection System Module

(1) Methane communication protocol module
The methane communication protocol module is mainly responsible for the serial communication with the methane laser concentration sensor. This is the implementation of certain API packages, such as sensor switches and methane concentration collection. The methane communication protocol module is also responsible for maintaining the serial port resources of the laser methane concentration sensor, such as the number of serial ports, connection status, baud rate, parity, and serial message buffer queues.

(2) Heterogeneous data upload module
The heterogeneous data transmission module is mainly responsible for uploading data resources such as methane collection history text, remotely captured images, and locally stored video records. According to different file types, there are different editing methods before uploading. Before uploading text files, use ZipOutputStream to decompress files with high compression rates. Use BitmapFactory to adjust the resolution of the image before upload to reduce the size of the image file, use MediaCodec to compress and encode the video before uploading to reduce the resolution, and use the bit rate to reduce the Reduce file size of the video. The above features can reduce the network and storage, and improve the stability of data transmission. For particularly large files, the method of splitting and sending is adopted, and the files are divided into multiple parts and numbered. Then split the device + multi-thread to upload, and use the data receiving unit to integrate and assemble parts to speed up the upload speed and perform the function of continuing transmission.

3.2 Remote Monitoring Center Module

The computer is also equipped with a laser monitoring application. The application process can not only display the methane gas content curve and real-time monitoring images taken by the panoramic camera, but also control the pan/tilt rotation and the camera zoom and send a warning when the telemetry alarm is sent. At the same time, manually turn on the camera to record the picture and video. After the inspection is completed, the APP can input, save and reproduce the track collection and alarm information of the inspection. After verifying the data, the inspectors use the four-G module to send the inspection data, pictures and short videos to the remote monitoring center.

The functions of the remote monitoring center include the collection of historical data such as inspection trajectories and leaks, the use of repeated inspection trajectories to understand the inspection process, the integration of the inspection trajectory with the GIS pipeline network, and the gas inspection. Perform statistics, classification and data mining, make inspection reports for leak locations, leak frequencies and leak categories, and guide inspectors.

3.3   Rights Management Module

(1) Tenant authority management
   The tenant authority management module is mainly responsible for the management of tenant authority. In principle, tenants are restricted from using the functions of the cloud platform. Most restrictions apply to various services provided by the cloud platform.
(2) Permission management of tenant sub-accounts
   The tenant sub-account authority management module is mainly responsible for managing the borrower authority of the sub-account. The sub-accounts use the functions belonging to the tenant under their respective licenses. For example, if tenants have limited login permissions for sub-accounts, sub-accounts will not be able to connect to the cloud platform. For example, tenants restrict the ability to remotely control sub-accounts, and sub-accounts cannot remotely control IoT devices.
(3) Function authorization management
   The operation authorization management module is mainly responsible for managing the licenses that the terminal can apply for. When certain functions are not authorized for the IoT terminal, even if the tenant or the tenant's sub-account has the corresponding permissions, the relevant functions cannot be performed in the terminal. This module is generally used for the locking function under special circumstances, such as a specific version of the equipment terminal is damaged, you can use this function.

4   Discussion

System repeatability refers to whether the output variable derived from the input of a single variable is consistent under the same conditions. System repeatability is an important indicator to measure system indicators, and the system performance with good repeatability is more stable.

Repeat the measurement of 5.05% of methane gas, make 5.05% of the gas into the gas chamber, check the mass flowmeter and continuously fill the gas chamber, if the gas chamber has been filled with measuring gas, check the mass flowmeter to make the gas escape rate It is roughly equivalent to the gas rate of filling, and the data value of this time is recorded with the host computer. Repeat the above operation several times, and the recorded density value and the corresponding digital value are shown in the Table 1:

Table 1. System repeatability experiment

First measurement	Second measurement
5.036	5.040
5.037	5.044
5.034	5.047
5.032	5.051
5.036	5.048
5.039	5.047
5.039	5.045
5.049	5.047
5.052	5.049
5.051	5.051

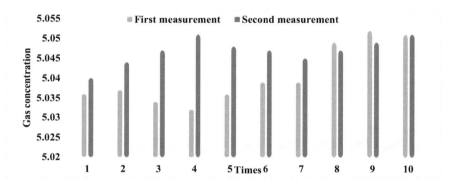

Fig. 1. System repeatability experiment

It can be seen from Fig. 1 that the measured value of 5.05% gas is about 5.05%, and the accuracy is better. Comparing the two measurements, the trend of the numerical value is similar, the numerical precision is the same, and the system has good repeatability.

The long-term stability of the system refers to the change of the detection results in the long-term measurement process when the same gas content is measured. It is an important indicator in the evaluation system indicators.

The system uses 5.05% standard gas for stability test, fills 5.05% of the gas into the gas chamber, and controls the mass flow meter to continuously fill the gas in the gas chamber. When the gas chamber is filled with the measured gas, the flowmeter makes the gas escape speed approximately equal to the inflation speed, and the host computer records the data value at this time. Measure and record full gas for a long time, and its stability is shown in Fig. 2.

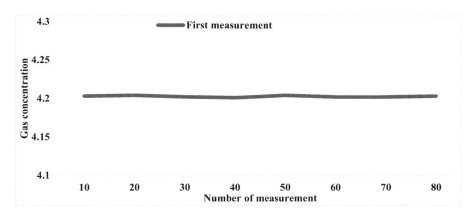

Fig. 2. System stability test chart

It can be seen from the figure that a straight line is basically processed at 5.05%, with small fluctuations and good stability, which can meet the requirements of actual measurement in the detection.

5   Conclusions

The development and research of high-efficiency and high-accuracy methane detection devices in complex environments play an important role in the safe development of coal mines, pipe corridors and dangerous areas in our country. The safety inspection in our country has important practical significance.

References

1. Wu, F.: Contactless distribution path optimization based on improved ant colony algorithm. Math. Probl. Eng. **2021**(7), 1–11 (2021)
2. Lv, G., Chen, S.: Routing optimizationin wireless sensor network based on improved ant colony algorithm. Int. Core J. Eng. **6**(2), 1–11 (2020)
3. Yu, C., Zha, W.: Detection and application of breaking of automobile mechanical transmission rod based on ant colony algorithm. Concurr. Comput.: Pract. Exp. **31**(10), e4759.1-e4759.7 (2019)
4. Jiang, Y.: Computer application of game map path finding based on fuzzy logic dynamic hierarchical ant colony algorithm. J. Phys.: Conf. Series **1992**(3), 032092 (2021)
5. Arivarasan, S.: A new modified mutation based ant colony algorithm for optimized fault tolerant routing protocol in MANET. J. Adv. Res. Dyn. Control Syst. **12**(SP3), 242–255 (2020)
6. Lima, V., Lima, E., Sherafat, H.: Roteirization of vehicles in the delivery/collection problems - application of a modifed ant colony algorithm. Revista Brasileira de Computação Aplicada **12**(1), 44–53 (2020)

7. Srinivasan, R., Jayaraman, M.: Experimentation on product and service life cycle on drive shaft using ant colony algorithm. J. Balk. Tribol. Assoc. **26**(4), 729–735 (2020)
8. Song, F., et al.: Interband cascade laser-based ppbv-level mid-infrared methane detection using two digital lock-in amplifier schemes. Appl. Phys. B **124**(3), 1–9 (2018)
9. Allsop, T., et al.: Methane detection scheme based upon the changing optical constants of a zinc oxide/platinum matrix created by a redox reaction and their effect upon surface plasmons. Sens. Actuators **b255**(1), 843–853 (2018)
10. Yang, J., Zhou, J., Lin, P.T.: Real-time isotopic methane detection using mid-infrared spectroscopy. Appl. Opt. **59**(34), 10801 (2020)
11. Syah, R., Nasution, M., Nababan, E.B., Efendi, S.: Sensitivity of shortest distance search in the ant colony algorithm with varying normalized distance formulas. TELKOMNIKA Indones. J. Electr. Eng. **19**(4), 1251–1259 (2021)
12. Sekiner, S.U., Shumye, A., Geer, S.: Minimizing solid waste collection routes using ant colony algorithm: a case study in gaziantep district. J. Transport. Logistics **6**(1), 29–47 (2021)

Low-Carbon Economic Life and Landscape Design Analysis and Research

Xiaoyu Zhu[✉]

College of Plant Science and Technology, Beijing Agricultural University,
Beijing 100096, China
zl7865310101@163.com

Abstract. With the rapid development of the economy, people's living standard is constantly improving, the pursuit of low-carbon economic life and healthy life is more and more obvious. As a basic project in the construction of urbanization, urban garden landscape is closely related to the daily life of urban residents. The purpose of this paper is to apply the low-carbon concept to urban landscape design, to provide a good environment for people's low-carbon life, and to provide ideas for many current landscape designers. This paper mainly analyzes the concept of low carbon and takes qiushui park in A city as A practical case to actively explore the research on China's low-carbon garden landscape design. With low carbon ecological technology, low-carbon landscape construction and construction related theory as the foundation, to autumn landscape engineering status as the breakthrough point, for A city region of ecological green space landscape construction by the thorough analysis, to explore A low-carbon landscape in related theory and technology problems in practical application, and through the investigation and analysis from the theoretical level, technical level and promote research and analytical design practices. It is concluded that the design and construction of low-carbon garden landscape in region A has good resource conditions and highly feasible technical means. In addition, compared with the previous landscape design, the low-carbon landscape design scheme obtained 93% satisfaction.

Keywords: Low-carbon life · Garden landscape · Garden design · Low-carbon recycling materials · Low-carbon landscape evaluation

1 Introduction

With the rapid development of economy and the expansion of population, the carbon emission of all countries in the world is also growing rapidly, and the accompanying global warming, fossil energy increasingly exhausted and other problems become more and more prominent. China is developing rapidly, and its total energy consumption and carbon emissions are among the highest in the world. The major carbon emission areas are concentrated in the eastern coastal areas, Inner Mongolia, henan and a few other inland cities. In this context, adjusting industrial structure, changing traditional economic development model and changing traditional social life style to reduce greenhouse gas emissions have become necessary choices to curb global warming and climate change.

© The Author(s), under exclusive license to Springer Nature Switzerland AG 2022
V. Sugumaran et al. (Eds.): ICMMIA 2022, LNDECT 138, pp. 28–36, 2022.
https://doi.org/10.1007/978-3-031-05484-6_4

With constant research and exploration in recent years related to learn, in terms of building low-carbon gardens landscape architecture field in China already have a certain theoretical basis, however, in terms of macroscopic, low-carbon landscape in our country is still a new concept, studies and discusses the theory, mostly in the concrete construction to build links we inevitably have to there is a lack of professional in the industry guidance, domestic common low carbon consciousness lack, lack of construction capital chain, lack of effective policies and measures to support the government. Therefore, the popularization of the concept of low-carbon landscape architecture in the field of landscape architecture in China is still a heavy task.

Mohamed Yusoff Abbas's research will help researchers in proposing design development situations in which people live in harsh living conditions and will help designers in designing and protecting the environment to better and better people's lives at the same time, people will be able to realize that culture can promote the environment and transform poor land to sustain life. These types of gardens attempt to balance the structure with greenery. The study examined ancient Persian gardens and harmony with nature as a means of achieving quality of life. It was to make this wish a reality that Mohamed Yusoff Abbas created such a garden, combining the beauty of tranquility with living space [1]. The optimization of regional landscape pattern plays an important role in improving the function and value of ecosystem and restraining the expansion of urban layout. Taking chengdu city as an example, LIUXu applied RS and GIS technology, landscape index and ecological service function evaluation, further analyzed the spatial and temporal characteristics of landscape pattern and spatial regional ecological function differences, and on this basis, determined the spatial distribution of ecological source land. Starting from the long-term goal of building a modern garden city in chengdu, this paper applies the cumulative cost distance model, introduces the theory of garden city, constructs the regional ecological corridor and ecological node, and explores the way to optimize the landscape pattern of modern garden city. Table urbanization process, a large number of arable land to construction land transfer; There are significant spatial differences in the strength of regional ecological functions [2].

Through extensive inquiry and collection of all kinds of information related to low-carbon gardens, this paper has a detailed understanding of the era background, development stage and process of low-carbon gardens at home and abroad, as well as the future planning and construction direction. This paper analyzes and compares the similarities and differences of typical low-carbon garden cases at home and abroad, points out the problems and imperfections in the development and construction of low-carbon garden landscape in today's world, and summarizes the focus and problems in the current and future development of low-carbon garden landscape. The overall environment of the experimental project site was investigated and studied in detail, and the status of natural, cultural, historical and other resources in A city was deeply understood and excavated. The location environment, climate and hydrological conditions, functional radiation range, vegetation and soil environment of qiushui park were analyzed and summarized.

2 Proposed Method

2.1 Landscape Overview of Low-Carbon Life

(1) The concept of low-carbon garden

At present, there is still no accurate definition of the concept of low-carbon garden. The concept of low-carbon originated from the increasingly exhausted fossil energy and abnormal global climate change in the context of the current era. Its core concept is to reduce carbon emissions and pursue the sustainable development goal of low emissions, high energy efficiency and high efficiency [3, 4].

(2) Characteristics of low-carbon gardens

Low-carbon landscape concept is: in the process of landscape architecture landscape planning and construction, follow the principle of ecology and the sustainable development idea, the use of energy conservation and environmental protection of landscape materials, maximize reduce CO_2 emissions, make to maximize the use of fossil energy efficiency improving, create the real environment friendly and carbon friendly landscape environment. According to this, low-carbon landscape is a new form of landscape that integrates ecology and sustainable development on the basis of traditional landscape planning and design [5].

Compared with traditional garden landscape, low-carbon garden has the following more prominent features:

1) Low energy consumption for construction

Low-carbon landscape on the construction contents and methods to choose more energy saving material and technical means, as much as possible by reducing the use of mass concrete construction and later have to reduce carbon emissions in the process of using, at the same time in the garden aesthetics on the basis of the existing natural conditions as much as possible to reduce planning land of earthwork construction, and make full use of rainwater collection and recycling means to improve the sustainable development of the garden [6, 7].

2) Have stronger carbon sink function

Landscaping can improve the climate in two ways: saving energy, reducing emissions and sequestering carbon. However, low-carbon gardens pay more attention to the utilization and improvement of the carbon sink capacity of green plants in the landscape than ordinary gardens [8].

In terms of improving landscape carbon sequestration ability, low carbon landscape pay more attention to the plant configuration of carbon sink capacity, the existing survey, evergreen shrubs deciduous trees and has a stronger ability of carbon sequestration, campus microclimate will improve with the increase of the vertical greening coverage, therefore, in garden plant configuration aesthetics on the basis of reasonable quantity increase evergreen shrubs deciduous trees and planting to improve campus carbon sink capacity is very important [4].

3) More significant ecological benefits

Low-carbon landscape in planning and design, material selection, plant configuration, the types of equipment, construction technology, curing management of each link is further merged the concepts of ecological low carbon, this makes the low-carbon landscape gardens have stronger than normal regulating benefits of ecology, the ecological regulation, improve microclimate, etc. have more outstanding performance [9].

4) Longer service life

Low-carbon landscape more respect the ecological and sustainable nature of the garden environment, pay more attention to the combination of short-term effect and long-term planning in planning and design [10].

5) More energy conservation and emission reduction

Traditional garden construction usually in order to meet the requirements of landscape aesthetics and using the large lawn, it is to a certain extent, caused some waste of land resources, and pay more attention to the use of three-dimensional plant low-carbon landscape, not only to avoid the unnecessary waste of land resources, also leaves out the produced by the late maintenance irrigation water and waste of human resources [11, 12].

2.2 Research Contents and Methods of Garden Landscape for Low-Carbon Economic Life

(1) Research content

In the current low carbon method based on the theory and technology, this article selects suitable for mature content of low-carbon landscape design, the design of the planning, construction, operation, maintenance and reuse, this paper wants to make of the landscape plan planning process fully operational, construction process, operation efficiency, and curing process rationalization, recycling process may change.

1) Define low-carbon landscape

For the specific requirements of low carbon, from reducing the energy consumption in landscape design; Enhance the carbon function of plant configuration; The recycling and reuse of landscape architecture resources; Reduce the landscape architecture of the four requirements of maintenance costs, we defined as: the landscape of low carbon in landscape architecture planning and design, construction, operation and maintenance management and reuse of the cycle, reduce the use of fossil fuels is harmful to the environment, improve the efficiency of energy use, reduce the emissions of carbon dioxide, formation is characterized by energy conservation and environmental protection of the ecological landscape architecture. As shown in Fig. 1.

Fig. 1. Life cycle of landscape architecture

2) Identification of low-carbon landscape concept

Material and energy are two basic elements of garden landscape. Only recyclable and renewable landscape can maintain healthy and vigorous vitality and serve human and nature for a long time. Low-carbon landscape needs to improve the species configuration and spatial form of green plants, promote the material circulation and energy flow of green plants, and drive the adjustment and renewal of the whole system [13].

3) Some misunderstandings of low-carbon landscape research

There is nothing wrong with advocating low-carbon life, but if the relationship between low-carbon and the realization of economic benefits, it will inevitably go astray.

On the basis of the research results of this discipline, the relevant contents of other disciplines should be properly integrated to achieve the cross-discipline, so as to ensure that the problem can be viewed in a more comprehensive way and more convincing arguments can be put forward. As show in Fig. 2.

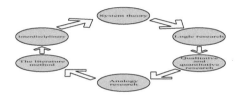

Fig. 2. Research methods

On the basis of the research on the achievements of this discipline, the relevant content of other disciplines should be properly integrated to achieve the cross-discipline, so as to ensure that the problem can be viewed in a more comprehensive way and more convincing arguments can be put forward.

2.3   Low-Carbon Performance of Landscape Design Elements

In the absence of necessary decompictors, many metabolic wastes are produced, and the substances in the system are difficult to circulate. It is characterized by simple structure of landscape elements and weak purification capacity, which usually needs to be maintained by humans. The low-carbon performance of landscape design elements is to follow the natural ecological process, increase the purification effect of the system, improve the species and quantity of animals and plants, and establish the circulation flow between substances, so as to reduce external carbon emissions and dependence on external resources.

3 Experiments

3.1 Experimental Data—Take Qiushui Park in a City as an Example

The qiushui park in a city covers an area of about 1,900 mu, which is a good place for people to relax and play with historical and cultural atmosphere. The construction of the new district in a city is derived from the policy strategy of expanding the urban development space in the provincial capital city. This region has very advantaged geographical location conditions. It is only separated from the provincial capital by a river, and many Bridges across the river make the traffic between the new area of a city and the provincial capital city more close. Due to its unique location and strategic factors, the new district of a city will assume more important regional responsibilities in the near future. At present, the new district of a city has been preliminarily planned and constructed, forming a scientific research and teaching area, a central business center, a central landscape area and a residential area. And colchicine status is in a new city district and its important central JingDian District, autumn park built for plain district not only provide a more perfect function area, more took on a new city district area microclimate regulation, improve the regional ecological environment quality, exert ecological benefit and improve the life quality of area residents.

3.2 Experimental Environment

The construction land of qiushui park in a city was originally an abandoned wasteland that had not been developed and used, and its vegetation conditions were relatively simple. The east, west, south and north sides of the project are urban road land and second-class residential land.

Qiushui park, located in plain new area of a city, has the natural environment characteristic of typical northern cities, and the climate, vegetation and hydrology conditions are very comfortable and pleasant. Therefore, this area has the natural resources for the planning and construction of urban gardens. In terms of temperature and climate, the summer in the autumn water area is relatively warm and the rainfall is relatively large, while the overall climate in the autumn park and winter is relatively dry. In terms of water quality of the soil close to the Yellow River, the water resource is very rich, so the colchicine has better water resources, in the autumn waters mainly good water permeability and sand loam, the use of plant growth and health, therefore, autumn waters sitting on vegetation resources are relatively rich in the natural environment, and quality is higher.

4 Discussion

4.1 Node Analysis of the Low-Carbon Design of Qiushui Park in a City

Technology and landscape interlaced, is a good place to experience the charm of modern science and technology. In the recycling area, there are mainly recycling space, green time and other nodes, where visitors can enjoy the fun and charm of recycling waste. The rainwater collection and utilization area and the low-carbon area of science and technology also have rich and colorful landscape nodes, such as the dew fragrance in the rain garden area, the stems cherish the pearls, the abundant scenery in the low-carbon science and technology park, and the science and technology world, etc. These two functional zones are good places for the interleaving of science and technology with the garden landscape and experiencing the charm of modern science and technology. Specific proportions are shown in Fig. 3:

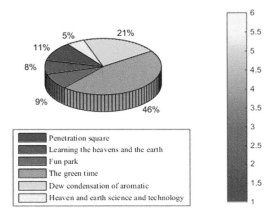

Fig. 3. Node proportion of low-carbon design of qiushui park

(1) Permeate the square
   Most of the squares in qiusui park are planned to use permeable pavement materials, and reasonable permeable structural layers are arranged in the underground part to facilitate the collection and treatment of rainwater during the use of the garden. In terms of the specific layout, there are various types of squares in the park, most of which are semicircle and square. They are mainly responsible for gathering and distributing crowds, cultural exhibitions, recreation and scenery appreciation.
(2) Learning horizon
   This landscape node is mainly used for people to communicate and learn here. As there are many universities built near the qiushui park, it is mainly considered to be a perfect place for students to communicate and learn, so that they can experience the happiness of knowledge harvest in the beautiful natural landscape environment.

4.2   Analysis of Planting Planning for the Low-Carbon Design of Qiushui Park in a City

Low-carbon landscape planting in autumn park planning, the main consideration of low carbon garden in carbon sequestration ability request, choice of solid carbon and releasing oxygen ability strong native tree species, at the same time pay attention to use deciduous shrubs, evergreen trees and other plants more stereo mix and match, to play a role of the park plant carbon sink capacity of the biggest. The planting plan is shown in Table 1:

Table 1. List of suitable plants in qiushui park

Watch the season	Varieties	Family name	Ornamental characteristics
Spring (February, march, April)	The lilacs	Oleaceae	Flowers purple
	Taeniasis and willow	Salicaceae	View branch
	Primrose	Oleaceae	Flowers yellow
	Virginia creeper	Grape family	Foliage
Summer (may, June, July)	Wild rose	Rosaceae	Look at flowers and fruits
	Henna	Henna division	Flowers white or red
	Wen crown fruit	No child family	White flowers with tree
	Flowers calamus	Irides	Design and color is rich
Autumn (August, September, October)	Day lilies	Lily	Flowers orange red to orange yellow
	The torch	Cashew	The autumn leaves red
	Lotus	Nymphaeaceae	A lot of colors, june-september
	Water lily	Nymphaeaceae	The number of lots of
Winter (November, December, January)	Yucca	Lily	Drought tolerance
	Sandy lam	Beckwell	Grow well
	Ningxia wolfberry	Solanaceae	Flowers and fruits may to October

5   Conclusions

At present, the planning and construction of urban low-carbon garden landscape is still in the preliminary stage of exploration, and the researches on low-carbon garden at home and abroad are not very in-depth, and most of them just stay at the theoretical level. In the actual construction process, low-carbon landscape is faced with such problems as backward technical level, lack of relevant policy support, insufficient attention from the government and the public, etc. This reminds us that the majority of landscape design planners need to constantly improve their learning ability, improve professional skills, grasp the development direction of the industry, establish a good professional ethics, adhere to the rigorous attitude of research, actively participate in and promote the concept and technical progress of the landscape industry. Only in this way can we truly create garden works that benefit the public and even future generations.

References

1. Abbas, M.Y., Nafisi, N., Nafisi, S.: Persian garden, cultural sustainability and environmental design case study shazdeh garden. Procedia – Soc. Behav. Sci. **222**(6), 510–517 (2016)
2. Liu, X., Li, Y.H., Wang, K.: Landscape pattern optimization of Chengdu based on modern garden city. J. Landsc. Res. **68**(4), 321–345 (2016)
3. Li, W., Duan, W., Shi, Q.: Song dynasty garden art and its inspiration on the modern urban landscape. Appl. Mech. Mater. **744–746**(5), 2206–2211 (2015)
4. Zhang, C.: Garden empire or the sublime politics of the Chinese-gothic style. Goethe Yearbook **25**(1), 77–96 (2018)
5. Burks, J.M., Philpott, S.M.: Local and landscape drivers of parasitoid abundance, richness, and composition in urban gardens. Environ. Entomol. **46**(2), 201–209 (2017)
6. Tor-ngern, P., Unawong, W., Tancharoenlarp, T.: Comparison of water-use characteristics of landscape tree (Tabebuia argentea) and palm (Ptychosperma macarthurii) species in a tropical roof garden with implications for urban water management. Urban Ecosyst. **21**(3), 479–487 (2018)
7. Major, J., Smith, C., Mackay, R.: Reconstructing landscape: archaeological investigations of the royal exhibition buildings western forecourt Melbourne. Int. J. Hist. Archaeol. **22**(1), 1–24 (2017)
8. Kulmatiski, A.: Community-level plant-soil feedbacks explain landscape distribution of native and non-native plants. Ecol. Evol. **8**(1), 2041–2049 (2018)
9. Chen, W.: Urban landscape architecture design under the view of sustainable development. IOP Conf. Series Earth and Environ. Sci. **81**(1), 012121 (2017)
10. Cong, W.-F., Jing, J., Rasmussen, J.: Forbs enhance productivity of unfertilised grass-clover leys and support low-carbon bioenergy. Sci. Rep. **7**(1), 1422 (2017)
11. Chen, C., Zhang, Y., Li, Y.: Highly conductive, lightweight, low-tortuosity carbon frameworks as Ultrathick 3D current collectors. Adv. Energy Mater. **7**(17), 1700595 (2017)
12. Chew, T.L., Jeng, S.L., Van Yee, F.: Enabling low-carbon emissions for sustainable development in Asia and beyond. J. Clean. Prod. **176**(11), 726–735 (2017)
13. Vandepaer, L., Gibon, T.: The integration of energy scenarios into LCA: LCM2017 conference workshop, Luxembourg, September 5, 2017. The Int. J. Life Cycle Assess. **23**(4), 970–977 (2018)

Fault Monitoring Technology of Electrical Automation Equipment Based on Decision Tree Algorithm

Lu Zhou[1(✉)], Yu Cui[2], and Amar Jain[3]

[1] School of Electrical and Information Engineering, Liaoning Institute
of Science and Technology, Benxi 117004, Liaoning, China
ZhouluLKX@126.com
[2] Siemens Ltd., Beijing 110000, China
[3] Department of Civil Engineering, Faculty of Engineering and Technology,
Madhyanchal Professional University, Bhopal, India

Abstract. With the continuous progress and development of human economy and society, the demand of human society for the stability of the power system is also increasing. The occurrence of sudden failures in the power system will cause significant economic losses and bad social consequences. Therefore, it is necessary to monitor the status of all electrical equipment in the entire power system in real time, and fully grasp the working status of electrical equipment at all times. This paper aims to study the fault monitoring technology of electrical automation equipment based on the decision tree algorithm. Based on the analysis of the basic process of data mining, the comparison of decision tree algorithms and the system performance requirements, a fault diagnosis method based on the C4.5 algorithm is proposed. The fault monitoring system of electrical automation equipment is designed. Experiments show that the algorithm can improve the classification accuracy, so to a certain extent the effectiveness of the algorithm in fault monitoring is proved.

Keywords: Decision tree algorithm · Electrical equipment · C4.5 algorithm · Fault monitoring system

1 Introduction

With the continuous development of power technology, substation equipment is an important part of the power grid. If defective electrical equipment cannot be discovered and discovered in time, local damage will spread, affecting the entire power grid, or even destroying the power grid, completely disrupting the power system, causing unnecessary economic losses and casualties. Therefore, the power operation department regards the monitoring and evaluation of electrical equipment as the top priority [1, 2].

At present, the United States, Europe, Japan and other developed countries and regions are at the forefront of technology in power equipment management and fault management, because these countries and regions have advantages and technology accumulation as pioneers in the field of information technology, and the power systems of these countries are mainly composed of primary. The composition of electrical equipment, the overall topology of the power grid is relatively simple, and there is no need for more complicated electrical equipment connection or control [3]. Therefore, taking the United States as an example, most of the power equipment management information systems in the United States now have corresponding internal power supply companies, which are mainly based on the automatic control capabilities of the automation equipment suppliers and the grid data communication service platform [4, 5]. The functional architecture of these business management platforms is relatively simple. Most of the operation and maintenance of power equipment and its fault management can be completed through intelligent service components or control equipment integrated into the equipment itself. The focus of the business management platform is batch maintenance and Centralized management [6, 7]. In addition, some foreign power companies also provide power management and maintenance equipment platforms for their power equipment. At present, the most popular applications abroad are mostly similar products from other well-known electrical equipment manufacturers such as Schneider, but these products are usually provided as independent application kits [8, 9]. True integrated management requires the development of a corresponding integrated management service platform based on the management suites provided by these vendors.

On the basis of consulting a large number of related references, this paper designs a fault monitoring system for electrical automation equipment based on the basic process of data mining, comparison of decision tree algorithms and system performance requirements. The system mainly includes four modules, namely the oil chromatography measurement module, oil micro-water and oil temperature measurement module, partial discharge measurement module and system fault diagnosis module, and apply the C4.5 algorithm proposed in this paper to the designed system to achieve the purpose of equipment fault monitoring.

2   Fault Monitoring Technology of Electrical Automation Equipment Based on Decision Tree Algorithm

2.1   The Basic Process of Data Mining

Figure 1 shows the general process in the basic process of data mining.

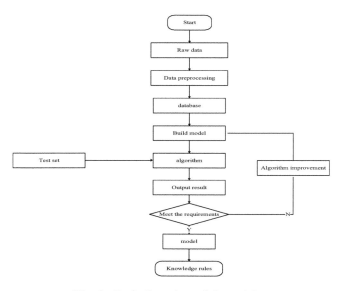

Fig. 1. Basic flow chart of data mining

(1) Original data source

The data source is the data that the computer can process. It can be used after general processing. For example, transaction data purchased by Treasure Net, flight reservations, student meal cards, hypertext and multimedia data of specific websites and sites, and most of the metadata stored in the database.

(2) Test set

This set of test results in the data mining flowchart can be used to detect and optimize the characteristics of the selected module. If the performance advantages of the algorithm are not prominent, optimization calculations can be performed [10, 11]. The price of the test set method is to reduce the amount of training.

(3) Data warehouse

The so-called data warehouse is a database that users can use to access and retrieve data. Most libraries are composed of many libraries and stored in the same way.

(4) Knowledge rules

Transform data to formulate knowledge or rules that inspire businesses and users. For example, supermarket decision-makers can carry out specific promotions and groupings of products based on their knowledge to increase turnover. In the process of data mining, the knowledge of analysis, induction and summary in statistics is used to process the data source, but it does not lack its own uniqueness. Analyze from the following aspects:

1) The problem is large. The data sources for mining are generally relatively large. For data mining, it is actually extracting rules from complex and changeable data. However, the organization and presentation of data is often dominated by corporate databases, and over-reliance affects the results and efficiency of mining. In general, the result set mined is huge.

2) Discreteness of variables. In practical applications, data can be divided into two categories: discrete data and continuous data. However, most statistical data is regarded as continuous data, which is not suitable for practical applications and requires special processing. In some cases, individual techniques are used to distinguish variables.

3) The effectiveness of evaluation standards. When evaluating method performance, old statistical methods need to consider many aspects. In theory, it is difficult to prove that the algorithm is optimal. The requirement to improve the performance of the algorithm is to trade time for space.

2.2   Comparison of Decision Tree Algorithms

(1) ID3 algorithm

The ID3 algorithm is a process of modeling and classifying the mathematical model of the decision tree. The decision-making process using the ID3 algorithm uses machine learning to build a decision tree. Use natural language to analyze the predicted database data, and extract samples from it for prediction, classification and results. The degree of information acquisition is the data difference between the two entropy information. Its definition can be expressed as:

The concept of information entropy H can be expressed as formula (1), where C_i is the attribute value of a given sample, P_i is the probability of the sample taking the attribute value, and the total probability is specified as 1.

$$H(p_1, p_2, \ldots, p_s) = \sum_{i=1}^{s} -(p_i \log p_i) \tag{1}$$

The higher the entropy value, the stronger the disturbance of feature change and the higher the uncertainty. The idea of constructing a decision tree is to gradually reduce the entropy value of the node.

Information gain represents the data difference between two types of entropy information, that is, the difference between the amount of information initially needed and the demand for new information after classification. The information gain of the attribute S related to the sample set D can be expressed by Eq. (2).

$$G(D, S) = H(D) - \sum P(D_i) H(D_i) \tag{2}$$

The calculation of ID3 algorithm is biased towards selecting attributes with more attribute values. For example, attribute M has two attribute values, and attribute N has four attribute values, and the calculation result is biased toward attribute N.

(2) C4.5 algorithm

C4.5 algorithm solves the defect of ID3 algorithm for feature selection with more attribute values, inherits all the advantages of ID3 algorithm, and can handle incomplete data and data sets with continuous attributes. Another advantage of using the C4.5 algorithm is that it has a higher classification rate than the ID3 algorithm, faster modeling speed, and does not affect the understanding of classification prediction rules after data classification. In this paper, the electrical automation equipment fault detection system is implemented using algorithm C4.5.

In the decision tree process, each time a node is created, the C4.5 algorithm will select the most effective data separation feature to classify a specific sample and classify it into one category or another. The current split function selects the item with the highest normalized information gain. Algorithm C4.5 then continues to calculate the highest normalized information gain in the next subset. The normalized information gain is the percentage of the information gain.

(3) CART algorithm

The CART algorithm uses the Gini coefficient value as the basic feature of node selection, and divides the current sample set into two modes through the retrospective division technique of dichotomy. Each time, the attribute branch is split into two sample subsets, that is, halved retrograde split. Each non-leaf node in the generated decision tree has two node branches, and the overall structure of the tree is a binary tree. The CART algorithm assumes that all nodes may be leaf nodes, and assigns a type to each node, which is divided into leaf nodes and non-leaf nodes. As for the classification method, the type with the highest number of events on the current node can be used, or the current node type error or other more complicated methods can be used.

2.3   System Performance Requirements

(1) Security requirements

The system must be configured according to the internal location of the power management operation and maintenance department, corresponding to the power equipment status data and debugging data access permissions in the system, and all users must log in and verify through the built-in authentication mechanism. After successful login verification, all internal access information and statistical functions should be restricted and configured according to the user's actual workplace. At the same time, the system must perform background log operations on all internal functions of successfully logged in and authorized users, and the system administrator is responsible for electronically displaying, exporting and analyzing log data.

(2) Normative requirements for functional design

The system design must be completed in accordance with the relevant software engineering specifications. The system adopts the idea of modular design. The interface between the units is clearly defined, and the coupling meets the interface requirements. The database design is table structure and logic, and follows the standard database design. The module programming code is standardized, and the code is readable, which is conducive to subsequent code retention.

3 Experiment

3.1 Oil Chromatography Measurement Module

Oil chromatography monitoring device is a widely used method to detect internal faults in transformers. The transformer is closed internally and cannot be detected immediately when an error occurs. The oil chromatogram can closely monitor the local overheating caused by the aging of the insulator. Different insulation problems inside the transformer will lead to different gas ratio distributions. By analyzing a set of gas ratios, the total ratio can be used to determine the internal fault of the transformer, it has occurred or is currently in a potential fault state. Therefore, dissolved gas is an indispensable part of transformer detection, which can accurately reflect the operating status of the transformer, identify the location of the fault, and provide it to maintenance personnel, who can maintain the fault and be responsible for avoiding accidents.

3.2 Oil Micro-water and Oil Temperature Measurement Module

The measuring principle of oil and water is similar to the way of detecting water in the laboratory, and no additional testing equipment is needed. Temperature monitoring requires the use of a thermistor to complete the experiment. The resistors are attached to different parts of the transformer, and the current signal is received through the signal transformer, and the oil temperature data related to the transformer is received.

3.3 Partial Discharge Measurement Module

The partial discharge module mainly monitors the partial discharge error of the transformer. Partial discharge is a common obstacle to transformers, but its harmfulness is not low. When a partial discharge occurs inside the transformer, transient sound will be produced. If the internal discharge of the transformer is less than 1000 times of charging, it will not affect the operation of the transformer. However, if the discharge exceeds 20,000 times, the transformer will not be displayed. Unable to resume normal operation. After that, the discharge continues to increase, damaging the insulator. Therefore, by monitoring the amount of partial discharge, the insulation state inside the transformer can be more clearly understood. The Mir method is a threshold alarm method. By adjusting and monitoring the amplitude of the high-frequency signal and the number of periodic pulses, when the monitored data exceeds the limit, the alarm device is activated. Use a sensitive voltage and current sensor to measure the voltage and current value of the legal discharge point, and process the measured data to obtain high-precision dielectric loss and capacitance value, so as to determine whether it is a discharge fault.

3.4 System Fault Diagnosis Module

The system troubleshooting module can make full use of the existing monitoring data to determine the type of transformer fault corresponding to the data as a diagnostic expert database, and store the content of the database on a dedicated diagnostic system platform. If the data currently conforms to the type of the error data, a dedicated diagnosis. The system determines that the transformer status is currently faulty.

4 Discussion

In this article, we will first implement the algorithm in Java and run it on the Weka platform. The test data set comes from the UCI machine learning database. In the experiment, 2/3 of the data in each data set is randomly selected for decision tree training, and the remaining 1/3 of the data is checked for decision tree. The experiment is compared with the traditional ID3 algorithm in terms of the scale and classification accuracy of the generated decision tree.

Choose four Tic-tac-Toe, Nursery, Car, and Kr-vs-sp data sets in the UCI database, and use β = 0.667 and β = 0.75 to create a decision tree, respectively. The test results are as follows:

Table 1. Analysis of results

Data set	Classification accuracy			Number of leaf nodes/total nodes		
	ID3	The algorithm of this paper (β = 0.667)	The algorithm of this paper (β = 0.75)	ID3	The algorithm of this paper (β = 0.667)	The algorithm of this paper (β = 0.75)
Tic-tac-Toe	77.6%	86.5%	81%	196/490	22/34	6/9
Nursery	86.5%	87.2%	85.5%	730/1746	26/36	17/23
Car	80%	85.1%	84.2%	234/558	47/67	18/26
Kr-vs-sp	89.2%	92.7%	90%	46/136	4/7	2/3

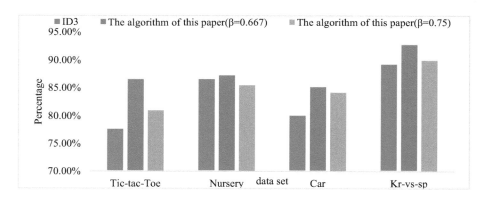

Fig. 2. Analysis of results

It can be seen from Table 1 and Fig. 2 that with the decrease of β, the construction scale of the decision tree seems to gradually decrease, and the classification accuracy of the decision tree has improved. Although the experimental results cannot clearly indicate that the classification accuracy changes linearly with β, for some data sets, it can be seen that the algorithm proposed in this paper improves the classification accuracy of the decision tree. The decision tree created using this algorithm is significantly smaller than the size of the decision tree created by the ID3 algorithm, which is conducive to understanding and using rules. At the same time, for some data sets, the algorithm proposed in this paper can improve the accuracy of classification, so the effectiveness of the algorithm has been proved to a certain extent.

5   Conclusions

Power system electrical equipment status detection is the main development direction of electrical equipment maintenance. The prerequisite for status detection is to achieve online detection and fault judgment, through the use of modern intelligent measurement technology to implement online detection of electrical equipment, and the use of artificial intelligence data processing technology. The realization of fault diagnosis is an inevitable trend in the development of modern intelligent measurement technology and fault diagnosis technology.

References

1. Darian, L.A., et al.: X-ray testing of high voltage oil-filled electrical equipment: physical background and technical requirements. IEEE Trans. Dielectr. Electr. Insul. **27**(1), 172–180 (2020)
2. Wen, J., Xie, Q.: A separation-based analytical framework for seismic responses of weakly-coupled electrical equipment. J. Sound Vib. **491**(3), 115768 (2021)
3. Yi, S., et al.: Upcycling strategies for waste electronic and electrical equipment based on material flow analysis. Environ. Eng. Res. **24**(1), 74–81 (2019)
4. Xie, Q., et al.: Influence of flexible conductors on the seismic responses of interconnected electrical equipment. Eng. Struct. **191**, 148–161 (2019)
5. Hoxha, E., Maierhofer, D., Saade, M.R.M., Passer, A.: Influence of technical and electrical equipment in life cycle assessments of buildings: case of a laboratory and research building. Int. J. Life Cycle Assess. **26**(5), 852–863 (2021). https://doi.org/10.1007/s11367-021-01919-9
6. Ma, J., et al.: Weakly supervised instance segmentation of electrical equipment based on RGB-T automatic annotation. IEEE Trans. Instrum. Meas. **PP**(99), 1–1 (2020)
7. Fabricio, M.A., Behrens, F.H., Bianchini, D.: Monitoring of industrial electrical equipment using IoT. IEEE Latin Am. Trans. **18**(8), 1425–1432 (2020)
8. Niu, Z., et al.: Electrical equipment identification method with synthetic data using edge-oriented generative adversarial network. IEEE Access **PP**(99), 1 (2020)
9. Tulloch, A., et al.: A decision tree for assessing the risks and benefits of publishing biodiversity data. Nat. Ecol. Evol. **2**(8), 1209–1217 (2018)
10. Esmaily, H., et al.: A comparison between decision tree and random forest in determining the risk factors associated with type 2 diabetes. J. Res. Health Sci. **18**(2), e00412 (2018)
11. Huang, Q., Zhang, F., Li, X.: A new breast tumor ultrasonography CAD system based on decision tree and BI-RADS features. World Wide Web **21**(6), 1491–1504 (2018). https://doi.org/10.1007/s11280-017-0522-5

Coal Mine Safety Risk Intelligent Control and Information Decision Analysis System

Xiao Wang[1,2(✉)]

[1] China Coal Technology and Engineering Group Shenyang Research Institute,
Fushun 113000, Liaoning, China
minozo2021@163.com
[2] State Key Laboratory of Coal Mine Safety Technology,
Fushun 113122, Liaoning, China

Abstract. In recent years, in order to strengthen the supervision of coal mine safety production, the Chinese government has formulated and promulgated a series of laws and regulations. At the same time, it has strengthened safety management and increased investment in coal mine safety, which also reflects that coal enterprises are gradually paying attention to safety production. The purpose of this paper is to study the coal mine safety risk intelligent management and control and information decision analysis system. This article is to build a safety production early warning system, analyze the risk sources of safety production in different time and space and environmental dimensions, cut the coal mine production system into independent evaluation elements, and overcome the natural environment and the variability, complexity and complexity of the natural environment and risk factors faced by coal mine production. Based on the network evaluation elements, the main factors of coal mine safety production risk are summarized, combined with qualitative and quantitative methods to determine the weights, and the coal mine safety production risk evaluation index system is constructed. This article has tested the system three times and found that the test result is consistent with the actual situation on site.

Keywords: Coal mine safety · Intelligent management and control · Information decision-making

1 Introduction

As one of the key elements of safety system technology, understanding and managing coal mine safety risks and information intelligence analysis is a key mechanism for managing, monitoring and evaluating safety, and it can effectively ensure the continuous improvement of coal safety production [1, 2]. The evaluation process takes the negative impact of coal mining as the starting point, analyzes and calculates potential losses and damages, and uses the early warning system to predict potential coal mining hazards, and targeted preventive measures can be taken.

In the research on the coal mine safety risk intelligent control and information decision analysis system, many scholars have conducted research on it and achieved good results. For example, MATHEY, M used neural network technology when evaluating all aspects of coal mine safety production [3]. Mu W conducted a comprehensive review of non-compliance characteristics when assessing safety systems, and discussed the necessity of applying neural network theory and technology in mine risk assessment [4]. It can be seen that the study of a coal mine safety risk intelligent control and information decision analysis system is of great significance to national mining safety.

This paper establishes a safety production early warning system, analyzes the sources of safety production risks from different time and space environmental dimensions, and divides the coal production system into independent evaluation elements to overcome natural, evolving, and complex problems. And the complex environmental hazardous substances encountered in coal production. Based on the network assessment data, this paper summarizes the main risk factors of mining area safety, and combines the power and magnitude method of determining the weight to define the coal mine safety risk assessment system.

2   Research on Coal Mine Safety Risk Intelligent Management and Control and Information Decision Analysis System

2.1   The Design of the Intelligent Management and Control Module for Coal Mine Safety Risks

(1) Real-time data display module
Real-time data projection transfers the information collected by the sensor to the database in an intuitive way such as tables, history, and line graphs, so that the control personnel can solve the downhole production problems in real time [5]. The real-time data displays the real-time data of the current time and the data trend of the most recent time.
(2) Smart map module
The intelligent map unit integrates the information of incoming and outgoing workers and the geospatial information of coal sources during the production process. Then store the information from the database. In the smart map module, when the mouse navigates to a specific carbon area, the information box will float, read the database in real time, and display the number of employees currently stored in the area, the number of positions, and the time entered. Each employee is underground, so you can control. The layer can understand the production situation of the bottom layer more intelligently and in real time [6].
(3) Safety information analysis module
The safety information analysis module is a sub-module under the accident prevention management module. It applies data mining technology to the system production management. After obtaining the coal mine safety production parameters, analyze them in the safety information analysis module to obtain the safety information of the coal mine.

(4) Emergency plan management module

In the safety information analysis module, the data mining technology is used to mine and analyze the safety parameters in the coal mine production process. According to the correlation between the obtained safety parameters, suggestions for coal mine safety governance are given, which are added to the emergency plan management module to facilitate the coal mine Employees can check it at any time and guide coal mine safety production more accurately [7–9].

Coal mine risk coupling may increase or decrease the risk of accidents, or the risk level may remain unchanged. According to the changes in the degree of risk caused by the coupling, coal mine risk coupling is divided into positive coupling, zero-degree coupling and negative coupling. When the influencing factors are positively coupled, it will increase the risk or expand the scope of the accident. Therefore, certain measures must be taken to avoid the positive coupling between the various influencing factors, to transform it to zero coupling or negative coupling, or to reduce each impact. The strength of the coupling between the factors.

(5) Alarm information module

Distress alarm: When underground workers encounter a dangerous situation, the mobile terminal presses for help to generate distress alarm information.

Over-limit alarm: The downhole environmental monitoring devices (sensors, etc.) are all set with upper and lower limit standard thresholds. If the monitored value exceeds this range and the data is abnormal, it will be judged as an over-limit alarm.

Equipment failure alarm: the underground production equipment and the ground master station regularly send signals to each other. If the master station does not receive the signal from the equipment within a period, it will be judged as equipment failure and an alarm will be issued, and the equipment failure alarm information table will be written into [10, 11].

Personnel overtime alarm: Each underground worker has its own prescribed working time, and personnel overtime alarm occurs when this time is exceeded.

Regional overcrowding alarm: If the number of people in a certain area of the mine exceeds the number of people that can be accommodated in the area, then an area overcrowding alarm will occur.

2.2 Theoretical Design of Decision Support System

(1) Classification of decision-making problems

1) According to the severity (or level) of the decision-making problem, it can be the decision-making process, strategic decision-making and project decision-making. The decision-making process refers to decisions involving general or specific issues. The decision-making process, also known as the decision-making process, refers to the evaluation of decision-making activities on the decision-making process of some local problems.

2) According to whether the decision-making problem is still reversible, it can be divided into customized decision-making and informal decision-making. Periodic decision-making, also known as repetitive decision-making, refers to the handling of problems that frequently arise in the production and operation

of enterprises. The rules are easy to compile, can run conventional systems, compiled into programs, computer generated, also known as system decision [12].

   3) According to the degree of understanding of the state of the decision-making process, it can be divided into decision-making, risk-based decision-making, non-critical decision-making and competitive decision-making. Decision-making refers to a decision that can accurately calculate all the facts in the decision-making process. Techniques can be expressed in mathematical expressions, with clear field functions and mathematical terms that can find the best solution. The choice of ideas is to find the best value for the problem. Risk-based decision-making refers to a decision that may have certain risks when implementing the idea.

(2) Decision support system and its characteristics

The focus of an information management system is to process large amounts of data. Performance research uses templates to aid decision making, which is obviously important in model-based decision making. With the development of technology and the complexity of the problems to be solved, more and more models are effective, and these models have also expanded from mathematical models to data processing models. For multi-model decision problems, before the advent of decision support systems, humans were used to determine the integration and coordination of models.

(3) Architecture based on X library

   1) Data parts

Data is the basis of assisting decision-making, and the function of decision-making is no different from data. Data elements are an important part of DSS. The database system includes a database and a data management system. It is responsible for storing and managing most of the data used by DSS and data outside the system, and identifying compatibility switching between multiple data sources.

   2) Model parts

The data represents events that have occurred in the past. The model can be used to transform previous data into current or future information that can be used for decision analysis. The model library system is the foundation of the decision support system, and is the most important and difficult part to do, including the model library and the model library management system.

   3) Dialog widget

The internal communication system is the man-machine interface for establishing and using the model system for decision support, development, analysis and statistical services. Through innovation, decision makers can use a wide range of DSS support services to conduct research, analysis and re-learning based on personal experience to select the best decision-making concept.

2.3   Basic Principles of Monitoring Information Fusion

The standard description function assignment to the monitoring information object in the mining area is the value of the object reflected by the monitoring information. A describes the size of the evidence support value. In the evidence support, the Bel function of the monitoring information of the mining area is defined, and the following relationship is set:

$$\text{Bel}(\Phi) = 0, \text{Bel}(\Theta) = 1 \tag{1}$$

$$\text{Bel}(A) = \sum\nolimits_{B \in A} m(B) \forall B \in A \tag{2}$$

Then Bel is called the Belief Function, and Bel(A) is called the degree of trust description of the monitored information object. And Bel(A) is the general description of the trust support degree of the object.

3   Experimental Research on Coal Mine Safety Risk Intelligent Control and Information Decision Analysis System

3.1   Use of the System

The evaluation record can be reproduced in the database-based data entry function, and the evaluation record can be recalculated in the gray calculation unit. In addition, the model index system and evaluation files can also be imported from external databases through the data management system at the same time, and the model solution provides model index systems and expert evaluation and certification services.

3.2   Design of the Master Control Program

The master control program mainly completes the following functions:

(1) Manage the performance of each model. Before activating the template, the man-machine interface is controlled by the main control system, and the required data is input into a specific data file or data set, and the data file or data processing is also prepared by the model. Follow the steps in the main test procedure to check the performance of the model.
(2) Data processing between templates. Each template only completes its own function, and the data transmission between the templates can only be completed by one main control system. If the data processing value is high, a data processing model can be designed for data processing to simplify the operation of the main control system. If the amount of data processing is not large, it is also done by the main control system.
(3) Human-computer interaction design. In order to control the performance of the model or determine the performance of the system, intermediate statistical results or temporary data entry need to design human-computer interaction functions.

4 Experimental Research and Analysis of Coal Mine Safety Risk Intelligent Control and Information Decision Analysis System

4.1 The Calculation Results and Analysis of the Gray Level Comprehensive Analysis of the Hazard of Gas Explosion

This paper conducts simulation test research on the intelligent control and information decision analysis system of a certain mining area. The calculation results of the program are shown in Table 1.

Table 1. Calculation results of gray category weight coefficients of gas explosion hazard

Risk level	Test 1	Test 2	Test 3
Level 1	0.173	0.063	0.022
Level 2	0.205	0.085	0.023
Level 3	0.214	0.073	0.021
Level 4	0.153	0.057	0.016
Level 5	0.109	0.033	0.009

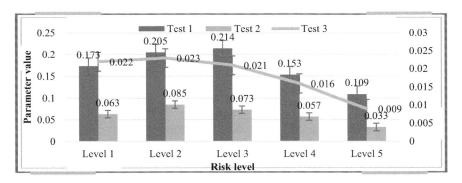

Fig. 1. Calculation results of gray category weight coefficients of gas explosion hazard

The analysis shown in Fig. 1 shows that the hazard of the gas explosion disaster system X belongs to level I–II, but the hazard of the gas explosion hazard system X belongs to level II, that is, it belongs to the safer category, which is consistent with the actual situation, so the safety situation is not optimistic. This is mainly due to gas accumulation hidden danger factors and high temperature fire source hidden danger factors belonging to the second level of influence, and in the influence factors of gas accumulation hidden danger factors, the air duct is too long from the working face, the air leakage of closed facilities, and the impact of ground pressure cause gas abnormalities. Gushing belongs to II–III or III grade and should be prevented immediately. Among the influence factors of high-temperature fire source hidden danger factors, there are transformer fault sparks, metal impact sparks, sparks from mining equipment cutting, and goaf fire sources belong to III Level, it should be treated immediately.

4.2 Examples of the Applicability of the System Model to the Index System

In view of the model index system established for different accidents, the system can analyze the specific accidents by applying the gray-level correlation comprehensive analysis model. Take this mine as an example, apply the gray-level correlation comprehensive analysis model to conduct a gray-level comprehensive analysis of its "coal spontaneous combustion fire hazard" as an example. The data is shown in Table 2.

Table 2. Summary of calculation results of coal spontaneous combustion ash weight coefficient

Risk level	Test 1	Test 2	Test 3
Level 1	0.034	0.049	0.063
Level 2	0.036	0.051	0.036
Level 3	0.035	0.046	0.060
Level 4	0.030	0.034	0.046
Level 5	0.019	0.018	0.014

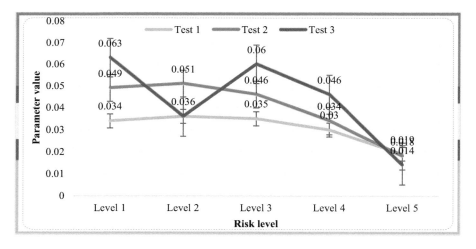

Fig. 2. Summary of calculation results of coal spontaneous combustion ash weight coefficient

As shown in Fig. 2, it can be seen that after 3 tests of the system, the test results are consistent with the actual situation on site, so the prevention and control of coal spontaneous combustion should be strengthened. This is mainly because the hidden danger factor of coal spontaneous combustion tendency belongs to grade III, the hidden danger factor of early warning cannot be dealt with in time and the hidden danger factor of coal seam oxidation and heat production belong to grade I–II.

5   Conclusions

Realize monitoring signal offline analysis and monitoring signal online disposal decision-making application, deal with and analyze faults, defects and accidents in units of events, and let the system guide the power grid regulation and operation work in reverse; realize intelligent mining area monitoring information specification and application, which is intelligent Power grid construction provides signal specifications and technical support; realizes closed-loop management between monitoring and management signals, monitoring fixed-value signals, and monitoring real-time signals, and uses technical means to eliminate problems caused by inconsistent signals in various production links. Through a high degree of analysis and benefit of signal data such as OMS, EMS, and comprehensive automation systems in mining areas, it provides accurate auxiliary decision-making for power grid construction, power grid state maintenance, and power grid operation.

References

1. Chen, J., et al.: Analysis of mine safety performance evaluation law based on matter-element analysis and rough set of concept lattice reduction. IEEE Access **PP**(99), 1 (2021)
2. Reed, G., et al.: An assessment of coal pillar system stability criteria based on a mechanistic evaluation of the interaction between coal pillars and the overburden. Int. J. Mining Sci. Technol. **1**(27), 11–17 (2017)
3. Mathey, M.: Modelling coal pillar stability from mine survey plans in a geographic information system. J. South. Afr. Inst. Mining Metall. **118**(2), 157–164 (2018)
4. Mu, W, et al.: Using numerical simulation for the prediction of mine dewatering from a karst water system underlying the coal seam in the Yuxian Basin, Northern China. Environ. Earth Sci. **77**(5), 215.1–215.19 (2018)
5. Taylor, K.: Arch closing in on permit for met coal mine expansion, but decision not made yet. SNL Energy Coal Rep. **14**(7), 8–9 (2018)
6. Wang, Y.: Prediction of rockburst risk in coal mines based on a locally weighted C4.5 algorithm. IEEE Access **PP**(99), 1 (2021)
7. Xu, J., Sun, C.: The design of coal mine water disasters prevention and control assistance decision system. IOP Conf. Ser. Earth Environ. Sci. **450**(1), 012051 (5pp) (2020)
8. Chen, W., Wang, X.: Coal mine safety intelligent monitoring based on wireless sensor network. IEEE Sens. J. **PP**(99), 1 (2020)
9. Wang, K., et al.: Roof pressure prediction in coal mine based on grey neural network. IEEE Access **PP**(99), 1 (2020)
10. Oliveira, W.D., et al. Power system security assessment for multiple contingencies using multiway decision tree. Electr. Power Syst. Res. **148**, 264–272 (2017)
11. Loiselle, M.E.: The penholder system and the rule of law in the security council decision-making: setback or improvement? Leiden J. Int. Law **33**(1), 1–18 (2019)
12. Ferjani, A., Mann, S., Zimmermann, A.: An evaluation of Swiss agriculture's contribution to food security with decision support system for food security strategy. Br. Food J. **120**(9), 2116–2128 (2018)

Microbial Detection Technology Based on Intelligent Optimization Algorithm

Yu Liu, Huizi Sun(✉), and Xiaoming Dong

The Second Affiliated Hospital of Qiqihar Medical University,
Qiqihaer 161006, Heilongjiang, China
Sunhuizi0110@163.com

Abstract. In recent years, with the large and disorderly widespread use of antibiotics by humans, a large number of drug-resistant strains have appeared and spread widely, and they have been increasing year by year. Their resistance to one drug is often characterized by multiple resistance to one drug, which results in nosocomial infections. Bringing extremely adverse effects, rapid and accurate pathogen identification and drug resistance monitoring are of great significance for the effective control of nosocomial infections and the prevention of the spread of drug-resistant strains. The purpose of this article is to study microbial detection technology based on intelligent optimization algorithms. This article analyzes the basic theoretical knowledge of holographic microscopy principles, holographic imaging optical paths and intelligent optimization algorithms, and introduces intelligent optimization algorithms and embedded related principles and technologies. Then the hardware platform, software architecture and intelligent optimization algorithm of the system are designed. In this paper, a digital holographic microscope is used to collect pathogen microbial images, and intelligent optimization algorithms are introduced to perform classification calculations to achieve classification and detection of pathogen microbial holographic image data sets. Experimental research shows that when the number of hidden layers of the intelligent optimization algorithm is 2, and the neural network structure with 10 neurons in each hidden layer has the smallest error in the estimation of the number of microorganisms.

Keywords: Intelligent optimization algorithm · Microbial detection · Contour extraction · Quantitative evaluation

1 Introduction

With the prosperity of the economy and the improvement of medical standards, modern medical biotechnology is also developing rapidly. From the completion of the Human Genome Project, to the continuous deepening of research in disciplines such as biochemistry, molecular biology, immunology, modern instrumental analysis, and computers [1, 2]. New pathogen detection technologies continue to emerge and are widely used in the detection of pathogenic microorganisms, making the detection technology of pathogenic microorganisms mature. The existing pathogen detection technologies are not limited to time-consuming and labor-intensive traditional methods. At present, a variety of efficient and sensitive pathogen detection technologies have been established, and their practicability is getting higher and higher.

In the research of microbial detection technology based on intelligent optimization algorithm, many scholars have studied it and achieved good results. For example, in the wastewater treatment system, Zhou T found biomass by analyzing the imaging characteristics of activated sludge [3]. A A Croxatto uses optical microscopes and color video recorders to capture microbial images and store and analyze digital images [4]. It can be seen that the research on microbial detection technology based on intelligent optimization algorithms is of great significance to the development of national science and technology.

This paper designs an algorithm for extracting pathogen microbial contours based on intelligent optimization. First, design the required network structure and training method based on the principle of traditional intelligent optimization algorithm. Finally, the region filling and morphological filtering are combined to realize the extraction of weak edge contours.

2 Research on Microbial Detection Technology Based on Intelligent Optimization Algorithm

2.1 The Design of the Typical Module of the System

(1) Microbiology report form module
   The dust report module is an important module in this system. Since the MIC method is selected for the drug sensitivity test used in the current system in the hospital, the drugs are fixed on the drug sensitivity plate by the manufacturer and cannot be changed. This way, some drugs are not clinically available, but some clinical drug sensitivity plates Nothing on it. Therefore, when we report to provide clinical reference for medication, the scope is reduced, so the original system has certain limitations for clinical reference.

(2) Drug library maintenance sub-module
   The drug library maintenance sub-module belongs to the foreground operation part. All personnel in the laboratory can maintain drug information. After the user enters the drug letter on the left, the sergeant adds a button, and the corresponding appointment information will be displayed in the table on the right. Click the save button to save the drug information. You can set whether the medicine expires or not through the radio button Expiration Set Kidney Series.

2.2 Computational Research on Rapid Microbial Detection

(1) Computational evaluation method for microbial detection
   The main application scenarios of the microbial detection algorithm based on next-generation sequencing are as described above, and the purpose is to quickly and accurately detect the known or unknown microbial gene sequence in the next-generation sequencing sample mixed with the background noise of the human

gene sequence. In practical applications, the evaluation indicators of the algorithm may include detection speed, detection accuracy, scalable speedup, false positive rate and false negative rate, etc. There are many types of data sets that may appear in actual scenarios. First, the sample to be tested may come from the blood, sputum or intestinal flora of the subject, and the microbial gene sequence and background noise sequence contained will be very different; secondly, different sequencing the length of reads brought by the instrument and the sequencing method is different. Generally speaking, the length of next-generation sequencing reads ranges from 35 bp to 250 bp. Third, different samples and different sequencing depths may bring data sizes from tens of Mb to tens of Gb. And so on; the base mutation rate of different pathogenic causes may also have a great influence on the detection accuracy results, etc.

(2) Optimization of microbial detection algorithm

The first is the process-level optimization, including the method used in RINS, which uses prior knowledge to determine the general type of microorganisms, and then directly compares with the microorganism reference database, which is an optimization point to improve the detection speed; and the method used in PathSeq, multiple cycles Comparison, providing different comparison algorithms and parameters, and completely removing the host gene sequence is an optimization point to improve detection accuracy.

The second is the comparison algorithm level optimization, including the precise k-mer matching used in Kraken, ignoring SNV, simplifying the algorithm structure, and labeling the reference database to match the k-mer comparison method, which is an optimization point to improve the detection speed; And the LDA comparison method used in Kraken. Because it uses k-mer for comparison, it may be successfully compared with multiple references. In this way, LDA is used to classify to the upper level to determine the species, which is an improvement in detection accuracy. Optimization points; there is also the method of using the CS-SCORE value in CS-SCORE to calculate and use the value to divide the data, and finally the method of comparison, which belongs to the optimization point to improve the detection speed; also includes the data division in Readscan, And then use heterogeneous systems such as multi-core CPU or LGE to perform multi-process/thread analysis, which is an optimization point to improve the detection speed [5].

The third is the optimization of the reference database level, including the establishment of multiple reference gene databases in Readscan, the use of prior knowledge to determine the general category of organisms before running, and then the method of specifying a single microbial reference database for comparison, which is an optimization point to improve the detection speed; And the method of preprocessing the reference database in Kraken and CS-SCORE, or k-mer division, or division by CS-SCORE value, and the same processing with sample data, and the method of comparison, which is to improve the detection speed Optimization point.

2.3   Research on Network in Network Network Image Recognition Algorithm Based on Particle Swarm Optimization

(1) Global average pooling

Traditional CNN is calculated in the lower layer. In the final convolutional layer, the feature map usually needs to be vectorized for classification tasks and sent to the fully connected layer and the subsequent Softmax logistic regression layer. This kind of fully connected layer easily leads to the problem of overfitting. To solve such problems, the global average pooling method can be used to replace the traditional fully connected layer in CNN, thereby improving the generalization ability of the network. In the last convolutional layer of MLP, each feature map corresponds to each classification task category. Then obtain the mean value of each feature map, and then send the generated vector directly into the Softmax layer [4, 6]. The advantage of this method is that no parameter optimization is required, and the feature mapping relationship is simple. The spatial information can be used to treat the global average pooling as a structural regularization matrix, so as to force the feature mapping to correspond to the confidence map category in an explicit manner.

CNN Tradition is at the bottom. At the last convergence level, the feature map must always be vectorized for the performance function and sent to the fully connected level and the next Softmax regression level. This fully bonded layer can easily lead to cracking problems. In order to solve these problems, the method of global network can be used instead of CNN, thereby improving the overall capability of the network. At the final convergence level of MLP, each feature map corresponds to each specific functional category [7, 8]. After that, the average value of each feature map is taken, and then the original vector is sent directly to the Softmax layer. The advantage of this method is that no parameter optimization is required, and the relative structure of the image is simple. Field information can be used to maintain the global average as a system update matrix to force feature maps to be transparently suitable for reliable map categories.

(2) Intelligent optimization algorithm improvement method

Compared with the basic intelligent optimization model, the difference of the method in this paper is that the intelligent optimization model is optimized in two stages, instead of only the optimization model based on the gradient descent method. In the first stage, this paper uses a gradient-based method to pre-train the intelligent optimization model [9, 10]. In this stage, the network structure, objective function, hyperparameter settings used and basic intelligent optimization model are used; in the second stage, it is mainly to further explore the solution space, in order to fully explore the solution space.

2.4   Design of Holographic Microscopic Image Recognition System

(1) Optical path and image acquisition

The digital holographic optical path has two forms, namely the coaxial holographic optical path and the off-axis holographic optical path. Compared with the off-axis holographic optical path, the coaxial optical path has a simple structure, requires less optical components, and takes up less space. For digital holographic microscopy technology, the coaxial optical path can better reflect its portability and stability than off-axis. The coaxial holographic optical path only needs part of the coherent light to work normally, and the relative position accuracy of the components in the optical path is not high, which makes the built system more robust [5]. The coaxial optical path can obtain clear holograms for tiny particles and transparent objects. The off-axis holographic optical path needs to use a relatively high standard coherent light source, a large number of optical components are required to build the optical path, and the optical path components are required to have high precision to meet their phase difference requirements, so that the twin images can be separated, the conjugate image can be eliminated, and a clear and interference-free image can be obtained. Due to the complexity of the off-axis holographic optical path, portability is relatively difficult and the stability is low.

Therefore, the coaxial holographic optical path is more suitable for this system. The light source used in this system is a red LED lamp, which emits 650 nm partially coherent light, and 200 μm micro-holes are used to shield the light source to enhance its coherence. Part of the coherent light emitted by the light source is divided into two parts. Among them, the measured object is directly illuminated. The light folded by the measured object is called the object light wave, which contains the characteristic information of the measured object, and the unrefracted light is called the reference wave. The object light wave interferes with the reference wave in the imaging plane, and its interference fringes are captured by the image sensor CCD. The interference image captured by the CCD is the hologram with the feature information of the object, which is transmitted upward via the data bus and processed by the embedded platform RK3399 processor. In order to make the holographic image more accurate, this system uses a 3.75 μm pixel size black and white CCD sensor (SONY ICX445) as the image sensor.

(2) User information interaction
    Due to the portability requirements of this system, the display output, mouse and keyboard input methods in the traditional digital instrument interaction method are not suitable for the digital holographic microscopy system in this article. This system uses an LCD liquid crystal screen to output data to the user, and the touch screen collects user input instructions. This article uses a 720P resolution LCD screen to output data, its size is 5.5 in., and a touch screen of the same size is attached to it for user input. Due to the popularity of smart phones, the touch screen interaction method has lower learning costs for users and is more usable. The LCD screen is connected to the RK3399 core through the LCD bus, and the system controls the screen display through the display driver interface. The user input data is connected to the 2IC bus through the touch screen, and the processing core defines the input device through the touch screen driver, and sends the input information to the application program through the device interface.

2.5   Determination of the Immigration Rate of Bacterial Flora

First, NP habitats x are arranged in descending order of their corresponding habitat suitability index HSI values, and then the number of species Si is determined by formula (1). In the formula, the variable i is the label of x after sorting.

$$S_i = S_{max} - i, i = 1, 2, \ldots, NP \tag{1}$$

Secondly, the immigration rate λi and the emigration rate μi of the habitat xi can be calculated by formula (2).

$$\begin{cases} \lambda_i = I\left(1 - \dfrac{S_i}{S_{max}}\right) \\ \mu_i = \dfrac{E \times S_i}{S_{max}} \end{cases} \tag{2}$$

Among them, I and E are the parameters that affect the maximum emigration rate and the maximum emigration rate of the habitat.

3   Experimental Research on Microbial Detection Technology Based on Intelligent Optimization Algorithm

3.1   Repeatability Test

Select 2 samples, the number of bacteria contained in each gram or milliliter in these two samples is different in order of magnitude, and the same sample is repeatedly tested with the rapid detection system 10 times.

3.2   Estimation Method

Since what is obtained in the national standard is the maximum possible value (MPN value) of living coliforms, the total number of bacteria and the total number of bacilli used for estimation should also be the number of live bacteria. These two sets of data can be determined by the aforementioned rapid detection system based on computer vision Obtained, because the image digitization may produce corner effects, the calculated eccentricity of the cocci will be slightly greater than 1, so the bacteria with the eccentricity greater than 1.10 are counted as bacilli.

4   Experimental Research and Analysis of Microbial Detection Technology Based on Intelligent Optimization Algorithm

4.1   Repeatability Test Analysis

The coefficient of variation is used to describe the relative dispersion of the data. The ratio of the standard deviation to the average becomes the coefficient of variation. The coefficient of variation can eliminate the influence of different units and (or) averages on the comparison of the degree of variation of two or more data. The experimental investigation results are shown in Table 1.

Table 1. Repeatability test analysis

Sample number	Bottled mineral water	Cereal flour products
1	71	287
2	73	308
3	71	286
4	70	279
5	66	263
6	69	282

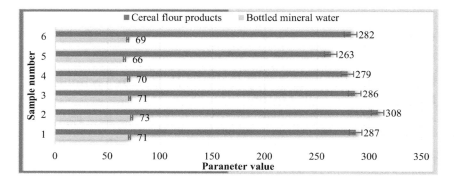

Fig. 1. Repeatability test analysis

As shown in Fig. 1, it can be seen that the method in this paper has good reproducibility, and the higher the bacterial concentration of the sample, the smaller the degree of variation of the method. When the concentration of the sample is low, the reproducibility of the results is slightly worse, and the degree of variation of the method increases.

4.2   Determination of Network Structure

Here we mainly optimize the number of hidden layers and the number of neurons contained in each hidden layer, and establish artificial neural networks with different structures. The root mean error (RMSE) and absolute mean error (MAE) and the correlation coefficient R ($0 \leq R2 \leq 1$) between the estimated value and the actual value are used to measure the estimation ability of different network structures. The experimental results are shown in Table 2.

Table 2. Analysis of the estimation results of the number of coliforms of artificial neural networks with different structures (MPN/100 g)

Hidden layers	Number of neurons per layer	RMSE	MAE
1	10	17.52	10.24
1	20	17.50	10.32
1	30	17.32	9.52
2	10	17.46	10.67
2	20	17.89	10.25
2	30	17.92	9.37

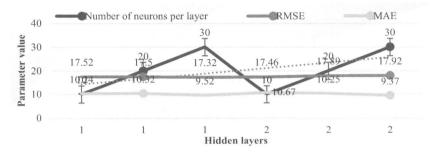

Fig. 2. Analysis of the estimation results of the number of coliforms with different structures of artificial neural networks (MPN/100 g)

It can be seen from Fig. 2 that the number of hidden layers is 2, and the neural network structure with 10 neurons in each hidden layer has the smallest error in the estimation results. Since the number of input and output neurons of the network is small and the calculation time is short, the accuracy of the estimation result should be considered first. Therefore, a neural network with a structure of 2-10-10-1 is selected. The logarithmic sigmoid transfer function (logsig) is used between the input layer and the hidden layer and between the two hidden layers, and the linear transfer function is used between the hidden layer and the output layer. The training function uses the Levenberg-Marquardt algorithm (trainlm) that is commonly used when approximating the BP network function.

5   Conclusions

This paper designs and implements a microbial holographic microscopic image classification system based on convolutional neural network, deploys a streamlined GoogLeNet-Lite network model under the embedded platform, and detects and recognizes the original holographic image of microorganisms. The self-collected data set test results are not ideal. It is necessary to improve the coaxial holographic optical path to improve the imaging quality, and increase the CCD heat dissipation system to make its work stable, so as to obtain more high-quality field data to join the training, so that the system can be used in the industrial field.

Acknowledgements. In 2021, Qiqihar science and Technology Bureau, innovation incentive project Project Name: Study on the application effect of molecular biology technology in clinical pathogenic microorganism detection Project NO. CSFGG-2021335.

References

1. Lee, H., et al.: Microbial respiration-based detection of enrofloxacin in milk using capillary-tube indicators. Sens. Actuators B: Chem. **244**, 559–564 (2017)
2. Wang, J., Hao, N., Wu, W.: Detection of toxic substances in microbial fuel cells. Sheng wu gong cheng xue bao = Chin. J. Biotechnol. **33**(5), 720–729 (2017)
3. Zhou, T., et al.: Microbial fuels cell-based biosensor for toxicity detection: a review. Sensors **17**(10), 2230 (2017)
4. Croxatto, A.A., et al.: Towards automated detection, semi-quantification and identification of microbial growth in clinical bacteriology: a proof of concept. Biomed. J. **40**(6), 317–328 (2017)
5. Khan, M., Oh, S.W., Kim, Y.J.: Power of scanning electron microscopy and energy dispersive X-ray analysis in rapid microbial detection and identification at the single cell level. Sci. Rep. **10**(1), 2368 (2020)
6. Uusitalo, S., et al.: Stability optimization of microbial surface-enhanced Raman spectroscopy detection with immunomagnetic separation beads. Opt. Eng. **56**(3), 037102 (2017)
7. Dm, A., et al.: Evaluation of 16S rRNA broad range PCR assay for microbial detection in serum specimens in sepsis patients. J. Infect. Public Health **13**(7), 998–1002 (2020)
8. O'Dwyer, M., et al.: The detection of microbial DNA but not cultured bacteria is associated with increased mortality in patients with suspected sepsis—a prospective multi-centre European observational study. Clin. Microbiol. Infect. **23**(3), 208.e1-208.e6 (2017)
9. Mritunjay, S.K., Kumar, V.: Microbial quality, safety, and pathogen detection by using quantitative PCR of raw salad vegetables sold in Dhanbad City, India. J. Food Prot. **80**(1), 121 (2017)
10. Zhang, X.: Study on microbial detection technology in food safety. E3S Web Conf. **233**(10), 02029 (2021)

Artificial Intelligence Applied in Electrical Engineering Automation

Changsheng Bi[1(✉)], Libin Liu[1], Ming Zhou[1], Shangyi Liu[1],
Zhiyuan Zhao[1], and Lei Li[2]

[1] Dalian Power Supply Company, State Grid Corporation of China,
Dalian, Liaoning, China
1060610108@alu.hit.edu.cn
[2] Urban Rail Technical Development Department,
CRRC Dalian Locomotive and Rolling Stock Co., Ltd., Dalian, Liaoning, China

Abstract. Electrical engineering automation is an independent discipline, which contains many research projects, such as electrical automation control systems, electronic information applications, and computer intelligent design. The experiments in this paper found that the speed and angular velocity of the intelligent robot gradually stabilized and were close to the expected values; the position deviation and angle of the intelligent robot gradually became consistent and stable in the later stage, indicating that during the operation of the intelligent robot, fuzzy controllers were used to perform the trajectory. The feasibility of real-time tracking and its ability to control the operation path of the intelligent robot close to the ideal trajectory.

Keywords: Electrical engineering · Artificial intelligence · Electric automation · Intelligent robot · Path planning

1 Introduction

The birth of robots is undoubtedly one of the greatest achievements of humankind in the 20th century. The so-called robots, as the name implies, are mechanical devices that simulate human behavior and have the ability to automate [1, 2]. In order to achieve the goal that artificial intelligence technology can be applied to the production process of electrical engineering automation, we should work harder to find the best route [3, 4].

Wu Z. studied the covariance intersection algorithm of random vector distribution estimation and information fusion. His research focuses on partially related random vectors. His results will make the already popular covariance crossover more applicable and more accurate for distributed estimation and information fusion problems [5]. Baranov LA considers reviewing methods to improve energy efficiency to automatically control the movement of subway trains: choosing a train control mode with an optimal sequence of energy, allocating time between the tourist substations along the line, and using a regenerative braking mode. In the case of heavy traffic, the allowable error for a given train running time is required to be ±2.5 s [6]. Mudunuri S.P. proposed a method for fully automatic recognition of low-resolution face images in uncontrolled environments. Learning common methods uses a multidimensional

© The Author(s), under exclusive license to Springer Nature Switzerland AG 2022
V. Sugumaran et al. (Eds.): ICMMIA 2022, LNDECT 138, pp. 62–69, 2022.
https://doi.org/10.1007/978-3-031-05484-6_8

scaling transformation matrix to transform low-resolution [7]. Ondrej Kovac proposed an improved bee algorithm for multi-target search and coverage of robotic "bees" [8].

In this paper, the ant colony algorithm is improved to solve the defects of mobile robot path planning problem. An improved ant colony algorithm based on ant colony classification and search strategy is proposed, and a new path evaluation function is constructed, and dual feedback based on the introduction.

2 Proposed Method

2.1 Convolutional Neural Network

(1) Convolutional layer

The convolution kernel and the input image convolution operation are used to extract features from the input image [9, 10]. Therefore, the convolution layer is also called a feature in terms of function. The extraction layer, and the output generated by the convolution layer is called the output map (Feature Map) [11, 12]. The k-th feature map Y_k in the same convolutional layer can be expressed as [13, 14]:

$$Y_k = f(W_k * x) \tag{1}$$

In the formula, f represents a non-linear activation function, the input image is represented by x. The calculation formula for 2D convolution is:

$$S(i,j) = (W_k * x)(i,j) = \sum_m \sum_n I(i+m, j+nW_K(m,n)) \tag{2}$$

The process of convolution operation performed by the convolution layer through the convolution kernel is shown in Fig. 1 [15]. If a convolution error occurs due to the step size or the size of the convolution kernel, you need to zero-pad the input image.

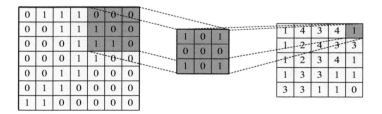

Fig. 1. Convolution kernel calculation process

The calculation of the convolutional layer is shown in the following formula:

$$X_j^i = f\left(\sum_{j \in M_j} (X_j^{l-1} * k_{j,i}^l + b_j^l)\right) \tag{3}$$

X^l_j represents the j-th feature map of the l-th convolution layer; f(x) represents the activation function; M_j represents the output set of the previous layer; $k^l_{j,i}$ represents the convolution kernel; b^l_j represents the additive bias of the convolution feature map X^i_j; * Means convolution operation.

(2) Pooling layer

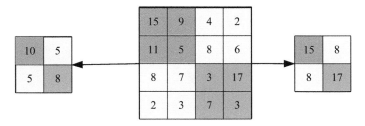

Fig. 2. Average pooling and maximum pooling

Maximum value pooling can make the highest element in the receptive field output to the next layer:

$$Y_{kij} = \max_{(p,q)\in R_{i,j}} x_{kpg} R_{ij} \qquad (4)$$

Y_{kij} is the largest pooled output of the k-th feature map; x_{kpg} is the element size on (p, q) in the pooled area R_{ij}. As shown in Fig. 2, the maximum pooling and average pooling are visualized respectively.

(3) Activation function layer
    The Sigmord function increases monotonically over the interval [0,1] and is calculated by the following formula:

$$f(x) = \frac{1}{1 + e^{-x}} \qquad (5)$$

The Tanh function is a modification of the Sigmord function. The following formula calculates:

$$f(x) = \frac{e^x - e^{-x}}{e^x + e^{-x}} \qquad (6)$$

In the development after Relu, the deformed Leaky Relu and other functions of Relu are derived. The specific form of Relu is shown below:

$$f(x) = \begin{matrix} x & x < 0 \\ 0 & x > 0 \end{matrix} \qquad (7)$$

(4) Fully connected layer and classifier

The fully connected layer can convert the output two-dimensional feature map into a one-dimensional vector. The calculation formula is shown in Eq. (8):

$$f(x) = W \times x + b \tag{8}$$

(5) Local receptive field

Such a neuron connection will have the best response to the receptive field of the input picture, as shown in Fig. 3.

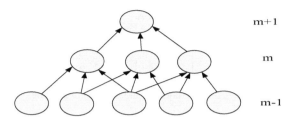

Fig. 3. Schematic diagram of local receptive field neuron connections

Assume that the m − 1 layer is an input, the input is generally an image, and the receptive field of the m layer is 3.

2.2 Testing Standards

Intersection over union (IoU) is often used as a reference standard for evaluating target detection accuracy in target detection tasks.

The intersection ratio of rectangular frame A and rectangular frame B is:

$$IoU = (A \cap B)/(A/B) \tag{9}$$

That is, the overlapping area of the two rectangular frames and the two rectangular frames are combined to form the proportion of the graphic area.

3 Experiments

3.1 Experimental Design

The environmental model is to be able to better plan the path of the robot. The quality of the environmental modeling method also has a great impact on the quality of the path planning. Therefore, the quality of the path planning is inseparable from the method of environmental modeling of.

3.2   Data Acquisition

Step 1: Construct the environmental information matrix of the robot's working space, and use the grid method to construct a mathematical model.

Step 2: Parameter initialization, including the weight coefficients of the path evaluation function as w_1, w_2, w_3, the safety threshold $d_{threshold}$ of the distance between the two turns of the robot, the number of two types of ants m_1 and m_2;

4   Discussion

4.1   Analysis of Simulation Results of the Algorithm

(1)   Optimization of the algorithm

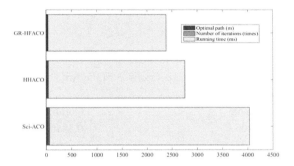

Fig. 4. Algorithm results when the obstacle is globally known

As shown in Fig. 4, the classification ant colony algorithm can plan an optimal path for the robot even in a more complex space environment.

(2)   Convergence speed of the algorithm

Table 1. Operation results of the algorithm with known obstacles

Algorithm	Optimal path (m)	Number of iterations (times)	Running time (ms)	Success rate
Sci-ACO	62	40	5073	90%
HHACO	40	17	4237	20%
GR-HFACO	30	12	3427	100%

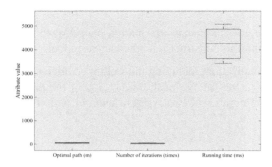

Fig. 5. The results of the algorithm when the obstacle part is known

As shown in Table 1 and Fig. 5, the path planning simulation research of the improved ant colony algorithm and the basic ant colony algorithm is performed in a variety of grid environments.
(3) Obstacle avoidance of the algorithm

Table 2. Performance comparison of the two algorithms

	Conventional artificial fish school algorithm	Conventional artificial fish school algorithm
Optimal path	30.9706	30.3848
Average path length	32.9232	32.5563
Longest path length	35.0711	34.3848

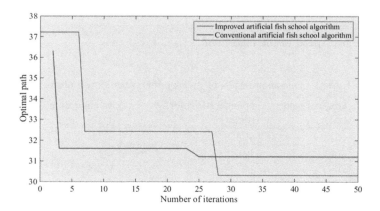

Fig. 6. Convergence simulation curve

As shown in Table 2 and Fig. 6, in the robot's working space, when obstacles are densely distributed, the number of U-shaped traps is large, and the depth is deep, the basic ant colony algorithm is often used for path planning research.

4.2 Application of Artificial Intelligence Technology in Electrical Automation

Application of artificial intelligence technology in equipment diagnosis in electrical automation.

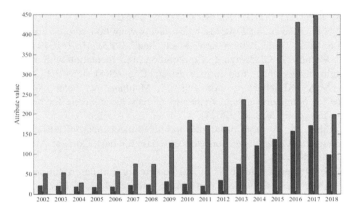

Fig. 7. Artificial intelligence enterprise development

As shown in Fig. 7, in the traditional work, when the equipment fails, it mainly depends on manual maintenance. These repair workers mainly rely on their work experience and their familiarity with the work to repair.

5   Conclusions

Path planning research is an important prerequisite to realize the autonomous movement of electric automation robots and improve the robot's intelligence level. This paper uses ant colony algorithm to solve the problem of robot path planning. Traditional ant colony algorithm has the disadvantages of slow convergence and easy fall into local optimum when solving path planning problems, and the planned path is poor in smoothness and low security, which is not good for robot Precise tracking control.

References

1. Zhao, C.: Application of virtual reality and artificial intelligence technology in fitness clubs. Math. Probl. Eng. **2021**(20), 1–11 (2021)

2. Mao, K., et al.: An artificial intelligence platform for the diagnosis and surgical planning of strabismus using corneal light-reflection photos. Ann. Transl. Med. **9**(5), 374 (2021)
3. Fornasier, M.: The regulation of the use of artificial intelligence (AI) in warfare: between International Humanitarian Law (IHL) and Meaningful Human Control. Revista Jurídica da Presidência **23**(129), 67 (2021)
4. Palomino, R., Low, K.-B., Ji, C., Petrovic, I., Waltz, F., Schmitz, T.: Micro computed tomography analysis of four-way conversion catalysts using artificial intelligence-enabled image processing. Microsc. Microanal. **27**(S1), 1028–1029 (2021)
5. Wu, Z., Cai, Q., Fu, M.: Covariance intersection for partially correlated random vectors. IEEE Trans. Automat. Contr. **63**(3), 619–629 (2018)
6. Baranov, L.A., Maksimov, V.M.: The energy efficiency of an automatic control system of subway train movement and requirements for its subsystems. Russ. Electr. Eng. **89**(9), 546–549 (2018). https://doi.org/10.3103/S1068371218090031
7. Mudunuri, S.P., Biswas, S.: Low resolution face recognition across variations in pose and illumination. IEEE Trans. Pattern Anal. Mach. Intell. **38**(5), 1034–1040 (2016)
8. Kovac, O., Mihalik, J., Gladisova, I.: Convolution implementation with a novel approach of DGHM multiwavelet image transform. J. Electr. Eng. **68**(6), 455–462 (2018)
9. Lashin, M.M.A., Khokhar, A., Alrowais, F., Malibari, A., Saleh, W.: Using artificial intelligence for optimizing natural frequency of recycled concrete for mechanical machine foundation. Recycling **6**(3), 43 (2021)
10. Yang, Z., Talha, M.: A coordinated and optimized mechanism of artificial intelligence for student management by college counselors based on big data. Comput. Math. Methods Med. **2021**(6), 1–11 (2021)
11. Boyraz, P., Bayraktar, E.: Analysis of feature detector and descriptor combinations with a localization experiment for various performance metrics. Turk. J. Electr. Eng. Comput. Sci. **25**(3), 2444–2454 (2017)
12. Lourdes, V.G.: New challenges for ethics: the social impact of posthumanism, robots, and artificial intelligence. J. Healthc. Eng. **2021**(6), 1–8 (2021)
13. Hariharan, R., Rahul, I., Darshanam, M.D.: MPPT based on artificial intelligence system for photovoltaic system using virtual instrumentation. J. Crit. Rev. **7**(5), 1284–1290 (2021)
14. Garingan, D., Pickard, A.J.: Artificial intelligence in legal practice: exploring theoretical frameworks for algorithmic literacy in the legal information profession. Leg. Inf. Manag. **21**(2), 97–117 (2021)
15. Kuric, I., et al.: Approach to automated visual inspection of objects based on artificial intelligence. Appl. Sci. **12**(2), 864 (2022)

Line Loss Allocation of Distribution Network Capacity Based on Optimal Multi-objective Programming with Annealing Algorithm

Zhe Chen[1], Bingcheng Cen[1], Dandan Zhu[1], Qian Zhou[1],
Zhouyue Ling[2(✉)], and Xiaolong Yang[3]

[1] Electric Power Research Institute of State Grid Jiangsu Electric Power Co.,
Ltd., Nanjing, Jiangsu, China
[2] School of Economics and Management, North China Electric Power
University, Beijing, Beijing, China
lingzhouyue@163.com
[3] Northeast Electric Power University, Jilin, Jilin, China

Abstract. Considering the large light no-load loss of the transformer on the power side and the potential of cost sharing on the user side, based on the distribution network line loss sharing, this paper establishes a multi-objective programming model for distribution network line loss sharing from the goal of minimum network loss and minimum carbon emission, and considering the factors such as transformer capacity, line loss cost and line loss rate. Finally, the annealing algorithm is used to solve the model, the penalty function analysis of the constraint conditions is carried out, and the convergent feasible optimization solution is obtained. The comparison of multiple schemes shows that the load rate of the transformer is between 60%–70%, and the line loss can be further planned according to the actual demand. The effectiveness of the model and the correctness of the conclusion are verified by simulation.

Keywords: Line loss allocation · Multi-objective programming · Annealing algorithm

1 Introduction

With the deepening of the construction of new power system, the grid connection of DGs has become the development trend of distribution network [1]. However, with the increase of the proportion of new energy grid connection, the problem of network loss allocation will be paid attention to by the power supply side and the user side [2].

There have been a lot of research on distribution network loss allocation at home and abroad, reference [3] simplified the difficulties associated with the decomposition of cross-term of power loss equation mathematically. In addition, the network loss of distribution network is also affected by many factors such as transformer location and grid connected operation mode [4]. Therefore, it is necessary to study the influence of transformer in distribution network loss, so as to encourage the development of distribution network to reduce loss and carbon.

© The Author(s), under exclusive license to Springer Nature Switzerland AG 2022
V. Sugumaran et al. (Eds.): ICMMIA 2022, LNDECT 138, pp. 70–78, 2022.
https://doi.org/10.1007/978-3-031-05484-6_9

In terms of solving line loss allocation by multi-objective programming, reference [5] introduced carbon emission constraints into the model. Reference [6] proposes that the goal should be to reduce the installation cost of transmission lines and the annual operation cost of conventional generator units. Reference [7, 8] uses the multi-objective programming method to analyze the cost-benefit of power grid. Reference [9] establishes a multi-objective programming model with minimum investment cost and minimum network loss cost. In the field of DG, reference [10] establishes a multi-objective and phased planning model by considering distributed generation access. Reference [11] proposed a multi-objective planning method for distributed generation distribution network considering the correlation between uncertainties such as wind speed and load demand. Literature [12] proposed a multi-objective PV planning model, active power losses and static voltage steady index are considered and discussed in the operation of distribution network.

To solve the above problems, this paper intends to establish a distribution network line loss allocation model based on multi-objective programming. Taking the minimum network loss and carbon emission as the objective function, and taking transformer capacity, line loss cost, line loss rate, load demand, voltage level and branch power as constraints, the multi-objective programming scheme is optimized by annealing algorithm, The better solution under the comparison of multiple schemes is obtained. Finally, a line distribution network is taken as an example to verify the practicability of the method.

2 Model Construction

2.1 Multi-objective Programming Objective Function

The mathematical expression of the objective function is:

$$
\begin{cases}
\min F = \{f_1(x), f_2(x)\} \\
\text{s.t.h } (x) = 0 \\
\quad g(x) \leq 0
\end{cases}
\tag{1}
$$

where $f_1(x)$ is the minimum line loss target of distribution network line:

$$
f_1(x) = \lambda_g P + \Delta P
\tag{2}
$$

$$
\Delta P = P_1 + \left(\frac{S}{S_N}\right)^2 P_2
\tag{3}
$$

In Eqs. (2) and (3), P is the total power loss of the line; ΔP is the comprehensive power loss of transformer; P_1 is the no-load loss of transformer; P_2 is the short-circuit loss of the transformer; S is the actual load of the transformer; S_N is the rated capacity of the transformer. Using the improved rage loss allocation method, the network loss coefficient of the distributed generation λ_G is:

$$\lambda_G = \frac{1}{2} \times \frac{W_{loss}}{\sum_{j=0}^{m} W_{Gj} + \sum_{k=0}^{z} W_{Gk}} \tag{4}$$

where j is the node number, m is the number of nodes, W_{Gj} is the distribution output of node j, k is the number of distributed generation node, z is the number of distributed generators, W_{Gk} is the distribution output of node k.

$f_2(x)$ is the minimum carbon emission target:

$$f_2(x) = \sum_{i=1}^{N_b} \gamma_1 (P_{G_i}^t)^2 + \gamma_2 P_{G_i}^t + \gamma_3 + \gamma_4 e^{\varepsilon P_{G_i}^t} \tag{5}$$

In Eq. (5), $P_{G_i}^t$ is the output of distributed power generation in unit i in the t year (MW); γ_1, γ_2, γ_3, γ_4, ε is the discharge coefficient.

2.2   Constraints of Multi-objective Programming

$g_1(x)$ is the transformer capacity constraint:

$$g_1(x) = P_{it} - P_{it}^{\max} = 0 \tag{6}$$

where $P_{it}^{\max}$ is the maximum installed capacity of transformer i in year t(KVA).

$g_2(x)$ is the line loss cost constraint, which is constrained according to the actual situation of local electricity price:

$$g_2(x) = P_{it} - P_{it}^{\max} = 0 \tag{7}$$

In Eq. (7), P is the total power loss of the line, D takes the local electricity price (¥/ kWh).

The percentage of line loss in power supply is called line loss rate $g_3(x)$:

$$g_3(x) = \frac{E_g - E_s}{E_g} \times 100\% \leq 12\% \tag{8}$$

where E_g is the generating capacity (MWh); E_s is the electricity sales (MWh). According to the general line loss rate constraint, the line loss rate is generally controlled within 12%.

In year t, the sum of the installed capacity of distributed energy and the load supplied by the distribution network shall not be less than the maximum load demand of the system in year t, i.e. $g_4(x)$:

$$g_4(x) = (1 - S) \sum_{i=1}^{N} P_{it} + P_j \geq (1 + \eta) P_2 \tag{9}$$

In Eq. (9), S is the line loss rate; P_t is the maximum load of the system in year t (MW); N is the number of distributed generation units in year t; P_j is the load supplied by the distribution network; η is the reserve factor.

Among the voltage samples obtained in the online loss allocation optimization model, those that meet the deterministic voltage constraint should meet a certain degree of confidence:

$$\Pr\{ V_i^{min} \leq V_{i,n} \leq V_i^{max} \} \geq \gamma_U \tag{10}$$

where Pr is the probability of event occurrence; γ_U is the confidence that the voltage meets the constraint; V_i^{min}, V_i^{max} are the lower and upper limits of node i voltage.

The branch power shall not exceed the allowable limit, the same as the voltage. For the uncertainty problem in this paper, the branch power shall meet a certain confidence, which is an opportunity constraint, as shown in the following formula:

$$\Pr\{0 \leq P_k \leq P_k^{max}\} \geq \gamma_P \tag{11}$$

where γ_P is the degree of confidence that the current meets the constraint; P_k^{max} is the upper limit of network loss of branch k.

2.3   Annealing Algorithm Optimization

In this section, the penalty function method is selected to deal with the constraints. By imposing some penalty on the infeasible solution in the optimization calculation, the solution gradually converges to the feasible optimization solution after repeated iteration.

The fitness function expression is:

$$F_{fit}(x,\sigma_k) = \frac{1}{F_1 + F_2 + P(x,\sigma_k)} = \frac{1}{F_1 + F_2 + \sigma_k \sum_{i=1}^{G} (q_p \psi_{Pi} + q_V \psi_{Vi})} \tag{12}$$

where F_{fit} is the fitness function; F_1 is the objective function of line loss; F_2 is the objective function of carbon emission; G is the number of inequality constraints; σ_k is the dynamic penalty factor in the annealing algorithm, which is defined as:

$$\sigma_k = \frac{1}{H_p} \tag{13}$$

$$H_{p+1} = \alpha H_p, \alpha \in [0,1] \tag{14}$$

With the increase of evolutionary algebra, gradually decreases and gradually increases (σ_k is controlled by the total evolutionary algebra to prevent infinite increase), which makes the solution group of the optimized individual tend to the feasible region.

$$\psi_{Pi} = \begin{cases} \frac{P_i - P_{imax}}{P_{imax}} & P_i \geq P_{imax} \\ 0 & P_i < P_{imax} \end{cases} \tag{15}$$

$$\psi_{Vi} = \begin{cases} \frac{V_i - V_{imax}}{V_{imax}} & V_i \geq V_{imax} \\ 0 & V_i < V_{imax} \end{cases} \tag{16}$$

q_P, q_V are power and voltage constraint penalty factors; ψ_{Pi}, ψ_{Vi} are penalty functions of power and voltage constraints, which are used to balance the influence of different power and voltage values and dimensions of each node.

The selected selection probability P_{si} is expressed as:

$$P_{si} = \frac{f_i}{\sum_{i \in N} f_i} n > 0.5\,N \tag{17}$$

where f_i is the fitness value of the individual on chromosome i; i is the serial number of chromosome; N is the size of chromosome population; n is the number of individuals whose fitness value is greater than the average fitness value in the population. The adaptive adjustment mode is as follows:

$$P_c = \begin{cases} P_{cmax} - \frac{(P_{cmax} - P_{cmin})(f' - \overline{f})}{f_{max} - \overline{f}} & f' \geq \overline{f} \\ P_{cmax} & f' < \overline{f} \end{cases} \tag{18}$$

$$P_m = \begin{cases} P_{mmax} - \frac{(P_{mmax} - P_{mmin})(f' - \overline{f})}{f_{max} - \overline{f}} & f' \geq \overline{f} \\ P_{mmax} & f' < \overline{f} \end{cases} \tag{19}$$

where P_c is the crossover probability; P_m is the probability of variation; P_{cmax} and P_{mmax} respectively represent the maximum value of crossover rate; P_{mmax} and P_{mmin} represent the minimum value of variation rate respectively; P_{mmax} is the maximum fitness of the population; $\overline{f}$ is the average fitness of the population; f' is the larger fitness of the two individuals participating in the cross operation; f is the individual fitness of variation.

2.4 Construction of Distribution Network Capacity and Line Loss Allocation Model Based on Annealing Algorithm Optimization Multi-objective Programming

The construction of distribution network line loss allocation model based on multi-objective programming is divided into the following steps:

1) Input the original data of distribution network: obtain the transformer characteristic data and line loss data. Then, the non-technical part of line loss is calculated by statistical line loss and technical line loss, and the line loss data is normalized;

2) Initialization of multi-objective programming model: select the conventional crossover probability, mutation probability range, population size and the value of each objective function;

3) Confirm the size of each matrix: construct the power consumption matrix and its corresponding power supply matrix size through the initial power of the line and the position of the power supply point;

4) A set of initial populations of multi-objective optimization of distribution network are randomly generated: the power supply node matrix is transformed into a self identification decoding method of genetic operation power grid path and power flow;

5) Construct the fitness function of multi-objective optimization of distribution network with penalty function: convert the objective function into fitness function and punish the infeasible solution.

6) Calculate population fitness value: set the number of iterations of hyper parametric optimization. If the model evaluation result does not meet the optimization requirements, go back to step 5) re initialize the fitness function, otherwise go to step 7);

7) Output the multi-objective optimization scheme of distribution network: input the data into the adaptive genetic algorithm model to calculate the objective function value of line loss allocation scheme. The specific calculation process is shown in Fig. 1.

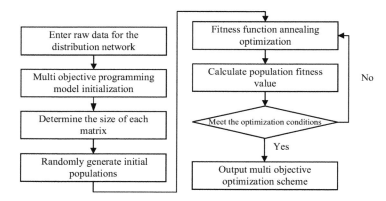

Fig. 1. Calculation flow of multi-objective optimization scheme of distribution network

3   Empirical Analysis

In order to verify the effectiveness and feasibility of multi-objective loss reduction planning of distribution network, a distribution line of a 10kV distribution network is taken as an example to test its effectiveness and operability. The capacity structure of each transformer on the line is shown in Fig. 2.

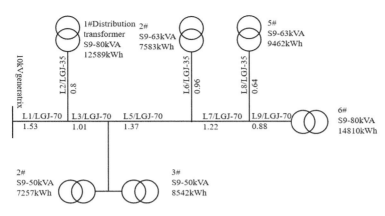

Fig. 2. A distribution line of a 10 kV distribution network

A 10 kV line with a total length of 9.12 km mainly supplies commercial power and residential power. The line scale includes lgj-70, lgj-50 and lgj-35; The whole line is equipped with 6 distribution transformers, all of which are S9 series, with a total capacity of 386 kVA. The total operation time in a month is 720 h, the total active power delivered is 60991 kwh, the reactive power delivered is 34556 kvar, the power factor $\cos q$ is 0.87, and the total reading power of distribution transformer is 60243 kWh.

Considering the minimum network loss and carbon emission, the control parameters of genetic algorithm are selected. According to the selection range of conventional crossover probability and mutation probability and the adaptive adjustment of genetic operator, the population size N = 50 is selected in the optimization, the maximum and minimum values of crossover probability are taken as P_{cmax} = 0.95, P_{cmin} = 0.50, The maximum and minimum values of variation probability are P_{mmax} = 0.1, P_{mmin} = 0.001. The unit construction investment cost is 1 million yuan/km, and the power reference value S_B = 100MVA, the annual network loss hours or maximum load utilization hours are still selected as 2000h according to the current urban reliability requirements, and the current unit price is 0.5 yuan/kWh, the maximum and minimum crossing probability values are P_{cmax} = 0.90, P_{cmin} = 0.60, the maximum and minimum variation probabilities are P_{mmax} = 0.1, P_{mmin} = 0.0005.

The results are shown in Table 1 and Table 2.

Table 1. Result analysis of multi-objective programming

Scheme	Average load rate of transformer	Annual line loss cost, 10000 yuan	Annual carbon emission, t	Line loss, MWh
Scheme1	60%	384.6	9.06×10^6	7692
Scheme2	65%	386.9	9.32×10^6	7738
Scheme3	62%	369.5	9.66×10^6	7390
Single objective scheme	68%	708.6	11.08×10^6	14172

Table 2. Allocation proportion of multi-objective scheme

Transformer	1#	2#	3#	4#	5#	6#
Scheme1	23%	13%	12%	15%	15%	22%
Scheme2	22%	15%	13%	11%	14%	25%
Scheme3	20%	11%	17%	11%	15%	26%

The allocation of line loss can be further planned according to the actual demand. The results show that there are three similar schemes, the comprehensive load rate of the transformer is 60%, 65% and 62%, the annual network loss cost is 384.6, 386.9 and 3.695 million yuan respectively, and the annual carbon emission is 9.06×10^6, 9.32×10^6 and 9.66×10^6 t. The allocation of all transformers on the line is shown in Table 2. It can be seen that with the access of distributed generation, the allocation of line loss in distribution network should also take into account the actual load of transformer. When taking the line loss and carbon emission as the minimum target, the load rate of the transformer in the given reasonable scheme is between 60%–70%, and an optimal solution cannot be found for the minimum line loss and carbon emission. The allocation of line loss can be further planned according to the actual demand.

4   Conclusions

In this paper, the mathematical model of multi-objective distribution network capacity allocation and its minimum emission function are proposed. Its characteristics are as follows: the dynamic penalty factor in the annealing algorithm is used to construct the penalty function to obtain the better solution under the comparison of multiple schemes. Finally, through the simulation calculation of some areas of a 10kV distribution network, it is verified that the method is effective in solving the multi-objective programming optimization problem.

Acknowledgements. This work was supported by loss reduction and carbon reduction consulting service of new power system light load transformer of Electric Power Research Institute of State Grid Jiangsu Electric Power Co., Ltd. (SGJSDK00XTJS2100354).

References

1. Niu, H., Wang, S., Pan, N., Peng, Z., Liu, X.: The influence of grid-connected distributed generation on the power supply reliability of distribution network. J. Phys.: Conf. Ser. **1920** (1), 012033 (2021)
2. Moret, F., Tosatto, A., Baroche, T., Pinson, P.: Loss allocation in joint transmission and distribution peer-to-peer markets. IEEE Trans. Power Syst. **36**(3), 1833–1842 (2020)
3. Hota, A.P., Mishra, S., Mishra, D.P.: Active power loss allocation in radial distribution networks with different load models and DGs. Electr. Power Syst. Res. **205**, 107764 (2022)
4. Jian, L., Shenghong, L., Yong, X.: Research on the influence of distributed generation to low voltage distribution network line loss. Electr. Energy Manag. Technol. **5**, 49–52 (2015)

5. Luo, J., Lu, C., Meng, F.: Generation expansion planning and its benefit evaluation considering carbon emission and coal supply constraints. Autom. Electr. Syst. **40**(11), 47–52 (2016)

6. Ugranli, F., Karatepe, E., Nielsen, A.H.: MILP approach for bilevel transmission and reactive power planning considering wind curtailment. IEEE Trans. Power Syst. **32**(1), 652–661 (2016)

7. Celli, G., Pilo, F., Pisano, G., Soma, G.G.: Cost-benefit analysis for energy storage exploitation in distribution systems. CIRED-Open Access Proc. J. **2017**(1), 2197–2200 (2017)

8. Koutsoukis, N., Georgilakis, P., Hatziargyriou, N.: Active distribution network planning based on a hybrid genetic algorithm-nonlinear programming method. CIRED-Open Access Proc. J. **2017**(1), 2065–2068 (2017)

9. Ziari, I., Ledwich, G., Ghosh, A., Platt, G.: Integrated distribution systems planning to improve reliability under load growth. IEEE Trans. Power Delivery **27**(2), 757–765 (2012)

10. Borges, C., Martins, V.F.: Multistage expansion planning for active distribution networks under demand and distributed generation uncertainties. Int. J. Electr. Power Energy Syst. **36**(1), 107–116 (2012)

11. Zhang, S., Cheng, H., Li, K., Tai, N., Wang, D., Li, F.: Multi-objective distributed generation planning in distribution network considering correlations among uncertainties. Appl. Energy **226**, 743–755 (2018)

12. Cai, C., Chen, J., Xi, M., Tao, Y., Deng, Z.: Multi-objective planning of distributed photovoltaic power generation based on multi-attribute decision making theory. IEEE Access **8**, 223021–223029 (2020)

Design of Full Life-Cycle Management System for Construction Project Based on "BIM+GIS"

Yang Sun[(✉)]

Liaoning Vocational University of Technology, Jinzhou, Liaoning, China
lnl239@outlook.com

Abstract. BIM technology is based on multi-dimensional figures, and by establishing a three-dimensional engineering information data model, it significantly improves the efficiency and quality of work, and reduces the probability of errors and risks. GIS represents the processing results in the form of maps, graphics or data, and meets the requirements of digitalisation and visualisation for construction management. "BIM+GIS" can facilitate the transformation of the management mode of construction projects through data integration, system integration or application integration. In this paper, we study the application scheme and spatial data storage model of "BIM+GIS", and design a system consisting of "3D visualization subsystem, Integrated management subsystem, Control management subsystem, monitoring management subsystem, information expansion subsystem, system maintenance subsystem" and other six subsystems. The system is designed to provide a solution for the development of a lifecycle management system for construction projects.

Keywords: BIM+GIS · Building information modeling · Geographic information system · Construction project · System design

1 Introduction

Construction projects generally have problems such as high energy consumption, serious waste of resources and low productivity. In order to change the status quo of the construction industry, developed countries in the West have started to carry out in-depth analysis and research on the various problems that currently exist from information technology, and promote changes in project management methods through information technology. BIM technology is based on multi-dimensional figures, and by establishing a three-dimensional engineering information data model, it enables different participants from the project to exchange and share information data through this model during the whole life cycle of the construction project, thus significantly improving work efficiency and project quality and reducing the probability of error and risk generation. BIM aims to use 3D digital technology to present an engineering model of all relevant information of a construction project, build a digital platform for the aggregation and interaction of information of all related parties of a construction project, and meet the needs of the whole life cycle management of the project [1, 2]. Modern construction is developing in the direction of complexity and systematization. GIS represents the processing results in the form of maps, graphics or data, which meets the requirements of construction

© The Author(s), under exclusive license to Springer Nature Switzerland AG 2022
V. Sugumaran et al. (Eds.): ICMMIA 2022, LNDECT 138, pp. 79–85, 2022.
https://doi.org/10.1007/978-3-031-05484-6_10

digitization and visualization and provides new ways for construction project management [3]. "BIM+GIS" technology will lead the future development direction of the construction industry and promote the transformation of the management mode of construction projects, which has great exploration value [4].

GIS-BIM integration can be used in all phases of the full lifecycle management of a construction project: firstly, in the planning and design phase. Secondly, the construction phase. Provides a methodological component for visual safety management, for project cost control, applied to GIS cost clusters for time analysis and cost scenario forecasting. Thirdly, the operations and maintenance phase. Enhanced BIM-GIS integration provides emergency response, prevention, disaster modelling, response strategies and heritage protection for solving complex problems, as well as large-scale project operations, indoor navigation and ecological rating to assist staff with issues such as sustainability assessment and asset management. Fourthly, demolition is the final stage of a construction project. This phase generates a large amount of waste and BIM can be used for waste estimation or efficient planning and GIS for analysis and optimization of the waste distribution process. Enhanced BIM-GIS integration optimizes waste reuse and recycling, minimizes waste materials and assesses demolition times and impacts on the surrounding environment.

2   Component Software of BIM

Dana K. Smitn, President of the building SMART Consortium, in his book "Building Information Modeling - A Strategic Implementation Guide for Architects, Engineers", states: "The days of relying on one software to solve all problems are gone". BIM is not a single piece of software, but a complete system of multiple pieces of software, as shown in Fig. 1.

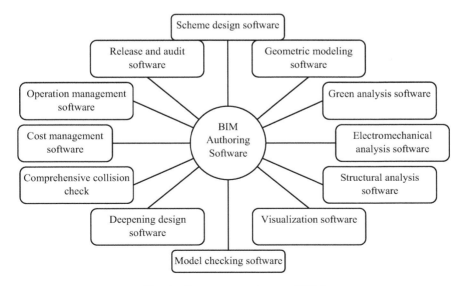

Fig. 1. Component software of BIM

2.1   Core Modeling Software in BIM

BIM core modelling software is the foundation of BIM and consists of four main components [5, 6]: firstly, Autodesk's Revit architectural, structural and M&E series, which has a good market presence in the civil construction market with the advantage of AutoCAD. Secondly, the Bentley Building, Structures and Equipment range, in the areas of plant design and infrastructure, has a clear advantage. Thirdly, Nemetschek Graphisoft includes ArchiCAD, which is part of the global market-oriented products and is the core BIM modelling software with the most market impact. Fourthly, Gery Technology Dassault is the world's most advanced mechanical design and manufacturing software, applied to the engineering and construction industry, modeling capabilities, presentation capabilities and information management capabilities, the advantages over traditional construction software are obvious.

2.2   Other Software Included in BIM

There are many other software packages included in BIM in Fig. 1, of which only four are briefly described in this paper: firstly, the Scheme design software, which translates the numerical project requirements in the design brief into a geometry-based building scheme for communication between all parties involved in the project and for research and demonstration of the scheme. Secondly, Geometric modelling software, which is easier and more efficient to use than the core BIM modelling software for the early stages of design or complex building shapes, and can even achieve functions that are not possible with the core BIM modelling software. Thirdly, Green analysis software, which uses BIM model information to analyze the project in terms of daylight, environment, heat, landscape visibility and noise. Fourthly, Structural analysis software, using information from the BIM core modelling software for structural analysis, the results of the analysis on the structure of the adjustment and feedback to the BIM core modelling software to automatically update the BIM model.

3   BIM and GIS Integration

3D GIS has been committed to the study of macro geographic environment, providing a variety of spatial queries and spatial analysis functions. The application scheme for the integration of BIM and GIS is shown in Fig. 2.

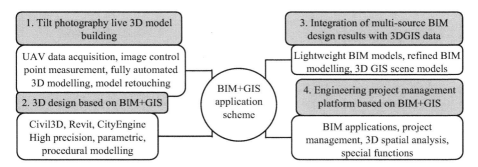

Fig. 2. BIM+GIS application scheme

BIM technology in geographic location accuracy, spatial geographic information analysis and the overall display of the building's surrounding environment are obvious deficiencies; three-dimensional GIS can complete the geographic location of the building and spatial analysis, more perfect large scene display, to ensure the integrity of information. 3D GIS is not accurate enough for the model of the building itself, while BIM model is the integration of 3D spatial information and building performance of the building. Therefore, BIM and GIS both have their own advantages and shortcomings, and the integration of BIM and GIS is just the right way to complement each other, integrating BIM technology and GIS technology and using them for engineering project management platform development [7, 8].

At present, the mainstream BIM and GIS data integration methods are divided into the following three categories: firstly, integrating GIS data into BIM; secondly, integrating BIM data into GIS; thirdly, integrating custom model architecture BIM and GIS information, while establishing the corresponding mapping relationships.

4   Spatial Data Storage

As the volume of engineering projects increases, the spatial relationships in the model become more complex and diverse, and the existing data storage and retrieval methods for building information modeling cannot meet the requirements of use. GIS can display geographic information more vividly with the help of multimedia data, so that users can understand geographic information more conveniently and clearly. Whether it is BIM or GIS, a large amount of spatial data needs to be stored. Spatial data includes map data, terrain data, attribute data, image data and metadata. Metadata is data about data. The content of metadata includes descriptions of data sets, descriptions of data quality, descriptions of data processing information, descriptions of data conversion methods, descriptions of data updates and integration, etc. [9, 10]. The spatial data storage model is shown in Fig. 3.

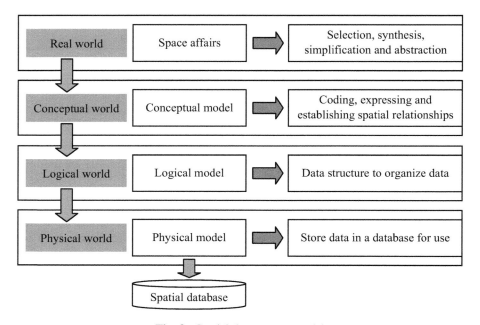

Fig. 3. Spatial data storage model

   The real world consists of spatial things or phenomena; the conceptual world is the cognition and abstraction of geographic space, forming a conceptual model, including the four basic objects of "point, line, surface and body"; the logical world corresponds to the logical model, expressing spatial objects and their relationships in a form that can be processed by computers; the physical world corresponds to the physical model, storing data in a database for use. The physical world corresponds to the physical model and stores data in a database for use.

5   System Functional Design

The whole life cycle management system of construction project based on "BIM+GIS" is designed according to the hierarchical structure of subsystems and modules, with reference to relevant literature, the system functions are designed as "3D visualization subsystem, Integrated management subsystem, Control management subsystem, Monitoring management subsystem, Information expansion subsystem, System maintenance subsystem. "Each subsystem consists of four functional modules", as shown in Fig. 4.

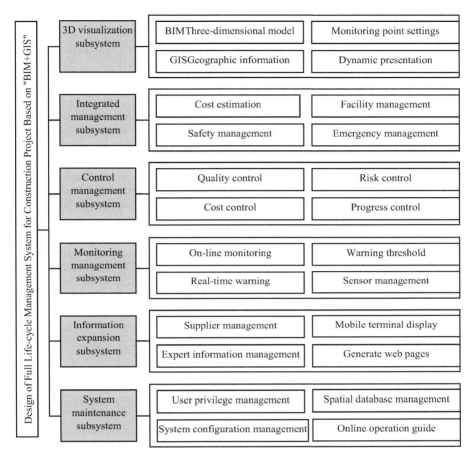

Fig. 4. System function

For the spatial data storage model shown in Fig. 4, 3D visualization subsystem, 3D visualization is a means of depicting and understanding the model, a form of representation of the data body. 3D visualization of geographic information contains graphical data visualization, data attribute information visualization, spatial analysis result visualization, map animation visualization and 3D spatial data visualization. Geospatial features are presented using visual coding. Control management subsystem, of which quality control is an important element, allows managers to implement control over construction personnel, materials and machinery, to allocate construction resources rationally, and to manage project construction progress and construction resources in a unified manner, thus providing reasonable control over the project progress.

6   Conclusion

As a modern digital information construction technology, BIM technology is a three-dimensional digital technology, while achieving the integration of various types of information in construction projects, on the basis of which a multi-dimensional data model is constructed, using digital methods to visually represent the functional characteristics and physical features of the construction project. GIS is a type of computer system that collects, stores, retrieves, manages, analyses and describes the location distribution of spatial objects, and their associated attribute data, with the support of computer hardware and software systems. BIM technology focuses on the microscopic representation of building information and is mainly applied to single buildings, while GIS technology focuses on spatial analysis and the representation of the macroscopic geographical environment. Through data integration, system integration or application integration, "BIM+GIS" integrates the advantages of each, so that each can take advantage of the strengths and avoid the weaknesses, and expand the application fields. The research results of this paper will help to arrange various construction resources in a reasonable and orderly manner, improve the cost control level of the constructors and builders, and promote the development of construction projects in the direction of industrialization and informationisation.

References

1. Abed, H.R., Hatem, W.A., Jasim, N.A.: Role of BIM technology in enhancing safety analysis of Iraqi oil projects. Asian J. Civ. Eng. **21**(3), 695–706 (2020)
2. Ham, N., Moon, S., Kim, J.-H., Kim, J.-J.: Optimal BIM staffing in construction projects using a queueing model. Autom. Constr. **113**(C), 103123 (2020)
3. Nikonorov, A., Badenko, V.L.: Flood events dynamics estimation methodology in a GIS environment. Computer **101**, 10110 (2021)
4. Ismail, M.H., Ishak, S.S.M., Osman, M.: Role of BIM+GIS checker for improvement of technology deployment in infrastructure projects. IOP Conf. Ser.: Mater. Sci. Eng. **512**(1), 12038–12049 (2019)
5. Valinejadshoubi, M., Bagchi, A., Moselhi, O.: Identifying at-risk non-structural elements in buildings using BIM: a case study application. J. Earthq. Eng. **24**(5), 869–880 (2020)
6. Tan, T., Chen, K., Xue, F., Weisheng, L.: Barriers to building information modeling (BIM) implementation in China's prefabricated construction: an interpretive structural modeling (ISM) approach. J. Clean. Prod. **219**(1), 949–959 (2019)
7. Gosling, P.C., Symeonakis, E.: Automated map projection selection for GIS. Cartogr. Geogr. Inf. Sci. **47**(3), 261–276 (2020)
8. Puno, G.R., Amper, R.A.L., Opiso, E.M., Cipriano, J.A.B.: Mapping and analysis of flood scenarios using numerical models and GIS techniques. Spat. Inf. Res. **28**(2), 215–226 (2019)
9. Karthi, S., Prabu, S.: Spatial data storage and retrieval in cloud computing environments using attribute based encryption algorithm. Int. J. Internet Technol. Secur. Trans. **9**(1/2), 163 (2019)
10. De'an, L., Liqiong, L.: A cadastral spatial data storage structure based on relational database. Geo-spat. Inf. Sci. **4**(3), 15–20 (2012)

Cost Prediction of Municipal Road Engineering Based on Optimization of SVM Parameters by RF-WPA Hybrid Algorithm

Feng Feng[✉]

Jiangxi College of Foreign Studies, Nanchang, Jiangxi, China
ff19790511@163.com

Abstract. In urban road construction, the most common is the urban internal traffic system. With the continuous improvement of people's living standards, travel needs and environmental awareness, higher requirements are put forward for urban road traffic safety. In order to meet the development needs of this new situation and new task, China began to implement a series of municipal engineering management system reform measures and achieved certain results, but there are still some problems to be solved. Therefore, based on the RF-WPA hybrid algorithm, this paper optimizes the SVM parameters and designs the municipal road project cost prediction model. Firstly, this paper expounds the significance of municipal road engineering cost, and then studies the RF-WPA hybrid algorithm and the optimized SVM parameter model. Based on this, the municipal road engineering cost prediction model framework is designed and developed, and the model is simulated and tested. Finally, the simulation results show that the prediction results of model parameters are greatly improved compared with those before and after optimization. The municipal road engineering cost prediction system based on RF-WPA hybrid algorithm to optimize SVM parameters has good prediction performance and high overall prediction accuracy.

Keywords: RF-WPA Hybrid algorithm · Optimization of SVM parameters · Municipal roads · Project cost

1 Introduction

With the rapid development of China's economy, the problem of traffic congestion is becoming more and more serious. In order to alleviate the pressure of urban road traffic, the state has issued many relevant policies and measures [1, 2]. However, in recent years, due to the large amount and long cycle of road construction and expansion, some sections have been overloaded. In some cases, due to the blind change of design in the construction, the vehicle limit time is prolonged, and even traffic accidents are caused. As a result, people's utilization of the road is reduced, the traffic congestion is becoming more and more serious, and safety accidents occur frequently, which seriously affects the normal travel of urban residents. Therefore, how to reduce the highway cost has become one of the key issues concerned by the government [3, 4].

Many scholars have studied hybrid algorithms. As an important research achievement in the field of artificial intelligence, hybrid algorithm has a lot of related research at home and abroad. China's hybrid intelligent transportation system is still in its infancy, and many technologies and theories are not fully mature [5, 6]. At the end of the 20th century, IEEE created a new artificial intelligence computing model - genetic algorithm (GA) to solve the traditional mathematical methods based on the problems existing in probability theory and simulation statistics. At the beginning of the 21st century, many intelligent algorithms proposed by scholars, such as artificial neural network method, chaotic entropy weight analysis method and random vector error optimization, were applied to hybrid intelligent transportation system [7, 8]. The above research has laid a research foundation for this paper.

Under the background of global economic integration, China's economy has developed rapidly and people's living standards have improved. People are increasingly pursuing more comfort, safety, environmental protection and energy conservation. Highway transportation is an indispensable part of the transportation system. Therefore, the scientific and rational planning of roads has become a topic of concern to the state and society. This paper takes an expressway as the research object, uses the RF-WPA hybrid algorithm to optimize the SVM parameter calculation model, compares it with the traditional investment estimation method, and draws a conclusion.

2   Discussion on Municipal Road Engineering Cost Based on RE-WPA Hybrid Algorithm to Optimize SVM Parameters

2.1   Significance of Municipal Road Project Cost

With the acceleration of China's urbanization process and the continuous improvement of people's living standards, China's urbanization is also deepening, which has higher requirements for infrastructure construction. At the same time, due to a series of factors such as serious traffic congestion, prominent environmental pollution and shortage of resources, urban development is facing great challenges. The cost of municipal road project directly affects the realization of the economic construction goal of the whole city [9, 10]. The cost of municipal road project refers to the construction scheme and the purchase price of materials and equipment determined according to the budget after estimating the construction cost of the whole city. All the expenses that the construction unit or other relevant enterprises need to spend in the construction process, that is, including the costs of early stage and later maintenance, management and repair of the project. The cost of municipal road project is, and then it is mainly composed of the following parts: (1) completion settlement. Estimate the investment of the whole project according to relevant national regulations and requirements. (2) Design change or adjustment scheme. After the bill of quantities and the contract price are determined, the cost change range in the construction drawing budget quota refers to the deviation between the actual work content and the data given in the bidding documents. At present, the main management contents of domestic municipal road project cost include: control and calculation in planning and design stage, construction and completion acceptance stage and maintenance work. In practical work, there may be deviations in varying degrees due to the influence of various factors. Therefore, it is necessary to reasonably predict the construction progress, raw material quality and other indicators, as well as the relationship between various costs and construction period for comprehensive analysis, so as to formulate the most scientific and feasible decision-making measures. The municipal road project cost is to estimate the economic construction cost of the whole city and provide reference basis for it, so it can accurately reflect the return on investment and investment risk value of the construction project [11, 12].

2.2   RF-WPA Hybrid Algorithm

Rf-wpa hybrid algorithm is a multi-objective optimization method based on asymmetric conditions. It uses the factors such as imbalance, system error and nonlinear constraints to determine the optimal parameters. This method decomposes a complex system into several small parts, and then optimizes these small units through self-learning and logic analysis. First, it is necessary to determine the role of each smallest component in the circuit. Secondly, the maximum working load of each component should be calculated according to the minimum component parameters. Finally, according to the known formula, the corresponding power density values of all components and their relationship with the impedance ratio (ESO) under the maximum load are obtained, the model equation is established, and the function coefficients and corresponding values are obtained. The model has the following advantages: (1) the RF-WPA hybrid algorithm does not need to establish the physical layer. Because each node is related to each other, each node can interconnect with each other through the physical layer. (2) The system composed of members in RF-WPA system is a complex nonlinear whole, and its structure and function become more cumbersome.

2.3   SVM Theory

SVM is a learning algorithm based on statistical learning theory. The selection of kernel function and kernel parameters plays a decisive role in the performance of SVM. Compared with kernel function, kernel parameter is the key factor affecting the performance of SVM. Because SVM has good machine learning ability, it has become an important research hotspot in the field of machine learning theory and technology. The basic idea of SVM is to design a certain structure of the function set to ensure that each subset can achieve the least empirical risk (making the training error 0). As long as the corresponding subset is selected and its confidence range is minimized, the function that minimizes the empirical risk for each subset is the optimal prediction function. The selection of optimal parameters is the most important and key link in model prediction. In the actual operation process, it will be affected by various factors. Therefore, in order to obtain accurate and reliable best data, we must have a deep enough understanding of the research problems. At the same time, we should also consider that there are differences in the results under different circumstances due to various reasons such as practical conditions and other relevant constraints. The advantages of SVM are:

1) SVM uses the principle of structural risk minimization to reduce the error of training set and the complexity of SVM Good generalization ability;
2) SVM algorithm is a convex optimization problem, while other learning algorithms such as neural network are easy to fall into local optimization. SVM algorithm is finally reduced to a quadratic optimization problem to calculate the global optimum;

3) Kernel function effectively overcomes the high-dimensional problem, that is, the complexity of SVM is not constrained and affected by the sample dimension, and improves the generalization ability of SVM;

4) The concept of VC dimension can reasonably control the bound of sample misclassification in SVM algorithm. The principle of SVM mainly includes SVM classification and SVM regression. SVM originates from the optimal classification surface in the case of linear separability. In reality, most of the problems are almost nonlinear separable. By means of nonlinear variation, the nonlinear problem is transformed into a linear problem in the feature space, and then the purpose of solving the optimal classification hyperplane in the feature space is realized.

2.4   Forecast Cost Index

When predicting the cost of municipal roads, various factors need to be considered, including engineering construction conditions, design parameters, etc. In order to better control the cost and income and effectively reduce the investment capital and labor cost, it is necessary to optimize the degree of interaction between various variables and the weight coefficient in the model to determine the weight of each index and the value of correlation function within the optimal value range, and then draw a conclusion and give corresponding suggestions and measures through data processing and analysis. The random forest method is used to analyze and detect the attribute information contained in the experimental data, and the cost level is selected as the decision variable and the attribute variable, so as to establish a preliminary numerical set. Using the random forest toolbox in MATLAB for statistical analysis, we can obtain the mean decrease Gini indicators of different attributes. The larger the mean decrease Gini index corresponding to each attribute index, the higher the importance of each attribute index can be proved. Root mean square error:

$$RMSE = \sqrt{\frac{1}{n}\sum_{i=1}^{n}(y-\hat{y})^2} \tag{1}$$

Average absolute percentage error:

$$MAPE = \frac{1}{n}\sum_{i=1}^{n}(y-\hat{y})/\frac{Y}{100} \tag{2}$$

Prediction accuracy of single project cost level:

$$H_p = \left[1-\sqrt{\frac{1}{n}\sum_{i=1}^{n}\left(\frac{y-\hat{y}}{y}\right)^2}\right]*100 \tag{3}$$

In the above formula, y is the real value of cost level, y is the predicted value of cost level, and N is the number of data groups. Among the above indicators, the smaller the values of RMSE and MAPE, the higher the prediction accuracy. HP in contrast, the higher the value, the higher the prediction accuracy.

3   Experiment

3.1   Municipal Road Engineering Cost Prediction Model

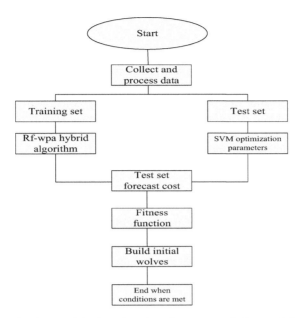

Fig. 1. Municipal road engineering cost prediction model

In the optimization model (as shown in Fig. 1), the RF-WPA hybrid algorithm can predict the system. Its principle is to use the network structure to simulate a complex nonlinear process. Firstly, the whole project is regarded as a linear combination network composed of two different levels, the same function and time series. Secondly, the mathematical expression is constructed according to the functional relationship contained in each level and calculated. Finally, the optimal scheme and parameters are determined according to the minimization principle, which realizes the maximum accuracy of the optimization model and provides a reliable basis for the actual system. The project cost prediction model based on RF-WPA hybrid algorithm to optimize SVM parameters mainly has the following characteristics: (1) this method can simplify the data acquisition, processing and analysis from the reality of the system. (2) This method can transform the nonlinear regression problem into linear equation or multivariate function.

3.2   Data Collection

In the project cost prediction, we must first collect, process and analyze the data. Rf-wpa hybrid algorithm is a multi-objective optimization model. Through the use of different parameter combinations to achieve the optimal control effect of system cost, and considering the possible interaction between different factors, resulting in errors and problems caused by large coupling between various factors. Secondly, due to the randomness of each parameter selection to a certain extent, it will lead to the deviation of untrue data, so it is necessary to analyze, process and store these information.

3.3   Prediction and Comparison Test Steps Before and After Optimizing Parameters

After the establishment of the model, it is necessary to analyze and predict the data, and the parameters obtained through the test are consistent with the actual situation. The first thing to be determined is that the initial error value is an important variable in the minimum fitting function. Secondly, the regression prediction model is established through the data set and brought into the regression equation. Finally, the least square method is used for fitting, and then the data are normalized to obtain the best value to complete the prediction test under the optimal scheme.

4   Discussion

4.1   Comparative Analysis of Prediction Before and After Optimization of Parameters

Table 1 shows the prediction results before and after the optimization parameters of the model.

Table 1. Prediction results before and after optimizing the parameters

Number of predictions	Before optimization	After optimization	D-value
1	1.569%	1.201%	0.368%
2	1.658%	1.350%	0.308%
3	1.578%	1.256%	0.322%
4	1.655%	1.362%	0.293%
5	1.697%	1.258%	0.439%

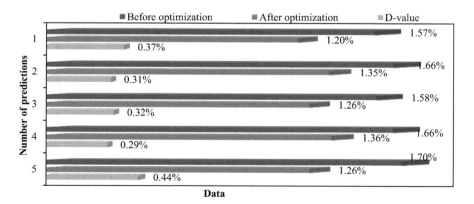

Fig. 2. Comparison of the prediction results before and after the optimization

Through the above analysis, it can be concluded that the comparison results between the optimized model parameters asnd the actual data are shown in Fig. 2. It is not difficult to see from the table that the prediction results of model parameter data have been greatly improved before and after optimization. Because the project cost prediction involves many aspects, and there is high correlation and large change range among various factors, it needs to be adjusted to get a more accurate and reasonable optimal scheme. In the optimization, the nonlinear programming theory and the idea of fuzzy mathematics are used to combine the parameters to determine the optimal structural size and variable value. Therefore, the municipal road engineering cost prediction system based on RF-WPA hybrid algorithm to optimize SVM parameters has good prediction performance and high overall prediction accuracy.

5   Conclusion

With the rapid development of China's economy, road construction has attracted more and more attention, and there are high requirements for the parameters required in the cost prediction of road engineering. At present, there have been many studies on hybrid optimization methods such as artificial neural network and genetic algorithm at home and abroad. This paper mainly obtains the optimal classification model through the analysis of the relationship between various data sets, and establishes a SVM system with more practical and nonlinear characteristics based on the RF-WPA combination principle.

References

1. Zhang, Y., Li, C., Zheng, X., Xu, H.: Fluctuating wind velocity prediction using LSSVM based on hybrid intelligent optimization of ABC and ABF. Zhendong yu Chongji/J. Vib. Shock **36**(15), 203–209 (2017)
2. Wang, X., Liu, S., Zhang, L.: Highway cost prediction based on LSSVM optimized by intial parameters. Comput. Syst. Sci. Eng. **36**(1), 259–269 (2021)
3. Su, C., Shi, Y., Dou, J.: Multi-objective optimization of buffer allocation for remanufacturing system based on TS-NSGAII hybrid algorithm - ScienceDirect. J. Clean. Prod. **166**, 756–770 (2017)
4. Song, R., Ni, L.: An intelligent fuzzy-based hybrid metaheuristic algorithm for analysis the strength, energy and cost optimization of building material in construction management. Eng. Comput. 1–18 (2021)
5. Mahdavi, S., Shaterzadeh, A., Jafari, M.: Determination of optimum effective parameters on thermal buckling of hybrid composite plates with quasi-square cut-out using a genetic algorithm. Eng. Optim. **52**(1), 106–121 (2020)
6. Schif, K.: Ant colony optimisation algorithm for the facility localisation problem. Tech. Trans. **11**(1), 103–112 (2018)
7. Chhipi-Shrestha, G., Rodriguez, M., Sadiq, R.: Framework for cost-effective prediction of unregulated disinfection by-products in drinking water distribution using differential free chlorine. Environ. Sci.: Water Res. Technol. **4**(10), 1564–1576 (2018)

8. Wang, B., Dai, J.: Discussion on the prediction of engineering cost based on improved BP neural network algorithm. J. Intell. Fuzzy Syst. **37**(7), 1–8 (2019)
9. Kaya, D.: Optimization of SVM parameters with hybrid CS-PSO algorithms for Parkinson's Disease in LabVIEW environment. Parkinson's Disease **2019**(18), 1–9 (2019)
10. Guangqian, D., Bekhrad, K., Azarikhah, P., Maleki, A.: A hybrid algorithm based optimization on modeling of grid independent biodiesel-based hybrid solar/wind systems. Renew. Energy **122**, 551–560 (2018)
11. Huang, Y.C., Huang, C.M., Chen, S.J., Yang, S.P.: Optimization of module parameters for PV power estimation using a hybrid algorithm. IEEE Trans. Sustain. Energ. **11**(4), 2210–2219 (2020)
12. Babbar, A., Prakash, C., Singh, S., Gupta, M.K., Mia, M., Pruncu, C.I.: Application of hybrid nature-inspired algorithm: single and bi-objective constrained optimization of magnetic abrasive finishing process parameters. J. Mater. Res. Technol. **9**(4), 7961–7974 (2020)

Three-Dimensional Scene Data Organization and Management of Indoor Positioning Based on Virtual Technology

Xi Chen[1(✉)] and H. Alsharif[2]

[1] Wuhan Railway Vocational College of Technology,
Wuhan 430205, Hubei, China
chenxi_wutie@163.com
[2] South Valley University, Qena 83523, Egypt

Abstract. Three-dimensional data is increasingly used in robotics, autonomous driving, virtual reality and augmented reality, and military defense. In the future, people will be able to acquire and process more three-dimensional data. The purpose of this paper is to study the organization and management of 3D scene data for indoor positioning based on virtual technology. First, the research background and research significance are introduced, and the concepts of virtual reality and 3D modeling technology are briefly introduced. Then, the research fields and research methods of 3D data processing are explained, and the research status of key technologies of 3D data processing at home and abroad is introduced. The research methods of the organization of 3D scene data and the management of 3D scene data involved in the following are introduced in detail, and the client of the 3D scene data management system for indoor positioning is designed. Through the experimental results, the frame rate of the quadtree frustum cropping is 112.3. The system implements an outdoor sand table model and supports scene browsing in first-person and third-person perspectives.

Keywords: Virtual technology · Indoor positioning · Three-dimensional scene · Data organization

1 Introduction

Virtual reality technology is an interdisciplinary technology developed on the basis of graphics, multimedia technology, sensor technology, computer simulation, and human-computer interaction technology [1, 2]. This is a hot spot of current computer technology research. 3D modeling technology is the foundation of all virtual reality systems and the primary technology of virtual reality technology [3]. Cloud data processing technology in 3D modeling technology has become an important research content of virtual reality technology [4]. 2D images are on the rise, but the analysis and understanding of searching for 3D perspectives is still in its infancy. Compared with 2D images, 3D data is more 3D, more detailed, more organized, and closer to the real world. Therefore, the analysis and understanding of the three-dimensional perspective is an important and important foundation for understanding and analyzing the real world [5, 6].

© The Author(s), under exclusive license to Springer Nature Switzerland AG 2022
V. Sugumaran et al. (Eds.): ICMMIA 2022, LNDECT 138, pp. 94–101, 2022.
https://doi.org/10.1007/978-3-031-05484-6_12

There are also more and more researches on deep learning based on three-dimensional data. Maples-Keller JL evaluates the literature on the effectiveness of incorporating virtual reality (VR) into the treatment of mental illness. The user experiences a sense of presence in a computer-generated 3D environment. Sensory information is transmitted through head-mounted displays and dedicated interface devices. These devices track head movement so that movement and images naturally change with head movement, creating a sense of immersion. VR allows the therapist to control the delivery of sensory stimulation and is a convenient and cost-effective treatment method [7]. Human-computer interaction in virtual reality and the presentation effect of virtual scenes are the two most important aspects of virtual reality experience. How to provide a good human-computer interaction method for virtual reality applications and how to improve the final presentation effect of virtual reality scenes has also become an important research direction. Zhao C uses the virtual fitness club experience system as the application background, analyzes the function and performance requirements of the virtual reality experience system in the virtual reality environment, and proposes to use Kinect as the video capture device to extract the user's somatosensory operation actions [8]. Therefore, how to locate a three-dimensional indoor scene with high quality to obtain a large amount of scene training data has become the focus of attention.

The innovations and results of a research presented in this article include: streaming media systems and storage devices for large-scale 3D visual data. Through in-depth research on the types of three-dimensional visual data, a distributed organization of three-dimensional event data based on image coding is proposed, and a distributed three-dimensional database system is developed on this basis; integrated data management is controlled through data recording and three-dimensional data processing company 3DSDML, Provide technical support for the rapid production of 3D scenes.

2   Research on 3D Scene Data Organization and Management of Indoor Positioning Based on Virtual Technology

2.1   Virtual Reality

Virtual reality is based on computer technology, combined with related science and technology, to create a digital community that is very similar to the real world in terms of visibility, listening, and touch [9, 10]. Users use special tools to interact and interact with objects in the digital environment, and can personally convey the feelings and experiences corresponding to the real community. The obvious truth is the kind of technology and technology that people create in the process of discovering and understanding nature. It gradually creates to understand nature, practice creation, and then improve and use nature.

Preliminary research content and virtual reality guidelines include three-dimensional modeling methods, operation technology, human-computer and application interaction, development platform and supporting fields, and applications. The VR model is the first step in virtual reality research, and it is also a very important step. Therefore, the data collection and processing of the three-dimensional model is the key to the realization of virtual reality technology [11, 12].

2.2   Organization of 3D Scene Data

Since the three-dimensional horizontal vector data of indoor conditions is divided into two categories: institutional data and future design data, in the refinement process, the two types of data are divided according to different characteristics.

Since the background structure data is some inherent characteristics of the 3D landscape, these inherent characteristics are fixed in most 3D scenes facing indoor conditions, so when recreating such data, you need to pay attention to the ceiling, floor, and walls.

The three-dimensional design data of future indoor positioning refers to the three-dimensional design data of different materials and objects, such as the design of tables and chairs. Therefore, when we divide the front-end design data, we can divide it into two areas of normal shape and irregular shape according to its geometric shape.

2.3   Management of 3D Scene Data

Three-dimensional visualization data has the characteristics of large data volume, complex system, multiple genres, and multiple variables. In these data, it also contains a lot of structured/unstructured data, power/static data. All these have brought problems to the control of 3D event data. The data-based architecture (DOA) mentioned in Sect. 2 focuses on the current quantity, type, and real-time data, including data-based, and unified management of data under different settings. It uses data identification such as data identification and location mark to develop data customization and access authority, recognizes data management and exchange through the registry and registry, and sets up several data processing functions, which can be modified from complex to complex and flexible. It is difficult to build complex application systems to determine data sharing, access and collaboration among multiple systems.

In response to the difficulty of 3D scene data management, a data-based architecture is adopted, and multiple 3D event data are identified and registered through metadata through the data registry, and the data is managed uniformly. Design 3DSDML three-dimensional data management language, combined with the results of three-dimensional mass indexing, according to the metadata method, integrate the registry into the data registry to form a metadata database, the registry stores data, and accesses the three-dimensional scene data database through metadata for unified management.

3   Investigation and Research on the Organization and Management of 3D Scene Data for Indoor Positioning Based on Virtual Technology

3.1   System Development and Test Environment

The operating system adopts Windows 8.164bit, and the software development of the data acquisition system is carried out under VS2010 +PCL+OpenNI and other software. It uses PCL source code and CMake compilation. Before CMake compiles the code, some third-party libraries need to be installed. Including: Boost, Eigen, Flann, VTK, OpenNI, Qhull. In the process of development and testing, first test on the computer, and then package it into an.apk file after reaching the expected goal, and install it on the mobile phone for testing.

3.2 Simulation Data

In the early stage of development, due to the limited amount of actual data, in order to test the massive data scheduling algorithm of the outdoor model, we used Auto-deskMaya to create a number of simulation data. The model is a high-rise building and a total of 3 layers of LOD were produced, of which the level of detail is the highest. Named Building_XXX_LOD0, it has a simple one-story lobby indoor environment, the others are named Building_XXX_LOD1, Building_XXX_LOD2 in turn, and then 1000 copies of it are created. The size and shape are completely determined at will, and only enough data is ensured.

3.3 Classification and Scheduling Strategy of 3D Scene Data

Three-dimensional raster dataset: When setting raster data, you need to first set the raster data in the current viewing port, and then set the data around the viewing area according to the preset algorithm. The viewing area refers to the area where the user is visible on a three-dimensional level at a specific viewing position and line of sight. Among them, point V represents the lane position, point O represents the intersection of the line of sight and the plane, the cone formed by points V, A, B, C, and D is the viewing cone. The viewing cone and the three-dimensional plane in the rectangular area are defined by the three-dimensional view A′, B′, C′, and D′ in the area constitute. Establish a three-dimensional coordinate system with point O as the origin of the coordinates, set the coordinates of point V as (xv, yv, zv), and the equation of the VO line can be expressed as Eq. 1.

$$\begin{cases} x = x_v + x_v t \\ y = y_v + y_v t \\ z = z_v + z_v t \end{cases} \tag{1}$$

Regarding the four lines of the viewing cone V-AC as rays emitted from the viewpoint, the ray equation is established, and the coordinates of the intersection point of the ray and the three-dimensional scene can be obtained through the ray equation, and the scope of the viewing zone can be further obtained.

The correlation coefficient of three-dimensional scene data is a value used to describe the correlation between various three-dimensional scene data. It is defined as: in the three-dimensional scene data scheduling, the correlation coefficient of data A and data B refers to when data A is called, The ratio of the number of times data B is called in an adjacent time interval to the total number of times data A is called, as shown in formula 2:

$$R(A \cap B) = \frac{Q_A \cap Q_B}{Q_A} \tag{2}$$

The calling correlation coefficient is in order. $R(A \cap B)$ and $R(B \cap A)$ are different because of their different denominators.

The correlation coefficient is stored in the data registration center. After the current 3D scene viewport scheduling is basically completed, the scheduler will decide which data needs to be scheduled in the cache in advance according to the correlation coefficient table of the data in the current scene, waiting to be used.

3.4   System Test

At the beginning of the project, we used simulated data to test the loading and scheduling experiments of outdoor models. The scheduling adopts regular grid + quadtree index. The number of experimental models is 100 and the number of LODs is 3. When constructing the grid and quadtree, refer to the Building object in the same linear table together, use the camera position to calculate the grid number of the model that needs to be loaded, and perform the loading operation on the model file that needs to be loaded. When rendering, use the frustum to crop the quadtree, and then further judge the relationship between the Building object in the node to be rendered and the frustum to determine whether to render the object. Later, the method was applied to actual data for testing, and an outdoor sand table model was realized, and it supports scene browsing from first-person and third-person perspectives.

4   Investigation and Analysis of 3D Scene Data Organization and Management of Indoor Positioning Based on Virtual Technology

4.1   3D Scene Data Management System Client for Indoor Positioning

The client is a graphical terminal program that integrates the functions of 3D scene display, creation, saving and loading. It not only covers all the functions of the current web browser, through this client, users can achieve efficient indoor 3D scenes Build on demand.

The 3D scene data management system client for indoor positioning can be divided into three parts functionally: the creation, saving and loading of 3D scenes, as shown in Fig. 1.

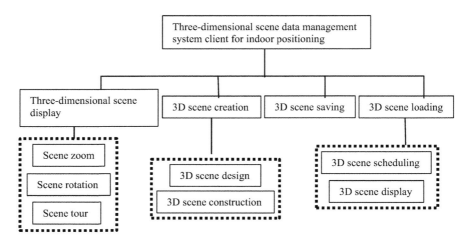

Fig. 1. The functional structure diagram of the client side of the 3D scene data management system for indoor positioning

3D scene display function: Provides visual display of 3D scene, can zoom and rotate the scene, and provide scene roaming function. 3D scene creation function: It provides users with a graphical 3D scene design and construction environment, which can realize on-demand creation of 3D scenes through various vector data and raster data stored in the background. 3D scene saving function: the created 3D scene is persisted in the storage cloud in 3DSDML format, and the 3D scene is registered in the data registration center for future recall. 3D scene loading function: It can load the saved 3D scene data and show it to users. When data is loaded, the data is scheduled and loaded according to the three-dimensional scene data scheduling method described to ensure the efficiency of data loading.

4.2   Indoor and Outdoor Scene Scheduling and Roaming Test

Turn off and turn on the rendering information in the case of quadtree frustum clipping, and the comparison result is shown in Fig. 2. When the function is turned off, all loaded objects are in the active state, and the result can also be seen from the Unity's scene object manager (the objects are all highlighted); on the contrary, only the objects located in the frustum are in the active state. The grayed-out objects in Unity's scene object manager represent inactive objects, as shown in Table 1.

Table 1. Comparison of clipping and rendering results of turning off/on quadtree frustum (both are peak conditions)

Comparison item	Turn off quadtree frustum clipping	Enable quadtree frustum clipping
Frame rate (FPS)	37.5	112.3
Rendering time consuming (ms)	18.4	1.6
Draw Calls	852	245

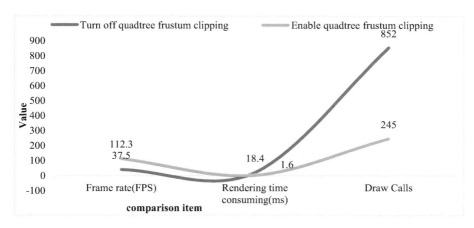

Fig. 2. Close/open quadtree frustum cropped rendering results comparison (all are peak conditions)

When roaming in an integrated indoor and outdoor scene, when the camera reaches a specific area of an outer frame model with an indoor model (judged by the record in the attribute description file), the outdoor scene camera is set to inactive and will be used in the indoor scene The camera is set to the active state, and at the same time all the loaded outdoor models are unloaded, and the indoor models are loaded.

The positioning function can be easily implemented in Unity. The path finding system needs to find the best route between two points. Mesh (Mesh, similar to the one used in 3D graphics) is an excellent choice to solve this problem. In the three-dimensional scene, the reachable area is represented by the polygon of the mesh surface. Therefore, the possible path through the mesh surface is naturally decomposed into polyline segments between two adjacent polygons. In theory, the mesh surface used to represent the ground can also be used for positioning.

5   Conclusions

In recent years, with the rapid development of 3D data acquisition technology and digital model generation technology, people have the ability to acquire more and larger 3D scene data. Therefore, more and more researchers are focusing on the analysis of the entire three-dimensional scene. In this paper, we have conducted in-depth research on the current problems in the field of 3D data indoor positioning technology. From the classification and scheduling strategy based on 3D scene data and the 3D scene of simulated data, a 3D scene indoor positioning solution is proposed, which is interactive and friendly. For the purpose of high construction efficiency and high compliance with requirements, indoor and outdoor scene scheduling and roaming experiments have been carried out, and the experiments have proved that the system is running well.

References

1. Esposito, C.: Interoperable, dynamic and privacy-preserving access control for cloud data storage when integrating heterogeneous organizations. J. Net. Comp. Appl. **108**(APR.), 124–136 (2018)
2. Rathnayaka, P., Baek, S.H., Park, S.Y.: Stereo vision-based gamma-ray imaging for 3d scene data fusion. IEEE Access **PP**(99), 1–1 (2019)
3. Werner, G.: The virtual world in focus at the digital factory leading trade fair: more room for virtual technology. Industrie Anzeiger **139**(9), 46 (2017)
4. Brenner, S.C., Sung, L.Y., Tan, Z.: AC1Virtual element method for an elliptic distributed optimal control problem with pointwise state constraints. Math. Models Methods Appl. Sci. **31**(14), 2887–2906 (2021)
5. Akdere, M., Ach Es On-Clai,r K., Jiang, Y.: An examination of the effectiveness of virtual reality technology for intercultural competence development. Int. J. Intercul. Relations **82**(1), 109–120 (2021)
6. Rouhani, M., Lafarge, F., Alliez, P.: Semantic segmentation of 3D textured meshes for urban scene analysis. Isprs J. Photogra. Remote Sensing **123**(JAN.), 124–139 (2017)
7. Maples-Keller, J.L., et al.: The use of virtual reality technology in the treatment of anxiety and other psychiatric disorders. Harv. Rev. Psychiatry **25**(3), 103–113 (2017)
8. Baker, S., et al.: Evaluating the use of interactive virtual reality technology with older adults living in residential aged care. Info. Proc. Manag. **57**(3), 102105.1–102105.13 (2020)
9. Ahmad, M., Eslam, M.S., Anshasi, H.A.: Virtual reality technology for pain and anxiety management among patients with cancer: a systematic review. Pain Manag. Nurs. **21**(6), 601–607 (2020)
10. Abpp-Gero, C., Abpp-Cn, N.: Promoting technology and virtual visits to improve older adult mental health in the face of COVID-19 - ScienceDirect. Am. J. Geriatr. Psychiatry **28**(8), 889–890 (2020)
11. Lee, J., Kim, J., Choi, J.Y.: The adoption of virtual reality devices: the technology acceptance model integrating enjoyment, social interaction, and strength of the social ties. Telematics and Informatics **39**(JUN.), 37–48 (2019)
12. Good, K.D.: Sight-seeing in school: visual technology, virtual experience, and world citizenship in American Education, 1900–1930. Technol. Cult. **60**(1), 98–131 (2019)

Hotel Intelligent System Design Based on Machine Learning Technology

Yanjie Yu[✉]

Liaoning Vocational College of Light Industry, Dalian 116100, Liaoning, China
yuyj2021@126.com

Abstract. With the rapid development of my country's economy and the continuous improvement of people's living standards, the hotel industry has developed rapidly, and more and more star-rated luxury modern high-end products have appeared. How to improve the hotel's service level and work efficiency is a very important issue. To this end, this article intends to start with machine learning technology to conduct intelligent research on hotels. The research purpose of this article is to improve the management service level and work efficiency of the hotel. This article mainly uses experimental testing and comparison methods to analyze the performance of the hotel's intelligent system. Compare the time and load capacity of the system with the user's expected value, highlighting the characteristics of the system's performance. The experimental results show that the time-consuming of the system is less than 0.7, and the carrying capacity is 29 more on average. This is enough to show the feasibility of the hotel's intelligent system.

Keywords: Machine learning · Hotel intelligence · System design · Technology research

1 Introduction

Hotel is an indispensable industry in modern urban life. It can not only meet people's material and cultural needs, but also provide spiritual and emotional convenience to a certain extent. Hotels are essential in this fluid society. How to improve the reputation of the hotel and attract people requires the hotel to adjust and improve its management service level. For this reason, keeping up with the times is a way to make hotels smart and convenient.

There are a lot of research theoretical results on hotel intelligent system design based on machine learning technology. For example, in order to ensure the safe use of data in various application scenarios, some people have proposed a design scheme of a data desensitization system based on machine learning [1, 2]. Some people also elaborated on the design principles of five-star hotel low-voltage system integrated on-board power supply, elaborated the design of five-star hotel low-voltage system integrated on-board power supply, and the content of the architecture, functional system and services [3, 4]. In addition, some people said that in the fiercely competitive environment, more and more hotel managers are paying more and more attention to hotel construction and its construction quality [5, 6]. Therefore, the research on the hotel's intelligent system design is very necessary. This article intends to use machine learning technology to study intelligent systems.

This article first studies the related technology of machine learning, and analyzes its structure and algorithm. Secondly, briefly describe the system integration of intelligent buildings. Then studied the hotel's intelligent system, including the hotel background management system, intelligent guest room center management system and so on. Finally, a system test is performed to test the performance of the intelligent hotel system, and the results are obtained.

2   Hotel Intelligent System Design Based on Machine Learning Technology

2.1   Machine Learning Related Technologies

Data preprocessing and evaluation methods for machine learning systems.

Due to the use of too many parameters, etc., learning artificial neural networks sometimes leads to very complex models. We usually refer to this situation as overfitting, which means that the function used by the model is too complex to be included in our training set, which creates the illusion that the model is particularly effective [7, 8].

SVM machine learning algorithm is widely used in classification problems and regression analysis problems, and its predictive ability performs well on small sample data. Perform linear regression analysis on the sample data set in this high-dimensional feature space to obtain the linear function of the high-dimensional feature space:

$$g(a) = (q \cdot \ell(a)) + y \tag{1}$$

Q is the weight to be calculated, $\ell(\cdot)$ is the nonlinear mapping function, and y is the bias.

According to the KKT condition of the quadratic programming, the fitting function can be obtained:

$$g(a) = \sum_{l=1}^{m} (x_l - x_l^*) J(a, a_l) + y \tag{2}$$

m is the number of support vectors and $J(a, a_l)$ is the kernel function. The SVM machine learning algorithm can replace the inner product calculation in the high-dimensional feature space through the kernel function, effectively reducing the amount of calculation.

2.2   Hotel Intelligent System

If you want to reduce costs, improve service and management quality, and promote economic benefits, you should implement intelligent hotel management. The management system fully integrates various interfaces such as switchboard, multimedia, Internet, VOD, voice mail, electronic lock, night control cabinet, security monitoring, etc., to provide customers with flexible and reliable third-party integration basic guarantee [9, 10].

(1) Intelligent guest room center management system

The intelligent room center management system can view the status of each room and various customer service requests in real time. If the customer requests an inspection, you can use the remote control to enter the room. Different systems can be displayed in different signal areas to know the service needs of different customers.

(2) System control object

The system adopts a two-layer communication network. The upper layer is the information field, which uses Ethernet communication to integrate third-party systems and expand customers; the lower layer is the control domain of the DDC controller for master-slave bus communication, which collects equipment signals and controls related equipment. The external system takes over the general configuration software, supports the communication of many equipment manufacturers, opens and exchanges data with the building management system through OPC, DDE and other methods. This project mainly includes the following building control technology subsystems: cold and heat source systems, air conditioning systems, fresh air systems, air supply and exhaust systems, air supply and drainage systems, power distribution systems, intelligent lighting systems, elevator systems, and web navigation functions.

The intelligent hotel check-in system is an intelligent hotel management system based on the Internet. Compared with traditional liquor management systems, this system focuses on mobility and convenience, while using smart devices to increase hotel business. Efficiency makes the customer's stay easier. The system is mainly divided into four major systems: smart hotel cloud management system, guest room mobile application, smart guest room system and background maintenance management platform.

(3) System working process

The process of perceiving the environment and collecting environmental data, through the installation of sensors and communication networks, converts home environmental information into digital information, and transmits it to the data processing module. Data processing, processing based on data collected by sensors, including filtering invalid data, aggregating environmental data, and device status data. In data learning, the built-in input vector is sent to the machine learning unit. Through the study of the nature of the data by the learning unit, useful information is obtained and the output vector is generated. The output vector is the basis for equipment condition monitoring or equipment planning. Auxiliary control, the auxiliary system controls the hotel accommodation status or sends hotel equipment according to the results of data learning.

2.3 Design of Smart Hotel System

The smart hotel system will generate a large amount of user usage data and store it on the server. The system needs to combine the smart hotel with machine learning algorithms, and use machine learning algorithms to learn and summarize the user usage history data generated by the smart hotel system. The mathematical theory model of system input data and system output data is developed to realize the energy-saving reminder function of home smart devices, and bring home users a smarter hotel experience.

The smart gateway part is the heart of the entire smart hotel system, the hub of system data instructions, and the manager of smart devices. The smart gateway manages the wireless network of multiple smart device nodes. At the same time, the Internet can be accessed through the home router, and the server can remotely establish a communication link with the smart gateway, thereby realizing functions such as remote control of smart home equipment and downloading of sensor data. The smart device node is an important part of the hotel's smart device. This system implements data communication between the smart device and the smart gateway by installing a high-frequency module on each smart device.

In order to make it easier for operators and building managers to operate, all information must be simple, clear, intuitive and standardized on a unified platform. On the other hand, to meet the user's human needs, if the user needs various services, the intelligent system can provide support for them conveniently and quickly. Its composition is shown in Fig. 1:

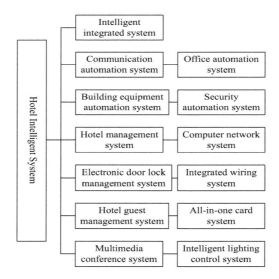

Fig. 1. Structure composition of hotel intelligent system

Advanced and intelligent design schemes should choose appropriate advanced technology to meet the various functional requirements of the hotel project itself, while ensuring that each subsystem can meet the needs of hotel facility management and various functions of the hotel. In terms of practicality, the design of intelligent systems should be based on the actual management needs of the hotel, rather than one-sided pursuit of the latest technology. The newer the technology, the greater the probability of failure, and the higher the cost of the system. Reduce the maintenance of the use effect caused by the use of new technologies. The scientific and intelligent design of the system should be carried out in strict accordance with the path of scientific management and technological development, and should not deviate from this direction. Otherwise, product functions will be outdated or severely redundant, resulting in failure to achieve the design and requirements or use of more than functions, resulting in technical waste. Open and intelligent system design must use mature, advanced and practical technologies. These technologies must be open and use international and national standardized products to optimize the integrated design of the system. Economically, on the basis of realizing the above principles, the economics of system design and equipment selection should also be considered, and the rationality and economic benefits of investment should be evaluated. Scalability, the system design should be future-oriented and expandable to meet the requirements of gradual system implementation and future system expansion. The design and construction of the system should be viewed from the perspective of the benefits of the entire life cycle of the hotel project. The existing system has sufficient scalability to handle possible functional changes and system updates.

3　System Test

3.1　Development Environment

The system web browser selects Internet Explorer 8, the developer tool plug-in selects MyEclipse developer tool to use Eclipse IDE, the system runs in the JDK development environment, the database uses MySQL5, and the web application service uses Matou. In the system hardware environment, the processor is preferred, Corei4 is selected, the system memory is 6GB, and the hard disk is 400G.

3.2　Test Method

The black-box test method must first build a black-box test system, and then start the test according to the designed scheme. After completing the test, analyze the test results. Usually, when analyzing the results, the actual test results are compared with the data set at the beginning. However, it is important to pay close attention to the fact that when defining data, you must consider the actual situation and consider many factors. The system test can be tested in two ways, one is the positive test, setting the correct data value. The other is a negative test, which sets wrong data. See if the system meets the data value by mistake, if there is an answer, it means there is no problem with the system. If there is no response to the wrong data, there is a problem with the system. In this way, the positive and negative parameters can more fully reflect the accuracy of the system.

3.3　System Function Test

System testing is difficult, involving a wide range of areas, a large amount of data, a long test cycle, and due to timing issues, the system only performs black box testing. Perform functional tests on the system according to the functional classification of users and modules. Regarding the input and output, error response, and exception management of the system after problems occur within the system, the main focus is on the measurement of exception management and system operation effects.

3.4   System Performance Test

From the perspective of access speed, number of concurrent users, data capacity requirements, etc., it is necessary to test the efficiency and correctness of hotel management system design, Javaization, and system performance.

4   Analysis of Test Results

4.1   Performance Analysis

In this system function test, all possible input conditions are used as test inputs. By comparing the test input and output, it can be seen whether the system can meet the ideal use standard in practice. The test result completely shows that the system can meet the expected demand in actual application. The details are shown in Table 1:

Table 1. System performance data analysis

	Actual performance	User index requirements
Access speed for an individual client	1.1	1.8
Data statistics speed	1.8	2.5
Individual browser page access speed	0.9	2.1
Concurrent number	31	28
The number of lines at the same time	76	62
Maximum number of users	300	240

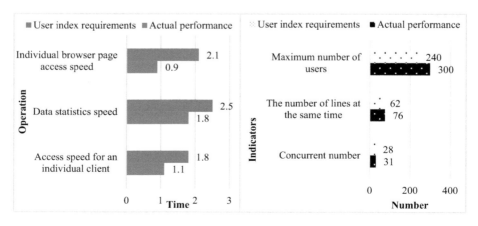

Fig. 2. System performance data analysis

As shown in Fig. 2, we can see that the actual speed achieved in various indicators is smaller than the expected speed. The data for the maximum number of users, the number of parallels, etc. are larger than expected. This shows that the system designed in this article can meet the basic needs of users.

5   Conclusion

Although smart hotel technology is constantly evolving and developing, most of the smart hotel solutions on the market are still very simple. They can only take on the function of remote control of hotel accommodation by customers, and the demand for hotel intelligence has not been fully implemented. Therefore, the research on hotel intelligence needs to be further in-depth. Through the research of this paper, it is found that the system designed in this paper has the characteristics of short time and large carrying capacity. The design of this article has certain reference significance to the hotel's intelligent system.

References

1. Yassine, A., Mohamed, L., Achhab, M.A.: Intelligent recommender system based on unsupervised machine learning and demographic attributes. Simul. Model. Pract. Theory **107** (4), 102198 (2021)
2. Carballo-Cruz, E., et al.: An intelligent system for sequencing product innovation activities in hotels. Latin America Trans. IEEE (Revista IEEE America Latina) **17**(2), 305–315 (2019)
3. Lee, W.J., Kwag, S.I., Ko, Y.D.: Optimal capacity and operation design of a robot logistics system for the hotel industry. Tour. Manag. **76**(Feb.), 103971.1–103971.10 (2020)
4. Josep, V. et al.: Prediction and prevention of hypoglycaemic events in type-1 diabetic patients using machine learning. Health Infor. J. **26**(1), 703–718 (2019)
5. Hernández, J.F., et al.: Machine learning and statistical techniques. An application to the prediction of insolvency in spanish non-life insurance companies. Inter. J. Digi. Acc. Res. **5** (9), 1–45 (2020)
6. Pavlova, A., et al.: Machine learning reveals the critical interactions for SARS-CoV-2 spike protein binding to ACE2. J. Phys. Chem. Lett. **12**(23), 5494–5502 (2021)
7. Saad, A., Mohamed, M.S., Hafez, E.H.: Coverless image steganography based on optical mark recognition and machine learning. IEEE Access **PP**(99), 1–1 (2021)
8. Amador-Angulo, L., Castillo, O.: Comparative study of metrics that affect in the performance of the Bee Colony Optimization Algorithm through interval type-2 fuzzy logic systems. In: Melin, P., Castillo, O., Kacprzyk, J., Reformat, M., Melek, W. (eds.) NAFIPS 2017. AISC, vol. 648, pp. 61–72. Springer, Cham (2018). https://doi.org/10.1007/978-3-319-67137-6_7
9. Thebelt, A., et al.: Maximizing information from chemical engineering data sets: applications to machine learning. Chem. Eng. Sci. **4**, 117469 (2022)
10. Carballo-Cruz, E., et al.: An intelligent system for sequencing product innovation activities in hotels. Latin America Trans. IEEE (Revista IEEE America Latina) **17**(2), 305–315 (2019)

The Application of Intelligent Meter Reading and Inspection in Substation Operation and Maintenance Management

Hongliang Zou[1]([✉]), Jingbin Qian[1], Xingxing Xiong[2],
and Jiawei Chen[1]

[1] Science and Technology Branch, Taizhou Hongchuang Electric Power Group
Co., Ltd., Taizhou, Zhejiang, China
zoucs086@yummail.cn
[2] Taizhou Power Supply Company, State Grid Zhejiang Electric Power Co.,
Ltd., Taizhou, Zhejiang, China

Abstract. The intelligent substation inspection robot system is the development direction of intelligent substation and unattended substation inspection technology. This article takes the intelligent substation as an example to describe the operation principle, system function and inspection route design method of the intelligent substation inspection system in detail. It reflects the broad development space and application prospects of the intelligent inspection system of substations in the daily inspection management of substations.

Keywords: Substation · Operation · Maintenance management · Intelligent meter reading · Inspection

1 Introduction

Routine inspection of substation equipment is an important management system to ensure the safe operation of substations. Up to now, due to the shortcomings of high labor intensity, low safety factor, and high labor cost in conventional manual inspection methods, intelligent inspection robot systems have emerged in the inspection management of substations III. The intelligent substation inspection robot system is developed by the State Grid Corporation to meet the development needs of intelligent substations and unattended substations, and to comprehensively improve the level of substation intelligence [1]. It has safe independent property rights and invention patents, and the overall technical level is at the international leading level.

The system takes intelligent inspection robots as the core, integrates robotics technology, power equipment non-contact detection technology, multi-sensor fusion technology, pattern recognition technology, navigation and positioning technology, and Internet of Things technology, which can realize all-weather, all-round intelligent inspection and monitoring of substations, Improve the automation and intelligence level of normal inspection operations and management, provide innovative technical testing methods and safety guarantees for intelligent substations and unattended substations, and accelerate the process of unattended substations [1]. The patrol site of the substation robot is shown in Fig. 1.

© The Author(s), under exclusive license to Springer Nature Switzerland AG 2022
V. Sugumaran et al. (Eds.): ICMMIA 2022, LNDECT 138, pp. 109–117, 2022.
https://doi.org/10.1007/978-3-031-05484-6_14

Fig. 1. Substation robot patrols the scene

This article takes the intelligent substation robot inspection system as an example to study the operation principle, system function and inspection route design of the robot. The use of robot technology for inspection is the development direction of intelligent substation and unattended substation inspection technology.

2 Intelligent Technology for Substation Operation and Maintenance

The application of intelligent technology for substation operation and maintenance has the characteristics of digitization of information on the whole site, networking of communication platforms, standardization of information sharing, and interaction of advanced applications. The various data collected through the optical fiber network are automatically analyzed through the information integration system. Complete functions such as control, protection, and monitoring, realize intelligent functions such as intelligent alarm, sequence control, accident analysis and decision-making, and seamlessly connect with the system [2]. The application of intelligent technology can implement the intelligent operation and maintenance mode of "online centralized monitoring, offline partition inspection", that is, while realizing remote centralized monitoring, a certain number of operators are retained in the station, and the platform is continuously monitored for 24 h. Once there are any abnormal alarms, immediately arrive at the site for processing immediately, achieve rapid response, and provide double guarantees for the safe operation of substations to meet the power demand of important industries or fields. The historical data collected by the system can be formed into corresponding reports, which not only eliminates the cumbersome and inconvenient manual meter reading, but also can be used for later advanced data analysis [2].

2.1 Technical Characteristics

(1) Maintain stability. Through the application of intelligent technology in the substation, digital control of mechanical equipment can be realized. During the operation of the substation, reasonable analysis of relevant dispatching

instructions and implementation of corresponding operations can ensure that the operation has good stability. For the application of intelligent technology in the substation system, the traditional equipment management mode has been innovated, and advanced intelligent mechanical equipment is introduced into the substation system to ensure the automatic operation and maintenance of the power system, so that the monitoring equipment has a certain scientific rationality and guarantees the equipment the corresponding functions in the management can be carried out reasonably, which lays a good foundation for the improvement of the overall stability of the power system [3].

(2) Automation. During the operation of the power system, in order to be able to do a good job in the automatic management of the substation equipment, it is generally necessary to start from the following two aspects: First, equipment management. Second, substation management. Equipment management is mainly to implement management of electrical equipment, using information management methods [3]. Relevant data and information in the substation are exchanged and communicated with the information platform to ensure that the whole can be integrated, intelligent technology is used to strengthen the construction of the digital information platform, so that the power system has a basic guarantee in actual operation to ensure the overall system the stability and safety of the substation can also be optimized to improve the quality and effectiveness of the power system.

2.2   Intelligent Inspection System

The substation equipment inspection robot system adopts a hierarchical control structure, which is divided into two layers: base station and mobile station control system layer and system layer. Substation inspection robots are mainly used for outdoor substations instead of patrol personnel to patrol and inspect [4]. The robot can carry infrared thermal imager power station equipment detection devices, visible light CCD, etc., autonomously and remotely, instead of monitoring outdoor high-voltage equipment, to detect internal thermal defects of power equipment, external mechanical or electrical problems such as foreign objects, damage, heat, leakage, etc., provide relevant data for the operator to diagnose the operation of power equipment and the precursors of failure accidents [4].

3   The Principle of Robot Intelligent Meter Reading and Inspection

The robot relies on the front-end magnetic sensor to identify the pre-designed patrol track and executes various commands of the parking position by identifying the buried RFID points, including turning, detecting, parking, charging, opening the door, and so on. Complete the established inspection task through the combination of the navigation of the magnetic track and the command of the stop. There are two main ways to start a robot task. One is timing start, which is mainly an autonomous operation task without human operation after the robot has been debugged. The other is a manual control start mode, which is mainly used in special emergency situations with visiting requirements [5].

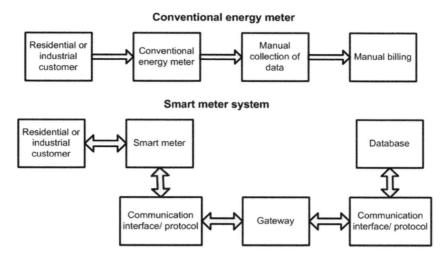

Fig. 2. Schematic diagram of intelligent inspection topology

Robot charging mainly includes automatic (recommended) and manual (the power of the robot is no longer sufficient for starting) (see Fig. 2). Automatic charging means that the robot can collect the RFID points for charging, so that the robot will execute the charging command. Also, if the robot is to be automatically turned on for charging the door control of the cabin is also the RFID point that allows the robot to collect the opening point.

4   System Function and Patrol Route Design

The substation intelligent inspection robot system is divided into two parts: the base station and the mobile station robot body. According to on-site work requirements, the main system functions of the robot system developed are as follows [6]:

(1) It is convenient for the user to configure the robot system and issue automatic planning tasks according to the configuration: According to the user configuration, the visible light image and infrared image are automatically collected and stored, analyzed and processed, and the intelligent collected data is displayed dynamically.

(2) Manually control the movement of the robot and the movement of the robot pan/tilt, display the current position and movement status of the robot in real time, and display the data information of the current position and speed of the robot.

(3) Display the power status of the robot, and perform analysis and abnormal alarms, display the signal status of the peripheral sensors of the robot, and give an alarm.

Query the historical data collected by the robot on-site and conduct comparative analysis. The intelligent inspection robot adopts the magnetic navigation method, which has the advantages of strong anti-interference ability, high reliability, and is not affected by rain and snow. It is the most reliable and mature navigation method for outdoor mobile robots [6]. The positioning point provides positioning and task information for the robot in the global path planning, including position coordinates, equipment interval, task content, etc.

5 The Application of Intelligent Meter Reading and Inspection in Substation Operation and Maintenance Management

5.1 Intelligent Meter Reading and Inspection is of Auxiliary Significance to the Maintenance of the Electric Meter Device

Using the remote inspection function of the smart meter reading system is of great significance to the operation and maintenance of the electric meter device. According to the type of electric meter failure, combined with the work experience of electricity inspection, several abnormal judgment rules such as electric meter loop, loss of voltage, and loss of current inspection are formulated. And solidify into the intelligent meter reading system to form the remote inspection function of the electric meter device [7].

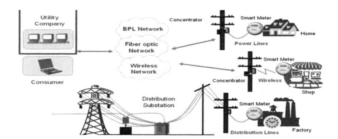

Fig. 3. Application of intelligent inspection in substation maintenance

The process of remote inspection is as follows: inspection and follow-up on-site inspection is determined by reusing the results of qualitative inspection, forming a closed-loop process to ensure fast and accurate troubleshooting. Remote inspections provide tools for comprehensive inspections of on-site meters and improve the management level of meter devices; the emergence of remote inspections makes on-site inspections more targeted, and the reliability of inspections is greatly improved. After the function is realized, the staff of the meter center will use the intelligent meter reading system to call the meter to collect the data of the terminal in real time [7]. Through the statistical analysis of various data, the system can grasp the operation status of the on-site meter device in time, as shown in Fig. 3.

5.2   The Intelligent Meter Reading System Has Auxiliary Significance for the First Inspection of the Electric Energy Meter Device

The first verification of the meter device before it is put into operation is of great significance. In the past, it was impossible to know the information of the meter device in real time. With the intelligent meter reading system, the staff can carry out operation and maintenance in strict accordance with the technical specifications to ensure that the first inspection of the meter device is in place after it is put into operation., And shut down the production of the meter device [8]. For example, clarify the first inspection test work process, development method and work requirements, prepare the first inspection work contact list, maintain contact with customers and power supply stations, follow up the operation information of the meter device in real time, and use the intelligent meter reading system to monitor the operation after the operation when the line is under load, ensure that the first inspection is carried out once the on-site verification conditions are met [8].

5.3   The Alarm Function of the Intelligent Meter Reading System Monitors the Fault

The use of the alarm function of the intelligent meter reading system to monitor the faults on-line effectively solves the problem of large space and long time for troubleshooting of the electric meter device. Due to the increase in electricity consumption, the number of meter devices has increased. It is quite difficult to use traditional methods to troubleshoot meter device failures and find the failures in time. If the device failures are not detected in time, it may bring more changes to users or power supply companies [9]. Failure to troubleshoot the electricity meter device for a long time will result in users not being able to use electricity normally, or equipment damage, and harm to the vital interests of users; for power supply companies, failure to troubleshoot will involve some electricity that cannot be recovered. Huge user base, even minor failures can cause immeasurable losses.

Using the intelligent meter reading system, as shown in Fig. 4, the staff can observe the alarm function of the system at any time. The system screens out more than 30 failure criteria and solidifies them into the energy data management platform. When the meter device fails or if the electric energy fails, the corresponding alarm event will be displayed. After the corresponding screening rules are automatically screened (such as light load and other alarm events), the work order flow will be automatically triggered to the marketing management system within the specified time limit [10].

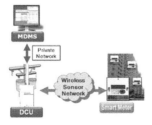

Fig. 4. Smart meter reading system

5.4 The Use of Smart Meter Reading Can Promptly and Accurately Maintain the Operation and Maintenance of the Electric Meter Device and Supplement the Electric Quantity

Intelligent meter reading can accurately record the real-time and historical values of three-phase current, voltage, power factor, power, etc. of the electric energy meter device; it can timely monitor the abnormal current and voltage of the electric meter device and issue an abnormal warning. In addition, through the intelligent meter reading system, the substation bus power imbalance statistics and abnormal analysis, the substation feeder line loss statistics and abnormal analysis, the main transformer loss statistics and abnormal analysis of the substation, the station area line loss statistics and abnormal analysis; can be based on these data the analysis accurately locates the inspection location, as shown in Fig. 5. Using the advanced intelligent meter reading system to assist the operation and maintenance management of the meter device can enable the meter personnel to flexibly arrange and deploy the work [10]. Compared with the traditional operation and maintenance management method, the efficiency is higher, and the troubleshooting accuracy rate is higher.

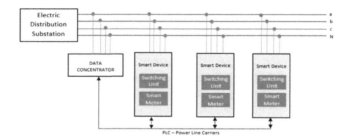

Fig. 5. Intelligent meter reading inspection data collection maintenance management

Using the intelligent meter reading system to upload the frozen power of the electric energy meter regularly and to record the current, voltage, power and other functions of the electric meter regularly, it can accurately record the electricity consumption of the electric energy meter device at each time period and provide a factual basis for the electric power recovery [11]. This method has been obtained. Widely recognized by users, greatly improving the efficiency of work. A fair and reasonable calculation of electricity consumption is a prerequisite to ensure that the interests of users and power suppliers are not damaged, and to protect the interests of both parties. In the past, due to the failure of the meter device, the inaccuracy of the meter was caused. Fairness, one's own possible interests are impaired; at present, the use of smart meter reading in a timely and accurate manner has implemented a policy of more refunds and less compensation for power errors caused by failures. Both parties can accept and maintain fairness [11].

5.5   Use Intelligent Inspections to Monitor Abnormal Electricity Consumption and Prevent Electricity Theft

Using the traditional on-site inspection method, it is impossible to monitor the abnormal electricity in real time, but the intelligent meter reading system can provide a variety of ways to monitor the abnormal electricity consumption. The alarm system for abnormal electric quantity is very sensitive. It can alarm in time for the abnormal response of electric energy meter, terminal, electric meter loop, current, voltage and so on [12]. Through the alarm, the staff can analyze the user's data in time and draw corresponding conclusions; At the same time, the system's alarm function will alarm for abnormalities in electricity, bus balance, and line loss. Through the analysis and calculation of the abnormalities, it can be analyzed whether the user has the possibility of electricity theft.

For users whose voltage or current data is zero for a long time, the system can quickly find out. If this kind of statistical work is to be manually excluded, it will take a lot of time, but it will be saved by the automatic troubleshooting function of smart meter reading. A lot of time, manpower and material resources. With the application of the automatic system of electric meters, the fairness and justice of electric energy meters are guaranteed to a certain extent, and whether users are stealing electricity can be detected in time, so that the interests of power supply companies are protected. Breaking the past due to limited manpower and the need for certain periodic restrictions for inspection and verification work, through the alarm function of the alarm system, the analysis of abnormal data can prevent the theft of electricity [12].

6   Summary

The use of intelligent technology for substation inspection not only has the flexibility and intelligence of manual inspection, but also overcomes some of the shortcomings and shortcomings of manual inspection. It is more suitable for the actual needs of the development of intelligent substations and unattended substations. This article explains the basic principles of intelligent meter reading and intelligent inspection system and focuses on the function of intelligent inspection system and inspection route design for intelligent substation, which fully reflects the substation intelligent meter reading inspection system in the daily inspection management of substations. The broad development space and application prospects.

References

1. Jin, K.: Application of artificial intelligence technology in power monitoring system. Integr. Circuit Appl. **37**(02), 84–85 (2020)
2. Abedi, H., et al.: AI-powered in-vehicle passenger monitoring using low-cost mm-wave radar. IEEE Access **10**, 18998–19012 (2022)
3. dos Santos, C.H.F., et.al.: Geometrical motion planning for cable-climbing robots applied to distribution power lines inspection. Int. J. Sys. Sci. **52**, 1646-1663 (2021).

4. Yun, A., et al.: Development of a robot arm link system embedded with a three-axis sensor with a simple structure capable of excellent external collision detection. Sensors **22**(3), 1222 (2022)
5. Kim, P.K., Kim, H.H., Kim, T.W., Lee, Y.N.: The intelligent system design for eliminating blind spots of PTZ controlled CCTV with raspberry pi and multiple pi cameras. Intell. Sys. **2**, 107–112 (2018)
6. De Kerf, T., et al.: Identification of corrosion minerals using shortwave infrared hyperspectral imaging. Sensors **22**(1), 407 (2022)
7. Thompson, S., Kagami, S., Okajima, M.: An autonomous mobile inspection robot for an electric power sub-station. Int. Conf. Info. Cont. Auto. Rob. **2**, 300–306 (2013)
8. Xu, M.: Intelligent technology in substation operation and maintenance technology. Elec. Technol. Soft. Eng. **9**(11), 237–239 (2020)
9. Jiang, W.J., Zhou, L.: Common problems and solutions for smart substation operation and maintenance. Decision Exploration **10**(08), 44–47 (2020)
10. Davtalab, O., et.al.: Automated inspection in robotic additive manufacturing using deep learning for layer deformation detection. J. Intell. Manuf. **33**(3), 771–784 (2022)
11. Zhang, Z.F.: Application research on intelligent inspection system of substation robot. North China Electric Power Univ. **15**(2), 147–149 (2017)
12. Badr, M.M., et al.: Detection of false-reading attacks in smart grid net-metering system. IEEE Inter. Things J. **9**(2), 1386–1401 (2022)

The Application of Visualized Monitoring of Partial Discharge in Power Equipment

Wei Chen[1(⊠)], Xueyan Wang[1], Hui Li[2], and Mingxiang Su[2]

[1] Science and Technology Branch, Taizhou Hongchuang Electric Power Group Co., Ltd., Taizhou, Zhejiang, China
chenwei1980@yummail.cn
[2] Taizhou Power Supply Company, State Grid Zhejiang Electric Power Co., Ltd., Taizhou, Zhejiang, China

Abstract. In the context of the continuous development of modern smart substation technology, related equipment is becoming more and more intelligent, and the application of online monitoring technology in smart substations has become mature. The application of online visual monitoring technology for power transmission and transformation equipment can lay a good foundation for the visualization of smart grid equipment status. This paper studies the characteristics of the on-line visual monitoring technology of power transmission and transformation, and its application in the detection of partial discharge of power equipment.

Keywords: Electricity · Equipment · Partial discharge · Visualization · Monitoring

1 Introduction

With the improvement of the voltage level of power transmission and transformation, the voltage of power equipment is getting higher and higher, and the requirements for the insulation performance of power equipment are getting higher and higher. In these devices, due to defects in the electrodes and insulation surfaces, damage or aging of the outer insulation of the wires, etc., partial discharges of the power equipment may be caused. If such faults are not handled in time, it may cause the final breakdown and failure of the insulation. And even cause damage to electrical equipment. To prevent the occurrence of such accidents, the on-line detection of partial discharge of power equipment becomes necessary without affecting the normal power transmission and transformation [1].

In the process of partial discharge, in addition to the transfer of charges and the loss of electric energy due to various ionization, excitation and recombination processes, electromagnetic radiation, ultrasound, luminescence, heat generation, and the appearance of new products are also generated. Therefore, in view of these phenomena of partial discharge, the basic methods of partial discharge detection are divided into two categories: electrical measurement methods and non-electric measurement methods. Electrical measurement methods mainly include pulse current method, ultra-high frequency detection method and so on. Partial discharge detection devices developed by

V. Sugumaran et al. (Eds.): ICMMIA 2022, LNDECT 138, pp. 118–126, 2022.
https://doi.org/10.1007/978-3-031-05484-6_15

many units in China generally use this method to extract the discharge signal, but the actual application effect is often not ideal. The main reason is that the field electrical noise interference is too large, which makes it difficult to obtain the real partial discharge signal. Although there are many methods and ideas for suppressing interference, there are not many that have been successfully used in monitoring systems. Therefore, further research and improvement in theory and application are needed.

Non-electric measurement methods mainly include ultrasonic detection method [2], optical measurement method, infrared detection method [3], chemical detection method, etc. The advantages of these methods are that they are not interfered by electrical signals during measurement and have strong anti-interference ability. These occasions have been widely used. Corresponding testing equipment includes ultrasonic detector, ultraviolet imager, infrared thermal imager, transformer oil analyzer, etc. Considering that in the process of partial discharge, since various ionization, excitation and recombination processes are accompanied by light radiation, detecting the light radiated by partial discharge is also a corona discharge detection method [4]. For example: A normal working high-voltage power equipment generally does not exist and will not generate ultraviolet light with a wavelength of 220–280 nm. When a partial discharge occurs inside the electrical equipment, it will radiate ultraviolet light. The ultraviolet sensor is used to detect the inside of the electrical equipment. In combination with optical imaging technology, it can sensitively and accurately detect partial discharge phenomena that cannot be directly detected by human eyes and conventional equipment in electrical equipment, to help on-site staff quickly locate the discharge point and discover the internal electrical equipment in time. Troubleshoot and eliminate the danger as early as possible to prevent problems before they occur.

2 Working Principle of Partial Discharge in High Voltage Equipment

When partial discharge occurs in high-voltage equipment, different forms of discharge will be generated according to the strength of the electric field. During the ionization process, the electrons in the air continuously gain and release energy. Through spectral analysis, it is found that light waves and sound waves are emitted when the electrons release energy. The radiated spectrum includes three spectrum bands of ultraviolet, visible and infrared. With the increase of voltage, the radiated spectrum of partial discharge also increases in the ultraviolet region [4–6]. Therefore, the use of ultraviolet signals as a means of detecting partial discharge has two main advantages:

2.1 High Sensitivity

The ultraviolet spectrum produced by partial discharge of high-voltage power equipment is mainly concentrated in the band below 200–400 nm, as shown in Fig. 1. The solar radiation in the air has a wide range of wavelengths, but the components with wavelengths between 220 and 280 nm are almost completely absorbed by the earth's ozone layer. The ultraviolet radiation in the "blind area of the solar spectrum" with a wavelength below 300 nm has become extremely weak [5]. This system selects the

UV-R2868 UV sensor that works in the 220–280 nm band as the detection sensor. Its sensitivity can reach 5 000 cpm, and it only responds to the UV signal in the 185–265 nm band.

2.2   Strong Anti-interference Ability

Since the inside of the power equipment is usually a closed black box environment, during normal operation, there is strong electromagnetic interference inside the equipment, and there is no ultraviolet signal. Since there is almost no ultraviolet light with a wavelength of 220 to 280 nm in the surrounding environment, the use of the ultraviolet radiation signal of this band as the detection object is hardly affected by the external ambient light. For example, in a normal working dry-type transformer, if an ultraviolet signal of this band is detected inside it, it can only be caused by partial discharge inside the power equipment [4]. Therefore, this method uses the ultraviolet signal of a specific wavelength band as the detection object. Since the sensor itself is not sensitive to various electromagnetic signals, the ultraviolet optical signal detection system has a strong anti-interference ability.

3   The Application Significance of Online Visual Monitoring Technology in Power Transmission and Transformation Equipment

The online visual monitoring of power transmission and transformation equipment uses advanced technology to automatically monitor the equipment, and the online visual monitoring technology of power transmission and transformation equipment continues to automatically monitor and monitor the status of power equipment to realize the collection, transmission, storage and forwarding of online detection status data.

Condition monitoring is equipment developed by using the life characteristics of equipment. Condition monitoring refers to the technology for monitoring the operating characteristics of machinery and equipment. Condition monitoring is divided into technology based on condition maintenance services and technology for predictive maintenance services [5]. CBM can obtain the status information of the equipment and inform people of equipment maintenance needs. TBM can prevent many accidents, but unexpected failures may still occur during maintenance.

As the energy shortage problem becomes severe, intelligence has become an inevitable trend in the development of power grids. At present, China Research has done a lot of research in the related fields of smart grid. State Grid Corporation's smart grid reform link provides complete solutions in terms of informatization access. A lot of research and practice have been carried out in the monitoring and evaluation of the condition of power transformers, arresters, and other equipment, as well as the failure mechanism. The detection technology has been greatly improved. Modern equipment management is an important part of enterprise asset management [6]. Equipment maintenance and repair is the core content of equipment management. The purpose is to maintain the reliability and availability of the equipment at the expected level, scientifically extend the maintenance interval, effectively manage the maintenance process, and achieve the goal of maximizing corporate value.

With the advancement of science and technology, the equipment maintenance strategy is improving. The condition maintenance system based on equipment condition evaluation is based on the development of management methods and detection technology. Condition overhaul is an advanced overhaul method [7]. Through advanced condition monitoring, the equipment status is judged, the fault location and its situation are judged, and the necessary overhaul is carried out. It provides a reliable guarantee for the safe and stable operation of power equipment.

4 Research on Online Visual Monitoring Technology of Power Transmission and Transformation Equipment

4.1 Online Visual Monitoring of Transformers

Online visual monitoring uses relevant equipment for real-time monitoring, as shown in Fig. 1. On-line visual monitoring of capacitive equipment borrowing damage and leakage current, on-line visual monitoring of partial discharges such as transformers, and on-line visual monitoring of power transmission lines have gradually reached the practical technical level. The key to the online visual monitoring technology of transformer oil chromatography is oil and gas separation technology and gas detection technology [7]. At present, the more mature and effective one is meteorological chromatography.

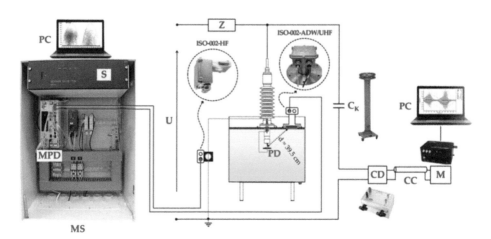

Fig. 1. Online visual monitoring of transformer

Oil and gas separation technology separates the gas dissolved in the oil. The main degassing methods currently used are membrane degassing, vacuum degassing, and dynamic headspace degassing. The membrane degassing method uses polymer membranes to realize the separation of oil and gas by using the unbalanced air pressure on both sides of the membrane. The gas content in the oil is obtained by calculation [3]. The disadvantage is that it takes a long time to reach the equilibrium state, and the polymer film needs to be replaced frequently.

Vacuum degassing method includes vacuum pump degassing method and bellows method. The vacuum pump degassing method has high degassing efficiency, but its disadvantage is low repeatability. The bellows method uses a motor to drive the bellows to compress oil and gas into vacuum separation gas. The remaining oil will affect the next measurement. The dynamic headspace degassing method is stirred by a stir bar in the sampling bottle, and the gas is separated out and returned to the sampling bottle through the detection device [8]. The headspace degassing technology has the characteristics of short oil-gas balance time and good repeatability and is widely used.

The excellent performance of the gas detection components determines the performance of the monitoring device. The gas-sensing semiconductor is in the gas chamber, and the gas sensor is used to detect the mixed gas content. This method is not scientific enough as a basis for diagnosis. The sensitivity to different gases is different, which can easily cause false alarms.

Meteorological chromatography transports the separated gas samples to the chromatographic column under the action of high-purity nitrogen. The advantage of this method is that it can accurately detect the liquidity of each gas, but the disadvantage is that the period is long. More suitable for regular testing [8].

The advantage of photoacoustic spectroscopy is that it can reduce the workload of calibration and can detect the concentration of each component gas with a small number of samples. This method does not need a chromatographic column and has good repeatability. This method has many advantages and is widely used.

4.2   On-Line Visual Monitoring of Transformer Partial Discharge

Partial discharges include pulsed flare discharges and non-pulsed glow discharges. According to the discharge characteristics of the transformer, pulse current method, ultrasonic method, radio frequency method, etc. are generated. The pulse current method measures pulse electrical signals when the transformer bushing is grounded, and high voltage occurs between the windings. It can reflect the characteristics of the number of discharge pulses, and it needs to extract the partial discharge pulse signals effectively and accurately [6]. The disadvantage of this method is the low measurement frequency. The radio frequency detection method usually uses a Rogowski coil type sensor to extract the signal from the neutral point of the detected equipment. It can work in a wide frequency range and is easy to install. This method is widely used in the field of generator online visual monitoring, as shown in Fig. 2.

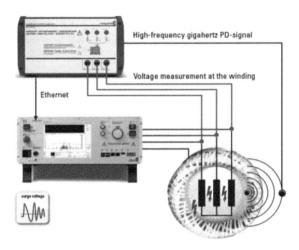

Fig. 2. Radio frequency detection method

The discharge energy detection method measures the power consumption of the total charge in each cycle. It can be used to measure the glow discharge, which is difficult to correspond to by the current method. The disadvantage is that it is difficult to separate the discharge loss. The ultra-low frequency local detection method uses 0.1 Hz voltage to measure the discharge of the equipment, which can reduce the capacity of the equipment and the volume can be smaller. The disadvantage is that it is difficult to prove the equivalence of 0.1 Hz and 50 Hz power frequency electricity.

The ultrasonic method uses the internal discharge of the piezoelectric sensor of the fuel tank to generate ultrasonic waves to determine the current. Its advantage is that it can greatly reduce the influence of interference and can qualitatively judge whether there is a discharge signal. Used as an auxiliary means in transformer testing. UHF detection method receives partial discharge through UHF sensor to generate UHF magnetic wave. Its biggest feature is strong anti-interference ability, high sensitivity, and good development prospects. In recent years, the UHF method has gradually been applied to transformers. The ultra-high frequency detection method is used to detect the received ultra-high frequency electromagnetic waves nowadays [7]. It has a large amount of information and high measurement frequency, which can comprehensively study the partial discharge characteristics in the transformer insulation system.

4.3   On-Line Visual Monitoring of Transformer Winding Deformation

Winding is a component with many faults, and most of the faults are caused by winding deformation. At present, the detection methods of transformer winding deformation mainly include frequency corresponding analysis, short-circuit reactance test and vibration signal analysis, as shown in Fig. 3. Frequency response analysis is the most advanced winding deformation diagnosis method in the world. It uses changes in capacitance before and after winding deformation to measure the winding transfer function and reflect the winding situation [8]. It has strong anti-interference ability, and the frequency response method has high sensitivity.

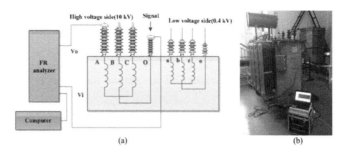

Fig. 3. Online visual monitoring of transformer winding deformation

The short-circuit reactance test uses the signals on the one and two sides of the transformer to obtain the short-circuit reactance of the transformer online. Realize the on-line visual monitoring of the transformer winding condition, but the poor transformer and voltage fluctuation will affect the strategy result. Vibration signal analysis uses sensors to measure the vibration signal during winding operation, which reflects the condition of the winding core [8]. It can also monitor the winding deformation and the loosening of the iron core and other faults at the same time.

4.4   Online Visual Monitoring of GIS Combiner

Various defects may occur during operation of GIS. Such as conductor burrs, contamination on the surface of insulators, deterioration of PT and CT insulation, etc. Defects in GIS operation will cause partial discharges of varying degrees, which will enlarge insulator defects for a long time. Partial discharge monitoring in GIS is an effective method to find faults.

Partial discharge monitoring methods are divided into electrical and non-electric methods. Non-electric methods include optical, chemical, and mechanical methods. The optical method uses a photomultiplier to measure the light emitted by the internal discharge of the GIS through the GIS observation window [8]. This method has low monitoring sensitivity and has a monitoring dead angle. The chemical method is not very sensitive, but it is effective for gas monitoring in small air chambers. The mechanical method can be monitored by accelerating sensors or ultrasonic probes.

The ultrasonic wave has a short wavelength and concentrated energy. The piezoelectric sensor can perform qualitative analysis on the discharge signal. The monitoring range of this method is limited, and it is currently combined with the UHF method for partial discharge positioning.

Electrical methods include conventional measurement methods and UHF methods. The external wave electrode method sensor has a simple structure, but the sensitivity is not high. The ground pulse current method has low detection sensitivity and can be used as a means of GIS inspection [9]. The embedded electrode method in the insulator has good anti-interference, but it is necessary to design the embedded electrode when the insulator is manufactured, which increases the difficulty of manufacturing the insulator.

Research and practice have proved that the detection of GIS partial discharge in the ultra-high frequency band has strong anti-interference, and the current detection system can perform partial discharge detection and positioning. The advantage of this monitoring method is that the sensor is independent of GIS, the measurement system is movable, and the sensor is light, and it is easy to use. GIS built-in sensors are not suitable for running GIS, which is conducive to external interference, but sensors need to be installed at intervals [9].

The SF6 in each gas cell of GIS has a certain amount of moisture, which comes from the volatilized moisture of the insulating material inside the gas, and the commercial gas contains a small amount of moisture. There are many methods to test the moisture content of SF6. The disadvantage of the commonly used coulomb solution is that the electrolysis efficiency will decrease with time. Some directly perform SF6 dew point test, which can more accurately reflect the hazard degree of SF6 level to the equipment than the relative volume value.

GIS equipment uses SF6 as insulation. Monitoring GIS equipment is SF6 gas meter. Supervision is a necessary means to monitor the operating status. Rely on the pressure contact on the density relay to send out the SF6 gas leakage to the distant place. It is prone to accidents such as damage to the main equipment due to SF6 leakage.

During production, the cross-linked battery will have residual bubbles, and the breakdown voltage of the bubbles will be lower than that, and partial discharge will occur first. Partial discharge in the cable will produce electromagnetic waves, ultrasonic waves, and other phenomena, and produce new products [10]. Commonly used detection methods include high-frequency detection methods, ultrasonic detection methods and many other methods.

The ultra-high frequency detection method uses the UFH sensor to detect the signal to determine whether it is partial discharge. Its advantage is that it can perform local positioning and is suitable for online detection, but its disadvantage is that it is susceptible to interference from broadcast signals. The high-frequency current method detects the two parts of the cable body and the ground wire in an orderly manner. The advantage is that the principle is simple, and the disadvantage is that it is easily interfered by broadcast and television signals [10].

5 Summary

The online visual monitoring system can ensure the stable operation of power transmission and transformation equipment. Improve the ability to monitor the operating status of power transmission and transformation equipment. Reasonably design the system structure, automatically collect, analyze and process data, find faults in time, and eliminate hidden troubles.

References

1. Khelifa, H., Vagnon, E., Beroual, A.: AC breakdown voltage and partial discharge activity in synthetic ester-based fullerene and graphene nanofluids. IEEE Access **10**, 5620–5634 (2022)
2. Uwiringiyimana, J.P., Khayam, U., Suwarno, Montanari, G.C.: Design and implementation of ultra-wide band antenna for partial discharge detection in high voltage power equipment. IEEE Access **10**, 10983–10994 (2022)
3. Hamelmann, P., Mischi, M., Kolen, A.F., Van Laar, J.O.E.H., Vullings, R., Bergmans, J.W. M.: Fetal heart rate monitoring implemented by dynamic adaptation of transmission power of a flexible ultrasound transducer array. Sensors **19**(5), 1195 (2019)
4. Ahmed, S., Lee, Y., Hyun, S.H., Koo, I.: A cognitive radio-based energy-efficient system for power transmission line monitoring in smart grids. J. Sensors **12**, 3862375:1–3862375 (2017)
5. Jadhav, A.R., Malyala, P.R.S.K., Rajalakshmi, P.: Development of a novel IoT-enabled power- monitoring architecture with real-time data visualization for use in domestic and industrial scenarios. IEEE Trans. Instrumen. Measur. **70**, 1–14 (2021)
6. Gallucci, L., et al.: An embedded wireless sensor network with wireless power transmission capability for the structural health monitoring of reinforced concrete structures. Sensors **17** (11), 2566 (2017)
7. Liu, Z., Wang, M.X., Sun, G.: Research on the application of internet of things technology in the condition monitoring of power transmission and transformation equipment. Auto. Instrumen. **12**(03), 171–173 (2017)
8. Wen, J.F., Sheng, H.H., Jiang, J.N.: Design and application of online monitoring system for relay protection of smart substation. Jiangsu Elec. Eng. **11**(01), 212–224 (2015)
9. Sadiqbatcha, S., et al.: Post-silicon heat-source identification and machine-learning-based thermal modeling using infrared thermal imaging. IEEE Trans. Comput. Aided Des. Integr. Circuits Syst. **40**(4), 694–707 (2021)
10. Feng, L.F., Sun, J., Zhou, J.H.: Implementation of ARM9-based network video surveillance system. Electric Power Auto. Equip. **26**(10), 95–97 (2016)

The City Information Visual Based on ArcGIS Technology

Yanlin Wu[✉]

Hubei University of Technology, Wuhan, Hubei, China
wy1989@bohao.org

Abstract. The 3D visual of urban information is the prerequisite and foundation for building a digital city, and the consistency of the second and 3D GIS platforms makes the digital city application system have a wider application prospect. Analyze various 3D visual software and take ArcGIS as an example to discuss the 3D visual of city information and the corresponding spatial analysis, and study the corresponding 3D modeling process and related technical issues.

Keywords: ArcGIS · Technology · City · Information · Visual

1 Introduction

With urban construction and the urban scale rapid expansion, the traditional management model urgently needs to be reformed. The digital earth and digital city are becoming more infiltrating [1]. As of the beginning of 2008, more than 30 cities across the country have carried out the pilot the construction work of the digital city geospatial framework, and have achieved preliminary results, and the 3D visual of the city is the prerequisite and foundation for the construction of a digital city or virtual city in urban planning [1]. In terms of construction and display, it is helpful to improve the city management level. This paper studies the methods of city 3D visual. The consistency of the 2D and 3D GIS platforms, this paper proposes a combination of Google GIS Plugin and ArcGIS for 3D modeling and visual and analyzes the corresponding workflow and existing problems [2].

2 Defects in the Visual Expression of Traditional Test Data

The traditional visual representation of test data analysis results is relatively simple, mostly in the form of statistical charts. Statistical charts are generally used to express the relationship of changes in a specific place over time [3]. Time series analysis or spatial interpolation analysis are often used to express the characteristics and changing trends of statistical data in the form of points, lines, and surfaces. "Currently, for large-scale testing work, the traditional data analysis software is mainly Excel, Origin and SPSS." Compared with a simple 2D data table, the expression method is more vivid and concrete, which makes people clear and impressive [2]. However, since most of the test data comes from discretely distributed test points in space, it has the characteristics

V. Sugumaran et al. (Eds.): ICMMIA 2022, LNDECT 138, pp. 127–135, 2022.
https://doi.org/10.1007/978-3-031-05484-6_16

of spatio-temporal data. In other words, the location of each measurement point is separated from the test data, and the analysis chart of each measurement point is relatively independent, and it cannot accurately show the evolution and development of the test data in space [3]. Therefore, only using time series to analyze the characteristics and changing trends of the data is completely not enough.

3   3D Modeling Based on ArcGIS

The key to the malleability and flexibility of ArcGIS model lies in the modeling ease and surface and volume editing. ArcGIS contains two basic cartographic elements: "line" and "area". The line constitutes the surface, and the surface constitutes the body [4].

3.1   Point Modeling

In the 3D environment, some ground objects are directly abstracted as point elements. Such as statues, flagpoles, bus stops, trees, etc. These point elements, such as flagpoles and street lights, have regular geometric shapes, and their modeling is no different from that of surface features [3].

3.2   Line Modeling

The modeling of linear elements mainly includes the modeling of roads, rivers, and 3D pipelines. It has no pure "line" modeling in ArcGIS. Lines are considered to be 3D surfaces with a certain area [5]. For example, roads are regarded as surfaces with a certain width. In addition, the suspension tool in the ArcGIS terrain tool can project the road onto the undulating terrain to generate 3D road lines.

3.3   Terrain Modeling

ArcGIS includes a unique set of tools called "sandboxes" for terrain processing. You can use the "Sandbox" tool to create smooth terrain from the imported contours; you can also use the "Sandbox" tool to change the shape of the terrain by creating berms, slopes, ridges and valleys, and add roads, paths, and building foundations [6]. Also can import terrain in external software format, such as TIN, DEM, contour, etc. Figure 1 is a simple 3D landscape model that includes points (streets, landscape trees, etc.), lines (roads), and surfaces (vegetation, buildings, etc.) in ArcGIS software, and is super-imposed on topography.

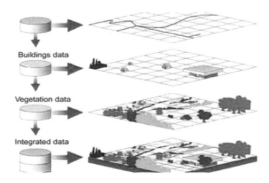

Fig. 1. Terrain modeling

3.4   Surface Modeling

Surfaces in SketchUp usually have the characteristics, such as modeling of buildings, lawns, ball fields, etc. It is mainly building modeling. One is geometric modeling. First, import the reference base map (remote sensing image, CAD data, shp format, etc.) in ArcGIS to obtain the outline of the building. Re-stretch to obtain a preliminary frame and make detailed modifications; the second is texture modeling, using a digital camera to take photos of all sides of the building, and correcting it in Photoshop, importing it as a material into the SketchUp software to map the model, and the software. It also comes with a large number of commonly used materials [7].

4   ArcGIS-Based Data Collection and Processing

The preliminary data preparation for the "Wuhan City Self-Examination Report" project includes site element data construction, building outline data construction, road traffic network status data construction, terrain data construction, etc.

4.1   Site Element Data

The acquisition and collection of current data of water system and green space mainly combines the "OpenStreetMap" and the Gaode map (https://lbs.amap.com) data collection; according to the scope of this physical examination, the scope of data collection is Wuhan Jiang'an District, Jianghan District, Qiaokou District, Wuchang District, Hongshan District, Qingshan District, Hanyang District, 7 central urban areas and Donghu Scenic Area, administrative. The total area of the region is about 800 km^2, shown as Fig. 2.

Fig. 2. Site element data

4.2   Building Profile Data

Mainly based on the number of building floors obtained from field research, combined with the geographic coordinates generated after satellite map correction; among them, the height of the building is roughly estimated according to the number of building floors and the height of the floor, and the satellite map generated by bigemap software determine the architectural projection; the scope of data collection is the scope of this physical examination, shown as Fig. 3.

Fig. 3.   Building profile data

4.3   Road Traffic Network Data

Mainly collect data based on the status quo of the urban road traffic network in Wuhan, confirm according to the field survey and Gaode map, generate osm data through

"OpenStreetMap", convert it into shp data and import it into arcgis for processing; the scope of data collection is the base area, shown as Fig. 4.

Fig. 4. Road traffic network data

4.4 Terrain Data

Mainly based on the Digital Elevation Model (DEM) of the "Geospatial Data Cloud", the raster accuracy of the data is 30m; the data collection range includes the base area and the surrounding area of 30–50 km area, shown as Fig. 5.

Fig. 5. Terrain data

5 City Information Visual Based on ArcGIS 3D Modeling

The system utilizes the touch operation, photographing, picture display, intelligent calculation and other functions of the mobile device to collect the appearance and texture characteristics of the ground objects through photographing and video recording, and display them smoothly on the device [8]. At the same time, the real-time positioning feature of the mobile device is used to obtain the current geographic location in time to realize the seamless association between the surface feature data of the ground object and the spatial data.

5.1 Model Processing

The model is processed in the 3D software, and the 3D Wuhan city model will be established. According to the corresponding longitude and latitude coordinates, the orthophoto map is pasted as the satellite map texture to form the true 3D geographic landscape of Wuhan city. In addition, such as water system green space, transportation network, etc., select or re-establish the corresponding surface material according to different feature attributes, and give it appropriate texture. Then choose an appropriate projection method, and project it on the geomorphic landscape according to the latitude and longitude coordinates corresponding to the ground surface model to obtain a geomorphic landscape with a sense of virtual technology.

5.2 Visual of Interactive Dynamic Information

The information visual design of the urban self-examination report in Wuhan is mainly divided into two sections: eight special physical examination indicators in the urban self-examination in Wuhan and three bright spots. Take the visual of the "eco-livable" indicator among the eight special physical examination indicators as an example. First, use Unreal Engine 4 software to create a visual user interface of the "Wuhan City Self-Physical Examination Report" and a visualized interface of the "eco-livable" indicator information; Secondly, create a blueprint class pawn, add a camera, and set the location of the Wuhan city 3D model to be displayed by "Ecological Livable"; finally edit the interactive blueprint, create a "button", click to generate an event, and trigger the click to generate "Ecological Livable" information visual Function, Fig. 6 shows the blueprint node.

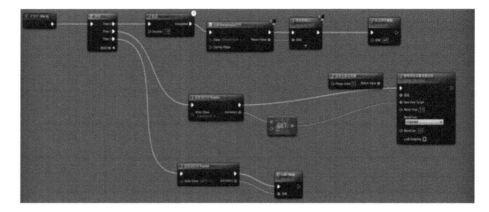

Fig. 6. "Eco-Livable" information visual interactive generation

Among them, the click event triggers the generation of the "Create Widget" and "Add to Viewport" nodes to call the designed "Eco-Livable" control blueprint, and the "Remove from Parent" node to specify the control blueprint of the initial interface as the target. The control blueprint of the interface is removed from the display. Use "Get Player Controller" and "Set View Target with Blend" to switch perspectives, and connect through "Sequence" to generate visual content of "eco-livable" information.

5.3   City Information Visual Based on ArcGIS

ArcGIS data flow is to enable two-way of 3D model data between ArcGIS and Google. Use the correct 2D data for 3D model. After modeling, the two and 3D data can work together. You must install the Google GIS Plugin control and add the Google GIS Plugin toolbar to the corresponding ArcGIS program interface [9]. Shown in Fig. 7, first use ArcMap to edit and optimize the 2D data to be modeled, especially the breakpoint problem of polygonal data. It can be connected in a straight line so that there is no breakpoint in the middle [10]. This can reduce the amount of data for subsequent modeling. Then use the Google GIS Plugin tool to export the elements that need to be modeled to ArcGIS, and model them in ArcGIS. Export the built model to data in ESRIMultiPatch (3.mdb) format. Finally, complete the visual of 3D geographic information in ArcScene or ArcGlobe, as shown in Fig. 8.

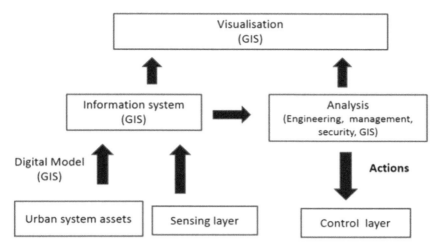

Fig. 7. Framework process of Google GIS plugin

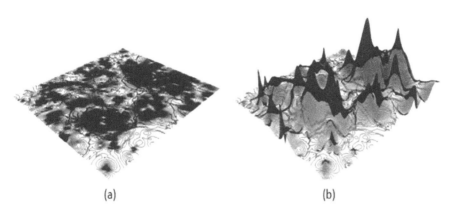

Fig. 8. Visual of ArcGIS

6 Summary

There are many methods to achieve the 3D visual of the city, but they all have their own shortcomings. ArcGIS, as a convenient, easy-to-use, and powerful 3D modeling software, coupled with the free and massive data sources in Google Earth as the basis, combined with the powerful ArcGIS spatial analysis, improves the efficiency of modeling and reduces the need for professional the limitation of modelers has greatly reduced the 3D modeling cost. It is a new bright spot in the city's 3D visual in recent years, and it has shown broad development prospects.

References

1. Bardsley, J.M., Howard, M., Lorang, M.: Matlab software for supervised habitat mapping of freshwater systems using image processing. Int. J. Remote Sens. **13**(23), 4906 (2021)
2. Wait, E., Winter, M.R., Cohen, A.R.: Hydra image processor: 5-D GPU image analysis library with MATLAB and python wrappers. Bioinformation **35**(24), 5393–5395 (2019)
3. Malczewski, K.: Image resolution enhancement of highly compressively sensed CT/PET signals. Algorithms **13**(5), 129 (2020)
4. Pugalenthi, R., Sheryl Oliver, A., Anuradha, M.: Impulse noise reduction using hybrid neuro-fuzzy filter with improved firefly algorithm from X-ray bio-images. Int. J. Imaging Syst. Technol. **30**(4), 1119–1131 (2020)
5. Jeelani, Z., Qadir, F., Gani, G.: Cellular automata-based digital image scrambling under JPEG compression attack. Multimedia Syst. **27**(6), 1025–1034 (2021)
6. Li, X.: Analysis of computer graphics and image processing technology. Wirel. Internet Technol. **11**, 123–125 (2016)
7. Santos, T., Rodrigues, A.M., Ramalhete, F.: Assessing patterns of urban transmutation through 3D geographical modelling and using historical micro-datasets. In: Gervasi, O., et al. (eds.) ICCSA 2015. LNCS, vol. 9155, pp. 32–44. Springer, Cham (2015). https://doi.org/10.1007/978-3-319-21404-7_3
8. Tong, Z.Y.: Sketchup architectural modeling detailed tutorial. Mod. Electron. Technol. **8**, 127–129 (2019)
9. Shen, D., Wong, D.W., Camelli, F., et al.: An ArcScene plug-in for volumetric data conversion, modeling and spatial analysis. Comput. Geosci. **61**, 104–115 (2013)
10. Jin, X.: Research on 3D visual technology of pipeline based on GIS. Mod. Inform. Technol. **13**(4), 57–60 (2018)

Intangible Cultural Heritage Protection Based on MAR Algorithm

Jian Rao and Xinyue Zheng[✉]

School of Arts and Design, Hubei University of Technology,
Wuhan, Hubei, China
zhengxinyuecs@aecd.cn

Abstract. Nvshu is an ancient gender script circulating in Jiangyong County, Hunan and its adjacent areas. At present, with the rapid development of national economy and culture, the inheritance and development of Nvshu culture is facing many difficulties, and the protection of Nvshu cultural resources is urgent. Under the background of the integration of science and technology, in order to protect the gradual loss of Nvshu intangible culture, a strategy for the protection of Nvshu culture using modern digital management technology is proposed. This paper analyzes the protection principles of Nvshu intangible cultural heritage, uses the MAR model to establish the Nvshu intangible cultural heritage database, analyzes the user's satisfaction with the database function through the MAR algorithm, and carries on the Nvshu cultural inheritance by means of digital collection and digital restoration, and strengthens The spread of Nvshu culture provides a certain reference for other intangible cultural heritage protection work.

Keywords: Nvshu intangible cultural heritage · Digital management · Intangible cultural heritage database · MAR algorithm

1 Introduction

Nvshu is a writing system born under the feudal society in ancient China, and it was spread in Jiangyong area of Hunan. It usually exists in the form of embroidery on fabrics such as fans, handkerchiefs, and straps, rather than traditional bamboo slips or paper. Therefore, in the 1950s, Nvshu was discovered by society and attracted wide attention from scholars at home and abroad. At the same time, the Nvshu culture is an important part of the Hanyao national culture in southern China. It has important value for studying the origin of human writing and civilization, female culture and human development.

There are abundant researches on the protection of intangible cultural heritage (ICH). For example, the Chinese government attaches great importance to the protection and inheritance of ICH, and accordingly provides funds to vigorously support related ICH areas, ICH projects and non-genetic inheritors, and has built many ICH studios and ICH Workshops, intangible cultural heritage exhibition halls, expositions, etc. It is precisely because of the government's emphasis on communication that ICH is becoming more familiar to the public, which enhances the public's interest in ICH, and

V. Sugumaran et al. (Eds.): ICMMIA 2022, LNDECT 138, pp. 136–144, 2022.
https://doi.org/10.1007/978-3-031-05484-6_17

thus more actively understands and learns about the content of ICH. The vitality of cultural heritage is constantly being activated [1]. Combining ICH with today's Internet technology, the use of the Internet can not only protect and inherit ICH, but at the same time carry out online sales of intangible cultural products; in addition, ICH has begun to integrate with tourism, and some ICH research trip projects have been launched welcomed by the public. On the other hand, the development of high and new technology has provided a variety of communication methods for the dissemination of ICH. The use of online and offline integration not only enriches the forms of communication, but also enhances the effectiveness of communication and expands the area of communication [2]. Although the early ICH studios, exhibition hall libraries and other places have a relatively simple display mode, they are only basic exhibitions of traditional text, pictures and videos, but now intangible cultural heritage exhibition halls are optimized in combination with digital technology on top of basic technology protection. The online or on-site interactive experience effect is enhanced in the display and communication.

Based on the current dilemma of the development of Nvshu ICH, this article explains the advantages of using digital technology to spread and protect ICH. According to the principles of intangible heritage protection, the Nvshu ICH database is designed and the database system functions are tested. It is hoped that users will pass the database. The interactive function of the system understands the cultural knowledge of Nvshu, and realizes the digital communication and protection of Nvshu culture.

2   Theories Related to Nvshu Intangible Cultural Heritage

2.1   Difficulties in the Development of Nvshu ICH

Nvshu is a kind of gender language for women to express emotions and communicate with each other. In the feudal society of China, women are a kind of spiritual resistance to the oppression of feudal male power. Since the discovery of Nvshu, it has only been spread in a small area in Jiangyong County and its surrounding areas. However, because of its unique shape and unique emotions contained in the text, scholars have successively arrived in Jiangyong, Hunan for cultural investigations. However, in the old society, because Nushu did not have standardized texts and teaching materials, it was copied by hand by generations of Nushu successors. In addition, after the death of the master, the works of Nushu were burned or buried as sacrificial objects tradition of. Therefore, only a few works of Nushu have survived as family heirlooms or souvenirs, and they were even on the verge of extinction. At the beginning of the 21st century, the discovery of scholars and the active implementation of the Nvshu cultural project by the Yongzhou government provided an excellent opportunity for the inheritance and protection of Nvshu's ICH. Under the combined action of many factors, the Nvshu ICH is facing the dilemma of modern development. The cost and effectiveness of the traditional communication method in the implementation process are open to question, and it is not suitable in the new era. Large-scale promotion. Therefore, the Nvshu culture is reproduced in a new form, and the Nvshu culture is "lively" inherited and protected by digital technology. For example, digital collection of Nvshu cultural relics, construction of Nvshu cultural heritage database and other measures are conducive to the wider spread of Nvshu culture, and it has practical significance for sustainable development [3].

2.2   Advantages of Digital Technology in the Application of ICH

(1) The integrity and reducibility of cultural protection is high
The primary value of digital technology in the inheritance and protection of intangible cultural heritage lies in the preservation of complete and accurate information, and it is not easily damaged or distorted. There are many intangible cultural heritage items in our country, and the content forms are complex and diverse. Digital technology can unify some cultural data and physical data in the database. Nowadays, with the rise of digital photography, three-dimensional modeling, virtual scenes and other related digital technologies, we can acquire, record, and synthesize comprehensive information content of intangible cultural heritage, such as sounds, movements, and expressions, to maximize complete records. On its basis, digital collection, modeling and restoration are carried out through digital images, videos, animations, renderings and other technical methods, to record the original appearance of intangible cultural heritage information, and to establish a data information database, so that it can be preserved to avoid inheritance In the process, the problem of missing data or incomplete inherited content has improved the accuracy of intangible cultural heritage content. At the same time, emerging digital technologies break through the traditional plane restrictions, and the preservation form is more three-dimensional and intuitive than in the past, which promotes the effectiveness of the inheritance of ICH [4, 5].

(2) Provide a wide range of platforms
The digital dissemination of intangible cultural heritage can be displayed on multiple platforms, thereby shortening the distance between intangible cultural heritage and the public. The digital communication platform is based on the traditional media display, using the Internet, multimedia and other means to carry out the digital display and dissemination of intangible cultural heritage. It includes digital exhibition platforms such as online digital exhibition halls, digital museums, and VR experience halls. There are some digital media platforms such as WeChat and Weibo. Compared with the past communication platforms, it breaks the traditional on-site exhibition activities and the limitation of exhibition space, and creates conditions for the promotion of intangible cultural heritage in different regions, so that the public can search and inquire remote intangible cultural heritage at any time. Culture has expanded the dissemination space of intangible cultural heritage. In addition, under the background of the continuous development of digital technology and multimedia technology, the digital display platform is no longer just a traditional static display, but a more dynamic, three-dimensional and interesting display form [6, 7].

(3) Strong interactive experience
Digital media has the function of reproducing intangible cultural heritage, perfectly combining vision and hearing, and enabling the public to obtain a complete

sensory experience. It has updated the way of displaying cultural heritage, created fresh presentation content, and is a new product in today's social progress. ICH is passed down from generation to generation, and the most essential feature is that it is human-centric, has unique communication vitality and creativity, and embodies the live inheritance process. Therefore, in the process of dissemination and display, more attention must be paid to the audience's immediate situational feelings, and digital means the use of is to create new ways and methods for the establishment of intangible cultural heritage interactive experience, build a platform for the joint perception and acceptance of cultural information from multiple sensory channels such as vision, hearing and touch, and help to establish an effective public evaluation and feedback system, and then change the form of cultural information transmission from one-way to two-way interaction [8].

2.3   Principles of Nvshu ICH Protection

(1) The principle of originality protection
The World Cultural Heritage Committee clearly stipulates that authenticity is an important principle for the protection of CH. The CH protected today is real, and it is not based on ancient documents and biographical similarities. In other words, the cultural heritage to be protected must be real in history. Only by protecting the authenticity of cultural heritage can we ensure the value and significance of all protected cultural heritage, and at the same time, the public resources invested by the whole society in protecting cultural heritage will not be wasted. When protecting intangible cultural heritage, we must prevent some government agencies from forcing some non-cultural heritage items as cultural heritage for local economic benefits, so as to increase the cultural heritage of the place and bring new tourist attractions. Stimulate local economic development. There are also some local governments who have applied for protection before they have sufficient capacity, resulting in damage to cultural heritage [9, 10].

(2) The principle of overall protection
The so-called integral protection is that the ICH itself as a whole is composed of many cultural units. Even if there are some representative characteristics, they cannot replace the entire cultural heritage. In order to fully protect the existing ICH, in addition to protecting the entire content of the ICH, it is also necessary to protect its surrounding environment. For some traditional handicrafts, the inheritors of these crafts must also be protected. All in all, to properly protect the integrity of ICH is to protect it from all aspects. In the protection of ICH, integrity is a very important point [11].

2.4   Mar Model

To establish MAR model for batch process data, it is also necessary to perform two-dimensional expansion processing on three-dimensional data. First, expand the three-dimensional data along the batch direction, extract the mean and standard deviation of each column, and perform standardization processing; then, re-divide the three-dimensional data into I modules, and establish the current time for each module (that is, each batch) data. The relationship with the previous L moments, and then the residual Ei of the MAR model of batch i is obtained. In the absence of special instructions, the superscript i indicates a certain batch [12]. The MAR model of each batch of i is as follows:

$$X_k^i = \sum_{l=1}^{L} \phi_l^i X_{k-l}^i + e_k^i \tag{1}$$

Among them, k = L,L + 1,.,K, L is determined by Akaike Information Criterion (AIC), $X_k^i, e_k^i \in \Re(J \times 1)$ is the measured variable and model residual of the i-th batch of data at time k, and X_{k-l}^i is the i-th batch of data The measured variable at the kl-th time, $\phi_l^i \in \Re(J \times J)$ is the model coefficient vector of the i-th batch of data at the kl-th time.

$$\begin{cases} AIC(L) = Nln\sigma_a^2 + 2L \\ \sigma_a^2 = \frac{\Delta}{N-L} \\ \Delta = \sum_{t=L+1}^{N} (\chi_k^i - \phi_1 \mathcal{X}_{t-1} - \phi_2 \mathcal{X}_{t-2} \cdots \phi_L \mathcal{X}_{t-L} \end{cases} \tag{2}$$

Among them, L is the model order, N is the number of sampling points, DD is the variance of the residuals, and DD is the sum of squares of the residuals.

3   Research on the Protection of Nvshu Intangible Cultural Heritage

3.1   Research Purpose

With the development of information globalization, DT can transform information into measurable digital data. In addition to various industrial fields, modern technology has also been widely used in intangible ICH. Utilizing the interactive, convenient, and immersive characteristics of digital media can provide more cultural expressions for the informatized expression of Nvshu ICH, and make the content of its dissemination more attractive. Establishing a digital management and protection strategy for Nvshu culture is conducive to better inheriting Nvshu ICH and provides a new development path for Nvshu ICH.

3.2   Research Content

This article mainly uses DT to design the Nvshu ICH database, test the database system function and the user's satisfaction with its function.

4 Nushu Intangible Cultural Heritage Digital Protection Management Design

4.1 Functional Design of Intangible Cultural Heritage Database

Figure 1 shows the architecture of the Nvshu ICH database. First, the participants of ICH digitization coordinate on the data integration platform according to the digital technology standards, and carry out project information data sorting, digital coding, integrated editing, data processing, and target citation of the Nvshu intangible cultural heritage. Secondly, the database manager loads the data into the archive library and release library to support users (ordinary users, academic research institutions, government agencies, cultural enterprises, libraries, archives, etc.) full text, keyword search, audio and video search, etc. Satisfy the user's basic purpose. Establish an interactive display platform again. Digitization is not only a display, but also a human-computer interaction function. Users can interact with the database (such as Nushu cultural knowledge interactive games). At the same time, the platform should be designed to be open, based on user permissions cultural heritage data can be added and supplemented by oneself, and ordinary users and local people can be encouraged and promoted to participate in the construction of the database content. Finally, a content downloading platform is constructed. The function of the Nvshu Intangible Heritage Database is to spread and inherit the excellent culture, so that the cultural value can be maximized. The platform functions include online learning, paid downloads, content customization, etc. It mainly provides cultural content and information data for cultural public welfare services and industrial applications.

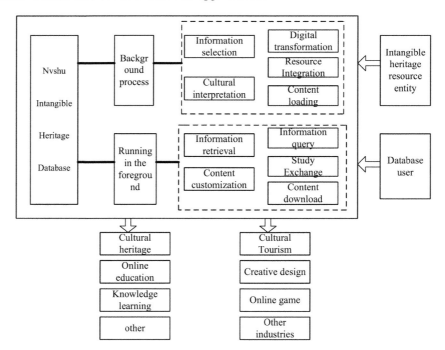

Fig. 1. Nvshu intangible cultural heritage database architecture

4.2   Functional Test of Nvshu Intangible Cultural Heritage Database

(1) Database function response speed test

Table 1. Functional test results

	Theoretical value	Test value
Retrieval speed	≤ 5 s	1.6 s
Content customization speed	≤ 5 s	3.4 s
Download speed	≤ 5 s	2.8 s

The functional response speed of the Nvshu intangible cultural heritage database was tested 10 times, and the average response speed was finally taken. The results are shown in Table 1. The theoretical value of retrieval speed, content customization speed, and download speed should not exceed 5 s, and the average response speed of the three functions in the actual test results were 1.6 s, 3.4 s, and 2.8 s, which did not exceed the theoretical values. Therefore, the function response speed of the Nvshu ICH database meets the needs of the system.

(2) User satisfaction survey on database functions

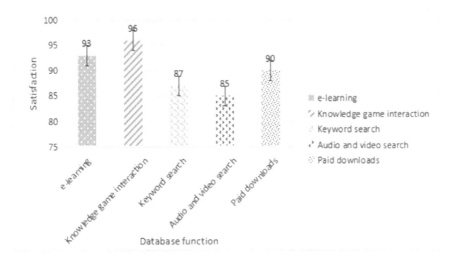

Fig. 2. Database function satisfaction

Figure 2 shows the results of a survey of users' satisfaction with the database functions after entering the Nvshu Intangible Cultural Heritage database. Among them, the online learning function means that users can use the Internet to directly learn the Nvshu culture online, and the user's satisfaction with this function is 93%; the knowledge game interactive function means that users can enter the game interface for human-computer interaction after learning the Nvshu knowledge, The satisfaction rating of this function is 96%; the satisfaction of keyword search and audio and video search functions are 87% and 85% respectively, which means that the user enters keywords related to the female book culture or audio and video search and it appears on the display screen The content you want; paid download refers to the Nvshu knowledge that needs to be unlocked to continue to understand or the audio and video that the user wants to save, etc., and the satisfaction of this feature is 90%. From this point of view, users are highly satisfied with the functions of the Nvshu intangible heritage database, and there is room for improvement. These functions help users learn Nvshu culture more deeply, which is conducive to the inheritance of ICH.

5   Conclusion

At present, DT plays an active role in the protection of my country's ICH. Through the digital processing of Nvshu culture, the establishment of a database of Nvshu ICH resources will protect and inherit Nvshu in a living form, can make Nvshu culture more colorful and spread the connotation and charm of Nvshu. Nvshu is a treasure of Chinese culture. In the process of digital protection of Nvshu culture, different original materials of cultural relics need to be classified and protected together. In addition, digital informatization of Nvshu resources is an innovative form of protection and inheritance of Nvshu culture. It is an important channel for further in-depth study of Nvshu cultural resources in the future, and at the same time provides a certain reference for the inheritance and development of other minority ICH.

References

1. Janisio-Pawlowska, D.: Analysis of the possibilities of using HBIM technology in the protection of cultural heritage, based on a review of the latest research carried out in Poland. Int. Soc. Photogramm. Remote Sens. **10**(10), 633 (2021)
2. Zhu, D.M.: General survey and protection of intangible cultural heritage in traditional medicine in Zhejiang Province. Zhonghua yi shi za zhi **47**(4), 226–229 (2017)
3. Faddis, A.: The digital transformation of healthcare technology management. Biomed. Instrum. Technol. **52**(s2), 34–38 (2018)
4. Oh, J.: Study of intangible cultural heritage community through social network analysis-focused on the item of Pansori. J. Cult. Policy **31**(1), 158–183 (2017)
5. Cuhlová, R., Nesiba, J.: GSR management in protection of cultural heritage in China: artistic processing of sugar. Listy Cukrovarnické a Řepařské **137**(2), 79–84 (2021)
6. Isa, W.M.W., et al.: Digital preservation of intangible cultural heritage. Indonesian J. Elect. Eng. Comput. Sci. **12**(3), 1373–1379 (2018)
7. Dou, J., et al.: Knowledge graph based on domain ontology and natural language processing technology for Chinese intangible cultural heritage. J. Visual Lang. Comput. **48**(Oct), 19–28 (2018)
8. Dharminder, D.: LWEDM: learning with error based secure mobile digital rights management system. Trans. Emerg. Telecommun. Technol. **32**(2), 232–235 (2021)

9. Sillanpaa, J., Lovelock, M., Mueller, B.: The effects of the orthopedic metal artifact reduction (O-MAR) algorithm on contouring and dosimetry of head and neck radiotherapy patients. Med. Dosim. **45**(1), 92–96 (2020)
10. Çelik, M., Keser, A., Yapici, Ü.K.: The relationship between cultural diversity and cultural intelligence: a cross-cultural research. Int. J. Adv. Intell. Paradig. **19**(3/4), 450–469 (2021)
11. Xu, B.C., et al.: Analysis of structural characteristics and spatial distribution of the national intangible cultural heritage in China and its policy implications. Sci. Cold Arid Reg. **11**(5), 59–76 (2019)
12. Love, P., Matthews, J.: The 'how' of benefits management for digital technology: from engineering to asset management. Autom. Constr. **107**(Nov.), 102930.1–102930.15 (2019)

Optimization of Prefabricated Concrete Frame Building Based on Genetic Algorithm

Zhenzhen Geng[✉]

Chongqing Chemical Industry Vocational College, Chongqing, China
gzzccivc@orgmail.cn

Abstract. With the development of construction industrialization technology, prefabricated construction technology overcomes the shortcomings of previous on-site pouring construction methods with its good economic and social benefits, and has become a research hotspot in the current construction industry. Using genetic algorithm to optimize the structure design of prefabricated concrete frame building (PCFB) can not only achieve better economic efficiency, but also ensure the safety of the structure. Based on this, this article will study the PCFB optimization based on genetic algorithm. This paper investigates the current application of prefabricated concrete frame structures and the application of component optimization based on Matlab genetic algorithm. This paper analyzes the advantages of PCFB and the obstacles to its development, and proposes a PCFB optimization scheme based on Matlab genetic algorithm. The survey data shows that with regard to the application of genetic algorithm and BIM technology in PCFB optimization, 94.94%, 86.71% and 83.23% are used for "component collision check", "prefabricated component drawing", and "project construction simulation" respectively. It can be seen that the genetic algorithm has considerable application effects in PCFB optimization.

Keywords: Building structure optimization · Genetic algorithm · Prefabricated building · Concrete construction

1 Introduction

With the acceleration of my country's economic development and urbanization into cities, more and more people are pouring into cities, and housing resources are becoming more and more tense [1, 2]. The production methods of the traditional construction industry continue to expose problems such as high energy consumption, high pollution, and long construction period [3, 4]. In addition, the increase in public awareness of environmental protection and the increase in housing quality requirements pose a major challenge to the production methods of traditional building on-site pouring [5]. Prefabricated buildings have become one of the technical means to solve the above problems, especially PCFB can save labor costs while meeting the requirements of living quality, and significantly improve construction efficiency [6]. Therefore, it is necessary to further study the optimized design of PCFB to realize the steady development of prefabricated buildings in my country's construction industry.

© The Author(s), under exclusive license to Springer Nature Switzerland AG 2022
V. Sugumaran et al. (Eds.): ICMMIA 2022, LNDECT 138, pp. 145–153, 2022.
https://doi.org/10.1007/978-3-031-05484-6_18

Regarding the research on genetic algorithm and prefabricated buildings, many scholars have conducted multi-angle explorations. For example, Sabharwal S uses genetic algorithms to construct t-way coverage arrays and uses them in construction engineering management [7]; Ngowtanasawan G applies BIM's causal model to Thailand's construction engineering and engineering design industry [8]; Potter studied the correlation between products, processes and customer preference matching in prefabricated house construction [9]. It can be seen that there are not many studies on PCFB optimization. This article combines genetic algorithm to explore the PCFB structure optimization program, which has theoretical and practical significance.

The main purpose of this paper is to study PCFB optimization based on genetic algorithm. This article first investigated the application of prefabricated concrete frame structures at the present stage and the application of component optimization based on Matlab genetic algorithm through the method of questionnaire survey. Then, this paper analyzes the advantages of PCFB and the obstacles to its development. Finally, this paper combines genetic algorithm with PCFB structure optimization, establishes the objective function of prefabricated columns and prefabricated composite beams and the constraint conditions of the components, and realizes the genetic algorithm based on Matlab.

2   Investigation on the Status Quo of PCFB Optimization

2.1   Purpose of the Investigation

The purpose of this investigation is to analyze the application of PCFB at the current stage and the application of component optimization based on Matlab genetic algorithm.

2.2   Questionnaire Design and Data Collection

(1) Questionnaire design

The questionnaire of this research is divided into two parts. On the one hand, it collects the interviewee's information, including the interviewee's work nature, working years and other basic information, as well as the survey interviewee's understanding of prefabricated buildings, so that some invalid data can be eliminated when the questionnaire is processed later. On the other hand, the survey respondents' opinions on the influencing factors restricting prefabricated concrete frame buildings, and the application of genetic algorithms in prefabricated building design at this stage are mainly surveyed.

(2) Data collection

The questionnaire adopts two methods: paper questionnaire and electronic questionnaire. In order to enable the results of the questionnaire to effectively reflect the actual situation, the interviewees' sources of influencing factors should be explained to a certain extent to ensure the credibility of this survey.

The issuing unit of this questionnaire mainly selects relevant units that have researched and practiced PCFB for distribution.

Since the main body responsible for the design and management of PCFB is the real estate developer, this article focuses on issuing questionnaires to real estate development units, including prefabricated building design and scientific research units.

A total of 324 questionnaires were distributed in this survey, and 319 were recovered. After excluding invalid questionnaires, there are a total of 316 valid questionnaires.

2.3 Questionnaire Reliability Test

This article uses Cronbach's coefficient to test. The reliability test result was 0.863, and the reliability test passed.

3 PCFB Optimization Status

3.1 Influencing Factors Restricting the Optimization of PCFB

From the perspective of prefabricated building designers, investigating the factors that restrict the optimization of PCFB are shown in Table 1: "component manufacturing costs" accounted for 79.43%; "on-site construction technology" accounted for 70.59%; select "component transportation and hoisting costs" accounted for 63.29%; "building project design and component deepening is difficult" accounted for 55.06%.

Table 1. Influencing factors restricting the optimization of prefabricated concrete frame buildings

Serial number	Constraints	Frequency	Proportion (%)
1	Component manufacturing cost	251	79.43
2	Site construction technology	223	70.57
3	Component transportation and hoisting costs	200	63.29
4	Project design and component deepening are difficult	174	55.06
5	Policy factors	121	38.29
6	Other	43	13.61

It can be seen from Fig. 1 that the current factors restricting PCFB mainly include component manufacturing costs, on-site construction technology, component transportation and hoisting costs, and the difficulty of design and component deepening. In terms of component manufacturing cost, due to the wide variety of PC components and the general use of steel for molds, the size of the mold is not scalable, and it is necessary to manufacture related molds for each PC (Precast Concrete) component. In

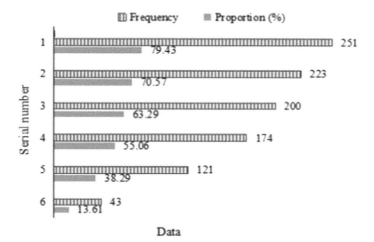

Fig. 1. Influencing factors restricting the optimization of prefabricated concrete frame buildings

the investigation, it was found that the general versatility of PC components is general. When each prefabricated component is due to design errors or manufacturing errors, there may be a phenomenon of lagging, waiting for new prefabricated components, which increases the economic cost of PCFB.

3.2   Application of Genetic Algorithm in PCFB Design

At present, the application of genetic algorithm and BIM technology in PCFB design emerges endlessly. The designer's application of genetic algorithm is shown in Table 2: It is used for "component collision check" accounting for 94.94%; for "prefabricated component drawing" accounted for 86.71%; used for "project construction simulation" accounted for 83.23%; used for "prefabricated component steel bar design" accounted for 82.28%; used for "statistics of engineering quantity and engineering cost" accounted for 63.61%.

Table 2. Application of genetic algorithm in the design of prefabricated concrete frame buildings

Serial number	Application	Frequency	Proportion (%)
1	Component collision check	300	94.94
2	Drawing of prefabricated components	274	86.71
3	Project construction simulation	263	83.23
4	Reinforcement design of prefabricated components	260	82.28
5	Complete library of prefabricated components	214	64.73
6	Statistics of engineering quantity and engineering cost	201	63.61
7	Other	81	25.63

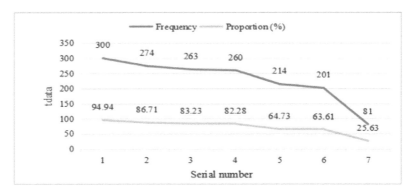

Fig. 2. Application of genetic algorithm in the design of prefabricated concrete frame buildings

It can be seen from Fig. 2 that the main application functions of genetic algorithm combined with BIM technology in deepening design are: component collision check, establishment of prefabricated component family library, prefabricated component steel design, project construction simulation. The genetic algorithm calculates the structure to make a BIM model, draws the relevant prefabricated component CAD deepening design drawings through the BIM model, and then calculates the engineering quantity according to the BIM model, which can partially reduce the weight, steel content and size requirements. The amount of steel bars used in the prefabricated structure can achieve the purpose of reducing the cost of construction raw materials for the prefabricated components.

4   PCFB Optimization Based on Genetic Algorithm

4.1   PCFB Advantage

(1) Save resources and energy. Compared with traditional cast-in-place concrete buildings, prefabricated concrete buildings have better resource optimization. Its components are easier to control resource and energy consumption in factory production, and its plaster-free feature can save energy and resources to a certain extent.

(2) Good environmental benefits. Compared with traditional cast-in-place concrete buildings, most or all of the PC components are produced in the factory. The production and protection equipment of the PC parts factory is relatively complete, and it has more standardized management, which can improve the efficiency of the use of materials, reduce resource waste and construction waste The generation of [10, 11]. There are fewer wet operations in component site construction, which can reduce water waste, reduce waste water discharge, improve the construction site environment, and produce good environmental benefits.

(3) Improve production efficiency. The production of prefabricated concrete buildings has a higher degree of standardization, a more complete industrial chain, and a

higher efficiency in the integration and utilization of social resources. The factory and mass production of PC components can help improve the proficiency of workers, thereby improving production efficiency, reducing the labor intensity of industrial workers, and releasing some labor resources.

(4) Control the construction quality. The characteristics of factory production of pre-fabricated components of prefabricated concrete buildings make the quality management system more perfect, and quality monitoring and quality acceptance easier. In addition, reducing the external wall insulation, plastering and other construction techniques, reducing the number of construction supply and demand can also reduce the risk of quality out of control, and reduce the workload of on-site quality management personnel for quality control.

4.2   PCFB Development Obstacles

Compared with traditional cast-in-place concrete construction methods, PCFB has better economic benefits, but there are also some problems.

(1) Building factories in different places, occupying a lot of land resources;
(2) PC components must be transported by large vehicles, causing great traffic pressure on the city;
(3) It is difficult to hoist large PCs on site.
(4) The joint structure is complicated, and the construction of on-site jointing and secondary pouring is difficult;
(5) The technical requirements for workers are high, but the current number of industrial workers is insufficient, and construction safety is difficult to guarantee;
(6) The construction cost increases with the increase in the assembly rate, and the enterprise lacks driving force.

4.3   PCFB Optimization Approach Based on Genetic Algorithm

(1) Establishment of objective function

The calculation formula for the cost of precast column concrete part C_1 is shown in (1):

$$C_1 = m_1 abl_b \times 10^{-9} \tag{1}$$

In the formula, a represents the height of column section (mm); m_1 represents the cost of unilateral concrete (yuan/m^3), b represents the width of the column section (mm), and lb represents the length of the column (mm).

The cost calculation of the concrete part of the precast laminated slab is shown in formula (2):

$$C_1 = m_1 abh \times 10^{-9} \tag{2}$$

In the formula, H represents the thickness of the prefabricated laminated board (mm).

(2) Establishment of constraint conditions for prefabricated components

The preliminary selection of the column section size of the prefabricated component must meet the influence of many factors such as the minimum section, the side displacement limit and the axial compression ratio at the same time. T usually adopts axial compression ratio control for preliminary estimation, and the aspect ratio of the column section should not be greater than 3, that is, a/b $\leq$ 3, and square columns should be used in the design. The calculation of axial compression ratio μ_c is shown in formula (3):

$$\mu_c = N/A_c f_c \qquad (3)$$

In the formula, A_c represents the column cross-sectional area; N represents the design value of the column axial force; f_c represents the design value of the concrete compressive strength. The calculation method of the column axial force design value N is shown in formula (4):

$$N = \beta F g m \qquad (4)$$

Among them, β represents the increase coefficient of the column axial force after considering the combination of earthquake action, F represents the load area of the column calculated according to the simply supported state, g represents the representative value of the gravity load per unit area after conversion, and n represents the number of floors above the check section.

In addition, it is necessary to constrain the PCFB optimized structure with the size of the steel bars, the weight of the prefabricated components, the steel content of the prefabricated components, the spacing of the stirrups, and the minimum reinforcement ratio of the components.

4.4 Operation Steps Based on Matlab Genetic Algorithm

Matlab language has abundant function resources, and the language is concise, providing users with an intuitive program development environment. The operation steps based on Matlab genetic algorithm are as follows:

(1) Selection. The research adopts the method of random traversal selection.
(2) Cross reorganization. Using binary coding, the number of initial populations. Due to the real number coding of the optimized parameters, the selected individual genes are randomly recombined. Determine the cross-recombination rate before optimization P_x.
(3) Variation. Within the scope of each optimization variable, the encoded variable is randomly selected, and then real-valued mutation is performed. Determine the mutation rate before optimization P_m. The mutation operation effectively ensures the diversity of individuals in the population. After the hybridization operation, the chromosomal genes in the hybridization space are randomly changed according to the mutation rate P_m, and there are about $P_m \times N \times L$ mutations in each generation.

(4) The algorithm termination condition. Determine the maximum evolution algebra, and terminate the calculation when the maximum evolution algebra is reached.

5   Conclusions

With the acceleration of my country's urbanization process, the country vigorously promotes the industrialization of construction in order to realize the transformation and upgrading of the construction industry. Assembled concrete structure has the advantages of high production efficiency, good product quality, small impact on the environment, and is conducive to sustainable development. In assembled concrete structure, assembled concrete frame structure has a wide range of applications. This paper is based on genetic algorithm to study the PCFB optimization, and completed the following work: Through the method of questionnaire survey, this paper found that the constraints of the optimization of the prefabricated concrete frame structure at the present stage are component manufacturing cost, on-site construction technology, component transportation and hoisting cost, It is difficult to deepen the design and component deepening, etc. The use of genetic algorithm to calculate the structure to make a BIM model can realize construction collision detection, project construction simulation, etc.; the advantages and development obstacles of PCFB are analyzed, and the genetic algorithm is combined with PCFB structure optimization. And finally established the objective function of prefabricated columns and prefabricated composite beams and the constraint conditions of the components, and realized the genetic algorithm based on Matlab.

References

1. Boscardin, J.T., Yepes, V., Kripka, M.: Optimization of reinforced concrete building frames with automated grouping of columns. Autom. Constr. **104**(Aug), 331–340 (2019)
2. Abey, S.T., Anand, K.B.: Embodied energy comparison of prefabricated and conventional building construction. J. Inst. Eng. (India) **100**(4), 777–790 (2019)
3. Tan, T., et al.: Barriers to Building Information Modeling (BIM) implementation in China's prefabricated construction: an interpretive structural modeling (ISM) approach. J. Clean. Prod. **219**(May 10), 949–959 (2019)
4. Baghdadi, A., Heristchian, M., Kloft, H.: Connections placement optimization approach toward new prefabricated building systems. Eng. Struct. **233**(2), 111648 (2021)
5. Razavialavi, S.R., Abourizk, S.: Genetic algorithm-simulation framework for decision making in construction site layout planning. J. Constr. Eng. Manag. **143**(1), 04016084.1–04016084.13 (2017)
6. Wang, L., Janssen, P., Ji, G.: SSIEA: a hybrid evolutionary algorithm for supporting conceptual architectural design. Artif. Intell. Eng. Des. Anal. Manuf. **34**(4), 458–476 (2020)
7. Sabharwal, S., Bansal, P., Mittal, N.: Construction of t-way covering arrays using genetic algorithm. Int. J. Syst. Assur. Eng. Manag. **8**(2), 264–274 (2017)
8. Ngowtanasawan, G.: A causal model of BIM adoption in the Thai architectural and engineering design industry. Proc. Eng. **180**, 793–803 (2017)

9. Potter, A., et al.: Product, process and customer preference alignment in prefabricated house building. Int. J. Prod. Econ. **183**(Jan. Pt.A), 79–90 (2017)
10. Sardone, L., et al.: A preliminary study on a variable section beam through Algorithm-Aided Design: a way to connect architectural shape and structural optimization. Proc. Manuf. **44**, 497–504 (2020)
11. Maljaars, E., Felici, F.: Actuator allocation for integrated control in tokamaks: architectural design and a mixed-integer programming algorithm. Fusion Eng. Des. **122**(Nov), 94–112 (2017)

The Site Layout and Optimization Research on BIM-Based Prefabricated Building Construction

Limei Wang, Yan Zhang, and Xiaoya Huang[⊠]

Chongqing College of Architecture and Technology, Chongqing, China
keyanhxy@163.com

Abstract. The development of various modern high-tech technologies has brought new opportunities for the development of various industries and has a good role in improving production quality and production efficiency. Building Information Modeling (BIM) is a commonly used technical method in modern architectural engineering design and management. This article mainly analyzes the application measures of BIM technology in the layout of prefabricated building construction sites.

Keywords: BIM · Prefabricated · Building construction · Site layout · Site optimization

1 Introduction

The requirements for site layout are getting higher and higher. The traditional methods of arranging temporary construction and construction machinery on site based on experience cannot meet the needs of modern construction. Traditional site layout methods are not only inefficient but are also affected by the size of the site and the structure of the project. It is often that the site layout cannot meet the construction needs and cannot provide convenience for subsequent adjustments [1]. The Revit modeling software is used to set up the site model, and then the roaming software is used to perform roaming simulation on the site. Simulate the site layout in advance and make optimization adjustments, to achieve the rationality and scientific of the site layout and provide strong data support for subsequent site change adjustments [1].

BIM is a brand-new tool applied in the fields of architecture, engineering, and civil engineering. It can realize the effective integration of building information and form a good guiding role for the building design and construction process. Through the three-dimensional simulation method, it is a good help for the improvement of work efficiency, resource saving, and cost reduction in the construction process of construction projects [2]. However, in the construction site layout of prefabricated buildings, how to make reasonable use of BIM technology is worthy of our deep consideration.

V. Sugumaran et al. (Eds.): ICMMIA 2022, LNDECT 138, pp. 154–161, 2022.
https://doi.org/10.1007/978-3-031-05484-6_19

2   The Application of BIM Technology in Construction Site Layout

2.1   Build a BIM Component Library for Construction Facilities

Combining BIM technology to perform modeling processing for various construction facilities in the construction site, marking their corresponding dimensions, materials and other information to form a unified construction facility management database [3]. For the commonly used electrical equipment, processing sheds, safety protection facilities, and construction passages in the construction site, the family functions configured by the BIM software can be used to construct the BIM family library of various construction facilities to provide for the production and management of construction facilities Data support, as shown in Fig. 1. For example, for the production of steel bar processing sheds, under the premise of ensuring structural stability, standard dimensions need to be planned uniformly so as to fully meet the space requirements of the project construction site and form a unified standard within the enterprise [3].

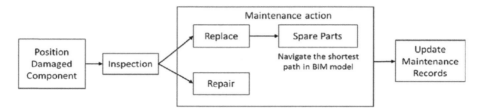

Fig. 1. Establishment of BIM component library

With the continuous enrichment and development of the enterprise BIM family library, the layout of various facilities on the construction site will be simpler. Classify and sort all family files to build a BIM component library [2]. When building a 3D model of a construction site, various components can be dragged into the 3D model to create a BIM model on site to provide a visual reference for the layout of the construction site.

2.2   Mechanical Equipment Management

2.2.1   Equipment Approach Simulation

The construction machinery has a huge volume and may be hindered by existing equipment and materials during the approach. Relying on BIM technology, it is possible to rationally design the mechanical approach path, find the collision point of the machinery during the approach process, and then re-plan the approach path (or make reasonable adjustments to the collision position) to ensure the machinery there will be no security issues during the approach [4].

2.2.2   Equipment Fixed Check

Under normal circumstances, construction units have high requirements for the fixed positions of various mechanical equipment on the construction site, and large equipment such as tower cranes must be subjected to force checking before being fixed, so as to ensure the stability of equipment operation during the construction process. In recent years, there have been frequent tower crane collapse accidents worldwide, posing a serious threat to people's lives and property [4]. The BIM technology is used to verify and inspect the fixing process of the tower crane at the construction site to ensure the installation quality of the foundation and fixing parts, and to provide a guarantee for the smooth progress of the construction process.

2.2.3   Cost Control

The main advantage of BIM technology is reflected in the flow of information. The BIM model not only contains the three-dimensional styles of various components, but also contains relevant information such as materials, dimensions, weights, and manufacturers. After the site modeling is carried out through BIM software, comprehensive statistics can be made on the number of mechanical equipment used in the layout process, the length of pipelines, and the amount of concrete engineering, forming a set of statistical data sets with good reliability, which is the cost of the project. Provide a basis for reasonable calculations [5].

2.2.4   Collision Detection

The contents of the construction site model that need to carry out collision detection mainly include [5]: ① The stacking position of materials and equipment. Through collision detection, analyze whether there are operational conflicts between various construction equipment, and at the same time make it clear whether the distance between the construction equipment and the raw material storage site is reasonable; ② Road planning and layout. Detect the road layout in the construction area and avoid crossing (or less crossing) between various construction roads as much as possible, so as to ensure the safety of the construction process; ③ Water and electricity layout. To prevent conflicts between the temporary hydropower layout and the on-site equipment layout, model demonstrations are carried out through BIM software to accurately identify the source of hazards, and corresponding measures are taken to optimize treatment in time.

3   Analysis of Construction Site Layout Based on BIM Information Model

3.1   Selection of Construction Site Layout Schemes Based on Analytic Hierarchy Process

(1) Establish a hierarchical structure model. In this article, the target is a reasonable site layout plan, which belongs to the target layer; the sub-target layer refers to various factors that affect the site layout, and the three solutions formed thereby belong to the plan layer.

(2) Construct a judgment matrix. At the same level, judge the relative importance of each factor according to the expert scoring method [7].

(3) For a certain standard, calculate the geometric average of the relative weights of the elements in each row vector of A, and then normalize, and the row vector obtained is the weight vector.

(4) Consistency inspection. According to the random consistency means CR in the formula, check whether the judgment matrix meets the requirements, that is, when CR is less than or equal to 0, the consistency requirements are met; otherwise, the judgment matrix needs to be adjusted until the requirements are met.

3.2   Fuzzy Mathematics Evaluation Construction Site Layout Plan Selection

The fuzzy comprehensive evaluation method is used to quantify the fuzzy factors in the selection of the plan in the decision-making of the layout plan of the prefabricated building construction site [7]:

Known evaluation index set W,

$$W = (W_1, W_2, W_3, \ldots, W_n) \tag{1}$$

W in the formula will give different weights to the judgement factors according to each W_i, and finally get the set of index weights expressed as W, which is:

$$U = (U_1, U_2, U_3, \ldots, U_i, \ldots U_n), \ U_i \geq 0, \sum_1^n U_i = 1 \tag{2}$$

U is the fuzzy subset on W, and the comment set is denoted as:

$$B = (B_1, B_2, B_3, \ldots, B_m) \tag{3}$$

Then, the membership degree of the i-th index for each comment level will be a fuzzy subset above the comment set B.

The comprehensive evaluation matrix R is:

$$R = \begin{bmatrix} r_{11} & r_{12} & \cdots & r_{1m} \\ r_{21} & r_{22} & \cdots & r_{2m} \\ \vdots & \vdots & \vdots & \vdots \\ r_{n1} & r_{n2} & \cdots & r_{nm} \end{bmatrix} \tag{4}$$

$$M = U \times R = (M_1, M_2, M_3, \ldots M_j \ldots M_n) \tag{5}$$

Take the evaluation results of the above judges as a weighted average, and use the arithmetic average as the final degree of membership of each evaluation level of the site layout plan, denoted by M,

$$M_t = (M_{t1}, M_{t2}, M_{t3}, M_{t4}, M_{t5}) \tag{6}$$

Calculate the different scores corresponding to the grading level, such as: X = (good, better, average, poor, bad) = (90, 70, 50, 30, 10); then get the final score, N = M_tX and finally sort out the scoring sequence of each construction site layout plan to determine the optimal site layout plan.

4   Measures to Optimize the Layout of BIM in the Construction Site of Prefabricated Buildings

The scientific division of various prefabricated components and specific construction areas is very important. The layout of the prefabricated building should be optimized from the following three aspects:

4.1   Tower Crane Layout Design

In the process of construction of prefabricated buildings, tower cranes are the most important construction machinery and equipment, and their application efficiency will have a direct impact on the overall construction process. It can be found that the unreasonable layout of the tower crane often causes the secondary transportation of the components, which will not only seriously affect the construction progress, but also pose a threat to the construction safety [8]. Therefore, it is necessary to rationally design the tower crane model and installation location. The specific content includes: ① Clarify whether the boom can meet the requirements of loading and unloading trucks of the components, so as to select the equipment model; ② Combine the requirements of equipment operation and coverage, and the safety distance requirements between transmission lines, etc., to refine the design of the tower crane installation location. After completing the above operations, carry out BIM simulation for each plan of the tower crane layout, and select the optimal plan through comparison and analysis.

4.2   Prefab Storage

The reserves of various components in the preset storage site need to fully meet the needs of building construction, and the storage site should also be optimized with reference to the actual situation [8]. In addition, whether equipment and material storage sites will cause traffic congestion during the construction process is also an issue that must be considered in construction design and management, and BIM software can be used for simulation and optimization [6].

4.3   Component Transportation Path Planning

After transporting the prefabricated components from the factory to the construction site, it is necessary to fully consider the transportation path, analyze whether it meets the unloading and hoisting requirements, and whether it will affect other operations. The use of BIM software can achieve a comprehensive simulation of the construction site to optimize the layout, so as to select the optimal warehouse location and transportation equipment driving route, which has a good effect on avoiding various construction problems [9] (see Fig. 2).

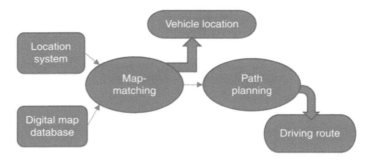

Fig. 2. BIM is used for transportation route planning

4.4 Optimization of Site Temporary Construction Layout

According to the size of the project, combined with the content and requirements of the construction manual and construction specifications, first determine the standards that need to be met for the site layout; then estimate the number of on-site staff based on the amount of work, construction period, etc., and then calculate the entire on-site office area, the scale and use area of the temporary construction of the living area and the production area; finally, according to the size of the site, use Revit software to simulate the layout of the entire construction area and office area [9]. The greening of the site shall be arranged. Finally, a complete three-dimensional field layout model is established [9]. Through the analysis of the roaming of the three-dimensional model, the facilities that do not meet the requirements or have a large optimization space are adjusted, and the optimal effect is finally achieved, as shown in Fig. 3.

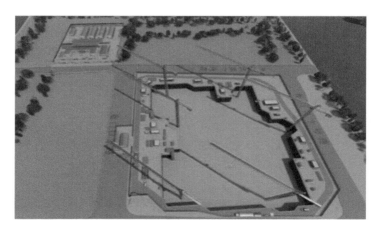

Fig. 3. Layout of temporary construction on site

4.5   Optimization of On-site Roadside

The construction site generally uses temporary circular roads surrounding the foundation pit. Due to the size of the site and the scale of construction, it is difficult to control the turning of the temporary road. If the angle is too large, it is not conducive to the construction. If the angle is too small, it will increase the use area of the site and increase the construction cost. Therefore, it is necessary to combine the vehicle technology and specification parameters used on site, and then guide the entire vehicle and road model to the roaming software for simulation analysis and use the roaming software to simulate the driving route of the construction vehicle in the software to obtain the normal operation of the vehicle required on the site the required minimum road width and optimal turning radius [10], See Fig. 4.

Fig. 4. Road layout on site

5   Conclusion

In summary, the full use of BIM technology in the layout of prefabricated construction projects can effectively improve the configuration efficiency of various resources and equipment within the construction area. Compared with the traditional method of site layout based on experience, the use of BIM technology It can fully integrate all kinds of on-site data for scientific analysis and three-dimensional effect display, and you can see the effect of temporary construction layout in advance; accurately obtain the optimal location and minimum use area of roads, work areas, and stacking areas in the site; more accurate Obtained to meet the relevant technical requirements of on-site construction machinery, and it is also convenient to realize the accurate layout of the tower crane and the material storage yard.

References

1. Murray, N., Fernando, T., Aouad, G.: A virtual environment for the design and simulated construction of prefabricated buildings. Virtual Real. **6**(4), 244–256 (2003)
2. Follini, C., et al.: BIM-integrated collaborative robotics for application in building construction and maintenance. Robotics **10**(1), 2 (2021)
3. Bassier, M., et al.: Drift invariant metric quality control of construction sites using BIM and point cloud data. Int. Soc. Photogramm. Remote Sens. **9**(9), 545 (2020)
4. García-Pereira, I., Portalés, C., Gimeno, J., Casas, S.: A collaborative augmented reality annotation tool for the inspection of prefabricated buildings. Multimed. Tools Appl. **79**(9–10), 6483–6501 (2020)
5. Mesáros, P., Mandicák, T., Behúnová, A.: Use of BIM technology and impact on productivity in construction project management. Wireless Netw. **28**(2), 855–862 (2022)
6. Chen, Y.: Analysis of the application of BIM technology in the construction of prefabricated buildings. Housing Industry **6**, 62–64 (2019)
7. Abdirad, H., Mathur, P.: Artificial intelligence for BIM content management and delivery: case study of association rule mining for construction detailing. Adv. Eng. Inform. **50**, 101414 (2021)
8. Hosseini, O., Maghrebi, M.: Risk of fire emergency evacuation in complex construction sites: Integration of 4D-BIM, social force modeling, and fire quantitative risk assessment. Adv. Eng. Inform. **50**, 101378 (2021)
9. Liu, W., Liang, M.: Research on BIM-based prefabricated construction site layout and optimization technology. New Silk Road: Mid **9**(10), 67–70 (2019)
10. Wan, A., Duan, A., Cai, B.: BIM-based site optimization method for prefabricated shear wall structure hoisting stage. Build. Struct. **10**(5), 110–113 (2019)

Computer Software Vulnerability Detection and Risk Assessment System Based on Feature Matching

Yan Chen[✉] and Ying He

Sichuan Vocational and Technical College, Suining 629000, Sichuan, China
chenyan9l2815@l63.com

Abstract. Although the use of information technology has brought countless conveniences and benefits to all aspects of people's lives and work, computer information technology is also a double-edged sword like other science and technology. The threats faced by information do not only come from computers and networks. Natural disasters, the environment, personnel misoperations, system vulnerabilities, etc. will all bring harm to the information system. The purpose of this paper is to study the design of computer software vulnerability detection and risk assessment system based on feature matching. A multi-stage method is proposed to detect vulnerabilities in cross-architecture binary software. This method combines deep learning technology and graph matching technology, and divides the vulnerability search task into phased processes. Specifically, first construct a 3-layer feature control flow graph for the functions in the binary software, capture the semantic information of the instruction through the word embedding and long-short-term memory network, and then generate the vector representation of the 3-layer feature control flow graph through the deep neural network and proceed. Function pre-filtering, and finally using an improved graph matching method to identify suspicious functions that are truly similar to vulnerable functions. The improved feature matching method in this paper has a number of correct matches of 355 in the Is binary file, which has achieved the best performance.

Keywords: Feature matching · Computer software · Vulnerability detection · Risk assessment

1 Introduction

Since human society entered the information age, information construction has achieved rapid development, improving people's production and life all the time [1, 2]. However, while enjoying the convenience and speed brought by informatization, people have gradually realized the importance of information security to everyone in today's social life [3]. In particular, various network security incidents such as theft of commercial documents, exposure of private documents, theft of personal accounts, and identity theft fraud have occurred frequently recently, making information security a major issue of common concern for the whole society [4]. Almost every user who uses a networked computer will worry about whether he will become the next target of

© The Author(s), under exclusive license to Springer Nature Switzerland AG 2022
V. Sugumaran et al. (Eds.): ICMMIA 2022, LNDECT 138, pp. 162–169, 2022.
https://doi.org/10.1007/978-3-031-05484-6_20

illegal hacker attacks [5]. Therefore, how to quickly detect software security vulnerabilities in networked computers, especially operating system software vulnerabilities, and successfully repair them before hackers use them to attack, has become an important topic for current information security workers [6].

At present, network information security has received widespread attention from all countries, and all countries have focused on it from a strategic height [7]. Network security detection and risk assessment technology occupies an important position in the field of network security technology, and research and development have become one of the hot spots. A scholar focuses on detecting three common types of vulnerabilities: Unused_Variable, Use_of Uninitialized Variable and Use_After_Free. We propose a software vulnerability detection method based on an improved control flow graph (ICFG) and several predicates of the vulnerability attributes of each type of vulnerability. We also defined a set of grammatical rules for analyzing and deriving the above three types of vulnerabilities, and designed three vulnerability detection algorithms to guide the vulnerability detection process [8]. Static analysis is an effective software assurance method. Shows that its most effective usage is to analyze interactively through software development processes that require high performance. A scholar focuses on the research of rule-based static analysis tools and proposes an optimized rule checking algorithm. Their technology improves the performance of static analysis tools by filtering vulnerability rules based on characteristic objects before inspecting source files. Since the source file always contains a small part of the rules of vulnerabilities, but not all of them, our method may achieve better performance [9]. Software vulnerabilities have become a major threat to computer security. Finding and repairing potential vulnerabilities in the software in time is of great significance for reducing system crashes and maintaining system security and integrity.

In order to solve the reference value of mining and using historical vulnerability data in the work of dealing with newly disclosed vulnerabilities, this paper proposes a security vulnerability similarity measurement method based on vulnerability characteristics. This method is based on the semantic richness of the vulnerability description text and the effectiveness of vulnerability risk assessment indicators for showing the characteristics of vulnerabilities, establishes a set of metrics to measure the similarity and proximity of security vulnerabilities, and realizes the analysis and retrieval of similar vulnerabilities. In order to comprehensively analyze the degree of similarity between vulnerabilities, this method combines the semantic characteristics of security vulnerabilities and the multi-dimensional reference coordinate system of risk characteristics, and proposes a two-stage similarity measurement method for security vulnerabilities.

2 Research on Design of Computer Software Vulnerability Detection and Risk Assessment System Based on Feature Matching

2.1 Design Scheme of Vulnerability Detection and Risk Assessment System

(1) Demand analysis

The system is used for weak detection and risk analysis of network information systems. Must be able to detect local and remote servers and perform a complete analysis and risk assessment of search results [10].

(2) Design ideas

The active test method is adopted, that is, the test system plays a role in recognition. First use multiple scanning technologies (to scan for open ports. Then, according to the services provided by the corresponding open ports, the patient search code corresponding to the function is stored in the vulnerability signature database, and the database is sent to the default port group to indicate the grouping result.) And make a judgment based on the response of the other party. Store the identified vulnerabilities in the result database, conduct risk analysis and generate an assessment report.

(3) Vulnerability signature database

Vulnerability characteristic database is an important part of the development of network security risk assessment system. Refers to the database that contains all the vulnerabilities in the vulnerability detection and risk assessment system. The final function is the function of the network database, so the binding code should be able to ensure the validity of the database search and the accuracy of the judgment after receiving the result database. Due to the wide variety of these types of vulnerabilities, eliminating the vulnerability code is an important task. Therefore, the implementation of vulnerability scanning is a risk analysis process, a process of establishing and maintaining knowledge and information capabilities over a period of time.

(4) Risk assessment

For the actual identified vulnerabilities, conduct risk analysis based on vulnerability assessment standards, risk supervision rules and risk tables, determine the vulnerability level and risk assessment level of the entire system, and further form and submit security recommendations.

2.2 Model Structure

In the Embedding Layer, the description text needs to be expressed in a way that the model can understand. Therefore, the embedding layer will use word embedding to obtain the GloVe word vector of each word in the vulnerability description text, thereby completing the vectorization of the input word sequence, and then passing it to the next layer of the model for text feature extraction and expression.

In the Bi-GRU layer (Bi-GRU Layer), the vulnerability description text will be processed by the embedding layer to obtain a text sequence represented by a word vector, and the Bi-GRU layer will complete the extraction of the text sequence feature and the characterization of the text semantics. The two-way GRU network can comprehensively consider the context information at each moment, and better extract text information. The number of Bi-GRU layers can generally be set to 1, 2, and 3, and the BiGRU output of each layer will be used as the input of the next layer. Too few layers of this layer may lead to insufficient feature information extraction, while too many layers may increase the amount of model calculation, but the actual effect has not been effectively improved. Therefore, the number of layers of BiGRU of the model proposed in this paper will be determined through actual experimental analysis.

In the attention layer (Atention Layer), the output of the Bi-GRU layer is used to calculate the weight vectors of different words in the text description. After obtaining the original feature vector of the text output by the Bi-GRU layer, the attention layer uses the Self-Atention mechanism to assign different attention weights to different vulnerability description words to highlight their importance, thereby improving the effect of model classification.

3   Investigation and Research on the Design of Computer Software Vulnerability Detection and Risk Assessment System Based on Feature Matching

3.1   Development Platform and Tools

System development uses VC++ programming language and Microsoft Visual Studio. NET2008VC.NET programming platform.

(1) The configuration of the computer used for programming is:
The operating system is Microsoft WindowsXPSP2;
CPU: Intel Pentium Dual CPU E2200@2.20GHz;
Memory: 1037420KB;
Hard disk: 640G;
(2) The running environment of the example program is configured as:
Client program: operating system is Microsoft WindowsXPSP2;
CPU: Intel Pentium Dual CPU E2200@2.20GHz;
Memory: 1037420KB;
Hard disk: 640GB;

3.2   Data Set

This data set contains the popular Linux software package CoreUtils of version v8.31 (the latest version) compiled by the compiler GCCv7.4. Two functions with the same name are regarded as similar, otherwise they are regarded as different. The reason for choosing Coreutils is that CoreUtils consists of multiple Unix shell utilities, there are many versions available, and the file size is suitable for evaluating feature matching methods.

3.3   Feature Matching Method

The feature matching method in this paper can be divided into two steps: initial matching and fuzzy matching. The former is based on the basic block layer features to find completely matched basic block pairs, and the latter is based on the former, based on the instruction layer features to find the optimal matching for unmatched basic blocks. The difference score between two basic blocks is a real value on the interval [0, 3], where 0 means they are the same, and 3 means they are completely dissimilar.

The graph embedding network in this paper is based on Structure2vec. Specifically, it will be expressed as the set of neighbor vertices of vertex v in graph g. Then the update formula of the vector of each vertex in each iteration is as follows:

$$\mu_v^{(t+1)} = F\left(x_v \sum_{v \in N(v)} \mu_u^{(t)}\right), \forall v \in V \qquad (1)$$

Next define this article; F function, the formula is as follows, where xv is the d1-dimensional vector of the basic block layer feature, p is the embedding size, pi(i = 1, …,n) is a p*p matrix, n Is the embedding depth:

$$\sigma(1) = P_1 * \mathrm{ReLU}(P_2 * \ldots \mathrm{ReLU}(P_n l)) \qquad (2)$$

Finally, accumulate and calculate the embedding vector of graph g.

4   Investigation and Analysis of Computer Software Vulnerability Detection and Risk Assessment System Design Based on Feature Matching

4.1   System Function Module

The vulnerability analysis system extracts text features from the vulnerability description text entered by the user, and then uses the text features to perform vulnerability risk assessment and similarity vulnerability measurement analysis and retrieval, and finally obtains 7 vulnerability risk assessment features and 5 most similar historical vulnerability data. According to system functional requirements, the system is divided into 5 modules: vulnerability risk assessment module, user login and management module, vulnerability data collection module, similar vulnerability measurement analysis module, front-end processing module.

Vulnerability data collection module: This module contains three parts: the collection of CVE vulnerability data, the processing of intermediate data, and the structured storage of vulnerability data. The module uses Python scripts to download and collect full data on the CVE vulnerability data download interface provided by the NVD website, and filter out the vulnerability attributes or indicators that are not used by the system in the vulnerability data. The structure and sequence are stored, and the old data in the system is updated and overwritten.

Vulnerability risk assessment module: This module is implemented based on the vulnerability risk assessment model of deep learning. The module uses the trained model as the core, and perfects the conversion processing function of the vulnerability description text input and model output results, and realizes the complete automation of the vulnerability risk assessment. At the same time, the module also provides the retraining and deployment functions of the vulnerability risk assessment model, which is convenient for administrators to update the learning parameters of the vulnerability risk assessment model in time, and effectively alleviate the decline of model classification effect that may be caused by new vulnerabilities. The program flow chart is shown in Fig. 1:

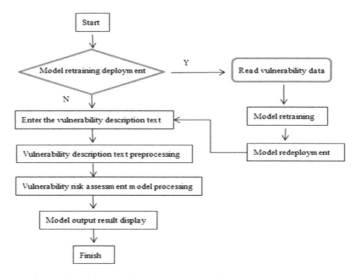

Fig. 1. Vulnerability risk assessment flowchart based on text features

Similarity vulnerability measurement and analysis module: This module mainly implements the similarity measurement method of security vulnerabilities based on vulnerability characteristics. In the first stage of the method, the module takes the vulnerability description text vectorized encoder model as the core, and uses the code to realize the calculation steps and formulas of the semantic feature similarity index, so that the system can quickly measure the semantically similar security vulnerabilities. In the second stage of the method, the module will call the vulnerability risk assessment module to analyze various risk assessment indicators for new vulnerabilities, and also use code to implement calculation steps and formulas for risk characteristics similar indicators, and combine the output results of the first stage of the method to calculate The historical vulnerability data that is most similar to the vulnerability to be analyzed is output.

4.2   Performance Evaluation of Function Similarity Measurement Module Based on Feature Matching

This paper uses the data set as the test set to explore the sensitivity of the feature matching algorithm in this paper to feature similarity. The experimental results are shown in Table 1. Taking cat, rm, Is and other 6 binary files as examples, the number of correct matches under different methods is explained in detail.

Method 1: The basic block distance calculation method based on the basic block layer features combined with the Hungarian algorithm;

Method 2: Basic block distance calculation method based on basic block layer features, minimum edit distance algorithm based on n-gram instruction sequence combined with Hungary algorithm (the graph matching algorithm in this article);

Method 3: The minimum edit distance algorithm based on the n-gram instruction sequence combined with the Hungarian algorithm;

Table 1. Number of correct matching functions

	Method 1	Method 2	Method 3
Cat	59	112	86
Rm	156	214	177
Is	222	355	301
CP	210	335	300
Dd	162	264	189
Chown	142	188	153

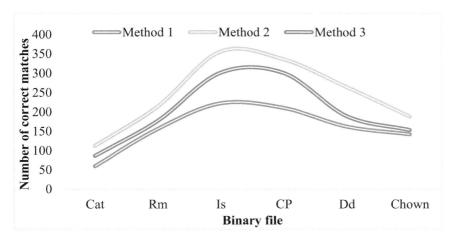

Fig. 2. The number of correct matching functions

The results show that method 3 has higher accuracy than method 1 and method 3. As shown in Fig. 2, it proves that the n-gram instruction sequence can describe the basic block better than the basic block layer features. In addition, the method that comprehensively considers the basic block layer characteristics and the n-gmm instruction sequence can find more correct matches than any other method. Therefore, it can be concluded that the improved feature matching method in this paper combines the advantages of the two basic block distance calculation methods and achieves the best performance in the evaluation.

5   Conclusions

With the development of computer technology, it is not only used in information technology and commercial development, but also in the aerospace and military fields. While it brings flexibility to people's lives, it also promotes social development. Due to the widespread use of open source software, potential vulnerabilities will also have a significant impact. This article investigates vulnerability detection and risk assessment technologies, analyzes and utilizes current well-known network security products at home and abroad, and manages the latest vulnerabilities reported through authoritative websites, and analyzes the principles of various vulnerabilities in detail. Discovery often uses the most advanced vulnerability detection and risk assessment technology to develop a vulnerability security system with independent property rights, which can provide technical support and guarantee for system testing.

References

1. Kazuki, K., et al.: Quantitative risk assessment of the first-hand / second-hand exposure occurred by vaping using computer simulated person integrated with virtual airway. J. Envir. Eng. (Transactions of AIJ) **83**(754), 1005–1013 (2018)
2. Masi, M.: ContractFuzzer: fuzzing smart contracts for vulnerability detection. Comput. Rev. **60**(12), 467–468 (2019)
3. Tndel, I.A., et al.: Collaborative security risk estimation in agile software development. Inf. Manag. Comput. Secur. **27**(4), 508–535 (2019)
4. Kartal, E., Balaban, M.E.: Machine learning techniques in cardiac risk assessment. Turkish J. Thora. Cardio. Surg. **26**(3), 394–401 (2018)
5. Given-Wilson, T., Jafri, N., Legay, A.: Combined software and hardware fault injection vulnerability detection. Innov. Syst. Softw. Eng. **16**(2), 101–120 (2020). https://doi.org/10.1007/s11334-020-00364-5
6. Shin, D., et al.: Human-centered risk assessment of an automated vehicle using vehicular wireless communication. IEEE Trans. Intell. Transp. Syst. **20**(2), 667–681 (2019)
7. Cardoso, P., Respício, A., Domingos, D.: riskaBPMN - a BPMN extension for risk assessment. Procedia Comp. Sci. **181**(4), 1247–1254 (2021)
8. Vila, P., et al.: Framework for a risk assessment model to apply in virtual / collaborative enterprises. Procedia Comp. Sci. **181**(34), 612–618 (2021)
9. Sicari, S., et al.: A risk assessment methodology for the internet of things. Comp. Commu. **129**(SEP.), 67–79 (2018)
10. Alawad, H., Kaewunruen, S., An, M.: A deep learning approach towards railway safety risk assessment. IEEE Access **PP**(99), 1–1 (2020)

Adaptive System Analysis of Coal Mine Spray Dust Reduction Based on Intelligent Analysis Algorithm

Baoyuan Chen[1,2(✉)], Xueming Wang[1,2], and Youliang Yuan[1,2]

[1] CCTEG Changzhou Research Institute, Changzhou 213015, Jiangsu, China
chenbaoyuan0825@126.com

[2] Tiandi (Changzhou) Automation Co., Ltd., Changzhou 213015, Jiangsu, China

Abstract. As a fast-developing and efficient analysis method, intelligent analysis algorithm plays an important role in the adaptive work of coal dust emission reduction, and is the key technology of coal dust emission reduction. The purpose of this work is to analyze an adaptive coal injection dust suppression system based on an intelligent analysis algorithm. On the basis of reading a large number of literatures, the research status of coal dust hazards and coal dust prevention technologies at home and abroad is investigated in detail. The mechanism of reducing spray dust and the factors affecting spray effect were studied and analyzed. On the basis of analyzing the spraying principle, the intelligent analysis algorithm is used for research, and the research conclusion is drawn. When the spray pressure was 4 MPa, the particle size was almost reduced by half, and increasing the pressure resulted in an increase in the pipeline pressure, the number of damaged parts and a decrease in the cost efficiency of the system.

Keywords: Intelligent analysis · Coal mine spray · Self-adaptive dust suppression · System analysis

1 Introduction

At present, my country's coal industry as a whole is not capable of sustainable development, and my country's coal production may continue to grow, but this will not help the destruction and loss of coal raw materials, pollution and environmental pollution, at the expense of health [1]. The sharp increase in the amount of dust generated in the industrial production process of coal mines will not only endanger the health of workers, but also endanger production safety [2]. The use of cutting machines to effectively control the dust generated by cutting plays a vital role in the prevention and control of dust in the entire region [3]. Measures are currently being taken to reduce dust spraying on charcoal cutting [4]. However, due to the low reliability of the sprayer itself, it did not play a significant role in reducing dust in actual production. In recent years, scientists in many countries have conducted extensive research on the design, configuration, printing and flow of water jet and vacuum equipment. However, due to the complexity of the system, high cost and low reliability, it is not suitable for popularization and implementation [5].

V. Sugumaran et al. (Eds.): ICMMIA 2022, LNDECT 138, pp. 170–177, 2022.
https://doi.org/10.1007/978-3-031-05484-6_21

The intelligent analysis algorithm lays the foundation for various management decisions, large and small, and realizes scientific management [6]. Aiming at the problem that the existing lump coal rate research methods cannot accurately express the behavior of coal and rock particles, Chen Li proposed a method for analyzing lump coal rate in fully mechanized mining face based on discrete element method. Taking MG450/1080-WD shearer as the research carrier, The coal mining process is studied and analyzed and the dynamic model of coal and rock particles is established; the analysis results show that when the drum speed of the MG450/1080-WD shearer is 48.74 r/min, the diameter of the drum hub is 233.79 mm, and the blade spiral When the lift angle is 14.09°, its performance is the best [7]. Parsakhoo A uses environmentally friendly dust suppressants such as cane molasses, polyacrylamide and bentonite to control fugitive dust emissions from forest pavements for 3, 9, 27 and 81 days. A rear spray system and a dust collector are used to implement treatment and dust emission records, respectively. The results showed that with the increase of the concentration of dust suppressant, the amount of fugitive dust showed a downward trend. In addition, the dust emissions started to decrease over time, with the lowest reduction efficiency at the end of the 81st day [8]. It is of practical significance to study the self-adaptive system of coal mine spray dust reduction based on intelligent analysis algorithm [9].

Aiming at the hazard of coal mine dust particles, this paper proposes a spray dust reduction technology, compares various spray dust reduction technologies, and formulates a specific and feasible scheme. By analyzing the harm of coal mine dust particles to human body, the main goal of this work: adaptive coal dust suppression system is studied. Based on this, the basic components are designed and calculated to obtain the basic size of the rotary pressure nozzle, the design of the air duct in the air supply system and the selection of the axial flow fan. On this basis, a simplified model of the air supply system is established by modeling software, and parametric fluid simulation is carried out to find the best airway structure.

2 Research on Self-adaptive System of Coal Mine Spray Dust Reduction Based on Intelligent Analysis Algorithm

2.1 Hazards of Dust

With the development of production technology, the mining industry tends to be large-scale and mechanized [10]. At present, most coal mines have adopted the extraction methods of fully mechanized mining and prospecting. This mining technology produces a large amount of dust and is difficult to control. Despite efforts to improve and perfect various dust removal equipment, a large amount of dust is still dumped in the air [11]. Coal dust in mines are all flammable substances. When it reaches a certain concentration in the air, it will burn and explode when it encounters external flames, electric sparks and high temperatures. While it is not uncommon for toner concentrations to reach the lower explosive limit, suspended toner often reaches the lower explosive limit when blasting coal seams. In addition, there is a large amount of continuous deposition of carbon dust on the mine road, which is a serious accident [12].

2.2   Spray Dust Reduction

Spray dust reduction is to use spray technology to atomize water and spray it out, collide with dust floating in the air, and achieve the purpose of dust reduction by increasing the weight of dust particles. The key to this technology is that the nozzle should be able to flow mist with good dust suppression. Some countries have developed a series of nozzles, and have carried out a lot of research on the establishment of nozzle testing centers to ensure the production quality and utilization of nozzles. Nozzle dust removal is a major development and is widely used not only for individuals, but also for turning and decorating.

Water spraying is one of the simplest and most effective methods to reduce the amount of dust in coal mines, such as drying, air purification, etc. When spraying on the excavation face, it can also eliminate the bong and shorten the ventilation time. At present, the dust removal technology adopted by smelters in the carbon cutting process is mainly used in the following fields: shale radial fog curtain and light hydraulic support to help reduce dust spray; coal machine high-pressure spray wall pressure secondary dust reduction.

2.3   Coal Mine Spray Dust Suppression Adaptive System

The adaptive coal mine dust removal system solves key technical problems such as real-time monitoring of coal mine dust removal and intelligent control of spray dust removal system. Draw the access specification of various sensors connected to the coal mine underground dust removal system, and draw the multiplexing shape of the sensor ports. The main control box of the underground dust removal system with intelligent control function can realize functions such as program storage and calling, sensor condition combination and other functions on the basis of simple control methods such as manual control and timing control of traditional mine dust suppressors. The entire system such as the adjustment of spray conditions can be adjusted according to the situation, enriching the operation of the dust removal system, so that the dust removal process can be carried out in a targeted and efficient manner.

In conventional nebulizers, the nebulizer valve is opened and closed manually. In this way, the sprayer control is not only functional, but also intelligently changes the sprayer and closes the sprayer according to the dust accumulation pattern through simulated operation. Therefore, the system must have the ability to turn on and off the sprinkler system in real time according to the dust concentration. This not only reduces the loss of human resources, but also saves the stagnant water in the well, and will not cause unnecessary waste. The most important thing is to adjust the amount of dust reduction correctly in order to achieve the best dust removal effect.

During carbon transfer and handling, a sprayer must be installed and used to spray dust during operation. However, most of the current sprayers manually control the transfer points and valves of the sprayer during transportation, which not only consumes a lot of manpower and material resources, but also struggles with the produced water in the well. Due to long-term wear and tear, it will cause water accumulation in the drying area, and in dangerous situations, it will endanger the normal operation of the drying material.

Therefore, in the drying zone and carbon transfer process, the need for a device that can automatically spray and reduce dust depends on whether there is material transfer.

3 Investigation and Research on the Self-adaptive System of Coal Mine Spray Dust Reduction Based on Intelligent Analysis Algorithm

3.1 Research Purpose

Due to the size, distribution and space requirements of dust particles in the air, compared with liquid dust reduction technology, the design scheme of dust reduction using high-pressure spray devices meets the design requirements of this era. In order to further improve the high-pressure injection system, it is necessary to obtain specific parameters through design and calculation to meet the actual needs, or to analyze the specific devices of the entire system to achieve the intended use.

3.2 Coal Mine Spray Dust Suppression Adaptive System Architecture

The interface of each sensor unit of the system is divided into two interfaces: switch quantity and frequency quantity, which can be respectively connected to the input quantity switch interface and input frequency interface of the main control box. The backup battery of the main control box is connected to the power control unit of the main control box. The appropriate working mode of the main control box can be adjusted through the remote control, and the connection status of the sensor and the type of the sensor can be determined at the same time. The output of the main control box drives the control box, controls the output signal and controls the solenoid valve switch, thereby controlling the operation of the spray line. The entire system configuration is simple and suitable for most application scenarios.

3.3 Matching Relationship Analysis Algorithm

The particle size of the mist is affected by the diameter of the nozzle and the spray pressure. The formula for the relationship between the three is:

$$D = K(1.79 \times d_0 - 1)/(d_0 P^{1.26}) \tag{1}$$

In the formula: D—cloud particle size, μm, K—proportional coefficient, take 6, 28×105, P—spray pressure, MPa, d0—nozzle outlet diameter, mm.

Different dust particle sizes correspond to different optimal dust reduction particle size ranges. By analyzing the dust particle trajectories captured by the fog particles, the formula for the particle size dmin of the precipitable fog droplets is obtained as:

$$d_{min} = \sqrt{18\eta D K_d/(\beta_p v_q)} \tag{2}$$

In the formula: β_p—dust density, Kg/m3; Vq—fluidity, m/s.

4   Analysis and Research of Coal Mine Spray Dust Self-adaptive System Based on Intelligent Analysis Algorithm

4.1   Analysis of Main Control Box Scheme

The main control box can cooperate with dust sensor, smoke sensor, touch sensor, infrared sensor, electronic ball valve and other activation sensors, spraying equipment and other sensors to realize various spraying operation modes. It includes automatic fire control, automatic touch control, automatic smoke exhaust, automatic dust removal and automatic sound control. It can well meet the basic requirements of dust reduction. After the sensor is connected, the corresponding parameters can be configured through the corresponding remote control, including system status, sensor access status, sensor type, sensor alarm level, etc., and the corresponding location information can be displayed through the LCD. The block diagram of the main control box system is shown in Fig. 1.

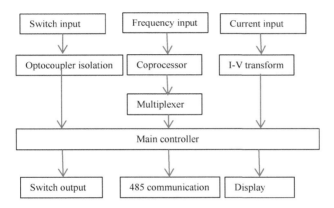

Fig. 1. Control box system structure

The main control box includes: power supply unit, main controller, sensor access circuit, power input unit, frequency input unit, optocoupler isolation circuit, communication 485 unit, display unit, etc. It is a complete and reliable motion management system. The control box system is powered by an intrinsically safe power supply, and the internal power supply passes through the power supply control circuit, and the power supply converts the voltage into the power supply voltage of each unit to supply power to each unit of the system. In the design of the sensor interface, the photoelectric isolation method is adopted to maintain the efficient control of the main control box and avoid interference. The central processing unit adopts the Cortex-M3 processor core, which has strong performance, high performance, and rich local interface and data processing capabilities.

In the man-machine interface, the LCD screen and LED indicators are used to indicate the current operating status, and the LCD monitor can display system status, sensor information, useful data, etc. The alarm information is provided by several LED lights on and off, and the operator can judge the corresponding alarm state of the sensor through the indicator lights. Therefore, it is easy to control sensor access and adjust system settings.

4.2   Intelligent Analysis of System Parameter Influence

The spray system in the air dust removal system is the process in which the aqueous solution is compressed into the nozzle and sprayed to form cloud particles of a certain size. The air supply system, that is, axial flow air, is pumped into the appropriate air chamber to form air pressure, and then the air flow passes through the spray system to transport the broken particles to the appropriate point at a certain speed. In order for the particles in the cloud space to play the most effective dust-reducing effect on the dust particles, factors such as the shape and size of the well-formed cloud particles and the optimal running speed of the particles need to be analyzed. However, dust suppression can only be performed effectively when there is a relative correlation between certain factors. When the nozzle diameter is 1.5 mm under different spray pressures, the values and particle sizes are shown in Table 1 and Fig. 2.

Table 1. Comparison of mist particle size under different pressures when the nozzle diameter is 1.5 mm

Spray pressure (MPa)	Calculated	Measured value
2	578	586
4	288	292
6	164	175
8	125	138
10	116	121

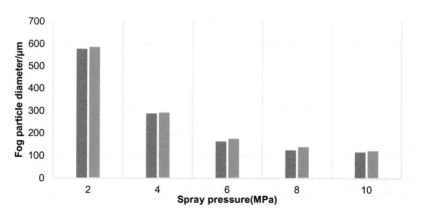

Fig. 2. Comparison of fog particle size under different pressures when the nozzle diameter is 1.5 mm

Due to the pressure loss in the pipeline, the actual spray pressure is less than the calculated amount, so the particle size of the mist is much larger than the calculated amount. The above results can be summarized as follows: when the spray pressure is 2 MPa, the cloud particle size is about 580 μm; when the spray pressure is 4 MPa, the particle size is almost reduced by half, about 290 μm. The spray density is 8 MPa and the particle size is about 130 μm. Then, the pressure inside the nozzle continued to increase and the particle size changed slightly.

With the increase of spray pressure, the particle size of the mist is greatly reduced, but when the pressure reaches 8 MPaW, the particle size also decreases, but the conversion rate is greatly reduced. As the pressure increases, so does the number of damaged parts. It shows that the pressure is not high enough, which leads to the decline of the economic benefits of the whole system.

Smaller droplet sizes can capture dust particles proportional to the square root of the particle size sprayed through the spray system. When the particle size of the mist is very small, the particle size of the dust you can collect is also very small, which is easy to remove, which is not conducive to vacuuming. It is very large and can collect dust larger than required. When the constant spray pressure is set to 8 MPa, using a nozzle with an output diameter of 1.5 mm, the attainable mist particle size is 120–140 μm, which can effectively capture 2–10 μm inhalable dust particles.

5   Conclusions

Coal mine dust refers to the small rocks and coal particles that can stay in the air for a long time, which are produced in the production process of our coal development, mining, coal mining and drying, drying, etc., referred to as my dust. The risks of coal mine dusts are mainly that they cause pneumoconiosis, reduced visibility and acceleration. Controlling coal dust pollution has become an important part of the coal industry. This paper firstly introduces the basic research and importance of dust reduction, as well as the current status of dust reduction technology at home and abroad. After analyzing the operation requirements of the spray dust suppression system in detail, the design concept and design life of each function of the system are given, and then the basic detection and realization circuit of each sensor unit is described in detail, and after the design of the sensor module circuit is completed, each sensor unit is completed. The connection and detection of the detection system and sensor operation.

Acknowledgements. Project no.: 2022TY2005 Project name: Development of intelligent dust removal module based on TD-Auto-OS platform.

References

1. Huang, Y., Chen, Z.: Ind. Mine Autom. **044**(009), 88–93 (2018)
2. Zhang, S., Hu, Q., Wang, G.: Ind. Mine Autom. **1**(1), 1 (2018)

3. Kilimci, Z.H.: Sentiment analysis based direction prediction in bitcoin using deep learning algorithms and word embedding models. Int. J. Intell. Syst. Appl. **8**(2), 60–65 (2020)
4. Chen, L., Du, W., Zeng, Z., et al.: Ind. Mine Autom. **044**(012), 60–64 (2018)
5. Kang, S.H., Song, H.E., Ahn, Y.C.: A study on the water spray system for coal dust reduction considering internal flow by belt operation. J. Power Syst. Eng. **24**(3), 14–22 (2020)
6. Córdoba, M.F., Ramírez-Romero, C., Cabrera, D., et al.: Measurement report: ice nucleating abilities of biomass burning, African dust, and sea spray aerosol particles over the Yucatán Peninsula. Atmos. Chem. Phys. **21**(6), 4453–4470 (2021). https://doi.org/10.5194/acp-21-4453-2021
7. Chen, L., Zhang, Y.: Analysis of lump coal rate in fully mechanized mining face based on discrete element method. Ind. Min. Autom. **45**(02), 57–62 (2019)
8. Parsakhoo, A., Hosseini, S.A., Lotfalian, M., et al.: Effects of molasses, polyacrylamide and bentonite on dust control in forest roads. J. For. Sci. **66**(5), 218–225 (2020)
9. Fatemidokht, H., Rafsanjani, M.K., Gupta, B.B., et al.: Efficient and secure routing protocol based on artificial intelligence algorithms with UAV-assisted for vehicular ad hoc networks in intelligent transportation systems. IEEE Trans. Intell. Transp. Syst. **22**, 4757–4769 (2021)
10. Al-Tarawneh, M.: Data stream classification algorithms for workload orchestration in vehicular edge computing: a comparative evaluation. Int. J. Fuzzy Logic Intell. Syst. **21**(2), 101–122 (2021)
11. Rababa, S., Al-Badarneh, A.: Optimizations for filter-based join algorithms in MapReduce. J. Intell. Fuzzy Syst. **40**(1), 1–18 (2021)
12. Sharma, A., Kumar, N., Kumar, A., et al.: Comparative investigation of machine learning algorithms for detection of epileptic seizures. Intell. Decis. Technol. **15**(1), 1–11 (2021)

A Study on the Multi-modal Interaction Pattern of "Culture + Museum"

Cen Guo[1], Ming Lv[1], Yuzhu He[1(✉)], and H. Alsharif[2]

[1] Design Art, Shenyang Jianzhu University, Shenyang, Liaoning, China
1142366015@qq.com
[2] South Valley University, Qena 83523, Egypt

Abstract. In order to explore the way of integration and innovative development of national culture and modern digital museum. Taking Liaoning Provincial Museum as an example, this study mainly analyzes the feasibility of multi-modal for the digital display form of museums. "Culture + Museum" has become the most intuitive model to promote the development of local culture. Using high and new technology, visitors and museums to carry out multidimensional interaction. Using the concept of multi-modal interaction to explore the way of museum development. Analyze the cultural values conveyed by museums in different ways and perspectives. At the same time, this study can provide a new reference value for the construction of modern museums and put forward new ideas for developing national culture in the future.

Keywords: Digital museum · Science and technology · Cultural and creativity · Multi-modal interaction

1 Introduction

1.1 Overview of "Culture +" Concept

People's material life is rich, and culture has jumped out of the previous level of added value and gradually evolved into the core value of society. Culture has evolved through thousands of years of human civilization. On the one hand, it is the result of people's long-term creation and catalysis. On the other hand, it is also a historical phenomenon. Scientific and technological means are constantly innovating and information channels are constantly opening up. Culture not only has the traditional carrier form of books, images and literary performances, but also expands cultural and creative products that carry beauty and literature to meet people's personalized and diversified needs. The concept of "Culture +" arise at the historic moment. Culture is not only limited to its own development, but also involves various fields such as economy, science and technology, tourism and so on. "+" is not only a simple sum of numbers, but also a mutual promotion to carry forward the spiritual connotation of culture to a greater extent [1].

1.2 "Culture + Museum" Mode Forms the Advantages of Cultural Development

Only by integrating culture with innovation, tourism, science and technology and other industries and better integrating them into people's daily life can the distance between people and culture be drawn closer. As a collection carrier of local natural and cultural heritage, the museum is an intuitive display of culture and a summation of cultural propaganda. In short, "Culture + Museum" has become the most intuitive model to promote the development of local culture.

While living a rich life, more and more people are keen on expanding their horizons with tourism. What is different from the past is that people begin to pay attention to experience and sense of acquisition, which makes the traditional "sightseeing tourism" evolve into a new "experiential tourism". Experience can increase the authenticity and interest of the tour, help the rapid spread of culture, enhance the sense of identity of our national culture, and establish cultural confidence. In the context of the integration of culture and tourism, museums have become a new choice for experiential tourism. Every year during "May Day", "National Day" and other holidays, tourists are enthusiastic about museums. During this period, visitors across the country soared, and tickets to the museum were hard to get. For example, the Palace Museum needs to book tickets a week in advance. Related data show that from January to June 2021, the number of tickets booked through the Internet and mobile phone APP platform increased by three-quarters compared with the previous two years.

1.3   Problems and Suggestions of the "Culture + Museum" Model

However, with the vigorous development of museums today, most museums have not grasped the opportunity and still have problems such as lack of development power and small audience. Based on the research of multi-modal interaction, this paper discusses such issues and puts forward the following suggestions based on excellent cases:

(1) Thinking solidification problem. First, we should innovate our ideas. Only by improving the vitality of museum development and changing the traditional model can there be greater development space.

(2) Cultural characteristics are not obvious. There are few thematic exhibitions. Take Liaoning Province as an example. The history of Liaoning province is characterized by the traditional Manchu culture. The unique culture has become the core competitiveness of the city, which is then transformed into productivity, driving economic development and improving the comprehensive strength of the city. Therefore, in addition to daily exhibitions, the museum should focus on promoting prominent culture [2].

(3) Single display form. In the display link, cultural characteristics should be presented in a popular way. Education through entertainment can integrate culture into people's life. Therefore, we should emphasize the sense of experience and pay attention to different types of interaction. Take different modes as the starting point for innovation. When interacting with visitors, we can enhance the experience of multidimensional senses such as vision, touch and hearing. For example, for the display of traditional food culture, we can add innovative forms of smell and taste. Promote cultural development with scientific and technological innovation [5].

(4) Lack of cultural added value. Take cultural and tourism products for example, most of our cultural and tourism products are not representative, sameness and

other problems. Therefore, it is necessary to study the cultural added value of products. Cultural and tourism products have the characteristics of wide audience and large sales. Enhance cultural and creative forms, attach cultural effects to them and promote mutual fusion. Apply the concept of multi-modal to the design of cultural and creative products. Design elements and modal symbols are organically combined to select creative forms that can reflect cultural vitality.

2   Overview of Multi-modal Interaction

2.1   Concept of Multi-modal Interaction

Generally speaking, modality is people's senses. So multi-modality is the integration of multiple senses. The Turing OS is capable of providing multi-modal interaction. It is capable of simulating human interaction in various ways, such as text, sound, image, behavior, environment, etc. So it gives people a strong sense of experience. From the professional level, multi-modal can be understood as the man-machine interactive information flow and communication media integrating more than two communication methods or carriers [3]. The fusion of single modes forms a multi-modal form, roughly forming five modes: visual, auditory, gustatory, olfactory and tactile mode.

2.2   Implementation Approach of Multi-modal Interaction

Multi-modal interaction integrates multiple senses, through eye, ear, nose, mouth, and skin contact. The practical applications of technology are also designed around these sensory interactions.

(1) Visual interaction: Seeing through the eyes is the simplest way to interact. VR, augmented reality and holograms are all examples of this. It focuses on displaying content through virtual screen and enhancing content immersion [4, 6].
(2) Auditory interaction: Conveying information through sound is the fastest and most direct way for humans to recognize products besides visual modes. Sound can help users feel the change of tone and mood in different scenes.
(3) Tactile interaction: Tactile mode can receive more delicate and real information, and establish the association between product attributes and user feelings. It can assist VR technology in human-computer interaction. The human brain is very good at using tactile, visual and auditory signals and fusing them together to make it believe that an object is in the hand. By collecting tactile information of real objects and simulating them, simulated data are expressed by the tactile information of virtual environments, to improve the interactive and realistic degree of the simulated environment. Naoto Fukazawa, a famous Japanese designer, used the concept of multi-modal metaphor to design a group of beverage boxes. The rubber material imitated the texture and color of fruits of different flavors, making users feel more real when touching the beverage packaging [7].

(4) Olfactory interaction: Smell is the most direct sense to awaken people's instinctive behavior and emotional memory. Olfactory interaction technology can enhance the sense of immersion in virtual environment. The Internet can transmit all kinds of scent information. Modern computers have been able to analyze and process smells and output various smells through the terminal.

(5) Taste interaction: By controlling electrodes and electrolytes in different gels, taste experience technology can stimulate the tongue to experience different tastes. In Australia, Sipahh straws are specially treated with flavoring particles of natural ingredients. When the milk passes through the straws the flavoring particles are dissolved, changing the milk taste different and guiding people's taste perception. The ability of the sensory organs to sense the right stimulus is called receptivity. The interval between the minimum and maximum amount of stimulus that can induce sensation is called sensory threshold. Receptivity is inversely proportional to sensory threshold. This is weber's law:

$$\Delta I/I = K. \tag{1}$$

ΔI: the difference threshold. I: the original stimulus intensity. K: a constant. The intensity of the sensation is proportional to the logarithm of the stimulus intensity. This is Fischer's law:

$$S = Klg\ R \tag{2}$$

S: Psychological quantity. K: constant. Lg: logarithmic. R: physical quantities.

3 Proposed the Concept of Multi-mode Interactive and Immersive Museum

Liaoning Provincial Museum is a comprehensive museum, with permanent exhibition halls and temporary exhibition halls, rich collections, among which the Manchu national exhibition hall is featured. The concept of "Culture + Museum" is proposed based on the exploration of local culture, and the innovation was carried out by extracting the Manchu culture with national identity and freshness. Therefore, this study will take the Exhibition hall of Manchu nationality as the breakthrough point for analysis.

3.1 Virtual Reality Integrates Multi-modal Interaction

In the new era, technology and culture are constantly blending and catalyzing, especially in the protection of cultural relics, digital exhibition halls and other aspects, the role of science and technology in promoting culture is increasingly strengthened. Using AR and VR high-tech technology can mobilize visitors' vision, hearing and touch, so that folk culture and even Chinese cultural achievements can be exhibited in a new way, to achieve an immersive feeling [8].

Manchu has a long history and distinctive traditional costumes. Manchu people wear traditional "Qi zhuang". In the digital museum, traditional costumes can be presented in three-dimensional and full-perspective form. The life scenes of people wearing traditional national costumes can be displayed to interact with visitors. The most stylized custom is Manchu marriage. On the basis of preserving Manchu traditional national characteristics, some Han cultural elements are integrated. The same is true of Manchu funeral rites. However, the traditional marriage and funeral customs have been forgotten by history, and such scenes are difficult to reproduce now, which has become a great regret of the present era. In order to better understand Manchu cultural traditions such as marriage and funeral, tourists can use the VR technology provided by the museum to have a dialogue with time and space. It's like "coming to the scene". Wearable devices can be used to interact with ancestors. Limb activity can be recorded during this period. Different behaviors and interactions will have different "emotional effects" on the ancients. After the viewing, tourists can shuttle through the "street shops" thousands of years ago, or bargain or select goods. Tactile gloves, a wearable device that uses air pressure to generate force, can be used to experience the real touch of grasping virtual objects through the chip. The gloves not only accurately feed the wearer's hand movements back to the computer, but also simulate a complex array of subtle sensations, pressure, texture and vibration of objects. The wearer's skin is gently pulled by the actuator to mimic the pull of gravity on the object it is holding. To achieve tactile, visual and auditory collision, this innovative technology will enable people to complete the series of experience [9] (see Fig. 1, 2 and 3).

Fig. 1. The manchu clothing

Fig. 2. Offer sacrifice to Heaven

Fig. 3. Offering XueShen

3.2   Folk Culture is Presented in Multi-modal Form

Chinese cuisine which has its unique value. Manchu's traditional diet is closely related to folk sacrifice rites, so it has strong particularity. Manchu people's dietary habits are noodles and pork, cooking to boil, stew and roast. Manchu pastries and the characteristics of the Qing Palace "Manchu and Han banquet" enjoy a high reputation.

To lead the trend of the country, China has launched "Masters in Forbidden City", "It's new, the Forbidden City", "National Treasure" and other creative programs. Previously solemn museum to find a new "open way". Cultural lovers and tourists enthusiasm are rising. Based on this experience, the display of Manchu diet culture can also be used for reference in the development of relevant programs. First, use the cookers and utensils collected in the museum to show the ancient methods of placing delicious food, and the process of making delicious food is further restored, to promote Manchu food culture. This form can be further promoted, using remote publicity to attract tourists from other places to experience on site. Use innovative visual integration technology to realize the transformation from single culture to multi culture. This can broaden cultural coverage and cultural connotation, enhance cultural added value and competitiveness of museums.

In addition, while presenting delicious food in visual form, smell and taste experience can be added. Different gels are used to control the intensity of the five basic tastes: sour, sweet, bitter, spicy and salty. When the device touches the tongue, it can sense the unique flavor carried by history, and the sense of smell can also be simulated through the terminal. Innovative experience methods can break through the limitations of traditional exhibitions. It can evoke human instinctive behavior and emotional memory and senses.

3.3   Multi-modal Design Concept of Cultural and Creative Products

Since the "13th Five-Year Plan", China has added one new museum every two days on average. By the end of 2020, the number of registered museums in China has reached 5788, and it keeps rising every year. The museum scale has grown steadily, providing a foundation for the "national craze" set off by museum tourism.

The Palace Museum has changed its previous serious image. It designed more than 10000 innovative cultural and creative products, covering all aspects of life. The products with high creativity and high quality are welcomed by people at home and abroad. The museum adopts online and offline sales channels, with annual sales exceeding one billion yuan.

The key to the design of cultural and creative products lies in the shape, color, material and function. The multi-modal concept is used to establish multiple modal symbols for the functions, forms and cultural connotations of products. Organic combination of the design elements and modal symbols.

During the International Museum Day, museums around the country have launched boutique exhibitions, cultural and creative products and offline activities to make the theme of museums more inclusive of all ages and immersive content experience. More people can walk into museums to feel the charm of Traditional Chinese culture [10].

4 Conclusions

Using the concept of multi-modal interaction to explore the way of museum development, the history and culture radiate more abundant imagination space. Interactive and experiential exhibition can enhance users' willingness to participate and culture identity. Build a digital museum that everyone can participate in, and let users become cultural re-creators. At the same time, the interactive mode is used to publicize national culture. To raise people's awareness of protecting cultural heritage. Digital innovative design injects fresh vitality into local traditional culture, enhances cultural visibility and promotes the development of local tourism and economy to a greater extent.

References

1. Liu, X., Jiang, Y.: Aesthetic assessment of website design based on multimodal fusion. Futur. Gener. Comput. Syst. **117**(164), 433–438 (2021)
2. Kim, D.S., Lee, J.Y., Chun, B.S., Hwang, G.Y., Paek, W.K., Byun, B.K.: SMEP (science museum exhibition platform) for sharing and exchange system of natural history collection. J. Asia-Pac. Biodivers. **14**(3), 299–301 (2021)
3. Peng, F., Geng, Y., Wang, J., et al.: Invited paper: liquid crystals for virtual reality (VR). In: SID Symposium Digest of Technical Papers, vol. 52, no. 1, pp. 427–430 (2021)
4. Mohring, K., Brendel, N.: Producing virtual reality (VR) field trips - aconcept for a sense-based and mindful geographic education. Geographica Helvetica **76**(3), 369–380 (2021)
5. Carvajal, D.A.L., Morita, M.M., Bilmes, G.M.: Virtualmuseums captured reality and 3D modeling. J. Cult. Herit. **45**, 234–239 (2020)
6. Lau, N., Hildebrandt, M., Jeon, M.: Ergonomics in AI: designing and interacting with machine learning and AI. Ergon. Des. **28**(3), 3 (2020)
7. Sun, S.: Cultural and creative industries and art education. Phys. Procedia **33**, 1652–1656 (2012)
8. Klein, M., Gerlitz, L., Spychalska-Wojtkiewicz, M.: Cultural and creative industries as boost for innovation and sustainabled evelopment of companies in cross innovatlon process. Procedia Comput. Sci. **192**, 4218–4226 (2021)
9. Goodman, A.: Karl Linn and the foundations of community design: from progressive models to the war on poverty. J. Urban Hist. **46**(4), 794–815 (2020)
10. Glenzer, K., Divecha, S.: Upscaling community transformation. Action Res. **18**(4), 407–413 (2020)

Design and Implementation of Dust-Free Process Environment System Based on STM32 Microprocessor

Jiang Wang[1], Jiuxia Li[1], and Lei Tian[2(✉)]

[1] Tianshui Huatian Technology Co., Ltd., Tianshui, Gansu, China
[2] School of Electronic Engineering, Xi'an University of Posts
and Telecommunications, Xi'an, Shaanxi, China
ttla02@163.com

Abstract. In recent years, intelligent dust-free process environment system has increasingly appeared in our daily life. Based on this, this paper designs and implements a more efficient and widely used intelligent dust-free process environment system, which plays a role in the development of IC manufactures. The system mainly completes three parts of functions: the first part uses the temperature and humidity module and light intensity module to display the collected temperature and humidity information and light intensity information on the screen through the STM32 single chip microcomputer FSMC interface, and upload these information to the terminal (upper computer) through the ESP8266 module for real-time viewing by the greenhouse owner. The second part is to automatically supplement the light source according to the change of light intensity through STM32 single chip microcomputer to promote the growth of crops in the greenhouse. The third part is to realize the function of monitoring and recording.

Keywords: Dust-free environment · Microprocessor · Intelligent monitoring

1 Introduction

In recent years, with the rapid development of China's industry, intelligent dust-free process environment has introduced advanced modern industrial technology. The integrated circuits manufactured have been greatly improved from the original pure human labor to today's semi mechanized. The workers can work more labor-saving and obtain more benefits at the same time. Through continuous research and development, the hardware quality level and supporting capacity of intelligent dust-free process environment are improved [1].

According to the status of the development of China's intelligent dust-free process environment, it will have a higher degree of integration, a wider range of application of intelligent dust-free process environment system is more conducive to promote the development of China's IC manufactory [2, 3].

Intelligent dust-free process environment has created opportunities for the development of IC manufactory in China. The intelligent IC technology will rely on the interconnection of all things under 5G to diagnose the crop growth environment and

© The Author(s), under exclusive license to Springer Nature Switzerland AG 2022
V. Sugumaran et al. (Eds.): ICMMIA 2022, LNDECT 138, pp. 185–192, 2022.
https://doi.org/10.1007/978-3-031-05484-6_23

growth conditions, such as detecting the environmental temperature, humidity, light intensity and other information, and then put forward strategies to make the intelligent system adjust a more appropriate temperature and let unmanned aerial vehicles spray fertilizers and pesticides.

2   System Hardware Design

2.1   Information Collection Module

DHT11 sensor is a temperature and humidity composite sensor with calibrated digital signal output [4]. Its features include low cost, good stability, can also be used in harsh environment, fast response to these aspects. DHT11 adopts the serial mechanism.

This part uses the photosensitive sensor to form the light intensity module. Photosensitive sensors convert the optical signals into electrical signals by using the characteristics of photosensitive elements, and their response wavelengths are near visible wavelengths.

The response speed of the sensor was mainly determined by the response time of the PD and TIA. For the novel structure was fabricated to the PD, the whole response time of the PD is controlled by the following three factors:

$$T_{RC} = 2.2C(R_s + R_L) \tag{1}$$

$$T_{drift} = d/v_d = d^2/(\mu V_R) \tag{2}$$

$$T = \sqrt{T_{\text{arift}}^2 + T_{\text{aifüsed}}^2 + T_{RC}^2} \tag{3}$$

Generally, when detecting light intensity, photosensitive sensor is also often used as a device. If other signals are converted into optical signals and then through the photosensitive sensor, the parameter values of other signals can also be obtained. The photosensitive sensor used in this design is a photosensitive diode, whose structure is similar to the semiconductor structure, with a PN junction inside, so the photosensitive sensor also has one-way conductivity.

Infrared detection module HC-SR501 is an automatic control module using infrared technology [5]. This module works in the dc voltage range of 4.5–20 V and can output TTL level. It has a working range of 7 m. It has two triggering modes: Non-triggering mode and repeatable trigger mode.

This module has inductive blocking time, that is, when the inductive output high level becomes low level, the developer can manually set a blocking time, during the blocking time, the infrared sensor does not receive any inductive signal. Therefore, it plays an important role in interval detection and effective suppression of various interferences in the process of load switching.

The OV2640 offers a single UXGA camera. Generally, OV2640 module is controlled by SCCB bus. The OV2640 can achieve a maximum frame rate of 15 frames per second if in output UXGA image mode. If through SCCB interface programming can achieve any image processing process [6]. The OV2640 module consists of the following modules: photosensitive array, analog signal processing, 10-bit A/D conversion, 8-bit microprocessor [7].

2.2   Information Transmission and Display

Because ESP8266 module supports TTL serial port, if the TTL level of MCU is 3.3V or 5V, this module can be used. After firmware burning, the module can support three connection modes to realize data transmission, which are: serial port to WIFI STA mode, serial port to AP mode, serial port to STA+AP mode [8].

This part is mainly used for the display of the intelligent IC manufactory system. It can refresh the information of temperature, humidity and light intensity in real time under normal circumstances, display the normal operation of the camera module during the capture, capture the picture and keep the latest capture on the screen.

TFTLCD screen is different from the simple matrix of passive TNLCD and STNLCD. Each pixel of TFTLCD corresponds to a thin film transistor. Therefore, the crosstalk of TFTLCD screen can be eliminated, and the static characteristics of the LCD screen can be independent of the number of scan lines, which greatly improves the image quality of the display.

2.3   Information Storage and Processing

This module is mainly used to store images taken. Considering to support SPI/SDIO driver, and large storage capacity, THE selection of SD card, and its MCU system used to do external memory is very convenient. Use the development board PC8, PC9, PC10, PC11, PC12, PD2 pins [9, 10].

The light source module automatically adds light intensity according to the change of light intensity, and is connected with the PA2 pin of the development board for multiplexing function. Multiplexing is the output comparison function of timer 9 to control the intensity of light source [11]. This part of external circuit is composed of light-emitting diode in series with a pull-up resistor. The alarm module mainly uses STM32 to control the buzzer to send sound alarm and WiFi module to send alarm signal to the server. The buzzer alarm is also relatively simple. It is connected to the PF8 pin of the development board for general output function.

3   System Software Design

3.1   Collection, Display and Transmission

Write the temperature and humidity collection code dht11.c according to the working time sequence diagram of DHT11 temperature and humidity sensor. DHT11_Rst() is used to reset the DHT11, DHT11_Check() is used to wait for the DHT11 response, and DHT11_Read_Data() is used to read the temperature and humidity data. In addition, it is also important to initialize the PG9 pin to pull up, normal output mode.

Temperature and humidity light intensity display will mainly use TFTLCD screen, here need to write the screen driver and Chinese characters display function.lcd.c is used to drive the screen and text.c is used to display Chinese characters. First, write lcd. c according to LCD usage flow chart, which is showed in Fig. 1.

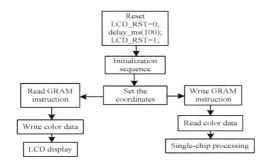

Fig. 1. The LCD function flow chart

Transmission is the key and difficult point in this design. In this part, esp8266. c is mainly written to establish the connection, usart.c is used to send data through the serial port. In short, this part is to control WiFi module to send and receive data through TCP/IP protocol in wireless local area network through AT instruction set.

The WiFi module can be set as TCP server, TCP client and UDP in STA mode, and the corresponding upper computer is set as TCP client, TCP server and UDP respectively. In this design, WiFi module in STA mode is used as TCP client and configured as TCP server in the upper computer software. Table 1 shows the configuration in TCP client mode.

Table 1. STA and TCP client configuration

Instructions	Functions
AT+CWMODE=1	Set the WiFi module to STA mode
AT+RST	Restart the WiFi to take effect
AT+CWJAP="xx", "xxxxxxx"	Add a WiFi hot spot: xx, the password is xxxxxxxx
AT+CIPMUX=0	Enabling single connection
AT+CIPSTART="TCP", "192.168.1. XXX", 8000	Establish a TCP connection to 192.168.1.xxx, 8000
AT+CIPMODE=1	Enable transparent transmission
AT+CIPSEND	Begin to transport

3.2   Monitoring and Storage

This part of the function is mainly realized by writing dt.c. The PA3 pin is first configured to initialize it, and the PA3 pin is set to normal input and drop-down mode. Through PA3 pin to infrared detection module output level detection, detection of rising edge trigger interrupt. External interrupts are implemented on interrupt line 3 and need to be configured for interrupt initialization. Configuration interrupt initialization includes initializing PA3, enabling interrupt clock configuration interrupt, connecting PA3 pins to interrupt line 3, and configuring NVIC interrupts for interrupt line 3 and interrupt line 3.

This part of the software is designed to capture and store images. malloc.c is the memory management driver code. The FATFS file system code is open source and only needs to be ported for use on the development board. sdio_sdcard.c is the code for storing data to an SD card. sccb.c, ov2640.c, and photo.c are used to capture JPG images. dcmi.c is used to transfer captured image data to SD card. The process flow chart of capturing process is shown in Fig. 2.

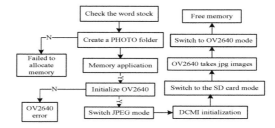

Fig. 2. Flowchart of program realization of capture

To accomplish this function, the driver code of SCCB interface should be written first. The second step is to write the OV2640.c code to turn on the OV2640 module. The OV2640 uses OV_SCL and OV_SDA to configure registers, as well as signals such as OV_PWDN and OV_RESET. Configure OV2640 initialization as shown in Fig. 3. Another important function here is to set the image output window function OV2640_Window_Set(), set the image output size function OV2640_OutSize_Set(), set the window function OV2640_ImageWin_Set(), set the image resolution size function OV2640_ImageSize_Set ().

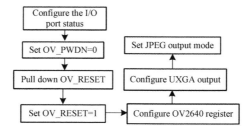

Fig. 3. OV2640 initial configuration flowchart

The third step is to compile dcmi.c which mainly completes four functions. The first is to enable the clock, configure the mode of the required pins and set the reuse function. Then complete the configuration of the DCMI. In this step, important parameters such as HSPOL/PCKPOL/VSPOL data width in the DCMI_CR register need to be configured. When frame interrupt is enabled, DCMI interrupt service function is written for data processing.

4   Function Implementation and Testing

The physical picture of the system is shown in Fig. 4 and the system startup interface is shown in Fig. 4. After entering the startup interface, you can get the function introduction of the system, the name of the system, the model of the development board used, and the department of the system.

Fig. 4. The whole system with the STM32 board

The test here is mainly on the screen and the upper computer temperature and humidity size, light intensity size test. The test environment was 8 a.m., 13 p.m., and 20 p.m. Figure 5(a) shows the display of temperature, humidity and light intensity on LCD screen and upper computer at 8:00 in the morning. This system can detect the animals or people in this room, the group of images obtained after testing are shown in Fig. 5(b).

(a)The data at 8:00   (b) The captured pictures

Fig. 5. The example of the display

5   Conclusions

By comparing the two captured images with the test indexes, the infrared monitoring alarm and storage functions have been verified. From here, we can also get the location information of the saved image after capturing: PHOTO/PIC00027.jpg; By comparing the two captured images with the test indexes, the infrared monitoring alarm and storage functions have been verified.

Acknowledgements. This work was partly supported by the Natural Science Basic Research Program of Shaanxi (Program No. 2021JM-460) and the Scientific Research Program Funded by Shaanxi Provincial Education Department (No. 21JC033); the higher education scientific research project of Shaanxi higher education society (No. XGH21161) and the undergraduate innovation and entrepreneurship plan (No. 202111664039, S202111664070 and 202111664019).

References

1. Viswanathan, S., Momand, S., Fruten, M., Alcantar, A.: A model for the assessment of energy-efficient smart street lighting-a case study. Energ. Effi. **14**(07), 1–20 (2021)
2. Yasuhisa, O.: Concept of an ideal PN junction. Bipolar-type insulated-gate transistors. IEEE Sens. 1–5 (2013)
3. Ping, H., Tang, C.: Indoor detection based on MLX90621 infrared sensor. Electron. Measur. Technol. **39**(11), 118–121 (2016)
4. Yan, A.I.: Intelligent lighting control system PWM dimming smoothing optimization. Times Agric. Mach. **44**(11), 114–115 (2017)
5. Sáenz-Peafiel, J.J., Poza-Lujan, J.L., Posadas-Yagüe, J.L.: Smart cities: a taxonomy for the efficient management of lighting in unpredicted environments. DCAI **6**(06), 63–70 (2019)
6. Sun, F., Yu, J.: Indoor intelligent lighting control method based on distributed multi-agent framework. Optik-Int. J. Light Electron Opt. **213**(07), 1–10 (2020)
7. Piao, S., Ciais, P., Huang, Y.: The impacts of climate change on water resources and agriculture in China. Nature **467**(16), 43–51 (2010)

8. Tolomio, M., Casa, R.: Dynamic crop models and remote sensing irrigation decision support systems: a review of water stress concepts for improved estimation of water requirements. Remote Sens. **12**(09), 39–45 (2020)

9. Jose, M.G., Pereira, L.S.: Decision support system for surface irrigation design. J. Irrig. Drainage Eng. **135**(11), 343–356 (2009)

10. Dadari, S., Ahmadi, S.H.: Calibration and evaluation of the FAO56-Penman-Monteith, radiation, and Priestly-Taylor reference evapotranspiration models using the spatially measured solar radiation across a large arid and semi-arid area in southern Iran. Theor. Appl. Climatol. **136**(26), 441–455 (2019)

11. Sun, Y.P., Lan, Y.P.: Research on self-learning fuzzy control of controllable excitation magnetic suspension linear synchronous motor. J. Electr. Eng. Technol. **15**(10), 843–854 (2020)

Distributed Genetic Algorithm in Energy and Power Blockchain System Design

Shaodan Zhou[✉]

Southern Power Grid Materials Co., Ltd.,
Guangzhou 510620, Guangdong, China
Zhou_shaoD@163.com

Abstract. The traditional energy and power system needs to solve the information security problem brought by the network. Once the system is attacked, the information of users in the power market will be leaked. The blockchain technology combines energy and power to improve the system and make the information in the system database public. It can be applied to the distributed power transaction of the distribution network. The distributed genetic algorithm (DGA) can be used to improve the transaction efficiency in the power data transaction, optimize the performance of the energy and power (EAP) blockchain system to detect the failure of power grid equipment, and build a reliable local area. The electricity market and ensuring the security of electricity data transactions are of great significance.

Keywords: Distributed genetic algorithm · Blockchain technology · Energy power system · Power data transaction

1 Introduction

One of the main tasks of distribution network trading is to design the power trading system in the power market mechanism, so that the trading information in the power market is symmetrical and the trading process is efficient and safe. Blockchain technology can make information transparent, and users can conduct electricity transactions fairly and equitably. This is why it is necessary to design an energy and power blockchain system to ensure the safety and reliability of transaction data.

Many scholars have analyzed and studied the application of DGA in the design of EAP blockchain system, and achieved good research results. For example, it is thought that the essence of the direct power purchase transaction of large users is decentralized decision-making, and the blockchain technology is introduced, and its application in technology matching, transaction mode, market settlement, etc. is discussed. By integrating virtual power plants and blockchain methods, the construction of the energy blockchain network model of distributed energy marketization effectively improves the performance of virtual power stations, and the cryptographic characteristics of blockchain also allow virtual power stations to achieve better security [1]. In terms of electricity transaction settlement, some scholars have applied smart contracts and DGA to electricity market transaction settlement, and verified the distributed accounting method of blockchain technology, the matching of consensus verification and transaction settlement, and the difficulties of key technologies with the realization, the

© The Author(s), under exclusive license to Springer Nature Switzerland AG 2022
V. Sugumaran et al. (Eds.): ICMMIA 2022, LNDECT 138, pp. 193–200, 2022.
https://doi.org/10.1007/978-3-031-05484-6_24

transaction efficiency is improved [2]. Some companies have arranged measurement equipment for grid nodes and established related communication mechanisms based on blockchain technology. The measurement equipment can detect the status of the current grid nodes, and all grid users can connect these devices through personal devices and view the information at any time, reducing daily maintenance costs and early detection of faults [3]. Although the DGA has a wide range of applications in the EAP blockchain system, it is still a key research direction to ensure the convenience of transactions and the security of system nodes when using the system for energy transactions.

This paper learns that the distributed genetic algorithm is a parallel algorithm, and the EAP blockchain system can realize power transactions. Therefore, this algorithm can be used to calculate the storage time of energy transactions in the blockchain system, obtain the optimal solution set in the population, and obtain the optimal solution set in the population. Optimize system performance. By integrating EAP blockchain technology, the designed system can effectively ensure the safe storage and management of information.

2   Overview of Distributed Genetic Algorithm and Energy and Power Blockchain System

2.1   Distributed Genetic Algorithm (DGA)

Genetic algorithm based on coarse-grained model, also known as DGA, is currently the most widely used parallel algorithm that can adapt to various calculations [4]. Coarse-grained models exchange data between subpopulations through a transfer function. Compared with the serial genetic algorithm, the parallel genetic algorithm has undergone a fundamental change in the calculation method. In implementation, there are several variants of coarse-grained models in applications [5]. For example, different subgroups can be assigned the same control parameters to obtain an evolutionary pattern that adapts to different computational programs. In addition, there are many different migration methods, you can choose the best one of the migration methods, you can choose only one of the methods that is not lower than the average of the subgroup, or just randomly choose from all the migration methods [6].

For genetic algorithms we expect to know when the optimal solution set is first reached. Since the genetic sequence is random, so is the time of first arrival. Here we are concerned with the expected value of the first arrival time. Let $\left\{ \overrightarrow{X}(n) \right\}$ be the genetic sequence, T be the first arrival time of the algorithm, and U be the optimal solution set of the algorithm, then:

$$T = min\left\{ n : \overrightarrow{X}(n) \cap U \neq \varphi \right\} \tag{1}$$

$$E(T) \leq \sum_{J=1}^{\infty} \frac{1}{p(1 - p^{n-1})} \tag{2}$$

where E(T) is the expected arrival time and p is the subpopulation.

2.2   The Role of Energy and Power Blockchain System

(1) Realize electricity trading
While the original use of blockchain technology was to facilitate cryptocurrency transactions, the technology has also greatly facilitated the development of electricity transactions. In this regard, there are mainly two types of transactions. One is equal trading. Several startups have launched a blockchain initiative that aims to revolutionize the energy industry by creating a transparent, immutable log of virtual transactions, allowing homes or businesses to sell distributed batteries or solar panels [7]. The second is network marketing. Blockchain technology is rapidly reducing grid transaction costs and reforming existing hybrid electricity markets. Furthermore, it can unite new markets for distributed energy, forcing them to work in tandem with existing energy systems [8]. Even with major changes in grid operations, these large-scale grid transactions remain inextricably linked and could ultimately have a major impact on the future of the power industry and its increasingly complex landscape.

(2) Energy financing
Certain startups offer blockchain applications, including the use of cryptocurrencies to source energy projects, the vast majority of which are clean energy sources. Blockchain networks can facilitate financing for renewable energy projects and expand the pool of potential investors [9]. However, it is unclear whether such a distributed network is really needed to finance the rapid growth of renewable energy production.

(3) Sustainable development of electric energy
One of the most immediate uses of blockchain in electricity is to record and sell electricity stability, including whether a unit of electricity can be recycled to create a single carbon footprint. This helps reduce trade, fraud and errors in regional renewable energy lending and expands the carbon offset market [10]. If these projects can be expanded, it will help the government control coal emissions and continue to use renewable energy.

(4) Equipment management
Other applications of blockchain technology in the power sector include the management of multiple networked devices connected to the grid to balance the network load to maximum power consumption [11]. Some utilities are now exploring models to better control network infrastructure assets and bring them into blockchain networks. These controls are designed to enable fast, uninterrupted transactions between consumers and electricity retailers. Finally, other initiatives are also trying to use blockchain technology to improve the cybersecurity of power systems.

2.3   Integration of Blockchain Technology and Energy Power System

In the EAP network, it has the characteristics of resource sharing, information openness, and availability for multiple users. As a standard distributed storage technology, blockchain has the characteristics of low concentration, non-intrusiveness and traceability. Of course, it is also very convenient, and provides a fast and convenient technology platform for the development and implementation of energy networks [12]. The combination of blockchain technology and the power grid is beneficial to improve its security, transparency and efficiency. It is integrated in the following aspects:

(1) Weak centralization

At present, my country's traditional energy and power trading methods are still dominated by on-site transactions. In exchanges, the energy is generally managed and controlled by the management and control center in a planned manner, and the safety and reliability of transactions are verified by a third party. The role of the third party cannot be ignored, which makes the third party have great authority in the transaction process, and also generates higher third-party costs. The blockchain is integrated with the grid, and each user can be either a generator or an energy consumer. Transactions between partners are done in real time. It is not based on a central transaction or governing body. Each node is only a part of the system, and there is no particularity in the node, and all have the same rights and functions.

(2) Intelligent scheduling

In traditional energy transmission, the central service frequently exchanges information with third parties, and the time cost is high, which is not conducive to processing user transaction operations at any time. The integration of energy power and blockchain, through the use of smart contracts, highly automates the entire transaction process of decision-making, execution and approval, which greatly reduces manual operations and time costs. At the same time, blockchain technology is being used to expand new power generation technologies and improve energy-saving technologies to connect and integrate different energy sources such as electricity, heat, water, and natural gas.

(3) Information security

In the traditional energy trading system, the trading center stores all the transaction information of the user. Due to the poor performance of the traditional trading system in defending against network attacks, the system is vulnerable to attacks, resulting in the loss or forgery of the data information in the system storage center. The center manages all the information of users. When conducting transactions, users do not know enough about the transaction information of the other party. It is difficult to ensure the accuracy of information by artificial management of information, and the interests of users are damaged from time to time. With the integration of blockchain and energy grid, each node has equal rights and stores the same information. Attacking network nodes will not cause damage to the overall performance of the system. At the same time, the blockchain uses encryption technology to encrypt data to prevent some human factors from attacking the data information in the system storage center.

3   Research on DGA in Energy and Power Blockchain System

3.1   Research Content

This paper mainly studies the influence of the introduction of distributed genetic algorithm in the EAP blockchain system on the system. This paper analyzes from two perspectives, one is that the transaction processing time can be optimized when the power data is stored, and the other is that the system is used to diagnose the fault safety of the power grid equipment.

3.2   Innovation Points

The traditional energy power system cannot guarantee the safe storage of power data, but after the introduction of DGA and blockchain technology, the data in the EAP system can be shared, and the power data information stored in the system can be made public and transparent. The DGA can also optimize the computing efficiency of the system during data transaction.

4   Application Analysis

4.1   Power Data Storage Application

As the country's basic energy industry, the power industry is the foundation for the sustainable development of the national economy. Power data storage and scheduling in the power industry is a top priority. The characteristics of blockchain can just solve the problem of safe storage of power data, design EAP blockchain system can effectively solve the problem of power data storage.

Table 1. System delay and throughput for 800 power data transactions

	Maximum delay(s)	Minimum delay(s)	Average delay(s)	Average throughput(tps)
Power data storage	4.93	1.21	2.85	314.36
Power data query	3.08	0.67	1.36	593.74
Electricity data transaction	6.77	2.15	3.92	185.26

Each request for power data will generate a transaction of power data in the blockchain network. Set the number of power data transactions to 800, and the EAP blockchain system stores, queries, and trades power data. The system delay and throughput (referring to the number of transactions that the blockchain system can process per second) are shown in Table 1. According to the data in the table, it can be found that the maximum delay, minimum delay, average delay and average throughput of power data transaction are significantly greater than the delay and throughput of power data storage and query. This is caused by the ordering of network nodes in the blockchain system.

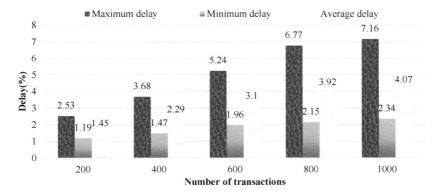

Fig. 1. Latency variation

As shown in Fig. 1, when the power data increases, the delay change of the blockchain system. It is easy to see from the figure that when the transaction volume of power data is 200, the maximum delay of the system is 2.53 s, the minimum delay is 1.19 s, and the average delay is 1.45 s. When the transaction volume of power data is 400, the maximum delay of the system is It is 3.68 s, the minimum delay is 1.47 s, and the average delay is 2.29 s. With the increase of transaction volume, the delay of a transaction will also increase. With the increase of transactions, after reaching the maximum processing speed, the blockchain system has no time to process subsequent requests, so the delay increases.

4.2   Application of Safety Diagnosis of Power Grid Equipment

The flow of equipment safety diagnosis scheme is shown in Fig. 2. Faulty devices are faulty nodes, and they can publish diagnostic request information in the consortium blockchain, which is visible to original supplier nodes and non-original supplier nodes in the chain. If the original supplier node and the non-original supplier node see the diagnosis request and respond to it, they are called diagnosis nodes. Depending on whether the faulty device is under warranty or not, this study is divided into two different responses. The faulty equipment is still in the warranty period, and the original supplier node will provide free diagnostic services after the fault occurs. At this time, the faulty device still needs to pay a certain amount of ether as a security deposit to the device safety diagnosis smart contract to prevent false requests from the faulty device. After the diagnosis, the deposit will be returned to the faulty node. When the faulty equipment exceeds the warranty period, the original supplier node and the non-original supplier node will compete fairly for the equipment, and whoever wins the bid will be diagnosed.

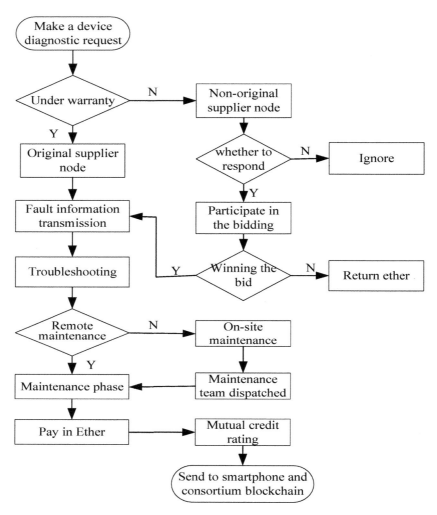

Fig. 2. The process of the energy and power blockchain system diagnosing the security of grid equipment

5   Conclusion

This paper analyzes the application of DGA in power blockchain system based on the characteristics of EAP industry. The analysis shows that the use of blockchain technology to build power system can solve the problems of power and energy transaction trust and data openness, and DGA can improve the overall computing power of the system optimizes the performance of the system to diagnose the safety of power grid equipment.

Acknowledgements. This work was supported by The research and application project of Southern Power Grid Materials Co., Ltd blockchain technology in the field of supply chain (ZBKJXM20200544).

References

1. Yan, M., Shahidehpour, M., Alabdulwahab, A., et al.: Blockchain for transacting energy and carbon allowance in networked microgrids. IEEE Trans. Smart Grid **12**, 4702–4714 (2021)
2. Shao, Y., Chen, B., Xiao, H., Qin, F.G.F.: Discussion on performance evaluation method of distributed combined cooling, heating, and power system. J. Therm. Sci. **28**(6), 1212–1220 (2019). https://doi.org/10.1007/s11630-019-1219-0
3. Kojonsaari, A.-R., Palm, J.: Distributed energy systems and energy communities under negotiation. Technol. Econ. Smart Grids Sustain. Energy **6**(1), 1–14 (2021). https://doi.org/10.1007/s40866-021-00116-9
4. Hussein, A.F., Arunkumar, N., Ramirez-Gonzalez, G., et al.: A medical records managing and securing blockchain based system supported by a genetic algorithm and discrete wavelet transform. Cogn. Syst. Res. **52**, 1–11 (2018). https://doi.org/10.1016/j.cogsys.2018.05.004
5. Wu, X.: Research on English online education platform based on genetic algorithm and blockchain technology. Wirel. Commun. Mob. Comput. **2020**(9), 1–7 (2020)
6. Xu, M., Feng, G., Ren, Y., et al.: On cloud storage optimization of blockchain with a clustering-based genetic algorithm. IEEE Internet of Things J. **7**, 8547–8558 (2020)
7. Chen, Q., Wang, K., Chen, S., et al.: Transactive energy system for distributed agents: architecture, mechanism design and key technologies. Dianli Xitong Zidonghua/Autom. Electr. Power Syst. **42**(3), 1–7 and 31 (2018)
8. Dmitrikov, V.F., Shushpanov, D.V., Petrochenko, A.Y., Alekseev, M.A.: Stability problems in designing aggregated and distributed systems of secondary power supplies. Russ. Electr. Eng. **91**(2), 108–114 (2020). https://doi.org/10.3103/S1068371220020030
9. Rahimi, F.A., Sasan, M.: Distribution management system for the grid of the future: a transactive system compensating for the rise in distributed energy resources. IEEE Electrification Mag. **6**(2), 84–94 (2018)
10. Hsu, P.C., Chen, W.J., Huang, B.J.: Distributed solar PV system for industrial application. J. Power Energy Eng. **11**(2), 78–84 (2017)
11. Javidtash, N., Jabbari, M., Niknam, T., et al.: A novel mixture of non-dominated sorting genetic algorithm and fuzzy method to multi-objective placement of distributed generations in Microgrids. J. Intell. Fuzzy Syst. **33**(3), 1–8 (2017)
12. Zhang, Y., Zhou, Y.: Distributed coordination control of traffic network flow using adaptive genetic algorithm based on cloud computing. J. Netw. Comput. Appl. **119**, 110–120 (2018). https://doi.org/10.1016/j.jnca.2018.07.001

Modular Information Fusion Model of Landscape Design Based on Genetic Algorithm

Fei Su[✉]

Nanchang Institute of Technology, Nanchang, Jiangxi, China
zhanglei81226@126.com

Abstract. With the acceleration of China's urbanization process, urban landscape design has also ushered in a new era, and it is an important link in the modular information management of landscape. This paper makes a comprehensive analysis and research based on genetic algorithm and ant colony theory. Firstly, it expounds the relevant contents of this subject and the development status at home and abroad. Then the concept, characteristics and functions of landscape design are introduced, and then the application of genetic algorithm is studied. Based on genetic algorithm, a modular information fusion model of landscape is designed, and the performance of the model is simulated. Finally, the experimental results show that the informatization degree of each module of the landscape design is relatively high, and the soft landscape and hard landscape are excellent in data fusion, data cleaning, data fusion and data synchronization. The full score of the experimental performance is 5, and the two main modules are more than 4, which shows that the performance of the landscape design module based on genetic algorithm is better, which can meet the needs.

Keywords: Genetic algorithm · Landscape architecture · Module design · Information fusion

1 Introduction

With the continuous progress of society, people's living standards have been greatly improved, and their material conditions and spiritual needs have been correspondingly improved [1, 2]. Landscape design is a very important part, which has a great impact on people's living environment and physical and mental health. Urban landscape design plays an important role in meeting residents' daily activities. Therefore, we must pay attention to this problem [3, 4].

Many scholars have conducted relevant research on landscape architecture. In foreign countries, the research on modular design of landscape started earlier. German scholars published a plan. If we regard all buildings as one, they will have the characteristics of non-interference, independence, unity and interconnection to a great extent [5, 6]. This is one of the methods that China first proposed to combine the functional space units in architectural design to meet the coordination requirements. Modular design method has been developed and widely used in foreign garden applications. After comparing and analyzing Chinese traditional gardens and foreign advanced urban greening concepts, Chinese scholars found that most areas in China did not fully realize that the relationship between urban green space system and plant configuration is mutual influence and supplement. This causes people to blindly pursue economic benefits and ignore ecological benefits. In addition, due to the low level of domestic landscape design and the lack of scientific and reasonable consideration in planning, there is a serious waste of green space [7, 8]. Therefore, based on genetic algorithm, this paper integrates landscape and information to optimize the landscape design model.

With the rapid development of society, people have higher and higher requirements for the quality of life. As a comprehensive system, urban landscape is an indispensable part of human survival and development. How to create more and greater value in a limited space and meet people's various needs has become a topic of concern for designers. Based on the modular information model of landscape architecture, this paper will study and propose a genetic algorithm to solve the above problems.

2   Discussion on Modular Information Fusion of Landscape Design Based on Genetic Algorithm

2.1   Landscape Design

1) Concept
   Urban landscape design refers to the creation of a garden space with certain characteristics, comfort and meeting people's spiritual needs through artistic means in a certain period of time in order to meet people's material and cultural needs. At the same time, it makes use of the biological and ecological characteristics of plants and according to people's requirements for environmental quality, through reasonable planning organization, arrangement, technology and economy to create an ideal ecological environment and carry out planning, construction and management activities to create a beautiful environment. With the continuous improvement of social and economic level and the accelerated development of urbanization, China's urban green space area is becoming larger and larger, and the per capita occupancy is also becoming higher and higher. At the same time, due to the increase of population, a large number of large-scale population gathered together in the process of urban development, resulting in increased traffic pressure and extended travel distance, people pay more and more attention to green land, which requires us to design a reasonable, scientific, functional and targeted landscape system to improve this situation and effectively improve its value. We should put people first in the process of urban construction. Because people are the most active social productive forces. Therefore, we should pay full attention to the harmonious development between man and nature. At the same time, we must consider the ecological effects and influencing factors produced by human activities. We should also pay attention to the artistic and practical value of landscape design to make it have certain ornamental value, aesthetic effect and economic benefits [9, 10].

2) Features
   (1) Diversity. With the continuous development of the city, people have higher and higher requirements for the living environment. Therefore, diversity should also be reflected in the construction of urban gardens. However, due to the large climate differences in different regions, there will be different types and styles, resulting in obvious distinctive characteristics among regions. For example, the South and North are warm temperate continental type, while the south is subtropical monsoon evergreen broad-leaved forest, coniferous forest and grassland landscape belt and tropical rain forest area.
   (2) Ornamental. From the perspective of urban development, human society is a complex system integrating politics, economy and culture. It not only refers to the material aspect, but also includes the spiritual aspect. In this process, there is an interdependence and mutual influence between man and nature, which exists in nature.
   (3) Social. It serves the people and the general public. Therefore, in the design, we must take into account the differentiated needs of different regions and people for the natural ecological environment, and pay attention to the requirements of the principle of ecological balance and the concept of sustainable development, so as to ensure that the whole ecological system can be fully utilized, And achieve the equilibrium between economic benefit and environmental benefit maximization [11, 12].
3) Function
   Landscape design is a cognitive activity of nature, which can also be said to be the product of human creation or expression in a certain time. As an art form, garden is obtained by perceiving all kinds of things in nature. In this case, we can better understand nature. These are products and material carriers that can be created only on the basis of the natural environment. Landscape architects plan and build according to different regions, climate, cultural background and other conditions. Through scientific and reasonable arrangement of plant landscaping, it provides a good and pleasant environment for the city. Landscape design can help people get more information in a limited time. For example, through the photosynthesis of plants and the inconsistent concentration of carbon dioxide in the air, urban air temperature and temperature can be controlled. Landscape planning is also an art. Designers divide different areas to form independent individuals and coordinate with each other to jointly complete the construction and development of the whole garden system. At the same time, they can also meet people's pursuit of beauty at a deeper level.

2.2   Genetic Algorithm

1) Basic idea
   In the process of designing landscape system, we must combine plant configuration with genetic algorithm. The theoretical basis of genetic algorithm is the theory of natural evolution. It looks for and uses information from the biological world to obtain the optimal solution. At present, it has been widely used in engineering

design, biology and other fields. Genetic algorithm is a random search method that simulates biological evolution and natural selection. It can optimize parameters in uncertain environment and obtain the global optimal solution. By comparing the fitness values of different individuals, we can judge whether each individual in the group conforms to the best combination model. If it is satisfied, the genes in the population will be replaced by a new population, and on the contrary, they will be re screened until they reach the most appropriate state. At the same time, genetic algorithm is a random search method based on the combination of natural selection and individual information processing ability. By simulating the biological evolution process in nature, we can process, learn and select information, and according to the relationship between these individuals. It can have a great impact on human beings in practical application.

2) Application

Genetic algorithm is a method to improve the degree of population evolution by selecting, crossing and mutation when the global search ability is poor. The technologies applied by genetic algorithm are as follows:

(1) Simple adaptation function

The construction of simple adaptation function is very simple. If f (x) is the objective function, the adaptation function can be taken as:

$$\text{fitnes}(x) = f(x) \tag{1}$$

At this time, the optimization target value is the maximum value.

$$\text{fitnes}(x) = M - f(x), M > \max_{X} f(x) \tag{2}$$

At this time, the optimization target value is the minimum. It can be seen that this construction method is directly related to the objective function and has the advantage of simple construction.

(2) Sorting adaptation function

The M chromosomes in the population are arranged in ascending order according to the objective function value, and their corresponding distribution probability is:

$$p(i) = \frac{2i}{m(m+1)}, 1 \leq i \leq m \tag{3}$$

In this way, the premature problem of some fitness functions is avoided, and the optimal individual of each generation is inherited with the greatest probability.

2.3   Relationship Between Landscape Design and Information Fusion

The modularization of landscape design is the rational distribution of plants, and in this process, there will be many uncertain factors. These problems may lead to great changes in the number and proportion of tree species planted in different regions, the same type or similar species at the same time. For this situation, we need to consider how to deal with it to adapt to environmental conditions and plant growth conditions, so as to design a garden landscape system that is more in line with people's needs and has a combination of diverse functional structure and landscape forms. In the process of landscape design, we often encounter many problems, but if we realize the information transmission and exchange between these different levels through integration. Then they can be effectively combined.

3 Experiment of Modular Information Fusion of Landscape Design Based on Genetic Algorithm

3.1 Landscape Design Module Based on Genetic Algorithm

In the planning and design of landscape, first of all, we should make a comprehensive analysis of the current situation of urban development and possible problems in the future. And the measures to be taken after these situations are scientifically and reasonably predicted. Then select the appropriate species and quantity according to different types of garden plants to meet the requirements of urban residents for green living environment. Finally, we should also consider that people's demand for ecological environment varies greatly in different seasons and climate conditions, and has certain periodic change characteristics. Therefore, we must fully consider and plan in combination with the actual situation. As can be seen from Fig. 1, there are two types of landscape design. One is hard landscape, which generally refers to artificial facilities, usually including pavement, sculpture, awning, seats, lights, peel boxes, etc. The second is soft landscape, which generally refers to the imitation of natural landscape, including artificial vegetation and rivers.

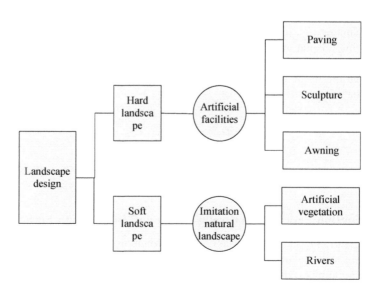

Fig. 1. Landscape design module

3.2   Operation and Development Environment of Landscape Design Module

During the operation of the modular information model of landscape design, it is necessary to divide each sub project area, and then integrate all biological resources in each area through certain methods. Genetic algorithm is used to realize some complex and important problems that are difficult to express intuitively and can not accurately describe their characteristics and functional requirements. Firstly, according to the landscape information, there is a certain degree of influence between plant growth, temperature and humidity and other environmental factors. Then, these factors are optimized with genetic algorithm to obtain the most appropriate combination scheme. Finally, the optimal scheme is selected according to the, which provides theoretical basis, practical reference data support and technical support for the garden design module, so as to achieve the work goal efficiently and finally achieve the expected functional requirements.

3.3   Information Fusion Simulation Test of Landscape Design Module Based on Genetic Algorithm

In the simulation design stage, the simulated plant model is established, and its characteristic parameters are set according to the actual situation. Firstly, the landscape information points should be determined. As landscape design is a complex system engineering project, which involves many aspects, and is highly interrelated, interactive and cross, we need to consider the coordination and integration relationship between different types and parts of the landscape in different regions. Therefore, we need to conduct a comprehensive analysis in the simulation stage, through comparison, the best combination scheme is found to achieve the best effect.

4   Experimental Analysis of Modular Information Fusion of Landscape Design Based on Genetic Algorithm

4.1   Modular Information Fusion Experiment of Landscape Design

Table 1 is the experimental data of modular information fusion of landscape design.

Table 1. Module informatization experimental data

Design module	Data migration	Data cleaning	Data fusion	Data synchronization
Hard landscape	5	4.5	4.7	4.9
Soft landscape	4.8	4.8	4.9	4.3

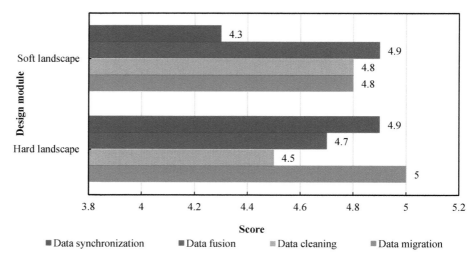

Fig. 2. Information comparison of the landscape design module

The experimental process of modular information fusion of landscape design is mainly to integrate different levels and types of data, and encode these data through genetic algorithm to get the corresponding results. In this system, the multi-objective optimization problem can be realized. As can be seen from Fig. 2, the informatization degree of each module of the landscape design is relatively high, and the soft landscape and hard landscape are excellent in data fusion, data cleaning, data fusion and data synchronization. The full score of the experimental performance is 5, and the two main modules are above 4, which shows that the landscape design module based on genetic algorithm has good performance, which can meet the needs.

5   Conclusion

With the continuous development of the city and the acceleration of urbanization, there are more and more related aspects of landscape design under the increasing material and cultural needs of people. As a landscape integrating nature, humanities and architecture, garden can create a more comfortable and pleasant space for human beings, and is highly comprehensive and has strong ornamental characteristics. Therefore, when carrying out the modular information fusion model of landscape planning, we need to consider its functional requirements, user habits and other factors, so as to realize the comprehensive utilization of these influencing factors, so that they can give full play to their respective advantages and values to the greatest extent.

References

1. Zhuang, J., Wang, Y., Zhang, S., et al.: A multi-antenna spectrum sensing scheme based on main information extraction and genetic algorithm clustering. IEEE Access **7**, 119620–119630 (2019)
2. Sui, S., Ma, H., Chang, H.W., et al.: Optimization design of metamaterial absorbers based on an improved adaptive genetic algorithm. Appl. Comput. Electromagn. Soc. J. **34**(8), 1198–1203 (2019)
3. Zhou, X., Jiang, P.: Variation source identification for deep hole boring process of cutting-hard workpiece based on multi-source information fusion using evidence theory. J. Intell. Manuf. **28**(2), 255–270 (2014). https://doi.org/10.1007/s10845-014-0975-7
4. Liu, W., Wang, L., Liao, C.: Estimation of Li-ion battery SOC based on model fusion and adaptive unscented Kalman filtering algorithm. Automot. Eng. **39**(9), 997–1003 (2017)
5. Arif, M., Wang, G.: Fast curvelet transform through genetic algorithm for multimodal medical image fusion. Soft. Comput. **24**(3), 1815–1836 (2019). https://doi.org/10.1007/s00500-019-04011-5
6. Darwish, S.M., Al-Khafaji, L.D.S.: Dual watermarking for color images: a new image copyright protection model based on the fusion of successive and segmented watermarking. Multimed. Tools Appl. **79**(9–10), 6503–6530 (2019). https://doi.org/10.1007/s11042-019-08290-w
7. Sulistyo, S.B., Woo, W.L., Dlay, S.S.: Regularized neural networks fusion and genetic algorithm based on-field nitrogen status estimation of wheat plants. IEEE Trans. Ind. Inf. **13**(1), 103–114 (2017)
8. Tavakkoli-Moghaddam, R., Safari, J., Sassani, F.: Reliability optimization of series-parallel systems with a choice of redundancy strategies using a genetic algorithm. Reliab. Eng. Syst. Saf. **93**(4), 550–556 (2017)
9. Dawid, H., Kopel, M.: on economic applications of the genetic algorithm: a model of the cobweb type*. J. Evol. Econ. **8**(3), 297–315 (2019)
10. Tsujimura, Y., Gen, M., Kubota, E.: Solving job-shop scheduling problem with fuzzy processing time using genetic algorithm. J. Jpn. Soc. Fuzzy Theory Syst. **7**(5), 1073–1083 (2017)
11. Qiang, L., Wu, C.: A hybrid method combining genetic algorithm and Hooke-Jeeves method for constrained global optimization. J. Ind. Manag. Optim. **10**(4), 1279–1296 (2017)
12. Arabasadi, Z., Alizadehsani, R., Roshanzamir, M., Moosaei, H., Yarifard, A.A.: Computer aided decision making for heart disease detection using hybrid neural network-Genetic algorithm. Comput. Methods Programs Biomed. **141**, 19–26 (2017)

Urban Intelligent Transportation Solution Based on Road Monitoring System

Haodong Fan[(✉)] and Y. I. Baldric

Shenzhen Urban Transport Planning Center Co., Ltd.,
Shenzhen 518000, Guangdong, China
93574881@qq.com

Abstract. With the development of cities, improving the efficiency of urban management is the inevitable trend of urban management. When a city's management development planning level is good, the city's economy is bound to develop rapidly. However, in today's social environment, the development of urbanization has encountered new problems and contradictions. Intelligent transportation system is an advanced system designed to provide a higher quality and efficient traffic management mode. It is committed to using more scientific algorithms and powerful opencv library to realize more accurate real-time detection of traffic flow passing through the road, so as to grasp the urban traffic conditions in real time and help the construction of smart city.

Keywords: Vehicle identification · Video processing · Intelligent transportation

1 Introduction

Intelligent video analysis technology is a new technology to solve the above traffic problems. It uses machine vision to monitor and judge the video picture, analyze the data, implant the refined feature formation algorithm into the machine, form a "machine brain" to automatically detect and analyze the video picture, and make corresponding actions, so as to make the camera become not only human eyes, but also human brain, saving human and material resources to a great extent, Effectively analyze road information and manage road environment.

"Digital city management" is a new urban management mode that integrates information technology, mobile communication technology and grid technology into specific urban management. "Digital city management" will improve the efficiency and scientificity of urban management by improving urban management space and rules, accurately finding the objects to be managed and collecting information. The establishment of a set of urban management video acquisition and monitoring system can realize accurate and real-time monitoring of the regulatory area and prevent the occurrence of monitoring blind spots. It can refine the management mode, build a sound and efficient management mechanism and improve the service level of the government [1].

Compared with the traditional analog monitoring method, when the digital video monitoring system is erected, the whole architecture is similar in the front end. The difference is that this digital video monitoring method can use coaxial cable to transmit the front-end signal, and the farthest transmission distance can reach 400 m. Optical

© The Author(s), under exclusive license to Springer Nature Switzerland AG 2022
V. Sugumaran et al. (Eds.): ICMMIA 2022, LNDECT 138, pp. 209–216, 2022.
https://doi.org/10.1007/978-3-031-05484-6_26

cable is used to connect the monitoring point with the video acquisition center, and then DVR digital hard disk video recorder is used for centralized video monitoring and management. In addition, the video receiving terminal can also upload the monitoring video to the video acquisition center through the network, which greatly simplifies the cumbersome erection of the whole monitoring system. As shown in Fig. 1 below, it is uploaded to the video acquisition center.

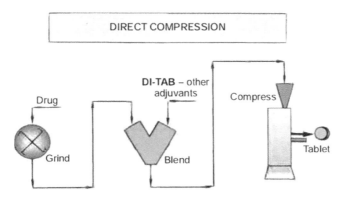

Fig. 1. Upload to video capture center

2   Related Work

2.1   New Problems Faced by Urban Traffic and Cause Analysis

With the development of economy, the degree and trend of urbanization are becoming more and more obvious in countries with rapid development. Urban traffic problems have become the main factors affecting socio-economic development and urbanization process. In recent years, China's civil economy has continued to grow at a high rate, the degree of social economy and urbanization has increased rapidly, and the demand for urban traffic has increased. However, China's cities, especially mega cities and large cities, generally have traffic problems, mainly reflected in traffic congestion, traffic accidents and environmental pollution.

Traffic congestion, as the bottleneck of traffic development, not only affects the normal economic and living order of urban society, reduces social production efficiency, but also makes the environmental quality deteriorate continuously. Most large and medium-sized cities in China have different degrees of traffic congestion. In 667 cities across the country, about 2/3 of the main roads are slow and even congested during urban traffic peak hours. In some large and medium-sized cities, the traffic environment is fragile, the traffic congestion is serious, the traffic flow on the main and secondary trunk roads is slow, and the road traffic efficiency is reduced, which often leads to long-term and large-area traffic congestion. Especially in mega cities, the traffic flow on the main and secondary trunk roads has reached saturation or over saturation during the traffic peak, and the travel time of residents and urban traffic cost have been seriously affected [2].

2.2   Research on Urban Intelligent Traffic Monitoring System

The application of intelligent traffic monitoring system in urban traffic to comprehensively monitor and control urban traffic plays a very important role in the normal operation of Expressway and its benefits [3]. Therefore, the main purpose of the design of urban traffic monitoring system is to produce a control scheme through the traffic flow detection, traffic condition monitoring, environmental meteorological detection and operation condition monitoring of the whole urban traffic line, so as to control the traffic flow, improve the traffic environment and reduce accidents, so as to make the urban traffic reach a higher service level. Comprehensive monitoring of urban traffic is carried out as shown in Fig. 2 below.

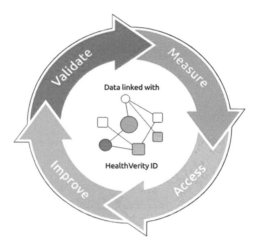

Fig. 2. Comprehensive monitoring of urban traffic

Through the urban traffic monitoring system, we can know the traffic operation status in advance, find problems as soon as possible, solve problems as soon as possible, avoid traffic jams, ensure traffic safety, and maintain the road operation at a specific service level. Through the monitoring and control of urban traffic to manage it, especially in ensuring traffic safety and smooth roads, maintaining a high service level plays a very important role [4].

The location of integrated supervisory and control system (ISCS) is similar to ATMs (advanced traffic management system) in its, but it is very different from ATMs in specific content. ISCS applies computer technology, network technology, system engineering theory and methods to varying degrees according to its own technical characteristics, so as to realize the operation and monitoring automation of urban road traffic, organically integrate the components of urban road traffic system, and make it present the overall functions that the components do not have. Its technical framework is shown in Fig. 3.

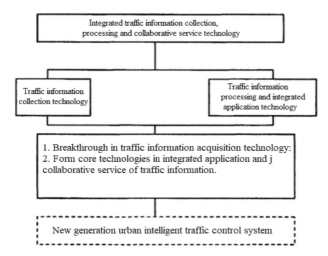

Fig. 3. Technical block diagram of intelligent integrated monitoring system

3   Research on Architecture of Integrated Intelligent Monitoring System

3.1   ISCS Development Based on Information Technology

Information technology, including computer technology, network technology and software engineering technology, is the technical core to realize ISCS [5]. It realizes the operation and monitoring automation of urban road traffic based on information technology, and organically integrates all components of urban road traffic system, so that it presents the overall function that each component element does not have. The core role of information technology is reflected in:

(1)   Through information technology, the individual decentralized traffic activities are guided and integrated to help individuals fully understand the relevant macro state.

(2)   Through information technology to enhance the management level, timely collection, transmission and analysis of information, managers can make scientific decisions according to the actual situation, and use the improvement of management level to improve the operation efficiency of the system [6].

(3)   Traffic information is closer to and integrated into residents' life, which will be conducive to the efficient utilization of various social resources [7].

3.2   Hierarchical Design of Intelligent Monitoring Platform Architecture

The basic technology used in determining the system framework scheme is object-oriented system structure analysis technology. The so-called object is the basic element of the system, and its characteristics include the information stored or generated therein and the basic operation of information processing under the determined boundary conditions. There are logical relations and information exchange relations between objects. Using the form of network diagram to describe the system structure on the basis of object concept will help us to establish a common language about system structure among many participants [8].

ISCS adopts hierarchical structure design, such as two-level structure, three-level structure and four-level structure. Two level structure: central level and intersection site level. Taking the intersection as the basic component unit, the traffic information of each terminal is uploaded to the database of the monitoring center through the wireless network and other network platforms. After centralized processing, the traffic control information is sent to each intersection level through the network platform, and the traffic flow is regulated through the signal lamp, information release system, etc. [9].

The disadvantages of the two-level division structure are mainly reflected in that the level is too fuzzy, the star network structure is simple, resulting in data link congestion, it is not easy to deal with complex traffic information in time, and it is difficult to achieve a high level of intelligence. Therefore, on this basis, engineering applications mostly adopt three or more levels of ISCS level division, introduce regional level, and adopt regional monitoring for medium and large urban traffic or large regional traffic [10]. Regional monitoring itself is equivalent to a secondary monitoring center, which has a certain independence of traffic information processing and is subordinate to ISCS monitoring center, so as to better straighten out the structure of traffic information communication, processing and control.

4 Technology and Its Realization

4.1 Extraction of Video Files

Video preprocessing uses the equal interval method to capture the video and make it a still image. The frame rate is 24 frames per second, which is the same as the original video and will not cause the loss of information.

4.2 Grayscale Processing

There are more than 16 million (255) pixels in a color image × two hundred and fifty-five × 255), while there are only 255 values for a gray image [11]. The time and space cost of processing color video is much higher than that of gray image. Gray image pixels are enough for vehicle recognition. We convert the monitoring video to gray level, using the following formula:

$$Cray = 0.299R + 0.587G + 0.114B \tag{1}$$

4.3 Extract Sample Features Using Haar Algorithm

Computer can simulate or realize human behavior through learning and induction, and replace people with efficient labor. A fully trained machine can recognize the traffic flow in the surveillance video. The training process is to select some features from Haar features and train AdaBoost classifier [12].

Haar like features, namely Haar features, is a common feature operator in the field of computer vision. It was first used for face description and is usually divided into four categories: linear feature, edge feature, point feature (i.e. center feature) and diagonal feature (see Fig. 4).

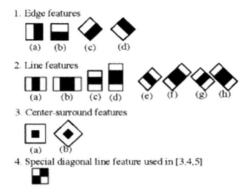

Fig. 4. Schematic diagram of four types of Haar features

(1) Calculation of rectangular features. In M × In a sub window of M size (see Fig. 5), you can calculate how many rectangular features exist in it. In pixel resolution M × M detector, for example, in M × In the sub window of M, this rectangle can be determined as long as the upper left vertex a (x1, Y1) and the lower right vertex B (X2, Y2) are determined [13].

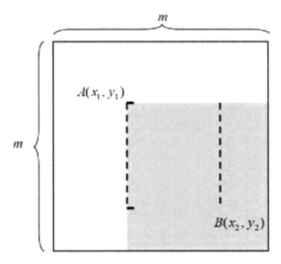

Fig. 5. Schematic diagram of rectangular feature calculation

After the classifier training of Haar feature is completed, you can save the process of establishing a join table by loading this file. With the cascade table, we can only get the collection of vehicles by passing the detected pictures and cascading tables to simultaneous interpreting of OpenCV's target detection algorithm [14].

5   Conclusion

Aiming at the multi-source heterogeneity and complexity of traffic information data, the key technologies of traffic information fusion and acquisition are studied. Complete the design of traffic information characteristics and data structure with good expansibility and compatibility, and study the application of modern signal processing technology in traffic information acquisition. Combined with the measured data, the signal detection method based on combined model method, wavelet decomposition and neural network is simulated and compared. Urban traffic problem is a worldwide problem with high complexity. The main research content of this paper, that is, the key technology of urban road traffic information acquisition, as the basis of intelligent transportation system, can promote the mitigation of urban traffic problems.

References

1. Nambajemariya, F., Wang, Y.: Excavation of the internet of things in urban areas based on an intelligent transportation management system. Adv. Internet of Things **11**(3), 10 (2021)
2. Shahgholi, T., Sheikhahmadi, A., Khamforoosh, K., et al.: LPWAN-based hybrid backhaul communication for intelligent transportation systems: architecture and performance evaluation. EURASIP J. Wirel. Commun. Netw. **2021**(1) (2021)
3. Dobrilovi, D., Brtka, V., Jotanovi, G., et al.: The urban traffic noise monitoring system based on LoRaWAN technology. Wirel. Netw. **28**, 1–18 (2021)
4. Meng, H.: Intelligent planning and research on urban traffic congestion. Future Internet **13**, 284 (2021)
5. Peyman, M., Copado, P.J., Tordecilla, R.D., et al.: Edge computing and IoT analytics for agile optimization in intelligent transportation systems. Energies **14**, 6309 (2021)
6. Zheng, K.: Design of intelligent transformer monitoring system based on GSM (2022)
7. Rezaei, M., Azarmi, M., Mir, F.: Traffic-net: 3D traffic monitoring using a single camera (2021)
8. Jain, S., Jain, S.S.: Development of intelligent transportation system and its applications for an urban corridor during COVID-19. J. Inst. Eng. (India) Ser. B **102**(6), 1191–1200 (2021)
9. Schulz, L.T., Dilworth, T.J., Rose, W.E.: Pragmatic application of AUC-based monitoring recommendations from the 2020 vancomycin consensus guidelines. Am. J. Health-Syst. Pharm. **78**, 1363–1364 (2021)
10. Sharma, P., Singh, A., Singh, K.K., et al.: Vehicle identification using modified region based convolution network for intelligent transportation system. Multimed. Tools Appl. (2021)
11. Jiang, J., Yang, Y., Li, Y., et al.: Lane-level vehicle counting based on V2X and centimeter-level positioning at urban intersections. Int. J. Intell. Transp. Syst. Res. **20**, 11–28 (2021)

12. Liu, Q.: Intelligent water quality monitoring system based on multi-sensor data fusion technology. Int. J. Ambient Comput. Intell. (IJACI) **12**, 43–63 (2021)
13. Dong, J., Meng, W., Liu, Y., et al.: A framework of pavement management system based on IoT and big data. Adv. Eng. Inform. **47**(2), 101226 (2021)
14. Lee, J., Yoon, Y.: Indicators development to support intelligent road infrastructure in urban cities. Transp. Policy **114**, 252–265 (2021)

Real Time Monitoring Method
of Comprehensive Energy Consumption
Based on Data Mining Algorithm

Gang Qian[1(✉)], Chenghong Tang[2,3,4], Yao Meng[5], Xiaomin Qi[5],
Jingtao Wang[6], and Jing Zhou[6]

[1] State Grid Shaoxing Power Supply Company,
Shaoxing 312099, Zhejiang, China
Jiangshangyou@sina.com
[2] State Grid Electric Power Research Institute, Nanjing 211106, China
[3] Nari Technology Development Limited Company, Nanjing 211106, China
[4] State Key Laboratory of Smart Grid Protection and Operation Control,
Nanjing 211106, China
[5] China Electric Power Planning and Engineering Institute,
Beijing 100120, China
[6] Beijing Nari Digital Technology Co., Ltd., Beijing 100085, China

Abstract. With the expansion of data centers, total consumption is growing
exponentially. The center is faced with problems such as reducing power consumption, reducing costs and improving resource utilization. Reducing energy
costs through monitoring is an important means of energy conservation in data
centers. In recent years, more and more real-time energy consumption data are
stored on the energy consumption management platform. There is a wealth of
knowledge hidden in the energy consumption data in recent years. A large
amount of data simplifies the traditional methods, and data mining methods have
achieved good results in many fields. Therefore, it is of great significance to
study a new energy consumption monitoring system and introduce data mining
method into the system.

Keywords: Energy consumption · Data mining · Real time monitoring

1 Introduction

Automatic monitoring of energy consumption and timely handling of abnormal energy
consumption by using control means is an important scheme for implementing energy
saving in data center. The energy consumption data monitoring platform is generally
equipped with a separate alarm module to feed back the abnormal energy consumption.
The current alarm management relies on setting threshold parameters manually to
determine the alarm level and alarm conditions [1]. However, the setting of threshold
basically depends on human subjective judgment, which is lack of comprehensive and
more accurate data support. On the other hand, this method can only view the abnormal
situation of historical energy consumption through data storage. For managers, it is a

V. Sugumaran et al. (Eds.): ICMMIA 2022, LNDECT 138, pp. 217–224, 2022.
https://doi.org/10.1007/978-3-031-05484-6_27

more concerned problem to find and deal with the abnormal energy consumption during operation in time [2].

These energy consumption data usually hide rich information and knowledge, but they also have high-dimensional and massive characteristics. For human limited processing, analysis and understanding ability, it is difficult for ordinary people to analyze and study the laws with conventional methods, the proposal of data mining technology provides us with a new perspective to explore big data. Data mining technology is a comprehensive discipline. It jumps out of complex formula calculation, so that people without professional knowledge can process and analyze massive data [3]. It has been applied in all walks of life and achieved good results. Therefore, it is of great practical significance to discuss the application scenario of data mining technology in energy consumption data monitoring system and study how to apply it to the system.

2   Data Mining

Data mining is an interdisciplinary subject, involving many fields such as database, informatics, statistics, machine learning, pattern recognition, artificial intelligence, data visualization and so on.

The contradiction between the explosive growth of data and the defect of human understanding ability promotes the emergence of data mining. Compared with expert system, data mining can not only learn the existing knowledge, but also find the unknown and hidden knowledge [4]. With the help of data mining technology, people can find potential patterns or knowledge information from massive data, so as to provide help for scientific research, commercial activities and other industries. In addition, data mining knowledge needs to be expressed intuitively and easily with the help of visualization technology. In short, data mining is a complex process, which needs repeated iteration, gradual implementation and human-computer interaction [5].

The data mining process consists of the following basic steps:

(1) Data collection: collect data from various sources and store them on suitable storage media.
(2) Data integration: unified storage of data from multiple sources for unified processing of data.
(3) Data preprocessing: various data storage media store data differently. Therefore, data preprocessing is required before data mining to improve the quality of data. There are many data preprocessing technologies. For example, data integration can unify multiple data sources; Data cleaning can filter out irregular data; Further processing of data by data transformation and data protocol; Sampling can obtain a subset of the data set.
(4) Data mining process: basic and core steps. Select appropriate algorithms to analyze data, such as clustering algorithm, classification algorithm, social network analysis, association rule algorithm, etc. The performance of the algorithm can be evaluated from the aspects of robustness, accuracy and understandability.
(5) Pattern evaluation: set evaluation criteria, evaluate the results of data mining process, and filter useful information [6].

(6) Knowledge representation: display the mining results visually in an intuitive and understandable way, or collect them as new knowledge into the knowledge base for future use.

The data mining process is a cyclic iterative process. After using the data mining algorithm to process the data, if the desired effect cannot be obtained, the input can be adjusted and the data can be mined again. Figure 1 shows the processing model of data mining.

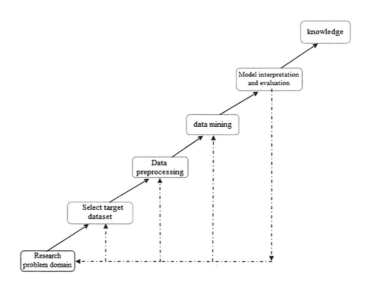

Fig. 1. General process of data mining

3   Research on Real-Time Detection Method of Abnormal Energy Consumption

3.1   Landscape Analysis

In the context of energy conservation and emission reduction, my laboratory has established a special fund project for the Internet of things in cooperation with a provincial company of China Mobile. The project relies on the dynamic environment of the data center to build an energy consumption analysis platform. I participated in the development of the platform. The energy consumption analysis platform integrates a variety of data mining algorithms, but no actual application scenario is found. The data mining module exists without the platform.

In addition, the platform also has a special alarm management module, as shown in Fig. 2. It is used to inform managers of abnormal energy consumption information. The premise of current alarm module operation is to manually set the energy consumption threshold at all levels [7]. When the operator clicks to view the historical alarm information, the selected data will be compared with the saved threshold, and the alarm information will be determined according to the difference between the actual energy consumption data and the threshold [8]. This traditional method has several problems. Firstly, the alarm only considers the historical data and does not realize the real-time data alarm; Secondly, the threshold is mainly obtained through manual observation or statistics, which is lack of accurate data support. Too high or too low threshold at all levels will affect the actual detection results; Third, this static detection method does not consider the changes of season and time. With the change of time, the energy consumption mode is also changing, but the threshold cannot change dynamically.

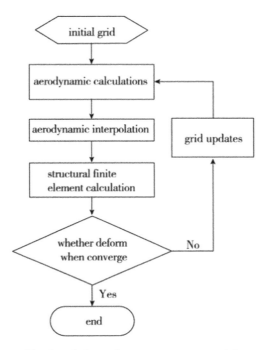

Fig. 2. Platform alarm management module

In the actual scene, people are more concerned about detecting the abnormal situation of energy consumption in time and automatically. In fact, the underlying dynamic environment continues to provide new energy consumption data for the energy consumption analysis platform. At present, the energy consumption analysis platform only carries out static analysis on historical data, and lacks real-time display of energy consumption data and real-time alarm function of abnormal energy consumption [9].

On the other hand, at present, the research of data mining in the field of energy consumption mainly focuses on energy consumption analysis and prediction, while there is relatively little research in energy consumption anomaly detection. The current situation of energy consumption analysis platform provides a good scenario for the application of data mining in energy consumption anomaly detection.

Based on the above analysis, this chapter proposes a real-time detection method of abnormal energy consumption. This method takes the energy consumption of the data center as the research object, analyzes the real-time data based on the massive historical energy consumption data, combines the energy consumption characteristics of the data center, and automatically identifies the abnormal energy consumption with the help of data mining technology.

3.2   Method Description

The abnormal attribute of real-time data is usually local. If it is placed in the global data for fuzzy processing, the result must have a great deviation. In addition, putting a piece of real-time data in a large amount of historical data is inefficient. For the computer room data center, the energy consuming equipment usually shows a stable and consistent energy consumption state in the same season, the same time period or other possible conditions. Whether the energy consumption is abnormal should also be discussed in this scope. Therefore, the analysis of energy consumption data under the same operation mode means the improvement of analysis efficiency and accuracy [10].

The process of identifying the energy consumption pattern of real-time energy consumption corresponds to the process of classification in data mining technology. There are many bases for classifying energy consumption categories. Energy consumption categories can be automatically identified by clustering algorithm, which is more common. However, because the energy consumption data is generally time series data with some regularity, the accuracy of using general clustering algorithm for time series data is not very high. Since the energy consumption data of the data center shows a certain law according to the time change, it is easy to obtain a certain energy consumption mode through statistical analysis.

4   Real Time Monitoring of Comprehensive Energy Consumption Based on Data Mining Algorithm

4.1   System Requirements Analysis

The original intention of the energy consumption analysis platform of the project I participated in, the Internet of things special fund project, is to analyze energy consumption data. The old architecture is no longer suitable for new business needs such as real-time data display and abnormal real-time alarm. In terms of monitoring energy consumption, new energy consumption analysis platform subsystem & amp; A real-time monitoring system for energy consumption data is established.

Based on the research background and theoretical analysis of the above chapter, the goal of realizing the real-time monitoring system of power consumption data is proposed. Combined with frequent data speculation and data IO working scenarios, the design scheme of power consumption data real-time monitoring system is proposed. New structures to meet business needs are being studied. On this basis, a perfect business module is designed to provide complete system functions. The anomaly detection method proposed in this paper is integrated with the alarm management module of the system. Intuitively display the energy consumption and data mining results. The system has friendly interface and good performance.The system can monitor and manage the energy consumption status (such as lighting, air conditioning, battery, etc.) of data centers such as computer room and base station, display the collected energy consumption information in real time in combination with the collection of existing energy consumption data, and realize energy consumption alarm in combination with data mining methods, so as to provide guidance and help for data center managers.

4.2   System Architecture Research

Combined with the realization goal of energy consumption data real-time monitoring system and the demand analysis of system function, this paper will study and design the system architecture in detail. Figure 3 shows the overall architecture of the system.

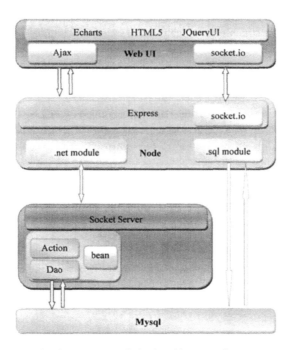

Fig. 3. System technical architecture diagram

As shown in the figure, the overall design of the system is divided into three levels, including client layer, server layer and data layer. Different from the traditional B/S architecture, the server layer is subdivided into middle layer and business server layer.

(1) Data layer. The data layer undertakes the tasks of data acquisition and data storage, and is the basic level of the whole energy consumption monitoring system. The bottom layer of the system relies on the dynamic environment of the data center, and provides hardware conditions for energy consumption data acquisition and timely storage through special acquisition equipment and transmission lines established by the Internet.

(2) Customer level. Because the system adopts B/S architecture, any device installed with browser can be used as a client, so that managers can view the energy consumption status anytime and anywhere. Due to the adoption of the web application mode, there is no need to install the client on the user's machine. As the client, the browser carries all the business logic of the original client.

(3) Middle tier server. One difference from the traditional B/S architecture is that a middle tier server is separated from the whole architecture. It does not have business processing capacity and only undertakes the responsibility of message distribution. In the traditional system, the business server usually needs to process both business and message forwarding. In addition to network IO, it also needs to process data IO, which greatly wastes the performance of the business server and is easy to cause a performance bottleneck. At the same time, a server with multiple functions is not suitable for development and maintenance.

5   Conclusion

With the increase of the size of the data center, the power consumption index also increases. The task of data center is to reduce power consumption, cut costs and improve resource utilization. Therefore, the construction of environmental protection and energy saving data center has attracted the attention of data center managers and data center manufacturers. Reducing startup energy consumption through monitoring is an important means of data center energy conservation. On the other hand, data mining methods have achieved good results in many fields. How to apply data mining method to power consumption data monitoring system has very important research significance.

Acknowledgements. Science and technology project of State Grid Corporation of China (Research on key technologies and modes of efficient operation of integrated energy system considering P2X and industry coupling).

References

1. Lines, A.M., Hall, G.B., Asmussen, S.E., et al.: Sensor fusion: comprehensive real-time, on-line monitoring for process control via visible, NIR, and Raman spectroscopy. ACS Sens. **5**, 2467–2475 (2020)
2. Zhao, Q., Gao, W., Gao, C., et al.: Comprehensive outage compensation of real-time orbit and clock corrections with broadcast ephemeris for ambiguity-fixed precise point positioning. Adv. Space Res. **67**(3), 1124–1142 (2020)

3. Lee, K., Jeong, H., Kim, S., et al.: Real-time seizure detection using EEG: a comprehensive comparison of recent approaches under a realistic setting (2022)
4. Ma, G.C., Delgado, M.G., Ramos, J.S., et al.: Mitigating damage on heritage structures by continuous conservation using thermal real-time monitoring. Case study of Ziri Wall, city of Granada, Spain. J. Cleaner Prod. **296**, 126522 (2021)
5. Lubken, R.M., Jong, A.M.D., Prins, M.: Real-time monitoring of biomolecules: dynamic response limits of affinity-based sensors (2022)
6. Ju, W., Dobson, I., Martin, K., et al.: Real-time monitoring of area angles with synchrophasor measurements (2020)
7. Jaroszewicz, M.J., Liu, M., Kim, J., et al.: Time- and site-resolved kinetic NMR: real-time monitoring of off-equilibrium chemical dynamics by 2D spectrotemporal correlations (2021)
8. Fu, C., Lv, Q., Badrnejad, R.G.: Fog computing in health management processing systems. Kybernetes (2020). ahead-of-print(ahead-of-print)
9. Wagner, W., Akowska, A., Aladi, C., et al.: Pilot investigation for a comprehensive taxonomy of autonomous entities (2021)
10. Park, J.H., Lee, B.: Holographic techniques for augmented reality and virtual reality near-eye displays. Light Adv. Manuf. **3**(1), 1–14 (2022)

Application Research on the Integration of Civil Engineering and Computer Aided Building System Under the Background of BIM Development

Chen Yang[1(✉)], Qi-miao Xie[1], and Yao Tong[2]

[1] Jinjiang College of Sichuan University, Meishan 620860, Sichuan, China
yang7670510@163.com
[2] Sichuan Tourism University, Chengdu 610100, Sichuan, China

Abstract. In the environment of population growth and energy crisis, the development of construction industry is facing severe challenges, with the continuous integration and penetration of Internet technology, information technology and computer technology in the field of construction engineering, BIM based construction engineering computer-aided technology came into being. This technology plays a great role in construction engineering design, construction, management and other aspects, and is of great significance to the development of modern construction industry. This paper mainly expounds the working principle of BIM, the advantages of computer-aided technology based on BIM and its specific application in construction engineering, in order to promote the further development, popularization and application of BIM.

Keywords: BIM Technology · Computer-aided · Architectural engineering

1 Introduction

From the development of the construction industry, the traditional architectural concepts and construction methods can not meet the needs of social development. From 2008 to 2020, China's construction industry has changed from initial high-speed development to medium-speed development (see Fig. 1). The slowdown in development has forced the construction industry to carry out corresponding industrial upgrading and technological transformation. In addition, the information requirements of construction industrialization also promote the integration of construction technology [1]. With the rapid development of information technology, many industries have entered the stage of accelerated development, and construction enterprises have higher and higher requirements for information technology, people have more and more strict requirements for the living environment, hoping to create a good living environment through green buildings [2]. Therefore, it is necessary to strengthen the integration of construction technology.

V. Sugumaran et al. (Eds.): ICMMIA 2022, LNDECT 138, pp. 225–232, 2022.
https://doi.org/10.1007/978-3-031-05484-6_28

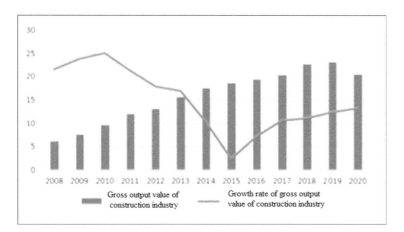

Fig. 1. Total output value and growth rate of national construction industry from 2008 to 2020

2   BIM Technology Advantage Analysis

2.1   Comprehensive Information Integration

Since ancient times, the huge amount of data, various drawings and different archives of the construction industry have made all participants miserable. As the soul of architecture, drawing is the decisive factor for the success or failure of a construction project. However, in the past, the revision of drawings in the construction industry occurred from time to time, and it affected the whole body. There is often one change that needs to be corrected and tested by various participants to ensure accuracy. It is time-consuming and laborious, and it is easy to miss key points, resulting in huge construction hidden dangers. But the emergence of BIM Technology solves this problem well [3]. All participants can summarize all building information in the same building model, and quickly correct the error points by using the powerful functions of BIM software, such as design, collision detection, global preview and so on. The BIM software changes the design function, and the relevant information can be changed automatically after the information is changed. Greatly save human and material resources, and effectively reduce the possibility of errors and omissions [4]. As shown in Fig. 2.

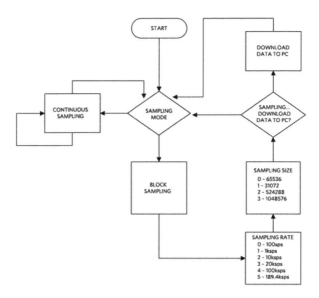

Fig. 2. Architectural design process based on BIM software

2.2   Integration of Participants

The whole life cycle theory has a long history. The five parts of preliminary planning, scheme design, construction, operation and maintenance, demolition and blasting constitute the whole process of a construction project from life to death. In the past, a construction project divided the work tasks of each stage very clearly, and there was little communication between the designer, the constructor, the owner and the supplier. Even, there is a lack of necessary communication between architectural designers and construction engineers [5]. Only when problems are found in construction will they negotiate and talk with them. The same is often true of the ubiquitous change design and poor communication in construction engineering. The birth of BIM Technology has successfully eliminated this problem. All participants of construction projects can communicate on the same BIM platform, with unified model and strong support of Internet technology, making the communication between all parties more smooth and correct. In this way, all participants will gradually integrate into one, work as a team and communicate well, which is more conducive to the rapid and effective implementation of construction projects [6]. Furthermore, the bean curd residue project can also be greatly reduced.

2.3   Real Time Information Update

The production process of construction engineering has always been complex and cumbersome. In the past, when the construction engineer found that the design drawing was wrong in the construction stage, he had to first apply to the owner for design drawing change, and then communicate with the architectural designer. After both parties make changes to the design, they need to report to the owner and submit the design change settlement. This series of complex processes is no wonder for construction practitioners. This is true for a change, not to mention the changes in a construction project can be seen everywhere. Among them, the problem of information retention has a deep impact [7]. The construction information platform constructed by BIM Technology solves this problem well. Based on the BIM Technology platform, all participants can quickly and effectively connect information according to the Internet. While communicating effectively, information is updated in real time to quickly solve various problems. Imagine the impact on the reduction of construction cycle if everyone in the construction industry is familiar with BIM Technology.

3   Application of Computer Aided Technology in Construction Engineering Based on BIM

3.1   Construction Site Analysis and Overall Building Planning

In the analysis of construction site, in order to ensure the comprehensive and reliable acquisition of building information, people often use BIM computer-aided technology in combination with a variety of other advanced measurement technologies, including geographic information system, remote sensing technology, GPS technology, three-dimensional laser scanning technology, etc. These advanced surveying and mapping technologies are used to measure, acquire and store the landform, surrounding building layout, spatial location and other information data of the construction site, and save them in the BIM database [8]. The software is used for analysis and calculation, and the construction project decision-making is scientifically evaluated according to the data and calculation results, so as to make reasonable planning for the construction site, Rational distribution of building layout. According to the planning standards and principles of complex space, according to the obtained spatial information and geographic information, use geographic information system to scientifically plan the building spatial structure, reasonably layout the control points, improve the efficiency of building analysis and planning, and provide data support and reference for project decision-making.

3.2   Construction Project Scheme Inspection and Performance Analysis

Before the formal construction of the construction project, the feasibility, scientificity, operability and economy of the construction scheme need to be tested to ensure that the construction scheme is feasible and meets the requirements of scientificity and economy. Computer aided technology based on BIM plays a great role in the inspection of construction engineering scheme. Through the BIM software to evaluate the designed construction scheme, including building space design, structural design and drawing design, we can accurately judge whether the scheme is feasible, whether the design of each part is scientific and reasonable, and whether the whole building structure is coordinated with each other. At the same time, on the BIM platform, all stakeholders of construction engineering can also pay common attention to key issues in construction engineering design and construction, and reach consensus in time, so as to effectively shorten the project decision-making time. In addition to the building scheme, the building performance analysis is also a key issue to be considered in the project construction. Only when the building performance meets the relevant standards, can it lay a solid foundation for the construction quality of the whole building project [9]. Therefore, BIM computer-aided technology is also widely used in building performance analysis. By importing the building model created in the design stage into the engineering analysis software, it can realize the objective analysis of various building performance, so as to effectively improve and improve the building performance.

3.3   3D Visual Design and Coordinated Design of Building Model

Computer aided technology based on BIM is widely used in the construction of three-dimensional visual model of architectural engineering, and gradually occupies a dominant position in the establishment of modern architectural model. Using this technology to build a three-dimensional building model and virtual visualization model can not only show the whole building structure intuitively and clearly, but also the effect of model design is very close to the real building effect. Through the simulation of fire fighting, escape and other situations by BIM Technology, we can also find the irrationality in architectural design in time, Uncoordinated parts of the building structure are found and automatically adjusted to promote further coordination among various parts and improve the overall coordination of the building, as shown in Fig. 3.

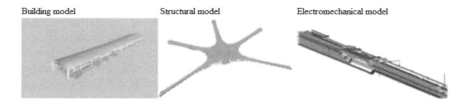

Fig. 3. Building BIM model

4   Countermeasures for the Integrated Development of Civil Engineering and Computer-Aided Building System

4.1   Improve Laws and Regulations and Industry Standards

Through the above analysis, it can be seen that the imperfections of laws, regulations and industry standards seriously restrict the integrated development of dress matching buildings and green buildings. Therefore, in order to achieve good development, relevant departments need to improve the current laws and regulations. In terms of specific operation, relevant state departments need to give policy support, such as tax preference, one-time technological transformation subsidy, etc. This can effectively benefit construction enterprises, encourage enterprises to strive to improve the level of production technology, and realize the integrated development of prefabricated buildings and green buildings [10]. At the same time, the construction industry should understand the problems existing in the current relevant technical system and improve it, so as to form a complete industry standard system to ensure the smooth construction of prefabricated buildings.

4.2   Improving Coordination Mechanism

In order to better integrate the development of green buildings and prefabricated buildings, relevant departments can improve the coordination mechanism from the following aspects. First, establish an integrated general contracting mode and form an integrated contracting mode of three links: assembly design, construction and construction production. Actively cultivate enterprises integrating the three and realize the benign development of prefabricated construction enterprises [11]. The second is to improve the production capacity of the equipment manufacturing industry, so as to provide corresponding development power for the integrated enterprises, vigorously promote the green building general contracting responsibility system, establish a team of high-quality professional and technical personnel, and comprehensively design and study the organization system to find out the defects and improve them.

4.3   Strengthen Logistics Management

In the process of integrated development of green building and prefabricated building, project managers should strengthen the application of BIM Technology and formulate component supply plan by applying BIM Technology in combination with the construction scheme, so as to ensure good cooperation between the two. During the production of prefabricated parts, the information exchange with the construction unit shall be done well, and the principle of "use layer by layer" shall be strictly implemented, so as to improve the connection between the site and transportation and avoid low transportation efficiency. At the same time, corresponding communication and coordination shall be done to ensure that the components are transported to the site for direct installation, which can not only improve the construction efficiency, but also effectively avoid on-site stacking and component damage during stacking. In addition, when installing prefabricated components, the construction personnel shall operate in strict accordance with relevant regulations and specifications to ensure the construction quality of prefabricated buildings.

4.4   Strengthen Talent Training

(1) Strengthen BIM technical personnel training
    BIM technical training can be carried out for all staff of the construction unit from all aspects of the construction unit, and BIM ability assessment is an important content of talent training. At the same time, for new employees, the construction unit shall carry out BIM technical training. After the training is completed and meets the post requirements, they can be allowed to work formally.

(2) Strengthen the training of management talents

Through regular training activities, improve the managers' understanding of prefabricated buildings, help the personnel of the construction unit to be familiar with the operation process of prefabricated construction, and realize the good management of prefabricated building construction. In addition, relevant technical engineers can be employed for on-site guidance to quickly improve the business level of management personnel.

5   Conclusion

In short, under the background of BIM development, the integrated development of civil engineering and computer-aided building system is the development direction of the future construction industry, which is of positive significance to the coordinated development of architecture and natural environment. The construction industry should actively change the traditional mode of production and strengthen the construction of green buildings. In order to ensure the coordinated and integrated development of the two, relevant national departments need to improve the current policies, regulations and industry standards, and promote the development of construction enterprises in the direction of prefabricated construction by formulating corresponding supporting policies. On this basis, improve the cooperation mechanism, form a contracting mode integrating assembly design, construction and production, and realize all-round good cooperation. In this process, the project leader shall strengthen the quality and logistics management of components, and do a good job in the communication between the prefabrication factory and the construction unit, so as to make the two cooperate closely. In addition, we should strengthen the training of professionals, improve employees' BIM skills and management level through regular training activities, and promote the integrated development of civil engineering and computer-aided building system.

References

1. Tasca, F.A., Goerl, R.F., Michel, G.P., et al.: Application of systems thinking to the assessment of an institutional development project of river restoration at a campus university in Southern Brazil. Environ. Sci. Pollut. Res. **27**(8), 14299–14317 (2020)
2. Naderi, A., Shakeri, E., Golroo, A.: Performance-based management for construction holdings by integration of measurement frameworks. Asian J. Civil Eng. **22**, 751–758 (2021)
3. Guo, J., Li, D., Wang, L., et al.: Experience and enlightenment on military, civil, and commercial integration in US space infrastructure development. Strateg. Study CAE (2020)
4. Li, S., Huang, T., Xia, Y.: Research on application value of traditional cultural elements in visual design. World Sci. Res. J. **6**(3), 176–179 (2020)
5. Torres, B., Völker, C., Nagel, S., Hanke, T., Kruschwitz, S.: An ontology-based approach to enable data-driven research in the field of ndt in civil engineering. Remote Sens. **13**(12), 2426 (2021). https://doi.org/10.3390/rs13122426

6. Zhang, S.: Research on the infiltration and integration of modern dance elements in folk dance teaching. Arts Stud. Criticism **1**(3) (2020)
7. Wang, X.M.: Research on the application of big data statistics in the field of economic management. Sci-tech Innov. Prod. (2020)
8. Boy, G.A.: Aerospace human system integration evolution over the last 40 years (2022)
9. Tong, J.: Research on application of multi-factor surrounding rock pressure calculation theory in engineering. KSCE J. Civil Eng. **25**, 2213–2224 (2021)
10. Zhang, Z., Li, Y., Wang, C., et al.: Research on cold-formed steel stiffened-web built-up I-section columns with complex edge stiffeners. Int. J. Steel Struct. **22**(1), 389–407 (2022). https://doi.org/10.1007/s13296-022-00580-8
11. Marvila, M.T., Azevedo, A.R.G., Monteiro, S.N.: Verification of the application potential of the mathematical models of lyse, abrams and molinari in mortars based on cement and lime. J. Mater. Res. Technol. **9**(4), 7327–7334 (2020). https://doi.org/10.1016/j.jmrt.2020.04.077

Study on Optimal Design of Urban Rainwater System Based on the Concept of Sponge City

Xiaoqiang Zhang[✉]

Shenzhen Urban Transport Planning Center Co., Ltd.,
Shenzhen 518000, Guangdong, China
935748819@qq.com

Abstract. With the development of urbanization, the drainage load borne by the traditional urban rainwater pipe network system gradually increases, and urban waterlogging occurs frequently. In order to ensure the safety of urban drainage and protect the ecological environment, the Ministry of housing and urban rural development began to promote the construction of sponge city and improved the relevant requirements for the planning, design, construction and management of urban rainwater system. Therefore, based on the construction concept of sponge City, this paper studies the optimal design method of urban rainwater system.

Keywords: Sponge city concept · Urban rainwater · System optimization design

1 Introduction

As the infrastructure of urban construction, urban rainwater system undertakes the important responsibility of waterlogging drainage and is an important index to evaluate urbanization. In the early stage of urbanization, rainwater was collected and discharged mainly by drainage pipe network; With the continuous development of urbanization and the increasing number of hardened roads, the surface runoff also gradually becomes larger. In case of severe rainstorm, only rainwater pipe network can collect and discharge rainwater without changing the design return period of pipe network, which can not effectively ensure the safety of urban waterlogging prevention [1].

In order to alleviate the load of drainage pipe network, China began to seek new ways of rainwater collection and utilization, combined the reduction of waterlogging disaster with the protection of ecological environment, and compiled the technical guide for the construction of sponge city with the goal of building ecological civilization [2]. Under this background, this paper studies the previous research results, collects relevant data, and analyzes the experience, research progress and main technical measures of sponge city construction; Determine the measure scale in combination with the technical measures for the construction of sponge City, so as to provide guidance for reasonably determining the measure scale of urban rainwater system [3]. The model is used to simulate and analyze the urban rainwater system, which provides a reliable basis for the optimal design of urban rainwater system; It has high theoretical value and practical significance to determine the optimization objectives and methods to make the optimization design of urban rainwater system more targeted.

V. Sugumaran et al. (Eds.): ICMMIA 2022, LNDECT 138, pp. 233–240, 2022.
https://doi.org/10.1007/978-3-031-05484-6_29

2 Basic Concept of Sponge City and Its Three Rainwater Systems

2.1 Basic Concept of Sponge City

Sponge City, like a sponge, has a certain elasticity and is better able to cope with environmental changes and natural disasters. Construction should be carried out in accordance with environmental priorities to ensure the safety of urban drainage and flood protection. Maximize the use of urban rainwater resources and protect the urban ecological environment through the organic combination of natural and man-made measures [4]. When constructing sponge city, it is necessary to coordinate the three systems of natural precipitation, surface water and groundwater, taking into account the complexity and long-term character of water supply and sewage systems, emphasizing coordination and management.

2.2 Three Rainwater Systems in Sponge City

Among them, the low impact development rainwater system treats rainwater through infiltration storage, regulation, transportation, sewage interception, purification and other measures to effectively control the total amount, peak value and pollution caused by runoff; Urban rainwater pipe and canal system is a traditional urban rainwater drainage system, which can collect, transport and discharge runoff rainwater together with the former; The over standard rainwater runoff discharge system is mainly used to deal with the rainwater runoff that cannot be treated by the rainwater pipe and canal system. The over standard rainwater runoff discharge system is a supplementary rescue system of the rainwater pipe and canal system, which is jointly constructed through various natural or artificial ways [5]. None of the three systems exist in isolation, let alone distinct boundaries. They are interdependent and complementary, which is an important basis for building a sponge city. As shown in Fig. 1.

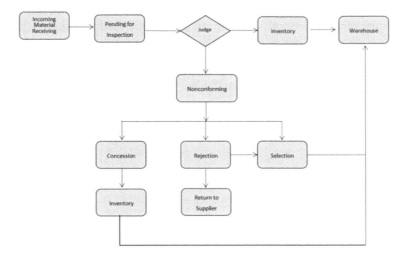

Fig. 1. Sponge city water system

3 Impact Analysis of Sponge City Construction on Urban Rainwater System

3.1 Analysis on the Impact of Sponge City Construction on the Composition of Urban Rainwater System

Compared with the traditional urban rainwater system, the construction of sponge city not only includes the traditional rainwater pipe and canal system, but also includes the source emission reduction and excessive rainwater runoff discharge system, which constitutes the urban rainwater system of modern rainwater and flood management. As the traditional urban rainwater system focuses on rainwater discharge, it mainly deals with the risk of urban waterlogging through the upgrading and reconstruction of pipe network, which can not comprehensively solve the problems of urban runoff pollution, saving water resources, protecting and improving the ecological environment. Figure 2 shows the composition of modern urban rainwater drainage system proposed by sponge city construction [6]. The source emission reduction system mainly aims at high-frequency small rainfall events. Through infiltration, water storage and other measures, it can absorb and reduce the runoff of urban rainwater drainage system from the source, which has good economic and social benefits. It makes up for the deficiencies of rainwater pipe and canal system in urban runoff pollution control, rainwater resource utilization, total runoff reduction and hydrological and ecological protection [7]. The rainwater pipe and channel system collects the collected rainwater into the rainwater main pipe or open channel through the rainwater pipeline, open channel and other facilities in the community or a plot, and finally transports the rainwater to the end discharge outlet, which mainly solves the problem of urban rainwater discharge with a design return period of less than 10 years [8]. Through the large-scale regulation and storage facilities at the end, the excessive rainwater runoff drainage system can effectively solve the heavy rainstorm or severe rainstorm exceeding the drainage capacity of pipes and canals, and reduce the risk of urban waterlogging. The construction of sponge city closely combines source emission reduction, rainwater drainage pipe and canal system and excessive rainwater runoff discharge system, breaks the pattern of traditional urban rainwater system dominated by drainage, and provides guidance for the construction of a modern rainwater and flood management system integrating ecological harmony, runoff pollution control, rainwater resource utilization and safe rainwater discharge [9].

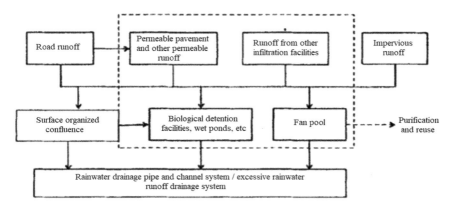

Fig. 2. Composition of urban stormwater system

3.2 Analysis on the Impact of Sponge City Construction on the Total Runoff of Urban Rainwater System

The traditional urban rainwater system mainly completes the discharge of about 80% of the total annual runoff through "fast discharge". The guideline for the construction of sponge city puts forward the total amount control index of annual runoff. Its principle is to reduce the annual discharge runoff of urban rainwater system by adding source emission reduction and collection, storage and utilization measures. Ideally, the total runoff control target should meet the runoff discharge standard before development and construction. If the natural landform before development and construction is considered as green space, the total annual runoff discharge rate is 15%–20%, that is, the total annual runoff control rate is preferably 80%–85%, so as to reduce the discharge load of traditional urban rainwater system [10].

The relationship between the remediation calculation time and precipitation calculation in Hefei Based on the standard annual method in different years, the total diameter setting rate is analyzed in detail. This thesis analyzes the general calculation method of the accumulation of construction measures in Schwammstadt, which the corresponding design settlement can generate under different types of accumulation. The total annual diameter fit rate according to recovery time is determined. Calculate the corresponding total annual runoff control rate under different total regulation and storage volumes, as shown in Fig. 3.

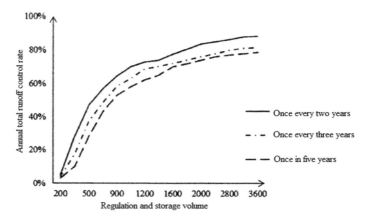

Fig. 3. Different storage volume corresponding to annual total runof volume control rate

It can be seen from Fig. 3 that the larger the total regulation and storage volume of sponge city construction measures, the more total annual runoff can be controlled; When the regulation and storage volume increases to a certain value, the control effect on the total annual runoff slows down. It shows that the construction scale of sponge city is not the bigger the better. The total annual runoff control target should be determined according to the actual situation and economic benefit analysis. When the total regulation and storage volume is certain, the smaller the design return period is, the higher the total annual runoff control rate can be achieved, indicating that the construction measures of sponge city are more targeted for the control of small rainfall events [11]. When the total annual runoff control rate is certain, the larger the design return period, the larger the scale of sponge city construction measures required. This is consistent with the increase of construction cost of urban rainwater system when the design return period increases, which further shows that the construction scale and total annual runoff control rate of sponge city should correspond to the design return period.

4   Research on Urban Rainwater System Design Under the Concept of Sponge City

4.1   Low Impact Development Rainwater System

(1) Do a good job in the overall urban planning
    Define the key urban areas to be developed, protect the urban scope of sensitive areas, and reasonably control the urban surface morphology. There are differences in hydrogeological conditions in various areas of the city [12]. In the development and design, we should prioritize and scientifically and reasonably control the key areas. For the ecologically sensitive areas of the city, the construction shall be prohibited or restricted, so as to realize the overall connection between the low impact development rainwater system and the urban rainwater pipe and canal

system, have a reasonable understanding of the urban land area, and understand the land characteristics and runoff characteristics of the corresponding areas".

(2) Implement special planning for water system

Clarify the scope of water system protection, make systematic planning of urban green space, and pay attention to the implementation of various drainage and waterlogging prevention measures. Carry out protection design according to urban water system, strengthen the connection between water area and coastline, and reasonably standardize the relationship between discharge and storage. Green space is a very important site in sponge City, which can not only provide enough reserved space for the city, but also ensure that the water system is in an elastic state. The design of rainwater system for low impact development shall be synchronized with waterlogging prevention to ensure that the urban drainage system can effectively adapt to the comprehensive planning of waterlogging prevention.

(3) Improve the detailed planning of repair

Clarify the classification of urban land, effectively deal with relevant constraints, and implement various connecting equipment and facilities. Before the design of low impact development rainwater system, we should understand the land classification of the city, avoid construction contradictions in advance, deal with the key and difficult points in development design, pay attention to the limitations of design and constraints, classify and analyze the sequence of system construction cycle, effectively deal with constraints and grasp the construction conditions. At the same time, according to the specific situation and existing differences of the city, do a good job in the connection between the low impact development rainwater system and various equipment and facilities to ensure that it meets the corresponding construction requirements.

4.2   Urban Rainwater Pipe and Canal System

(1) Make full use of terrain to discharge nearby

In the design of urban rainwater pipes and canals, the nearest drainage shall be considered. Combined with the actual situation, the rainwater shall be discharged into the nearby water body in the shortest distance by using the natural slope. Unless there is an impact on the ground, the rainwater main pipe is generally arranged at the lower part of the terrain when conditions permit, so that the ground rainwater can be led to the pipe channel. If the terrain is flat, the rainwater main pipe can be arranged in the middle of the drainage basin to improve the drainage efficiency.

(2) Rainwater pipes shall be arranged according to urban planning

The layout of rainwater pipes shall be considered in the process of urban planning. Rainwater pipes and canals shall be set according to the distribution of urban buildings and road layout, so as to ensure that most of the rainwater on the catchment surface of the street can be discharged into the rainwater pipes in the shortest distance. In addition, rainwater pipes and canals are mostly set under sidewalks and green belts to reduce the impact on roads and traffic under abnormal conditions such as ponding and maintenance. In addition, the rainwater inlet shall be set reasonably, and the coordination between rainwater pipe canal and other pipelines and buildings shall be considered, so as not to affect the function of other buildings and municipal facilities.

(3) Reasonably calculate the hydraulic of rainwater pipe and canal

In order to ensure the normal operation of rainwater pipes and canals, hydraulic calculation shall be carried out, and various basic data involved shall be scientifically calculated before calculation. The hydraulic calculation of rainwater pipes and canals is generally considered according to the uniform flow, and the formula is the same as that of sewage pipes, but the full flow state is selected for calculation, then the drainage basin is divided according to the calculation results, and the pipeline is located in combination with the urban master plan. Inspection wells shall be set according to the pipeline layout and the actual needs of the site to ensure that each area is not easy to be blocked.

5 Conclusion

After the economic and social development to a certain extent, the ecological environment will be restored. In order to reduce the adverse impact caused by such damage, sponge city construction makes full use of the city's own advantages, develops the rainwater system conducive to urban construction, establishes an urban development model that conforms to nature and respects nature, and completely solves the contradiction between urban waterlogging and water shortage through reasonable accumulation Infiltration and purification can improve urban water quality, lay a solid foundation for the recycling of water resources, and actively walk out of the road of sponge city construction with Chinese characteristics.

References

1. Bhat, H., Abraham, P.S.: Factors affecting water conservation potential of domestic rain water harvesting - a study on Bengaluru urban. BASE University Working Papers (2021)
2. Chapa, F., Krauss, M., Hack, J.: A multi-parameter method to quantify the potential of roof rainwater harvesting at regional levels in areas with limited rainfall data. Resour. Conserv. Recycl. **161**(2), 104959 (2020)
3. Munguía-López, A.D.C., Núez-López, J.M., Ponce-Ortega, J.M.: Identifying fair solutions in the optimal design of integrated residential complexes. Chem. Eng. Process. **157**, 108116 (2020)
4. Qian, L., Feng, W., Yang, Y., et al.: Comprehensive performance evaluation of LID practices for the sponge city construction: a case study in Guangxi, China (2021)
5. Zhou, Y., Sharma, A., Masud, M., et al.: Urban rain flood ecosystem design planning and feasibility study for the enrichment of smart cities. Sustainability **13**, 5205 (2021)
6. Lin, Y., Liu, S., Gao, S., et al.: Study on the optimal design of volume fracturing for shale gas based on evaluating the fracturing effect—a case study on the Zhao Tong shale gas demonstration zone in Sichuan, China. J. Pet. Explor. Prod. **11**(4), 1705–1714 (2021)
7. Qu, J., Chu, Z., Chen, C., et al.: Design of urban rain-waterlogging monitoring system based on video image recognition. Microcontrollers Embed. Syst. **38**, 1102–1115 (2020)

8. Toboso-Chavero, S., Villalba, G., Gabarrell Durany, X., Madrid-López, C.: More than the sum of the parts: system analysis of the usability of roofs in housing estates. J. Ind. Ecol. **25**, 1284–1299 (2021)

9. Watanabe, N., Setoguchi, T., Yokoyama, S., Guo, Z., Tsutsumi, T.: Urban-design process with snow and wind simulations: a study on the Kitami City hall. J. Civil Eng. Archit. **11**, 107–120 (2022)

10. Tansar, H., Duan, H.-F., Mark, O.: Catchment-scale and local-scale based evaluation of LID effectiveness on urban drainage system performance. Water Resour. Manag. **36**(2), 507–526 (2021). https://doi.org/10.1007/s11269-021-03036-6

11. Amiri, H., Azadi, S., Javadpour, S., Naghavi, A.A., Boczkaj, G.: Selecting wells for an optimal design of groundwater monitoring network based on monitoring priority map: a Kish Island case study (2022)

12. He, Z.: Study of comprehensive utilization of water resources of urban water distribution network. Water **13**(19), 2791 (2021)

Application of User Interface Design of Multimodal Information System in Blank System Simulation Model

Jianpeng Deng$^{(\boxtimes)}$, Wang Li, and Geng Pei

Shandong North Modern Chemical Industry Co., Ltd., Shandong 250033, China
jianpeng2002@163.com

Abstract. Nowadays, carbon fiber reinforced composites play an irreplaceable role in various fields. Facing the intensification of environmental pollution, energy conservation and emission reduction are the research focus, while thermoplastic is the reinforced matrix, which has light material, stable chemical properties, easy processing and molding and low cost, so it is widely used in all walks of life. With the development of simulation objects from the early simple engineering system to the current complex engineering system, object modeling technology presents the characteristics of strong coupling and complex and changeable interaction between the subsystems of the system and between the system and the environment. The simulation model has also developed from the basic function model, constraint model to the current spatial model and multi-level abstract model. Therefore, this paper studies the simulation model integration technology of carbon fiber blank system for wound material forming sabot based on computer technology.

Keywords: Winding material forming · Carbon fiber · System simulation

1 Introduction

In recent years, plastics have played an important role in almost all aspects of our life. In the context of the rapid development of industrialization, plastics are widely used in all walks of life all over the world because of their light material, stable chemical properties, low processing cost and easy processing and molding. In order to meet the needs of energy conservation and emission reduction, the current research focuses on light materials and strong mechanical properties.

Modeling and simulation technology is no longer limited to the behavior of a single organization or individual, but emphasizes the joint development of simulation system and model, so as to form a simulation system that can reflect the principles of different disciplines, different simulation environments and different simulation processes [1]. At the same time, it is also necessary to integrate different granularity and different levels of models developed by personnel in different fields with unified interface protocol specification, so as to meet the needs of the system for model complexity and ensure the correct interaction of data between models [2].

Carbon fiber plays an important role in the field of composites. The weight of high-performance materials produced by the combination of carbon fiber and polymer matrix is more than 20% lower than that of aluminum and more than 50% lower than that of steel. Therefore, carbon fiber, as a reinforcing material, plays an important role in the manufacture of high-performance composites. In addition, carbon fiber reinforced polymer composites (CFRP) not only have the advantages of light and strong, light and rigid, but also can be integrally formed in a large area [3]. The simulation model integration technology of carbon fiber blank system for wound material forming sabot based on computer technology has irreplaceable advantages, so it has a wide application prospect in national defense and civil use. The simulation model integration technology is shown in Fig. 1 below.

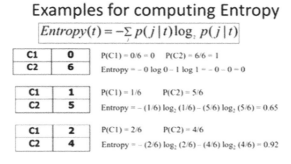

Fig. 1. Simulation model integration technology

2   Related Work

2.1   Overview of Carbon Fiber

Degradable leader, the largest product category and the most complete modified plastic production enterprise. High performance carbon fiber industry concept stocks include: Zhongjian technology, guangweifu materials, Hebang biology and blonde technology [4].

In the 1980s, high performance and ultra-high performance carbon fibers were successively developed. After more than 30 years of development, the industrial status of carbon fiber reinforced composites has been basically established.

The difference of structure and properties of carbon fiber is related to the production of precursor and carbonization process [5]. The following is the classification of different carbon fibers:

(1)   It can be divided into high elasticity module (HM), ultra-high elasticity module (uhm), high strength (HS) and ultra-high strength (VHS). Table 1 shows the corresponding attributes.

Table 1. Mechanical properties of different kinds of carbon fibers

Performance	Carbon fibre			
	UHM	HM	VHS	HS
Tensile modulus of elasticity	>400	300–400	200–350	200–250
Tensile strength	>1.7	>1.7	>2.76	2.0–2.75
Carbon content	99.8	99.0	96.5	94.5

(2) According to the type of precursor, it can be divided into the following categories:
 ① Viscose based carbon fiber. The output of viscose fiber obtained by oxidation exceeds 1% of the carbon fiber market.
 ② Polyacrylonitrile based carbon fiber. Standing call it man-made fiber.
 ③ Pitch based carbon fiber. It is made of isotropic and anisotropic asphalt.

2.2 Classification of Carbon Fiber Reinforced Composites

The composite consists of two parts: reinforcement phase and matrix phase. The transfer of stress is completed by the matrix, such as metal, resin, polymer, etc. The reinforced phase is mainly used to bear the load, so the mechanical properties are one of the factors to be considered in the selection of reinforced phase. Carbon fiber is not only relatively cheap, but also has excellent mechanical properties It is often used as an enhancement phase [6].

(1) Because carbon fiber reinforced resin matrix composites (CFRP) are excellent in strength and wear resistance, it is favored by many new energy vehicle enterprises, and its position is becoming more and more significant in the field of aerospace.
(2) Carbon fiber reinforced ceramic matrix composites belong to chemical synthetic materials, which are manufactured by a variety of chemicals through a series of reactions and processing. This material has better strength and toughness, high temperature resistance and corrosion resistance, so it is widely used in many fields [7]. Among them, carbon is used in the manufacture of new energy vehicle engines Fiber reinforced ceramic composites replace metal materials, which will greatly reduce the weight, not only save energy, but also have better environmental protection.
(3) Carbon fiber reinforced carbon composites. At higher temperatures, C/C composites still maintain excellent mechanical properties. Experiments show that the strength remains unchanged at temperatures above 2000c. Compared with steel braking materials, C/C composites as braking materials have the advantages of low density, large heat capacity, high high high temperature strength and better braking effect. They are commonly used in braking equipment of military aircraft Put it on.

3 Development Trend of Computer System Simulation Technology

Based on system theory, cybernetics, similarity principle and information technology, simulation technology integrates the knowledge of many high-tech fields such as computer technology, graphics and image technology, multimedia, network technology, object-oriented technology, software engineering, information processing and automatic control, and uses computer as a tool to simulate and reproduce the real environment, It is a comprehensive technology to conduct dynamic experimental research on the actual or envisaged system with the help of system model [8]. It has the characteristics of good controllability, non-destructive, safety, not limited by external conditions, allowing repeated times, high efficiency and wide application.

The development of simulation technology gradually shows new trends: (1) the scope of application is more and more extensive; (2) Simulation has gradually formed a high-tech industry; (3) Simulation technology has become a strong support for the development of national defense, economy and technology.

In recent 10 years, due to the rapid development of computer technology in all aspects and the intersection, integration and penetration of various technologies, has been expanding in breadth and depth, and a series of new topics have emerged, which are gradually closer to reality [9].

(1) Visual simulation

Visual simulation is to connect graphics and algorithm objects into a complete simulation system, and finally output the simulation results in a graphical way. The whole simulation process is clear and intuitive.

(2) Distributed simulation

Distributed simulation means that the whole simulation system adopts unified standards, protocols and databases, connects the scattered simulation equipment through LAN or WAN, and completes the simulation task through the interaction of each simulation equipment. With the continuous improvement of the application requirements of complex system simulation and the continuous development of the application field, the simulation system has also developed from a single computer system to a multi computer system. Many problems can not be solved by a single simulation system alone. It is necessary to rely on multiple simulation systems for collaborative simulation [10]. Distributed simulation has become one of the frontiers and hotspots in the field of complex system simulation, And continue to receive attention.

(3) Multimedia simulation

The introduction of multimedia technology in the simulation can make the whole simulation effect sound and color, and the simulation results are realistic and vivid.

(4) Virtual reality simulation

Virtual reality simulation technology can flexibly and effectively provide simulators with a feeling similar to the real scene. In addition, virtual design, virtual prototype and virtual manufacturing can be realized by using virtual reality technology.

4 Simulation Model Integration Technology of Carbon Fiber Blank System for Wound Material Forming Sabot Based on Computer Technology

Carbon fiber technology has strict technical barriers. Shanghai Petrochemical is one of the earliest companies in China to develop carbon fiber and carry out industrial production. As early as 2012, the company adopted the self-developed complete set of carbon fiber technology to produce 12K small tow carbon fiber. In January this year, the first 48K large tow carbon fiber project in China started construction in the company. After the project is put into operation, it will change the situation that all major tow carbon fibers in China depend on imports and are in short supply for a long time.

Filament winding technology is a molding technology with high molding efficiency, high utilization of fiber strength and good controllability of molding process. The fiber is introduced from the yarn frame. Through the linkage between the wire guide head and the machine tool spindle, the fiber is wound on the die surface according to the set winding law to complete the winding forming of the structure [11]. The winding principle is shown in Fig. 2. Winding technology can be divided into wet winding, semi dry winding and dry winding.

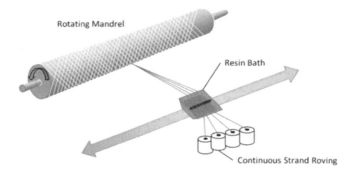

Fig. 2. Schematic diagram of filament winding

(1) Wet winding

After the untwisted roving (or cloth belt) is led out through the yarn frame, impregnated with resin through the rubber groove, and directly wound to the core mold under the control of the wire nozzle Come on. Wet winding is widely used in the production of winding products because of its simple operation and low requirements for equipment. However, the wet winding process also has some problems, such as difficult to control the glue content of products, high dispersion of product quality, poor forming environment and so on.

(2) Dry winding

In the traditional sense, dry winding mainly refers to a winding process in which the prepreg yarn (belt) is used to replace the resin impregnated fiber in the wet process, soften the viscous flow through the heating equipment on the winding machine, and then wound on the core mold. This process requires special fiber prepreg equipment, and the winding machine needs to be equipped with prepreg yarn heating equipment. Dry winding has the advantages of controllable glue content, stable product quality, high forming efficiency and clean processing environment. However, the process cost is higher than that of wet winding, so dry winding is mainly used in the fields of national defense, military industry and aerospace. In recent years, in the field of low temperature, medium and low pressure applications of wound products, the process of direct winding with dry fibers without impregnated resin has gradually attracted the attention of scholars at home and abroad because of its good impact resistance, excellent fatigue performance and recyclability in the later stage of service. It has been applied in some fields such as liquefied gas tank, rubber expansion joint, fiber air cushion crane and so on.

(3) Semi dry winding

Semi dry winding is a process between the first two winding processes. By wetting the fiber and resin before winding, then drying and making the resin react with the curing agent to a certain extent, and then winding this kind of fiber on the core mold. In this way, it does not need dry fiber prepreg equipment processing, but also can deal with the volatile matter in the resin and improve the quality of wound products. Moreover, the crosslinking degree of matrix in this method is lower than that of prepreg yarn, and winding can be carried out even at room temperature.

According to the unfolding results of filament winding trajectory, the winding modes are divided into geodesic and non geodesic winding modes. According to the differential geometry, after the geodesic is unfolded on the curved surface, it is a linear connection of two points, and its position is the most stable, as shown in Fig. 3. Therefore, the fiber will not slip under the winding tension. It is precisely because of the geodesic feature that the winding line type is uniquely determined after the core die size is determined. According to the definition of geodesic, when the fiber track deviates from the geodesic winding, it can be regarded as a non geodesic line type. A large number of winding practices have proved that the friction between the core mold and the wound fiber surface will inhibit the fiber slip to a certain extent, and the fiber is still in a stable position after deviating from the geodesic within a certain range. Compared with geodesic winding, non geodesic winding has a wider linear design space within the range of slip coefficient, and can optimize the winding products with variable angle according to the load condition, which is similar to the variable stiffness laying principle in fiber laying technology. Therefore, non geodesic winding can be applied to some products with complex shape and special load conditions. According to previous research experience, the slip coefficient of wet winding is 0.15–0.2 and that of dry winding is 0.39. Considering the influence of tension and fiber glue content in the winding process, non geodesic winding has a wide application space in dry winding.

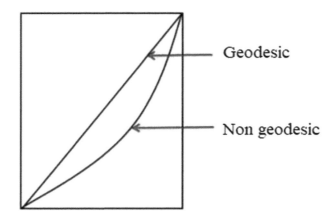

Fig. 3. Development results of geodesic and non geodesic on cylindrical surface

5   Conclusion

Carbon fiber composites based on carbon fiber or carbon fiber fabric are used as resin base materials, new composites with high specific strength, high specific stiffness, high temperature and good structure were made by specific composite processes. With its excellent mechanical properties and chemical stability, carbon fiber composites play a major mission of new product research and development and technological break-through in the application of major national strategic fields such as energy, trans-portation, aerospace, national defense and military industry. Especially in the field of national defense and military industry, the research on the simulation model integration technology of carbon fiber blank system for wound material forming sabot based on computer technology. The special materials represented by carbon fiber composites play an irreplaceable role in upgrading weapons and equipment and tackling key problems of new models.

References

1. Miraz, D., Ali, M., Excell, P.S.: Adaptive user interfaces and universal usability through plasticity of user interface design. Comput. Sci. Rev. **40**(100363), 1–26 (2021)
2. Hsu, H.P.: Design Communication and Media Usage in the Field of Cooperative Design: User-Interface Design Case (2020)
3. Tekinerdogan, B., Aktekin, N.: Model-based user interface design for generating E-forms in the context of an E-government project (2020)
4. Lucero, R.J., Sunmoo, Y., Niurka, S.T., et al.: Application of persuasive systems design principles to design a self-management application user interface for Hispanic informal dementia caregivers: user preferences and perceptions. JAMIA Open (1), 1 (2022)
5. Oehl, M.: Teleoperation of highly automated vehicles in public transport: user-centered design of a human-machine interface for remote-operation and its expert usability evaluation. Multimodal Technol. Interact. **5**, 26 (2021)

6. Nouri-Khorasani, A., Bonakdarpour, A., Fang, B., et al.: Rational design of multimodal porous carbon for the interfacial microporous layer of fuel cell oxygen electrodes (2022)
7. Jeon, M., Andreopoulou, A., Katz, B.F.G.: Auditory displays and auditory user interfaces: art, design, science, and research. J. Multimodal User Interfaces **14**(2), 139–141 (2020). https://doi.org/10.1007/s12193-020-00324-0
8. Chowdhury, A.: Design and Evaluation of User Interface of a Mobile Application for Aiding Entrepreneurship (2020)
9. Miraz, M.H., Excell, P.S., Ali, M.: Culturally inclusive adaptive user interface (CIAUI) framework: exploration of plasticity of user interface design. Int. J. Inf. Technol. Decis. Making **20**(1), 199–224 (2020)
10. Hamza-Lup, F.G., Goldbach, I.R.: Multimodal, visuo-haptic games for abstract theory instruction: grabbing charged particles. J. Multimodal User Interfaces **15**(1), 1–10 (2020). https://doi.org/10.1007/s12193-020-00327-x
11. Medjden, S., Ahmed, N., Lataifeh, M.: Design and analysis of an automatic UI adaptation framework from multimodal emotion recognition using an RGB-D sensor. Procedia Comput. Sci. **170**, 82–89 (2020)

Obstacle Avoidance Trajectory Planning Method for Space Manipulator Based on Genetic Algorithm

Lei Huang^(⊠)

Henan College of Industry & Information Technology, Jiaozuo 454000, China
fengyeshinow@163.com

Abstract. The robotic arm plays a very important role in the in-orbit service process of space, which can significantly improve work efficiency and save costs in various aspects. However, space manipulator system often has some special requirements. In the process of system operation, trajectory planning method should be adopted to avoid relevant obstacles, and genetic algorithm should be used to meet the requirements of ideal trajectory of motion time in joint space. In this study, task planning and obstacle avoidance planning methods will be proposed based on the specific requirements of space manipulator performance to improve the manipulator's operation accuracy.

Keywords: Genetic algorithm · Space manipulator · Obstacle avoidance trajectory planning method

1 Introduction

In the application stage, the space manipulator needs to carry the payload and carry out relevant space experiments, and even transport astronauts in space. Although the research on space manipulator in China started late, the further development of aerospace industry has made remarkable progress in many aspects in China in recent years. The task and assembly system of space in-orbit service has been effectively updated, which can guarantee the quality of in-orbit maintenance and in-orbit replacement work, and provide good logistics support and intelligent guarantee. In order to ensure the stable operation of the space manipulator, this paper proposes to use genetic algorithm to plan the ideal trajectory of joint space motion and avoid obstacles, so as to improve the operation accuracy and operation efficiency of the manipulator.

2 The Basic Content of Space Manipulator

(1) Kinematic dynamics of space manipulator. The research on mechanical arm is generally explored from three perspectives of structure, kinematics and dynamics, among which kinematics studies are the displacement relations and velocity relations between various links, while dynamics studies of mechanical arm are mainly analyzed around flexible dynamics and balance dynamics. Assuming that the base

of the space manipulator remains fixed, the space manipulator becomes a structure with multiple joints [1]. During trajectory planning, it is necessary to ensure the global obstacle avoidance function, control the motion of each joint Angle, and use corresponding algorithms to obtain the best proportion. In the current situation, the manipulator trajectory planning method has rich features and stable maturity, but there are still some defects in index correlation and performance, such as the vertical intersection interference of robot type determination plane. In some complex situations, interference and obstacle avoidance strategies may lead to no solution [2]. Therefore, mathematical description methods need to be selected to discuss the simplification process of mechanical arm, so as to make the simplification method of mechanical arm avoid various obstacles and the reliability of the domain itself. In the process of coordinate system design, in order to describe the displacement and posture state of the end coordinate system of the manipulator, the principle that the origin of the end coordinate system is the center point of the claw should be followed [3].

(2) Modeling of space manipulator. The current model assumptions of space manipulator are shown in Table 1.

Table 1. The space manipulator hypothesis requires

Requirements	Instructions
Mechanical arm	Connections are made by connecting rods and joints, and each joint is controlled by the system
System	The initial momentum and kinetic energy of the system are zero
Force	In the microgravity environment, the system of space manipulator ignores the influence of gravity and friction and maintains the conservation of relative momentum
Performance	The system is rigid

In the process of modeling, the problem of non-integrity constraint needs to be considered. Therefore, virtual manipulator method and generalized Jacobian matrix method can be adopted in the selection of modeling methods. The former sets the manipulator system free from external forces in free-floating space, and then follows the conservation laws of linear momentum and angular momentum to establish a virtual massless motion chain. In this case, kinematics modeling methods related to ground robots can be used in the virtual robot arm, but the early processing workload is large and large-scale calculation is required for modeling [4]. By contrast, the generalized Jacobian matrix method can combine the characteristic equation of the system with angular momentum information, and fully follow the requirements of kinematic characteristics and geometric parameters. The preliminary processing work is greatly reduced, but the later stage still involves the conversion and calculation problems in the inertial coordinate system [5].

3   Obstacle Avoidance and Task Planning for Space Manipulator

(1)   On-orbit task. Before understanding the mission and specific planning, we need to master the overall structure of the space manipulator system. Specifically, the system consists of three parts: payload cabin, service cabin and propellant cabin. The free flight process and the free floating process are the main components of the in-orbit service process. Once the tracking star is in the free flight state, the robot arm can move from the resting position to the ready to capture position. After the command is issued, the mechanical arm begins to visually capture the target star and enter the free-floating frame state. After the capture, the relative motion of the two stars must be kept within a certain range, or even 0, as shown in Fig. 1.

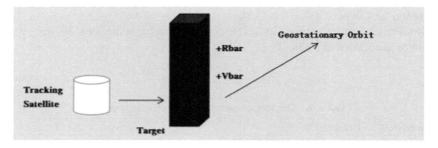

Fig. 1. The movement process of mechanical arm

Command post issued instructions, use the task of intelligent optimization algorithm analyzes the content and task object, get the target specific optimization program, based on spatial on-orbit task environment and other factors affecting task model is established, and the task instructions about the various content of overall planning, finally get the instructions of the complete, guide the satellite on-orbit service [6]. The basic flow is shown in Fig. 2.

The mission planning of in-orbit service will analyze the whole process of all services and find feasible in-orbit service schemes under different constraints [7]. At the same time according to the planning objects and tasks of technical analysis space manipulator, on-orbit service mission planning according to different types of tasks and the initial line of difference to find the optimal path and influence factors including environmental factors, target spacecraft, from multiple aspects, such as fuel factor, location factor to make the task execution process to achieve the best state, Therefore, considering the complexity of the mission environment and the complexity of the mission target, the content should be adjusted to establish a multi-objective on-orbit mission model. In other words, the corresponding model building method will be selected according to the different task types, and the hypothesis will be established first [8]. It is assumed that all mission objectives

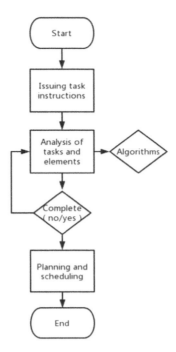

Fig. 2. Mission flow of space manipulator

have been determined completely before planning, and all observation tasks are regarded as point target observation tasks. Then, the goal task planning of space manipulator system can be transformed into a combinatorial optimization design problem of multi-objective [9].

(2) Classification and treatment of disorders. When performing in-orbit service tasks, the space manipulator system will divide all types of obstacles. One type is the obstacles generated by the service satellite or the target satellite, and the other type is the obstacles relative to the movement of the service satellite. Already on-orbit service movement, if the space manipulator never obstacle avoidance process, may lead to mechanical arm with the target satellite collision or other problems, the establishment of coordinate system, we establish coordinate axis centered on space manipulator system, can also be in the base coordinate system of the initial state as the inertial coordinate system for the analysis of dynamic equation.

For example in the case of collision detection, using axis section to represent the whole mechanical arm, surrounded by the cylindrical bounding box and ellipsoid and represent different types of obstacles, the collision detection problem of space manipulator can be converted into two bounding box collision detection, if the mechanical arm radius and surrounded and distance deviation will lead to collision [10].

(3) Obstacle avoidance planning. After finding the starting position and target position according to the obstacle information, a path without collision can be selected to meet the requirements of the starting and ending conditions, so that the space manipulator can maintain sufficient safety and stability in the process of task planning. According to the in-orbit service and the task characteristics of space robotic arm, the observation task has higher requirements for discovering obstacles, while space robotic arm itself has certain hardware limitations. The amount of calculation in autonomous obstacle avoidance planning should be kept at a low level, so known obstacles and real-time detection obstacles should be determined to reduce the amount of calculation. For point-to-point motion planning of space manipulator, collision detection should be carried out on all planned paths from the beginning of a certain moment to the end of another moment. If there is a collision, the previous step should be returned for re-analysis [11]. The main advantage of this obstacle avoidance planning is that it conforms to the working characteristics of the in-orbit service space manipulator itself, and can identify obstacles in a shorter time consumption, so as to select a bypass method and keep the manipulator in the correct trajectory. In addition, due to the hardware limitations of space satellites, the original inverse kinematics calculation of end-effector can be greatly reduced based on distribution.

4   Comprehensive Application of Genetic Algorithm

(1) Genetic algorithm and population establishment. In order to avoid unnecessary energy consumption, genetic evolution algorithm can be selected to initialize the population and evaluate the adaptability function according to different selection requirements of indicators. Some mutation operators can be systematically analyzed after evaluating individual fitness according to certain principles. In the establishing process of initial population and the need for space manipulator obstacle avoidance planning and continuous collision free path analysis, establish joint parameter function, then the generated path, determine whether each path is a collision occurs, the calculation of total time consumption each path movement process, the said position for the vector collection.

(2) Operator Design. Considering the collision-free design of space manipulator, it is required that in order to meet the realistic needs of diversified group trajectory, the roulette method and individual fitness function can be selected to determine the probability of being selected during the operation of replication operator. In addition, for mutation operator, mutation operator can be regarded as a new element in the group. Each mutation operator subject can represent the collusion-free trajectory and conditional constraints of the robot arm, thus forming the trajectory path change, and the new trajectory can become a new mutation individual after replacing the original trajectory.

(3) Fitness function design. The design accuracy of individual fitness function will directly affect the individual measurement requirements of different types of paths, and also determine the time efficiency of genetic evolution algorithm, which indicates that there is a close correlation between hardness and individual quality, and the selection of garlic wheel method is the main reference basis of individual selection probability. So the function of optimization goal can be used as fitness function and as the result of goal transformation. Also, in the process of reducing unnecessary energy consumption, the space manipulator in task planning should be considered, and the connection between motion path and fitness function should be established to establish energy consumption.

The function design stage should consider the starting condition and the termination condition, among which the termination condition is the most critical part. If several of the requirements are met, it can be judged as conditional termination.

First, each optimal individual can be recorded after the genetic algorithm reaches the number of iterations, which can be regarded as termination. Secondly, when the fitness target is known, the minimum deviation value can be specified. Finally, the individual fitness of the group remains flat, which can be regarded as producing the optimal fitness individual. When a complete process of replication exchange and mutation is completed, it indicates that the genetic programming has completed one generation, and the final termination condition can be reached only after repeated iteration. The piecewise evolutionary algorithm reduces the computation amount and obtains the optimization result of the manipulator trajectory. The piecewise algorithm and domain search algorithm are used to solve the original multi-objective task design problem of the space manipulator. Therefore, the space manipulator will always maintain the shortest time consumption to return to the original trajectory, and the amount of calculation is relatively small. However, the disturbance of the whole calculation process to the satellite, especially the energy consumption of the space manipulator system, should still be taken into account. In future work, fitness function based on energy consumption of space manipulator should be established to solve the disturbance caused by obstacle avoidance to other satellites in the base and other aspects from the root.

5   Conclusion

According to the system of space manipulator, the trajectory is evaluated by genetic algorithm in joint space. The purpose is to optimize the system dynamics performance of space manipulator, further embody the advantages of genetic algorithm, and maintain the stability and computational efficiency of the system. Subsequent work can consider to increase the corresponding constraint conditions, determine the meet the conditions of a certain optimal solution and the population and genetic algorithm iteration parameters, such as processing, eventually to improve the trajectory planning problems, makes the genetic algorithm can in any degree of freedom of obstacle avoidance of trajectory planning for promotion and application.

Acknowledgement. Jiaozuo Social Science Association project: Research on strategies for accelerating the high-quality development of Jiaozuo City by integrating into the "double circle" linkage priority development zone of Zheng Luo. Henan industry and Information Technology Vocational College extreme education reform project: Research on the optimization strategy of young teachers' ethics and style in Higher Vocational Colleges in the new era.

References

1. Miteva, L., Yovchev, K., Chavdarov, I.: Preliminary study on motion planning with obstacle avoidance for hard constrained redundant robotic manipulators. In: International Conference on Computer Systems and Technologies, pp. 71–75 (2021)
2. Noffke, B.W., Beckett, D., Li, L.S., et al.: Aromatic fragmentation based on a ring overlap scheme: an algorithm for large polycyclic aromatic hydrocarbons using the molecules-in-molecules fragmentation-based method. J. Chem. Theory Comput. **16**(4), 2160–2171 (2020)
3. Ashyralyev, A., Al Ammouri, A., Ashyralyyev, C.: On the absolute stable difference scheme for the space-wise dependent source identification problem for elliptic-telegraph equation. Numer. Methods Partial Differ. Equ. **37**(2), 962–986 (2021)
4. Receveur, J.-B., Victor, S., Melchior, P.: Autonomous car decision making and trajectory tracking based on genetic algorithms and fractional potential fields. Int. Serv. Robot. **13**(2), 315–330 (2020). https://doi.org/10.1007/s11370-020-00314-x
5. Krmer, M., Muster, F.I., Rsmann, C., et al.: An optimization-based approach for elasticity-aware trajectory planning of link-elastic manipulators. Mechatronics **75**(1), 102523 (2021)
6. Palmieri, G., Scoccia, C.: Motion planning and control of redundant manipulators for dynamical obstacle avoidance. Machines **9**(6), 121 (2021)
7. Ajayi, E.A., Obe, O.O.: Genetic algorithm based optimal trajectories planning for robot manipulators on assigned paths. Int. J Emerg. Trends Technol. Comput. Sci. **8**(8), 4888–4892 (2020)
8. Yokose, Y.: Energy-saving trajectory planning for robots using the genetic algorithm with assistant chromosomes. Artif. Life Robot. **25**(1), 89–93 (2019). https://doi.org/10.1007/s10015-019-00556-8
9. Shin, S.Y.: An improved deep convolutional neural network-based autonomous road inspection scheme using unmanned aerial vehicles. Electronics **10**(22), 2764 (2021)
10. Wanner, J., Brndle, F., Sawodny, O.: Trajectory planning with obstacle avoidance for a concrete pump using harmonic potentials. IFAC-Papers OnLine **53**(2), 9873–9878 (2020)
11. Nithya, P.K., Lal, P., Benjamin, G.E., et al.: Optimal path planning and static obstacle avoidance for a dual arm manipulator used in on-orbit satellite servicing. IFAC-Papers OnLine **53**(1), 189–194 (2020)

Physical Training Instruction Method Based on Data Mining

Bin Qian[✉]

Jiangxi University of Engineering, Xinyu 338000, China
qb15807901211@163.com

Abstract. In order to do a good job of physical training instruction and teaching through data mining technology, this paper will carry out relevant research. The research mainly discusses the basic concept of data mining and its role in sports training guidance, and then develops the design of data mining technology system, and puts it into practical application to test whether the system is effective. Through practical testing, the system can play the role of data mining technology, output better sports training guidance methods, can improve the quality of sports training, etc.

Keywords: Data mining · Physical training instruction · Effectiveness of physical education teaching

1 Introduction

Modern philosophy think that physical training guidance should be targeted, so as to make the level of the athletes improve, eliminate the athletes in the physical quality of "short board", and to do this we must understand the specific situation of the athletes, master status problem, the athletes to propose targeted training guidance. But practical perspective, the athlete body circumstance change frequency very fast, and have different forms, it can reflect the process in the large amount of information data, in the face of large scale, rapid update information data, the traditional manual mode of sports training guidance to comprehensive information collection, analysis data, training guidance advice may not be comprehensive and accurate enough. Related Suggestions based on the data mining technology into sports training guide work, using the technology to solve practical problems, improve the quality of sports training guidance, so the moment the main problem is how to use data mining technology in the athletic training guidance, in order to solve this problem to related research.

2 Basic Concepts of Data Mining and Its Role in Sports Training Guidance

2.1 Basic Concepts

Data mining is a kind of can carry on the "dig" to the huge amounts of data analysis technology, and common data analysis technology, the technology in addition to being able to analyze every message data, to identify, can also through the correlation, similarity and compatibility principle to obtain all information data connotation, such as in sports training to guide the work, Based on this technology can be athletes to vertical and horizontal coordinates of the upper limb and lower limb strength training data analysis, on the one hand, to understand the growth of the upper or lower limbs strength curve, on the other hand, know the gap between the upper limb and lower limb power, this is normal data analysis technology to do so, common the analysis results of data analysis technology can only tell people, how much is the onset of strength athletes, However, the gap between the two can not be shown, so it needs to be calculated manually. The difference between the data mining technology and common data analysis technology seems to have no too much value on the surface, but actually there is a huge difference, the reason is that in the face of complex information data analysis, data mining technology can be independent, clarify the logical relationship and ordinary technology does not have this function, need artificial to clarify complex logic, However, manual labor may not be able to complete the work well, so data mining technology has huge advantages compared with common data analysis technology. Figure 1 is the basic framework of data mining techniques [1–3].

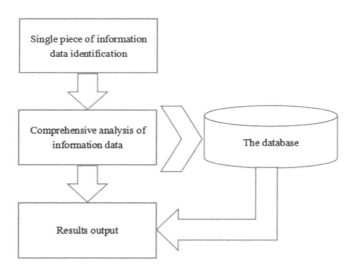

Fig. 1. Basic framework of data mining technology

2.2   Role in Sports Training Guidance

The main role of data mining technology in sports training guidance is reflected in two aspects, namely, simplifying training guidance and improving the pertinence of training guidance. The specific contents of the two aspects are as follows.

Simplify Training and Guidance

Along with the development of the era, the complexity of the sports training guidance is higher and higher, that staff working pressure is more and more big, this trend will continue to maintain in the future, so that the work pressure is likely to exceed the staff under the limit, and once this happens, certainly will cause a downturn of the quality of sports training guidance, so the sports training guide must be simplified. Look at this, the data mining technology can effectively simplify the sports training guide, the technology can independently complete the athletes sport information data collection, analysis, these two work is the most complex projects in sports training guide, therefore in the data mining technology, artificial don't need to work on complex focus too much on, You only need to make a decision based on the technical output [4–6].

Improve the Pertinence of Training Guidance

Targeted is the core of the sports training guidance quality indicators, to say the sports training guide targeted low, quality of work is in low level, but as the job complexity, difficulty and pressure, the higher frequency, artificial has targeted, it is difficult to guarantee the sports training guidance instructions manual mode does not meet the demand of the present sports training guidance. In the face of such situation, the data mining technology can provide effective help, namely the data mining technology to replace the artificial information data collection and analysis, and technical system operating efficiency than artificial, can in a very short period of time to tens of thousands, or even more data collection and analysis, so the results will be more comprehensive and in-depth, Can point out the problems of athletes, artificial nature can give specific training guidance suggestions [7].

3 Data Mining Technology System Design

Combined with the work of sports training guidance, the following will design a data mining technology system specially used for sports training guidance.

3.1 Overall Framework and Process of the System

In order to clarify the direction of system design, this paper gives priority to the overall framework and process of the system. The specific contents are as follows.

Overall Framework

Figure 2 is the overall framework of the system.

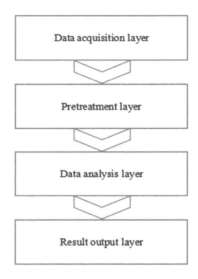

Fig. 2. General framework of data mining technology system

Process Design
Combined with Fig. 2, the process of the system is shown in Table 1.

Table 1. System flow

Process step no	Step content
1	Collect relevant information and data through manual records and equipment records
2	Preprocessing the collected information data to ensure the purity of information data
3	Single analysis and comprehensive analysis are carried out on the preprocessed information data
4	Output single analysis and comprehensive analysis results

According to Table 1, in the physical training and guidance work, the manual can directly understand the specific conditions of athletes according to the results, and then put forward targeted training and guidance suggestions. If the problem is complicated, it can also design targeted training and guidance programs. Therefore, this system is theoretically feasible.

3.2   System Implementation Method

Combined with Fig. 2, the system will adopt the hierarchical method to expand the design, and the implementation method and connection method of each layer are as follows.

Implementation Method of Data Acquisition Layer

Data acquisition layer is the basic layer of the system, and its main function is to collect the information data needed in sports training and guidance. Therefore, the first step of the design of this layer is to define the data acquisition method and integrate these methods into this layer., because physical training guidance work parts information data cannot be directly through the equipment acquisition, so the data collection method must be selected on the record, the equipment of the artificial two methods, of which the former is mainly responsible for collecting the athlete injury, age, and have no the information data of equipment training activities, etc., the latter is responsible for the record player in the data information in the equipment training activities, Such as through the sensor can record athletes "riding a bicycle" training cycle times, resistance size, etc., can judge athletes leg muscle endurance, average strength, explosive force value, etc. Therefore, two data collection methods can ensure the integrity and reliability of information data, paving the way for the follow-up work. In addition, because the magnitude of the collected information data is huge and needs to be stored, this paper chooses the SQL database with 1T capacity as the support, which is also an important part of the data acquisition layer [8–10].

Data Preprocessing Layer Implementation Method

Because there may be some invalid data or low-quality data in the process of data collection, such as repeated information data, these information data will not only lead to a slow follow-up analysis process, but also may interfere with the accuracy of the final results, so it must be processed, so it is necessary to realize the data pretreatment layer in the system. Method to realize the data preprocessing layer mainly through programming language preprocessing tool, need for this system, this paper chose the rechecking during the course of the Java language to write the data retrieval tool, this tool can detect each character of information data, if you find a set of information data of each character are exactly the same, will immediately delete one of them, Avoid the existence of fully informative data. It is worth noting that in the actual situation, there are many forms of data information that need to be preprocessed, among which repetition is only one of the more common ones. Therefore, in the actual work, tools need to be written according to the specific situation to ensure the pretreatment effect [11].

Data Analysis Layer Implementation Method

Data analysis is the main task of the first article of each layer to collect the information data analysis, so that the individual information data are defined, and then according to the definition of each data, combined with the analysis need to relevant information data to carry on the comprehensive analysis, comprehensive analysis of the results can reveal the relationship between the multiple information data, these relations can reveal the specific analysis, For example, in the work of sports training and guidance, the comprehensive analysis results show that the long-distance running speed of athletes decreases while the weight of athletes increases. The two information data are closely related, indicating that the reason for the decrease of long-distance running speed of athletes is the increasing weight of athletes. Two methods of the realization of analysis tools is the same, that is, the use of artificial intelligence tools to construct intelligent analysis process, and then to construct neural network and artificial intelligence logic, logic is completely joint sports training guidance needed, the process on the basis of each data node definition analysis, and then calculate the correlation between each basic nodes, similarity, In this way, the analysis results can be obtained.

3.3   Communication Design

In order to ensure the system can input the actual application, on the basis of the above designs need for communication system design, the purpose is through the channels of communication system of the physical layer with a terminal connected, the system is the physical equipment record part of the data collection layer (manual record can be directly connected to the terminal layer by means of cable transmission). The terminal layer refers to the framework of other components of the system except the data acquisition layer, and the two must be connected to analyze data. From this perspective, communication design aspects in this paper, we adopt the RFID technology in the Internet of things technology to construct the communication channels, with the help of RFID devices send infrared data to transmit information, at the same time because of RFID equipment signal format is electric signals, can't be read from a terminal layer, so it also increases the decoder in the communication channels, The function of decoder is to unlock the information data in the format of electrical signal, and then copy the data, and then send the replication group data to the terminal layer in digital format, which can be processed by the terminal layer, so the communication design is completed.

4   System Inspection

4.1   Inspection Scheme

With A one-year athletes training process as sample, with the athlete's weight, lower limb power mean, lower limbs explosive force as evaluation index, the first human to analyze these data and get the physical quality of the athletes before and after A year, and the corresponding training advice, and then input the information on the evaluation indexes into import system, Wait for the system analysis and output results. If the output results of the system are similar to or better than the results of manual analysis, it indicates that the system design is effective and data mining technology can improve the pertinence of sports training guidance; otherwise, it is the opposite.

4.2   Test Results

Table 2 shows the test results.

Table 2. Test results of this system

Indicators	A year ago	A year later
Weight	71 kg	83 kg
Mean strength of lower limbs	188 kg	205 kg
Lower limb explosive power	200 kg–203 kg	194 kg–211 kg
Targeted suggestions		The lower limit of lower limb explosive power of athlete A decreased after one year, because of the increase in body weight and the low average increase in lower limb strength. Therefore, it is suggested to lose weight and increase strength training

As can be seen from Table 2, athlete A's comprehensive physical quality has been improved after one year of sports training, but the lower limit of explosive power of lower limbs has been reduced. Therefore, the guidance advice of sports training in artificial mode is to increase strength training. Combined with the targeted suggestions in Table 2, it can be seen that the contents include suggestions in manual mode, and also indicate the reasons for the lower limit of explosive power of lower limbs and give suggestions for weight loss. Therefore, the application of this system is effective.

5   Conclusion

To sum up, sports training guidance is related to athletes' sports level, and even the probability of injury and injury of athletes, so it is very important to do a good job in sports training guidance. In the face of this requirement, data mining technology can provide people with help, with the help of this technology can improve the pertinence of sports training guidance suggestions, guarantee the quality of work.

References

1. Raut, A.R., Khandait, S.P.: Review on data mining techniques in wireless sensor networks. In: IEEE Sponsored 2nd International Conference on Electronics and Communication System (ICECS 2015). IEEE (2020)
2. Yu, M., Petrick, L.: Author correction: untargeted high-resolution paired mass distance data mining for retrieving general chemical relationships. Commun. Chem. 4(1), 40 (2021)
3. Javier, L.Z., Torralbo, J., Cristobal, R.: Early prediction of student learning performance through data mining: a systematic review. Psicothema 33(3), 456–465 (2021)
4. Shirono, T., Niizeki, T., Iwamoto, H., et al.: Therapeutic outcomes and prognostic factors of unresectable intrahepatic cholangiocarcinoma: a data mining analysis. J. Clin. Med. 10(5), 987 (2021)

5. Rodr Guezherrera, A., Reyesandrade, J., Rubioescudero, C.: Rationale for timing of follow-up visits to assess gluten-free diet in celiac disease patients based on data mining. Nutrients **13**(2), 357 (2021)
6. Wulandari, S.A., Kuswara, H., Palasara, N.: Analisis Penerapan Data Mining Pada Penjualan Kerupuk Rambak Menggunakan Metode Nave Bayes Classifer Untuk Optimasi Strategi Pemasaran. SITECH Jurnal Sistem Informasi dan Teknologi **3**(2), 83–94 (2021)
7. Romanov, F.D.: Discoveries of variable stars by amateur astronomers using data mining on the example of eclipsing binary romanov V20(Abstract). J. Am. Assoc. Variable Star Observer **49**(1), 115 (2021)
8. Schück, S., Roustamal, A., Gedik, A., et al.: Assessing patient perceptions and experiences of paracetamol in France: infodemiology study using social media data mining. J. Med. Internet Res. **23**(7), e25049 (2021)
9. Sarangi, R., Bahinipati, J., Pathak, M., et al.: Is data mining approach a best fit formula for estimation of low-density lipoprotein cholesterol. J. Family Med. Primary Care **10**(1), 327 (2021)
10. Avdagic-Golub, E., Begovic, M., Causevic, S., et al.: Profiling contact center customers for optimization of call routing using data mining techniques. In: 2021 20th International Symposium INFOTEH-JAHORINA (INFOTEH) (2021)
11. Peji, A., Molcer, P.S.: Predictive machine learning approach for complex problem solving process data mining. Acta Polytechnica Hungarica **18**(1), 45–63 (2021)

Optimization Model and Algorithm for Rail-Highway Combined Transportation of Dangerous Goods

Qingpin Ye$^{(\boxtimes)}$ and Wenying Wu

Wuhan Railway Vocational College of Technology Railway Transport
Management School, Wuhan 430205, China
redcherryqing@126.com

Abstract. In order to successfully complete the ironwork combined transportation of dangerous goods, this paper will carry out relevant research, the main purpose of the research is to establish the optimization model of the ironwork combined transportation of dangerous goods. In the research, the process and particularity of the rail-highway combined transportation of dangerous goods are discussed first, then the model hypothesis is put forward based on the main problems in the rail-highway combined transportation of dangerous goods, and then the optimization algorithm is proposed for the model hypothesis. By using the algorithm in this paper, the model of rail-highway combined transportation of dangerous goods can be linearly optimized. The optimized model can improve the stability of rail-railway combined transportation, reduce the risk probability and influence on the one hand, and guarantee the smooth completion of cargo transportation on the other hand.

Keywords: Dangerous goods · Rail-highway combined transportation of dangerous goods · Model optimization

1 Introduction

Under the background of global trade, Chinese and many other countries in the world to establish the trade relations of cooperation, so in order to support business activities, our country need to export the goods by means of blacksmith transport, delivered to the trade partners, but people gradually realized the blacksmith in transit transport and general cargo have very big difference, just from the point of view on the mode of transportation, the former is a greater risk of, In addition, the combined transportation of iron and steel industry often involves some dangerous goods, which will undoubtedly further increase the risk, leading to the increase of risk probability and the increase of risk influence. This background, people begin to attach importance to public transport of dangerous goods iron process, hoping to establish the most stable before shipping transportation scheme, in order to achieve this field put forward a train of thought, namely the first basic model is set up according to the original transportation condition, then adopts the algorithm is optimized for basic model, so you can get the optimized model, according to this model to adjust the transportation scheme, It can effectively improve the stability of the scheme and control the transportation risk. The proposal of the idea has been widely concerned by people, so how to implement this idea has become the focus of people's thinking, it is necessary to carry out relevant research.

2 The Process and Particularity of Rail-Highway Combined Transportation of Dangerous Goods

2.1 Process

Rail-highway combined transportation of dangerous goods is a long-distance land transportation mode, which involves a very broad geographical scope and is often used for trans-provincial transportation and national foreign trade transportation. The basic process of this combined transportation mode is divided into two steps: Firstly, highway short-distance connection, which can be divided into two stages: inbound short-distance connection and outbound short-distance connection. In the first stage, road transportation is the first link, and the goods are transported from the starting point to the transfer station, so as to reach the end of the road through continuous cycle, and connect with the second stage. In the second stage, the goods will be transported from the end of the road to the key point of the transit station through the road, and the cycle will continue to reach the destination. In this process, because some customers have special railway lines, the highway short-distance connection may only include inbound and outbound connection nodes; Second, long-distance railway transportation, the dangerous goods transported to the destination of the road will be loaded, starting from the railway starting point by means of railway transport, the dangerous goods will be transported to the terminal change station, continuous cycle until the goods arrive at the key delivery point. It is worth mentioning that in the process of iron and the dangerous goods transport of goods change operation is very important, that is because the iron public transport of dangerous goods need to constantly more exchange carrier, so there are a lot of transport the picture in site, every reach a shipped in site will conduct a complete operation of changing the outfit, operational requirements of the goods in good condition is transferred to other transport vehicles, And to ensure that the transferred goods can adapt to the new transport environment, to check the cargo protection measures [1–3].

2.2 Special

Compared with ordinary freight transportation, rail-highway combined transportation of dangerous goods is special, whose particularity is reflected in many aspects, such as long journey, many nodes, frequent reloading operation, etc., but the risk in transportation can best reflect its particularity. According to relevant theories, risks in the process of rail-highway combined transport of dangerous goods generally include road transport risks, goods transfer risks and railway long-distance transport risks, the details are as follows [4–7]. Figure 1 is the risk relation diagram of rail-highway combined transportation of dangerous goods.

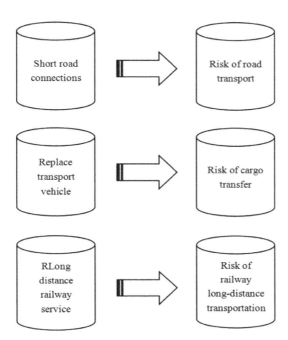

Fig. 1. Risk relation diagram of rail-highway combined transportation of dangerous goods

Risk of Road Transport

According to Fig. 1, road transport risks mainly occur in the stage of short-distance road connection, and there are many types of them. Here, only the common traffic accident risks are taken as an example for discussion. The main means of transport in the short-distance road connection stage is generally large trucks. The large volume of such vehicles indicates that the collision area is larger and there is a larger blind area of sight. Therefore, the probability of traffic risk in short-distance road connection is greater than that of ordinary cargo transport. At the same time, the transportation environment of large trains is mostly expressway, and the speed of transportation is relatively fast. In addition, the complicated road influence during transportation will further increase the incidence and influence of transportation risks. Figure 2 is the influence of traffic accident risk, once the outbreak of all objects within the scope of the impact and influence there are two types: first, the vehicles collided at high speeds, or hit the road surrounding facilities, can cause huge impact to the occupants, impact has the high fatality rate, can also lead to vehicle tilting, even knock the highway; Second, because of the large trucks to transport the goods as dangerous goods, so after has the traffic accident, dangerous goods may be leaked out, and then will cause new risks, such as the goods have flammability, leaked after affected by various environmental factors may cause a fire, or the goods belong to chemicals, after the leak can cause pollution to the surrounding air, it threatens the health of the surrounding people [8–10].

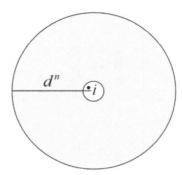

Fig. 2. Influence range of traffic accident risk (d is the radius of the range, and i is the location of the accident)

Risk of Cargo Transfer

The risk of cargo transfer mainly occurs in the process of reloading operation, and the main risk type is relatively single. It is usually the leakage of dangerous goods, or improper contact between reloading operator and dangerous goods, resulting in personal injury, or may lead to fire and other accidents when the goods are not leaked. Although there are few risk types of cargo transfer, the influence is very strong at that time. The reason is that the reloading operation needs a lot of manual work, and all the manual work is concentrated in the reloading station, indicating that the surrounding environment is relatively narrow and the personnel are relatively dense. At this time, the risk of cargo transfer occurs, and the surrounding staff may be affected. At the same time in site may exist in many vehicles carrying dangerous goods, risk of any vehicle in the transfer of goods, it may affect other dangerous goods vehicles, forming a chain reaction, the risk of this case fatality rate is very high, personnel for the first time around was affected, is likely to be killed, so the goods transfer risk must be aroused people's attention. Fortunately, in the case that reloading operation is carried out in strict accordance with the specifications, the probability of cargo transfer risk is very small, so the risk of cargo transfer is not considered as the main risk under normal circumstances.

Railway Long-Distance Transport Risks

When dangerous goods from highway short-distance transportation of large trucks, transferred to the railway vehicle on railway long-distance transportation risk is likely to occur, the risk of type two types of risk are similar to the above mentioned, the risks are more likely to occur in railway long-distance transportation risk is also a traffic accident and the leakage of dangerous goods, but different in that the risk of causing factors. The first railway in the long-distance transportation accident risk probability is small, the reason is that the safety of railway vehicle itself is higher, and railway standardization management, can effectively control risk, but the risk is unlikely to disappear entirely, if in the long distance transportation of railway vehicles have been some special factors, its security, the bottom line has been broken It will cause traffic accidents of unpredictable degree, such as vehicle tipping, sudden brake, etc. These accidents have great threats on the physical level, basically anyone affected will be severely impacted, the fatality rate is nearly 100%, and it may also lead to leakage of dangerous goods. Second everything is normal, even long distance transportation of railway vehicle is dangerous goods leakage situation is still possible, due to traffic when the wind is bigger, so strong wind could damage the goods protection facilities, cause the goods were leaked (in addition to the wind, and other environmental factors may lead to the risk, don't tell here). This risk is often difficult to be noticed by the occupants of the vehicle in the first time. Coupled with the extremely high speed of the vehicle, dangerous goods released within a short period of time will cause a wide range of impact, even affecting many cities. Figure 3 shows the scope of risk of leakage of dangerous goods in long-distance railway transportation (the scope of risk of traffic accident cannot be predicted, so it is not shown).

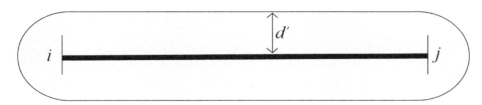

Fig. 3. Scope of risk of leakage of dangerous goods in long-distance railway transportation (i is the location of the accident, j is the end point, d is the radius)

3 Construction and Optimization of the Optimization Model for Rail-Highway Combined Transportation of Dangerous Goods

3.1 Model Construction

Problem Description: Iron the main problem with public transport of dangerous goods is a conflict between risk control and cost control, namely as a wide range of land transportation mode, which involves the cost is huge, so management naturally want to control the cost, but simply to control costs, could lead to increased risk, whereas blindly control risk, cannot be very good cost control, there is a conflict between. Faced with this problem, relevant fields believe that the best solution is to find a balance between risks and costs, and minimize costs and risks on the basis of the balance. Starting from this idea, the risk of rail-highway combined transportation of dangerous goods has been discussed above, so the cost of combined transportation is mainly introduced, as shown in Table 1.

Table 1. Cost of rail-highway combined transportation of dangerous goods

The cost of	Calculation method
Train operating expenses	Cost of individual trains × number of trains
Vehicle running cost	The sum of the product of the vehicle's individual operating cost × the number of vehicles running × the distance of vehicles running
Transshipment charges	Transshipment cost × transshipment amount of each transshipment station
Fees for road transport services	Individual road transport cost × transport quantity × road transport distance

Model hypothesis: Take the process of "public-iron, iron-iron, iron-public" as an example, where the set of risk R is R = $(r_1, r_2, \ldots, r_n)$, and the set of cost C is C = $(c_1, c_2, \ldots, c_n)$, the set model of the two is shown in Formula (1).

$$\sum_{i \in N | \phi(i) = o^h} \sum_{j \in N} x_{ij}^h = 1 \tag{1}$$

where i and j are the starting point and end point of a certain stage in the process of "public-iron, iron-iron, iron-public"; N is the collection of switching stations in the whole process; x_{ij}^h is a variable, it is defined as if the goods H $\in$ H (h is the set of I and J) $\in$ transit set, the variable is 1, otherwise 0.

3.2   Model Optimization Calculation

May, in accordance with the above model, through the correlation algorithm is optimized calculation, first of all, in the process of iron and dangerous goods transport model as a NP hard problem, aiming at the problem of minimum cost and risk of the smallest subproblems to parallel in series, in order to ensure follow-up calculation on (because the problem solving thinking is balancing the two subproblems, So the two problems cannot be calculated separately). Secondly using multi-objective mixed integer programming method, relying on polynomial matrix can be calculated, all non dominated solution in this case the multi-objective optimization method, to every single objective (a cost or risk factors) as model inputs, the generalization of non dominated solution can be converted to the typical non dominated solution, The representative non-dominated solution can optimize the corresponding single objective. Finally, by integrating all single objective optimization results, the optimal solution can be obtained, and the balance between cost and risk can be found on behalf of calculation. See Formula (2) for the specific calculation method.

$$|N||H| + |N||A| + |A_1| + |V_{ra}| + |H| \tag{2}$$

where V is the set of nodes; V_{ra} is a collection of railway and public transport handling stations; A is the set of alternate arcs.

In addition, the model of optimization calculation should pay attention to the integrity of data input, typically preferred column index system, to control the target to determine coefficient, such as velocity coefficients, risk factor, dangerous goods transportation cost coefficient, dangerous goods risk coefficient and so on, the details of these indicators as far as possible, otherwise the optimization calculation results is not complete.

4   Conclusion

In conclusion, as a special mode of land transportation, rail-highway combined transportation of dangerous goods has the characteristics of high risk and multiple risk types. Under these characteristics, people should pay attention to the planning work before rail-highway combined transportation of dangerous goods, and do a good job in planning can reduce risks and ensure the smooth completion of transportation. At the same time, the planning of railway-highway combined transportation of dangerous goods should not only consider the problem of risk control, but also pay attention to the cost. The optimal planning result should be exactly at the balance point of risk control and cost control, and minimize the risk and cost within a reasonable range.

References

1. Huang, W., et al.: Historical data-driven risk assessment of railway dangerous goods transportation system: comparisons between entropy weight method and scatter degree method. Reliab. Eng. Syst. Saf. **205**, 107236 (2021)
2. Andersson, S.E.: Safe transport of dangerous goods: road, rail or sea? A screening of technical and administrative factors. Eur. J. Oper. Res. **75**(3), 499–507 (1993)
3. Huang, W., Zhang, R., Minhao, X., Yaocheng, Y., Yifei, X., Dieu, G.J.D.: Risk state changes analysis of railway dangerous goods transportation system: based on the cusp catastrophe model. Reliab. Eng. Syst. Saf. **202**, 107059 (2020)
4. Cassini, P.: Road transportation of dangerous goods: quantitative risk assessment and route comparison. J. Hazard. Mater. **61**(1–3), 133–138 (1998)
5. Menyasz, P.: Transport Canada imposes city speed limit for rail transport of dangerous goods. Int. Environ. Reporter **38**(9), 538 (2015)
6. Rader, J.H.: Containers for transport of dangerous goods by rail: transport Canada considers adoption of transport Canada rail car standards. J. Hazmat Transp. **23**(6), 27–29 (2013)
7. Ebrahimi, H., Sattari, F., Lefsrud, L., et al.: Analysis of train derailments and collisions to identify leading causes of loss incidents in rail transport of dangerous goods in Canada. J. Loss Prev. Process Ind. **72**(1), 104517 (2021)
8. Li, X., Yang, Y.-F.: Research on operation and management of railway transport of dangerous goods in third-party logistics enterprises. In: Liu, B., Ma, M., Chang, J. (eds.) ICICA 2012. LNCS, vol. 7473, pp. 31–36. Springer, Heidelberg (2012). https://doi.org/10. 1007/978-3-642-34062-8_4
9. Purdy, G.: Risk analysis of the transportation of dangerous goods by road and rail. J. Hazard. Mater. **33**(2), 229–259 (1993)
10. Dvorak, Z., Leitner, B., Ballay, M., et al.: Environmental impact modeling for transportation of hazardous liquids. Sustainability **13**(20), 11367 (2021)

Analysis of Short Video Algorithm Marketing Strategy of Enterprises

Peng Wang[✉]

Jiangxi University of Engineering, Xinyu 338000, China
271504901@qq.com

Abstract. In order to better carry out enterprise marketing through short video, this paper will carry out relevant research, mainly discusses the basic concept and application value of short video algorithm, then conducts qualitative and quantitative analysis on short video algorithm based on SWOT analysis method, and finally puts forward enterprise short video marketing strategy based on algorithm results. The method in this paper can give full play to the role of short video algorithm, help enterprises find accurate marketing strategy, achieve accurate marketing and high-quality influence, which is conducive to the economic development of enterprises.

Keywords: Enterprise · Short video algorithm · Short video marketing

1 Introduction

Under the background of new media, short video has attracted widespread attention, so some short video platform users rapid growth, the performance for business to see business opportunities, or a short video as a kind of media, has the function of passing information to the audience, so if use short video marketing, can let more people know about their products, or even complete sales, So many enterprises began to enter the field of short video. But in the early part of the enterprise into the short video domain, many businesses found that short video marketing effect is not good, and then enterprises realized that the reasons for this phenomenon lies in the short video marketing lack of policy support, correct video content in the audience to accept degree is not high, it is hard to make transactions, the enterprise then began to emphasize short video marketing strategy, In this case, the short video algorithm has attracted the attention of the majority of enterprises. Using this algorithm can help enterprises find the right short video marketing strategy. Therefore, in order to achieve the goal, it is necessary to conduct research on the short video algorithm marketing strategy of enterprises.

2 Basic Concepts and Application Value of Short Video Algorithm

Short video algorithm is a macro concept, is a series of short video user behavior, interest orientation, video content acceptance, consumer demand and consumption characteristics calculation algorithm of collection, and these algorithms can help enterprises to improve the precision of short video marketing, thus to realize accurate, even super precision marketing, That is accurate or ultra precision marketing compared with the traditional marketing is not just the difference on the precision, the deeper point of view, accurate or super precision marketing more pertinence, purpose, and the traditional marketing is more subjective, is common in the thinking of "wide net, many fishing", explain the difference, there is also the basic train of thought for the two This aspect shows that short video algorithm has high application value [1–3].

In the present development trend of new media industry, expanding investment under the condition of good, good short video algorithm got people attach great importance to the application of performance, namely the modern new media industry mainly through advertising to reap the benefits, in the process of new media are able to get the favour of the enterprise, it is just because of the new media can effectively increase the advertising marketing factor, This is achieved through a variety of short video algorithm, short video platform with the aid of all kinds of algorithm technology, can carry on the control to the process of information transmission, and with the help of big data, each user's behavior, such as calculation, complete the user interest in painting, etc., according to the pictures can know what the users like to see, what you need, how the level of consumption, Therefore, they can accurately put into each user happy to see, acceptable advertising, so the conversion rate of advertising marketing naturally improve. At the same time, a short video algorithm in the application of short video marketing value more than that, also have the function of the increase user stickiness, return on investment, the short video algorithm as a kind of technology into a short video communication process, can cause the subjectivity to the user's access to information media, the influence of the assumption of short video platform for advertising to a user to push a certain enterprises for a long time, Then the user will become the user of the enterprise in a subtle way. The enterprise can also let the user download the APP of the mall through its own promotion, so the user becomes the user of the enterprise, and its viscosity increases. This performance also shows that the return rate of investment of the enterprise increases. Help enterprises to establish a stable marketing chain, obtain more market share, etc. Figure 1 is the basic logic of short video marketing under short video algorithm [4–7].

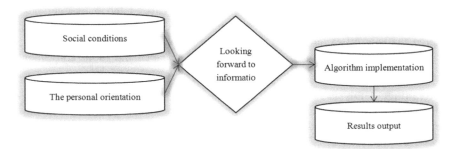

Fig. 1. Basic logic of short video marketing under short video algorithm

3 SWOT Short Video Algorithm Analysis

In order to better understand the short video algorithm, this paper will carry out relevant analysis based on SWOT method, referring to SWOT theory, including qualitative analysis and quantitative analysis of two steps, the specific analysis methods and contents of each step are as follows.

3.1 Qualitative Analysis

To facilitate analysis, this paper takes "Douyin", a popular short video APP, as an example to conduct qualitative analysis of this short video platform from the perspectives of internal production factors, external market environment and national policies. Qualitative analysis can be divided into four steps, including strength analysis, weakness analysis, opportunity analysis and threat analysis, as follows [8–10].

Advantage Analysis
According to the current situation of Douyin APP, compared with other similar short video apps, Douyin has the following main advantages: First, Douyin is one of the first new media enterprises in China to use algorithms to push videos or advertising information, which also helps Douyin attract more investment in advertising. Some enterprises in other fields have made huge profits through Douyin. This appearance to trill soon aware of the importance of the technology itself, it has been actively follow-up development in the technology development, since the algorithm has perfect technology system, relatively more mature, this is after the trill to firmly grasp the audience fragmentation of the main reason for the market, such as trill algorithm system has a strong independent learning ability, Through continuous learning and training, the information push path can be constantly optimized, so Douyin has advantages in algorithm technology. Second, thanks to the algorithm advantages, trill more precise positioning for the users, the mode of production for the trill new UGC pattern, this pattern is equivalent to trill user interaction with the link, it helps make the KOL marketing trill, under long-term let users more receptive to advertising, combined with the algorithm of positioning precision users, trill advertising content is more easily accepted by users, There is even a phenomenon that "users do not want to see advertisements on platforms other than Douyin". Third, trill platform or the nature of the media, so you need to update their own media forms, so the trill developed a broadcast medium in the near future, and quickly realized the marketing business integration, this makes the trill position in the market have new change, successfully gathered the enterprises and users, the two sides through information interaction, can more quickly complete marketing, This makes it cheaper and more convenient for enterprises to acquire market share.

Disadvantage Analysis
Although Douyin platform has made outstanding achievements in marketing and other aspects, it has some disadvantages in other aspects. The overall platform is not as perfect as it appears. Compared with other platforms, The main similarities of Douyin are as follows: First, according to trill among all users of more than 35% of users are minors, nearly half in the remaining 65% are from 20 to 25 people, these

younger users, values, attitudes, self control ability is low, trill marketing may lead to excessive consumption, even indirectly caused some malignant events, Such performance shows that Douyin has disadvantages in user management, and such disadvantages make Douyin face huge risks. Second, trill main short video creation slogan since its inception, attracted a large number of video creators, but because the content regulation, short video content is uneven, there is no lack of among them there are some vulgar, values orientation error, bad information of low quality content, so compared to other platforms, trill short video content controls exist disadvantages.

Opportunity Analysis
Douyin platform still has a very broad development prospects, but to seek the next development, we must find the right opportunity. From this perspective, There are two main opportunities for Douyin: First, trill has the good social attribute, can help users through decentralized content recommendation algorithm and the user, the user and enterprise establish a real network, according to the user's location, for example, recommend to the city within the scope of corporate advertising, so that users will be more trust trill and advertising enterprise, explain trill can rely on the social attribute of continuously forward good development; Second, trill development live business now favored by the vast number of companies, especially the electric business enterprise is the most respected, which led to the trill live and electrical business cooperation, and cooperation from the first beginning achieved good results, so the constant cooperation will generate more revenue, can constantly promoting the development of both sides.

Threat Analysis
At present, the success of Douyin has attracted the attention of people in the industry, so a variety of powerful competitors soon emerged in the industry, the first of which are Kuaishou, Tencent, etc. The existence of competitors is bound to bring threats, which are mainly reflected in two aspects: First, competitors also have strong capital, and many parties will compete to win more users. Therefore, Douyin may lose users in the face of competition, which is a big threat. Second, The success of Douyin depends on the short video algorithm. However, the algorithm will be tired in the long-term operation and can no longer provide fresh information to users. In this case, the user loss will be aggravated, so algorithm fatigue is a potential threat.

3.2   Quantitative Analysis

Quantitative analysis of short video algorithm can be divided into four steps: IFE and EFE matrix construction, strategic index calculation, strategic azimuth calculation and strategic intensity calculation, as follows.

IFE and EFE Matrix Construction
For convenience, taking the four indicators of SWOT algorithm, such as technical support advantages, inadequate content control disadvantages, opportunities of social attributes and threats of user loss, as examples, Delphi method is used to construct IFE and EFE matrices (IFE matrix contains strengths and weaknesses, while EFE matrix contains opportunities and threats), as shown in Table 1 and Table 2.

Table 1. Ife matrix

Indicators	The weight
Algorithm technology support advantages	0.214
Content control is not in place disadvantage	0.125

Table 2. The efe matrix

Indicators	The weight
Social attribute opportunity	0.115
Threat of user churn	0.185

Calculation of Strategic Indicators

The calculation method of strategic indicators is shown in Formula (1), which is the basic model and can be used to calculate indicators of different dimensions in IFE and EFE matrices. In the process, only the index dimension needs to be replaced.

$$S' = \sum_{i=1}^{n_s} S_i (1 \leq i \leq n_s) \tag{1}$$

In the formula, S is the strength, which can also be replaced by the weakness, opportunity and threat index. n is the dimension of the index, i is the index, and s is the weight of the index.

Calculation of Strategic Azimuth

The calculation of strategic azimuth can be expanded based on the model shown in Fig. 2.

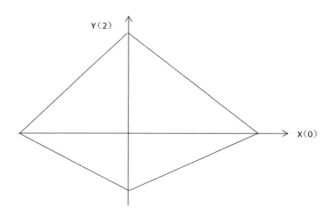

Fig. 2. Calculation model of strategic azimuth

According to the model, by substituting the function and the calculation results of all strategic indicators, we can confirm which quadrant is the center of gravity of the overall marketing, which is the direction of marketing.

Calculation of Strategic Intensity

The calculation of strategic strength is mainly divided into two steps: first, the calculation of strategic positive strength, the method is advantages × disadvantages; Second, the strategy negative intensity is calculated as opportunity × threat.

4 Enterprise Short Video Marketing Strategy

Through the above methods, this paper analyzes the marketing situation of Douyin short video enterprise, and concludes the marketing strategy most needed by the enterprise, namely, strengthening video content management, based on the marketing direction: For trill short video platform can be seen, analyzing the disadvantages and threats of trill platform if not as soon as possible in order to strengthen the management of video content, continue to spread the short low quality video, is bound to cause the attention of governments, could lead to a platform is banned, show video content on the issue is not only a disadvantage, is also a kind of threat, at the same time video content also has relationship with other threats, In addition, strengthening video content management is more conducive to Douyin to seize opportunities and continue to develop. For example, if Douyin's video content is managed, it can better play the advantages of algorithm technology and seize opportunities of social attributes. Therefore, strengthening the video content is advantageous to trill in many aspects, and strengthen the work there is urgency, so companies to focus on this, improve the quality of video content, avoid information spreading vulgar, false ideas, other enterprises shall cooperate on strategy trill short video content reform requirement, pay attention to their own content creation, so that to do a good job of marketing.

5 Conclusion

In conclusion, trill, short video platform in the modern social environment development speed very quickly, but rapid development lead to a short video platform to internal reform, so many problems, these problems not only results in a short video platform itself is affected, can also lead to other companies can not through the short video platform for high quality marketing. From this point of view, we should actively use the short video marketing algorithm to find the accurate reform target and gradually complete the reform. Enterprises should coordinate with the platform to synchronize the reform and clarify the correct marketing strategy.

References

1. Soria Morillo, L.M., Alvarez-Garcia, J.A., Gonzalez-Abril, L., et al.: Discrete classification technique applied to TV advertisements liking recognition system based on low-cost EEG headsets. Biomed. Eng. Online **15**(1), 75 (2016)
2. Korovyanko, O.J., Reydecastro, R., Elles, C.G., et al.: Optimization of a femtosecond Ti: sapphire amplifier using an acousto-optic programmable dispersive filter and a genetic algorithm. In: Lasers & Applications in Science & Engineering. International Society for Optics and Photonics (2006)
3. Onuma, K., Tong, H., et al.: TANGENT: a novel, 'Surprise me', recommendation algorithm. In: Proceedings of the 15th ACM SIGKDD International Conference on Knowledge Discovery and Data Mining, pp. 657–666 (2009)
4. Li, J., Ye, Z.: Course recommendations in online education based on collaborative filtering recommendation algorithm. Complex **2020**, 6619249:1–6619249:10 (2020)
5. Qian, Y.: Application of collaborative filtering algorithm in mathematical expressions of user personalized information recommendation. Int. J. Comput. Intell. Syst. **12**(2), 1446–1453 (2019)
6. Choi, S.M., Ko, S.K., Han, Y.S.: A movie recommendation algorithm based on genre correlations. Expert Syst. Appl. **39**(9), 8079–8085 (2012)
7. Rong, H.G., Huo, S.X., Hu, C.H., et al.: User similarity-based collaborative filtering recommendation algorithm. J. Commun. **35**(2), 16–24 (2014)
8. Klann, J., Schadow, G., Mccoy, J.: A recommendation algorithm for automating corollary order generation. In: AMIA. Annual Symposium proceedings/AMIA Symposium. AMIA Symposium, vol. 2009, p. 333 (2009)
9. Ge, Z., Liu, X., Li, Q., Li, Y., Guo, D.: PrivItem2Vec: a privacy-preserving algorithm for top-N recommendation. Int. J. Distrib. Sens. Netw. **17**(12), 155014772110612 (2021)
10. Chen, H.: Discussion on new marketing mode based on collaborative recommendation algorithm in big data era. In: ICISCAE (ACM), pp. 454–457 (2021)

Integrated Power Information Scheduling Algorithm Based on Middleware Technology

Mingyang Yu[✉], Zhiyong Zha, Bo Jin, Xing He, and Jie Wang

Information and Communication Branch, State Grid Hubei Electric Power Co., Ltd., Wuhan 430077, China
24870875@qq.com

Abstract. In order to improve the quality of power information integrated scheduling, this paper studies the power information integrated scheduling algorithm based on middleware technology. In this study, the significance of power information integrated scheduling is introduced first, then the algorithm is discussed by taking the node location and screening of power information integrated scheduling as an example, and the complete cycle of power information integrated scheduling is constructed through the algorithm. Finally, comparative experiments are carried out to verify the effectiveness of the algorithm. Through the research, the power information integrated scheduling algorithm is applied effectively under the support of middleware technology, which can improve the quality of scheduling work.

Keywords: Middleware technology · Power system · Power information integrated scheduling algorithm

1 Introduction

Middleware technology is a kind of technology applied between operating system and application program. The main application embodiment is called "middleware". Under the long-term development, this technology is gradually integrated into various operating systems, so it is also considered as an important part of operating system. The application form of middleware technology is generally integration, that is, a group of related middleware is integrated together, so as to obtain a platform, which is composed of development and operation of two sub-sections, to meet the basic needs of operating system operation. Middleware technology in the practical application of inevitably contains communication middleware, or other middleware is unable to communicate, not form a whole system, it is difficult to put into practical application, and because of the existence of the communication middleware, middleware technology has also had the build platform, supporting communication two big functions, namely the middleware + communication platform, according to the description, In a narrow sense, middleware specifically refers to middleware in distributed system, that is to say, any middleware can not be called middleware as long as it is not applied to distributed system. This definition also effectively distinguishes middleware, support software and functional software, which will make the concept more clear.

V. Sugumaran et al. (Eds.): ICMMIA 2022, LNDECT 138, pp. 278–285, 2022.
https://doi.org/10.1007/978-3-031-05484-6_35

Middleware in distributed system is mainly applied to the underlying operating system, middleware and intervention can significantly reduced the complexity of the underlying system, so that developers can in a standardized, simplistic environment development, program design and development of more efficient, it also can to a certain extent, optimize costs, Therefore, middleware technology has been widely concerned. Under this background, middleware technology has been gradually applied to different fields, including power information management. For example, people can integrate and dispatch power information using relevant algorithms based on middleware technology, which can greatly improve work efficiency and quality. However, it should be noted that middleware can only solve the complexity of the environment, and cannot determine the algorithm itself. Therefore, related researches are needed to realize the optimization of power information integrated scheduling based on middleware technology.

2 Significance of Integrated Power Information Scheduling

Electric power information integration is refers to the related organizations in the operation of power system, the system operation is caused by all kinds of information integration, the scheduling of a work, namely the scale of modern power system in China is very large, has achieved the national basic coverage, as well as very complex internal structure, contains all kinds of electric power engineering, electric power facilities, This case management of power system is hard to through artificial mode, due to the large scale and complex structure will not only lead to management work harder, and contact with more realistic factors, between osmosis and artificial ability is limited, unable to cope with work difficulty, also can't work in a complex environment. This point of view, there is an urgent need to through other ways to perform the power system management, so after a period of groping forward power information management, its main is installed in the power system, all kinds of information collecting device, so that the internal communication networks are formed, in all kinds of power engineering and power facilities system can communicate with each other, And communication of information can be workstations to receive, artificial can be directly on the workstation based on information of power system operation situation, found that the problem in order to adjust, in the process of the electric power information communication mode is divided into two kinds: integration, scheduling first, integration is to specify the target information is concentrated to the terminal workstation; Second, dispatching is a communication mode that is dominated by dispatching center and transmitted with other power institutions to reflect system working conditions, or control and adjust working conditions information. You can see that the electric power information integration is of great significance, it can make up for the inadequacy of human model, makes the power management work can continue to operate in the modern environment, stable electric power energy services for the general population in China, at the same time also give full play to the power system maintenance, maintenance, management of resources, reduces the power system failure rate, Therefore, integrated power information scheduling must be realized. Figure 1 is the basic framework of power information integrated scheduling [1–3].

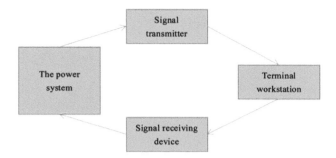

Fig. 1. Basic framework of power information integrated scheduling

3 Power Information Integrated Scheduling Algorithm Analysis – Taking Power Information Integrated Scheduling Node Location and Screening as an Example

3.1 Introduction of Power Information Integrated Scheduling Node Location and Screening

Actual effect for electric power information integration is to realize the communication of information, but in the process, how to accurately find the communication object is a key problem, the solution to the problem is before communication node positioning and selection, through node localization can confirm communication object location, but simply to node positioning is the positioning error may occur, So the node localization results is not the only, but several objects may be the best communication nodes, this case will be filtered, to confirm the best from the object node may be the best communication node communication object, is to make a choice, so the information integration scheduling node location and filter link is very important [4].

3.2 Algorithm Analysis

The following will analyze the power information integrated scheduling algorithm, mainly introduces the advantages of the algorithm, algorithm application process.

Advantages of the Algorithm

In electric power information integration scheduling node localization and screening of link, mainly using the traditional algorithm of positioning, although can realize node selection, but the response speed of performance problems, namely the traditional scheduling node localization algorithm and power information integration and screening link logic of compatibility is not good, so will happen in the practical application of the phenomenon of slow response. The algorithm proposed in this paper does not have such defects, which is the main advantage of the algorithm [5].

Algorithm Application Process

In the process of power information integrated scheduling node location and screening, the application process of the algorithm in this paper can be divided into four steps, as follows.

(1) Node positioning application

In the node location step, it is necessary to read the task information sent by the power enterprise before the power information integration scheduling, so as to adjust the subsequent information integration scheduling work and find the accurate node location. Based on this, find power system by means of middleware technology in wireless sensor network, as the source of integrated scheduling information, and then to conditional transformation of information data, select all eligible node as a target, at the same time to define the scheduling task executes instructions, instructions must be in priority integration status, if not in the state, should be adjusted, Or stop locating and resend. After the above operations are completed, combined with the information integration status, the grid information scheduling is taken as the priority execution condition to find the location, and the node location can be expressed by formula (1) [6–8].

$$y(n) = \sum_{j \geq 1}^{m+1} D_j(n+m) \tag{1}$$

In the formula, $y(n)$ is the locating node, n is the number of locating nodes, m is the number of information filtering transformation, j is the power information selection transmission path to perform the scheduling task, and D is the sum of squares of sequence distribution of locating nodes.

(3) Node reading and filtering

Through node positioning, several nodes that may be the best communication objects are found, but to confirm the best communication objects, it is necessary to conduct screening, and node screening must be established on the basis of understanding the actual situation of each node, so each node needs to be read first. Before reading the node, the staff should first confirm whether the integrated scheduling of power information is in the ready state. The node can be read only in the ready state, because the I/O channel of power information in the non-ready state cannot be converted. The method of node reading is relatively simple. It is mainly to comprehensively evaluate the communication quality of each node according to the relevant indicators of communication quality. Node after reading, if the node is larger, the communication between the quality gap you can choose the optimal node directly, but if the gap is small, it is difficult to distinguish between good or bad, then recommend the use of maximum likelihood estimation method for screening, namely in different nodes, which affects as the basis, when a node, which affects for PI, on behalf of the level of the node, In this case, nodes lower than the horizontal plane of the node are removed, so that a better choice can be made. When the collinearity of a node is $\pi/3$, the minimum error of nodes should be calculated according to the set node screening prefabrication, and accurate selection can be made according to the results [9, 10].

(3) Power information integrated scheduling time cycle planning.

To carry out the power information integration and scheduling work, the cycle planning of the power information integration scheduling time must be carried out combining with the node positioning and screening results of the power information integration scheduling. During the process, the work should be carried out according to the five states in Fig. 2 to obtain the complete cycle of the power information integration scheduling.

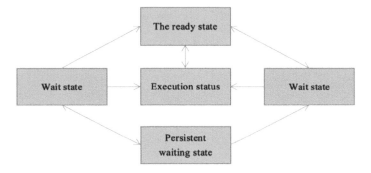

Fig. 2. Complete cycle of power information integrated scheduling (five states)

(4) Power information integrated scheduling process design

Scheduling time cycle of the algorithm and power information integration planning, integrated scheduling need in strict accordance with the scheduling time classifying information, in order to avoid in the process of scheduling information synchronous transmission information variables are too many problems, can work in using middleware technology to realize distributed heterogeneous computing, so to be able to confirm whether there is any unknown node scheduling information, If there are unknown nodes, the middleware technology can be used to locate them, but each time the unknown node is located, the cycle starts from the first step. After positioning, the information will be transmitted to the terminal workstation synchronously. See Fig. 3 for the transmission process.

The database in Fig. 3 is mainly used to store unknown node data and information, so it can be defined as unknown node database. Depending on the information in this database, the reliability of power information of cluster data middleware can be calculated with the help of Boolean operation method, and transmission can be carried out when the reliability reaches the standard. It also means that any information that is not reliable enough cannot be transmitted.

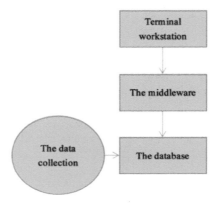

Fig. 3. Power information integrated scheduling process

4   Comparative Experiments

4.1   Preparation

In order to verify the effectiveness and advantages of the proposed algorithm, the response speed of the proposed algorithm in power information integrated scheduling is taken as the benchmark, and the proposed algorithm is compared with the traditional algorithm. At the same time, in order to ensure the practical experiments, this paper chose a certain electric power enterprises as the object, according to current situation of the electric power information integration experiments, collect the object of study for a month before and after all the scheduling information, and then the traditional algorithm and the algorithm is used to read information, information integration and scheduling and simulation process, the method of operation at the same time, After completing a cycle of integration and scheduling according to the same process, the response speed of the two algorithms can be judged and compared. In addition, scheduling tasks simulated in the experiment are shown in Table 1.

Table 1. Scheduling tasks simulated in the experiment

The task sequence	Task execution time required/ms	Priority
1	20	Higher
2	20	Higher
3	20	Higher

4.2   Result Analysis

Through comparative experiments, this paper realized two algorithms supported by electric power information integration scheduling tasks can be completed within task execution time requirements, therefore two algorithms have certain application value, but the response speed of the two algorithm in the usage of different task execution time, according to the results of this difference can be drawn. Firstly, the traditional algorithm occupies 5 s, 9 s and 7 s respectively in the three tasks, and the remaining time is mainly used for task execution. Secondly, the algorithm in this paper occupies 1 s, 1 s and 4 s respectively in the three tasks, and the remaining time is also used for task execution. This time of the algorithm takes the maximum < time takes up the minimum value of traditional algorithm, so the effective algorithm in this paper, and it has advantages, also shows that this algorithm in actual application of can better improve the efficiency of electric power information integration of task execution, faster response speed, shorter timescales and can be applied to the execution time demanding higher integration and scheduling tasks.

5 Conclusion

To sum up, with the support of middleware technology, power information integrated scheduling was strongly supported, with more convenient environment and simpler operation. At that time, to better realize power information integrated scheduling, algorithms with better compatibility with middleware technology were needed to provide help. Therefore, power enterprises should carefully choose the algorithm and do not recommend to use the traditional algorithm. The algorithm in this paper can provide reference help.

References

1. Shin, Y., Kim, S., Moon, I.: Integrated optimal scheduling of repair crew and relief vehicle after disaster. Comput. Oper. Res. **105**, 237–247 (2019)
2. Ebadi, R., Yazdankhah, A.S., Kazemzadeh, R., et al.: Techno-economic evaluation of transportable battery energy storage in robust day-ahead scheduling of integrated power and railway transportation networks. Int. J. Electr. Power Energy Syst. **126**(3), 106606 (2021)
3. Khalilpour, K.R., Gr Ossmann, I.E., Vassallo, A.: Integrated Power-to-gas and gas-to-power with air and natural-gas storage. Ind. Eng. Chem. Res. **58**(3), 1322–1340 (2019)
4. Dolatabadi, A., Jadidbonab, M., Mohammadi-Ivatloo, B.: Short-term scheduling strategy for wind-based energy hub: a hybrid stochastic/IGDT approach. IEEE Trans. Sustain. Energy **10**(1), 438–448 (2019)
5. Sohrabi, F., Jabari, F., Mohammadi-Ivatloo, B., et al.: Coordination of interdependent natural gas and electricity systems based on information gap decision theory. IET Gener. Transm. Distrib. **13**(15), 3362–3369 (2019)
6. Touboul, P.-J., Hennerici, M.G., Meairs, S., et al.: Mannheim carotid intima-media thickness and plaque consensus (2004–2006–2011). In: An Update on Behalf of the Advisory Board of the 3rd, 4th and 5th Watching the Risk Symposia, at the 13th, 15th and 20th European Stroke Conferences, Mannheim, Germany, 2004, Brussels, Belgium, 2006, and Hamburg, Germany, 2011. Cerebrovascular Diseases (Basel, Switzerland), vol. 34, no. 4, pp. 290–296 (2019)
7. Budhiraja, I., Kumar, N., Tyagi, S.: Deep reinforcement learning based proportional fair scheduling control scheme for underlay D2D communication. IEEE Internet of Things J. **8**, 3143–3156 (2020)

8. Mirzaei, M.A., Nazari-Heris, M., Mohammadi-Ivatloo, B., et al.: A novel hybrid framework for co-optimization of power and natural gas networks integrated with emerging technologies. IEEE Syst. J. **14**, 3598–3608 (2020)
9. Ma, A., Ji, J., Khayatnezhad, M.: Risk-constrained non-probabilistic scheduling of coordinated power-to-gas conversion facility and natural gas storage in power and gas based energy systems. Sustain. Energy Grids Netw. **26**, 100478 (2021)
10. Ouammi, A., Achour, Y., Zejli, D., Dagdougui, H.: Supervisory model predictive control for optimal energy management of networked smart greenhouses integrated microgrid. IEEE Trans. Autom. Sci. Eng. **17**(1), 117–128 (2020). https://doi.org/10.1109/TASE.2019.2910756

Research on English Translation Correction Algorithm

Yanyan Deng[✉]

Ganzhou Teachers College, Ganzhou 341000, China
yan2002412@163.com

Abstract. To improve the accuracy of the translation software English results, this article, taking the research of translation correction algorithms, this paper discusses mainly the present situation of the English translation software, demonstrated to improve the accuracy of translation in the necessity, after the correction algorithm were analyzed, and a basic algorithm model, finally for validation algorithm model, first for translation in the process of system design, and then on the basis of this system, The advantages of the proposed algorithm can be verified by comparison with the two mainstream translation systems. Through this study, the algorithm system can effectively improve the accuracy of English translation results, and has certain advantages compared with mainstream translation systems.

Keywords: English translation · Correction algorithm · Machine translation

1 Introduction

In the modern international environment, the exchanges between China and other countries become more and more frequent, so people often come into contact with foreign languages in daily life, and English is the international common language, making it the most common foreign language in People's Daily life. In this case, in order to break the language barrier, people need to use English translation software. With the help of software, English can be translated into National language, which is easy to read and understand. Therefore, English translation software has become an important tool in modern social life. However, the human language system is very complex, and the software translation logic cannot be completely imitated, resulting in translation errors to a certain extent, indicating that the accuracy of software translation needs to be improved. At this time, it is necessary to carry out relevant research on English translation correction algorithms, which is of practical significance.

2 Current Situation of Software English Translation

Software English translation is the core part of natural language translation, which has high application value in modern social life, so it has been widely used. However, the previous English translation software pays more attention to translation efficiency when it is designed, so the language translation method adopted is more dependent on

V. Sugumaran et al. (Eds.): ICMMIA 2022, LNDECT 138, pp. 286–292, 2022.
https://doi.org/10.1007/978-3-031-05484-6_36

experience and focuses on literal translation. Therefore, the translation does not need to consider language grammar and semantics, and the translation results are simple but the efficiency is fast. In this case because modern English translation have high request for case database software [1–3], the traditional way of translation software can only according to the part of speech logo to identify words literally, so translation often result in such aspects as the sentence structure, grammar errors, the reason is that translation logic is too pure, made in the translation process will produce wrong iteration pass increase phenomenon, Therefore, translation results may deviate from structured examples, and the deviation increases with the increase of iterative delivery, indicating that the accuracy of translation results is insufficient. In this case, it is necessary to correct the algorithm to improve the accuracy of translation results, otherwise it will weaken the value of English translation software and hinder social communication at the international level [4–6].

3   Analysis of Correction Algorithm

Facing the current situation of software English translation, how to improve the accuracy of software English translation results is a very urgent problem at present. Therefore, in order to solve the problem and achieve the goal, this paper chooses dependency tree to string model and corresponding correction algorithm. The algorithm content will be analyzed below.

3.1   Dependency Tree to String Model

The core of the dependency tree to string model is the triplet <d, s, a>, where <d, s> is the translation pair, D is the dependency tree of the source language,S is the target language string of the source language, and A is the description of word alignment between D and S. For D and S, A is equivalent to the logic of tandem. Figure 1 shows the dependency tree to string bilingual alignment model [7–9]. </d, s></d, s, a>.

Fig. 1. Dependency tree to string bilingualism alignment model

The upper part of the figure is the dependency tree D of the source language, which contains several nodes, and each node has two attributes of part of speech and word. Below each word is the description of related word properties of English words. Table 1 introduces common description relations (from Stanford Parser part-of-speech tag set). There are also black lines in the words to show the relationship between words. The lower end is the corresponding English string sequence S of Chinese language, and the dotted line between it and the upper end is mainly used to represent the alignment between Chinese word nodes and English words [10].

Table 1. Common description relationships

The part of speech	Describe
Noun	NN
The verb	VV
Adjectives	JJ

3.2   Analysis of Lexical Semantic Similarity

Lexical semantic similar degree is the key step in the English translation, the role is to determine whether English words and Chinese are similar, if accord with standard of similar degree, will the corresponding words as a primary scheme, similar degree analysis was carried out on the other words, after completion of all the analysis take the most similar words as the best answer. In this regard, this paper adopts the algorithm in GEC evaluation model, which regards corpus as a data set, and then directly converts it into the corresponding set according to the basic meaning of each indicator in combination with algebraic statistics for calculation. For details, see Formula (1).

$$Precision = \frac{\sum_{i=1}^{N} |e_i \cap ge_i|}{\sum_{i=1}^{N} |e_i|} \tag{1}$$

where, e is the text data set after model error correction, g_e is the standard data set, N is an evaluation index, and i is the standard value of the index represented by N.

3.3   Log-Linear Model Construction

Although the word semantic similarity analysis and related calculation can confirm the optimal answer, it does not mean that the optimal answer is the most accurate answer. Therefore, further analysis is required, and this step analysis mainly relies on log-linear model. Logarithm linear model is a kind of more characteristics identification model of thinking, operation will be in the original words to f, for translation is set to e, and then through the analysis of the translation of the maximum entropy model, if the result shows that there is really no higher than the optimal answer semantics approximation degree of the answer, then you will be the best answer is defined as the most accurate answer, instead will replace the answer, and cycle analysis, Until you find the right answer. In addition, log-linear model has strong extensibility, can set corresponding characteristics according to different goals and requirements, and can also be compatible with a variety of linguistic measures. Therefore, the model can effectively improve software English translation results and play a correction role. At the same time, with the help of the model software, English translation can also carry out the forward translation probability, reverse translation probability, translation language model and other feature functions, according to the function based on the translation system, combined with the actual requirements of automatic setting of feature functions, privilege weight, finally get the translation with the highest weight value.

4 Design and Verification of Algorithm Model System

4.1 System Design

In order to facilitate verification, this paper will carry out system design first. The design process is divided into two steps, namely structural design and logical design. The details of each step are as follows.

Structural Design

Structural design is mainly to ensure that the system can be put into use, so focus on the application process of the system, functional architecture, the whole design process is divided into three links: Firstly, the web page is selected as the carrier of the system structure, and the online English translation software platform and corresponding function buttons are set up through interface design and VB language development. Users only need to log in to the web page and enter the source language in the interface, and then the system will carry out analysis; Secondly, in terms of functional structure, taking The Chinese language as an example, the platform must have the function of source language input, English-Chinese corpus, translation display, and multilingual compatibility. All functions will be paired with the function buttons developed by VB language for easy use. Thirdly, the application process is designed according to the application process of the mainstream online translation software in the market, as shown in Table 2.

Table 2. Design results of system application flow in structural design

The step sequence	Operation
1	The login page
2	Input source language
3	Click the translate button
4	Generate a translation

Logical Design

Logic design is relatively complex, generally can be divided into four steps: first, the reference Sato&Nagao measures analysis of interdependent institutions, will remain tree to the string of alignment model of the source language dependent, and then through the matching principle, excerpts from corpus within a corpus as test object, according to the corpus to the source language semantic information, such as the complete close corpus translation; Second, the three supporting expression methods of adding, filtering and replacing are selected to support the operation of the first step. If there is no target word string in the process, the dependency tree relationship layer of the word string is changed into a new statement expression form. Third, reference source language dependent logic tree to string model, set up the system logic, in order to "she bought an English book," the statement as an example, the statements in the corpus of prefix Numbers as "e", so if the system input in the translation of Chinese "she bought an English book," prefix will confirm through the analysis of the language source language in Chinese label, if the same is "e", It will identify all the English corpus with the prefix "E" and finally find the best answer "she bought an English book". In addition, following this logic, She bought an English book in Chinese and English; Fourthly, in combination with the log-linear model, it is necessary to confirm the specific performance of its characteristic function, which usually has two, as shown in Table 3.

Table 3. Specific performance of feature function of log-linear model

The characteristic function	The connotation of the
Forward translation probability and reverse translation probability	If the number of words is the same, the semantic similarity between the translated sentence and the translated sentence is higher, and the same number of words is more, so the feature function can generate accurate translation results more easily
Language model	In order to generate high-quality translation, it is necessary to test the preset results of translation through functions, find out the deficiencies and correct them, so as to improve the fluency and colloquial degree of translation sentences. The translation language model is established based on the quality sentences, and the target language contains the probability of translation fragments

4.2   System Verification

In order to verify the effectiveness of the system, the test corpus is selected from the Chinese-English news corpus in CWMT2015 official evaluation, in which the number of English and Chinese corpus is 427700, and the number of Chinese corpus is 955. Therefore, the English corpus is imported into the initial corpus, and then the corresponding Chinese corpus is prepared to be input into the system. If the system can accurately translate the Chinese corpus into the corresponding English corpus, it shows that the application of the system is effective. In order to determine whether the system has the advantage at the same time, this article also chose the semantic language machine translation system, open source statistical machine translation system as the control group, three systems were tested in the same way, if the test results show that the system error rate is lower and has higher accuracy, this system also this system has the advantage.

First of all, in terms of system effectiveness test, this paper input "On November 19, according to the news on the website of the Ministry of Culture and Tourism, on November 18, the Ministry of Culture and Tourism issued a public announcement on the Management Measures for Performance Brokers (Draft for Comments)". According to the results, it can be seen that the accuracy of the system translation results is 100%, and the translation grammar and semantics are not wrong, so the application of the system is effective.

Secondly, in terms of strength test, three translation software systems are tested based on corpus. In order to more accurately compare the three and objectively judge their advantages and disadvantages, this paper will make judgment according to the evaluation results of BLEU value, that is, BLEU value is an indicator of accuracy matching value. If the accuracy of translation is higher, the BLEU value will be higher. BLEU value will be evaluated for each test. The results show that the BLEU values of semantic language machine translation system and open source statistical machine translation system are 25.419 and 25.179 respectively, while the BLEU values of this system are 25.529. Therefore, this system has certain advantages, at least in terms of translation quality, it is better than the other two translation systems. Meanwhile, taking "Information industry shows the trend of rapid development" as an example, the translation results of semantic language machine translation system and open source statistical machine translation system are "Keeping the momengtum going" and "Fast change" respectively. It can be seen from this that The translation results of The two software are quite different from The original text. In contrast, The translation of this system is "The Information Industry shows a trend of rapid Development", and The translation results are at least more accurate than those of The two software. Further, this system has advantages. In addition, by summarizing the results of ten thousand tests, the error rate of semantic language machine translation system, open source statistical machine translation system and our system is 27%, 15% and 1% respectively.

5   Conclusion

To sum up, English translation software plays a great role in modern social life, but its value depends on the accuracy of translation results. If the accuracy is correct, the software will not play an effective role and the value will naturally decline. From this point of view, it is necessary to improve the accuracy of software translation results by correcting the algorithm, which can solve the defects of traditional English translation software and greatly improve the accuracy of translation results.

References

1. Peng, J., Araki, K.: Correction of article errors in machine translation using web-based model. In: IEEE International Conference on Natural Language Processing & Knowledge Engineering. IEEE (2005)

2. Cui, G.: Design of translation accuracy correction algorithm for English translation software based on rough set. In: ICISCAE (ACM), pp. 558–562 (2021)

3. Souza, N., Costa, K., Dmitriev, V.: Development of computational 3D MoM algorithm for nanoplasmonics. J. Microwaves Optoelectron. Electromagn. Appl. **12**(2), 569–579 (2013)

4. Dräxler, P., Kögerler, K., et al.: An algorithm for finding all preprojective components of the Auslander-Reiten quiver. Math. Comput. **71**, 743–759 (2002)

5. Cai, Y., Velegkas, G.: How to sell information optimally: an algorithmic study. In: ITCS 2021, pp. 81:1–81:20 (2021)

6. Binder, K., Block, B.J., Virnau, P., et al.: Beyond the Van Der Waals loop: what can be learned from simulating Lennard-Jones fluids inside the region of phase coexistence. Am. J. Phys. **80**(12), 1099–1109 (2012)

7. Chen, Y.: Business English translation model based on BP neural network optimized by genetic algorithm. Comput. Intell. Neurosci. **2021**, 2837584:1–2837584:10 (2021)

8. Zhang, G.: Research on the efficiency of intelligent algorithm for English speech recognition and sentence translatione. Informatica (Slovenia) **45**(2) (2021)

9. Joseph, J.K., Chathurika, W.M.T., Nugaliyadde, A., Mallawarachchi, Y.: Evolutionary algorithm for sinhala to English translation. CoRR abs/1907.03202 (2019)

10. Li, W.: Fast retrieval algorithm of English translation core words based on simhash. In: Zheng, X., Parizi, R.M., Loyola-González, O., Zhang, X. (eds.) CSIA 2021. AISC, vol. 1343, pp. 822–827. Springer, Cham (2021). https://doi.org/10.1007/978-3-030-69999-4_115

CV Process and Data Mining
for Multi-modal Informatics Systems

Pattern Recognition in Convolutional Neural Network (CNN)

Zhengyu Sun[✉]

University of Surrey, Guildford, UK
sunzhengyu1230l@163.com

Abstract. One of the main topics of human science research is to explain the mechanism of brain activity and the essence of human intelligence, create intelligent machines similar to human intelligent activities, and develop the technology and intelligent application of human intelligence. The research on pattern recognition using CNN technology is a hot topic. Therefore, this paper studies the problems faced by pattern recognition and explores it. Firstly, this paper discusses the research background and significance of CNN, and briefly introduces the characteristics of CNN. Then it summarizes the relevant articles at home and abroad. Then it briefly describes the BP neural model used in eep and its existing reasons, and explains the better effects they may bring. Finally, a series of experimental tests are carried out on pattern recognition. It is concluded that the network with scaling parameter set to 4 and depth of about 26 has the best effect, which can effectively solve the problem of pattern recognition.

Keywords: Convolutional neural network · Neural network · Pattern recognition · Problem exploration

1 Introduction

Machine learning is an application of artificial intelligence: through repeated learning of data, an algorithm that can process specified data is obtained. In the supervised learning mode, unknown data can be predicted according to the formed prediction function. Unsupervised learning aims to solve various problems. By learning the unlabeled training set, we can find the potential information, structural knowledge and patterns in the data of the training sample set. In semi supervised learning, labeled data are used for training. Add labeled data to unlabeled data to enhance the effect of unsupervised pooling, so as to enhance the effect of supervised classification [1, 2]. At any given level, the input of each neuron comes from the local region of the previous layer, plus the weight determined by a set of weights, we can say that the current layer is associated with a kernel function. The output of a specific layer, i.e. the characteristic diagram, represents the input of the next layer [3, 4].

There are many related studies investigating pattern recognition in CNNs. Some researchers have proposed a multilayer feedforward network fault feedback propagation algorithm, called BP algorithm [5, 6]. In addition, some researchers at New York University use CNN to classify images and propose the famous lenet-5 structure. The network has seven layers (excluding input), first a convolution layer, then a lower sampling layer, then another convolution layer and a lower sampling layer, and finally three fully connected layers [7, 8]. The convolution layer uses local connection and weight distribution, and the lower sampling layer uses 2 * 2 regional average pooling. The above is the previous research results, which has laid a deep foundation for this study.

Firstly, this paper introduces the history and research status of neural network. On this basis, the BP artificial neuron model and its implementation are discussed, and its principle, structure and design flow are given. Secondly, the convolution algorithm is described and an example is given to illustrate its application in pattern recognition, including training set data extraction (SOM) method and the solution of learning and testing classification problem. Finally, the experimental case verification results and demonstration process are briefly described.

2   Discussion on Pattern Recognition Based on CNN

2.1   CNN

CNNs are similar to biological neural networks [9, 10]. It represents a multi-layer neural network, which has high interference suppression to image rotation, translation and scaling. Weight sharing can reduce network parameters, accelerate network convergence, and reduce network complexity to a certain extent [11, 12]. By learning and studying the relevant knowledge of CNN, the CNN has the following characteristics: (1) the larger the image sample database of the training model, the wider and deeper the CNN of the training model. (2) In order to avoid the complex image preprocessing operation in traditional algorithms, CNN can directly drive the original image. (3) The more complex the structure of the neural convolution network model, the greater the computational complexity of the network and the longer the time required for model training. This end-to-end learning algorithm avoids the errors caused by complex operations such as preprocessing, segmentation, noise reduction and feature extraction, and makes the image recognition process more scientific. The typical convolution network model architecture is mainly composed of input layer, convolution layer, pooling layer and full connection layer, as shown in Fig. 1.

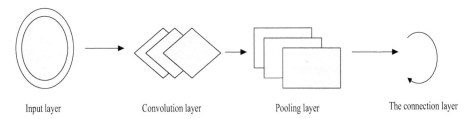

Input layer   Convolution layer   Pooling layer   The connection layer

Fig. 1. Typical neural network model

The parameters to be trained on the convolution layer represent a series of convolution kernels. Each convolution layer can contain many different convolution kernels. The recognition data searched by each convolution layer is a new image. The information learned by convolution layer neural network is applied to other parts of the recognition model, and the generated value is the weight of convolution kernel. The local connection and weight sharing of convolution layer significantly reduce the number of parameters in the network and reduce the complexity of training. The convolution layer is responsible for extracting the features of different local regions from the upper input image, and its output shape is:

$$C_j^{(l)} = g(\sum_{i \in U} C_I \times K + B) \tag{1}$$

The symbol * in formula (1) represents the convolution operation, and u represents the selection from the input characteristic diagram. All characteristic diagrams of layer (i − 1) connected with the jth characteristic diagram of layer L; Ci is the output value of the ith characteristic diagram of layer (i − 1); K is the convolution kernel connecting the jth characteristic graph and the ith characteristic graph of layer L. Layer (1-1) is represented by a two-dimensional matrix, and B is the offset of the j-th characteristic map of layer I.

2.2   Accretion Operation

In CNNs, a pool layer is usually inserted after the convolution layer. Its main function is to reduce the dimension of the output of the convolution layer, that is, it is also completed by moving a structure similar to the completed convolution kernel, but the filter of the pooling layer does not have to cover the whole depth in the sliding process, but can only move one node in depth. Of course, the pooling kernel also needs to manually set parameters such as size, fill method, slip step length, etc.

(1) Average pooling. This operation averages only the pixels in the neighborhood. There was an error during feature extraction. Due to the limited choice of neighborhood size, the variance of the final estimation will be greater. Using the average pooling operation can effectively reduce this error and preserve the image to a greater extent. Background information.

(2) Maximize pooling. The goal of this operation is to maximize the pixels of the neighborhood. Such errors always appear in feature extraction, that is, the error of convolution parameters, which eventually leads to the deviation from the estimated average value. Maximum pooling effectively reduces this error and further preserves the texture information in the image.

(3) Stack pooling. As the name suggests, this means that pooled operations overlap when performing operations, and studies have shown that this method can also reduce errors and reduce the error rate.

(4) Space pyramid pool. This is a very powerful integration operation. The spatial pyramid grouping operation can convert the convolution features of any scale image into features of the same dimension, because when fully connected, all the features to be connected must have the same dimension. Pyramid pool can achieve this goal.

The output form of the pooling layer is Eq. (2), where the down () function is the down sampling function.

$$A_j^i = \text{down}(A_j^{i-1}) \tag{2}$$

The traditional CNN usually connects one or more full connection layers after passing through multiple convolution layers and pooling layers. Its output form is:

$$a_j^i = f(\sum_i^s a_i(i-1)w_{ij} + b) \tag{3}$$

The main purpose of the full connection layer is to combine the classification learned in the convolution layer or pooling layer with the local information. However, the full connection layer will occupy most of the parameters in the network, resulting in over fitting. Therefore, in the 2014 googlenet model, the full connection layer is removed and replaced by the global average pooling layer, which makes the model training faster and reduces over fitting.

2.3   Identification of CNN

Deep learning requires a lot of data. The more tag data, the better the performance of the model. Researchers at home and abroad have failed to further improve the performance of deep learning model for a long time. There are many factors at work: first, a depth model with many representations requires a large amount of labeled data to perform better than other traditional methods based on small data sets. For example, the most credible studies are based on several data sets provided by UCI, many of which contain only hundreds to thousands of images. The situation has changed since Li contributed to the Imagenet database in 2009. In addition to the Imagenet database, cifar-10 and fashionmnist are the most commonly used data sets to verify the efficiency of the algorithm.

As the driving force behind deep learning, data is an indispensable part. At present, various defects contained in MNIST data sets have prompted many researchers to look for new data sets. Generally speaking, things are understood locally and then as a whole: for the pixels of the image, the surrounding pixels are more relevant, while the distant pixels are less relevant. This principle also benefits from human intelligence: in fact, some neurons in the biological visual system only respond to stimuli in specific areas. So much data is difficult to train and may even be impossible to achieve. In training, we must find a way to reduce the number of parameters.

2.4   Advantages of CNN

(1) Local perception. In general, it is easy to look at the local area first, but in general, a pixel in the image has greater correlation with the surrounding pixels, which refers

to the known pixels, which are far away, and the correlation is very weak or irrelevant. Therefore, it is not necessary for every neuron to perceive the big picture. The idea it contains is inspired by this idea.

(2) The division of weight is also called the division of parameters. Convolution uses this type of operation, which is very effective in reducing the number of parameters.

3 Experiment

3.1 Experimental Environment

Table 1 shows the machine configuration used in this experiment.

Table 1. Machine configuration

Processor	E5-2650 3.0 Ghz
Internal storage	128 GB
Fixed disk	2 * 156 GB 20k
GPU	TWO NVIDIA TESLA

Tensorflow used in this paper is the most active deep learning framework at present. It uses data flow graph for calculation, in which nodes represent mathematical operations and lines represent the interaction between tensors. At present, tensorflow can be deployed on one or more CPU hosts, and even on mobile devices.

3.2 Experimental Process

The random momentum gradient descent method was adopted in our experiment. The momentum values were set as 0 and 9, and the failure probability P was set as 0.3. The experiment was run on two NVIDIA Tesla K20C Gpus with a small batch of 50 samples on each GPU. Before transmitting the image data to the network, necessary preprocessing work can significantly improve the accuracy of the network. During the molding process, we filled each image with 0 in four directions to make it a 36 * 36 image, then randomly cut it into a 32 * 32 image, then randomly rotated left and right, and processed the image with ZCA lighting. It doesn't interfere with the sequence. So far, we have given the image preprocessing process and the complete network structure. The whole experiment process is shown in the Fig. 2.

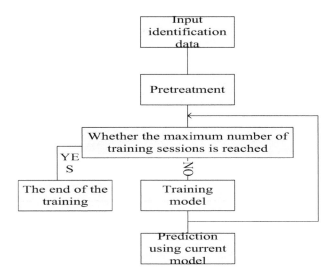

Fig. 2. Experimental procedure

4 Discussion

4.1 Identification Error Rate Analysis

Table 2 is the data table of model recognition error rate based on CNN.

Table 2. Model identifies error rate data

Model serial number	Scaling parameter m	Network depth	Number of participants	Test set error (%)
1	1	69	15.2M	4.49
2	3	81	177M	4.53
3	2	126	27.8M	4.43
4	1	57	37.1M	4.37
5	3	32	44.2M	4.96

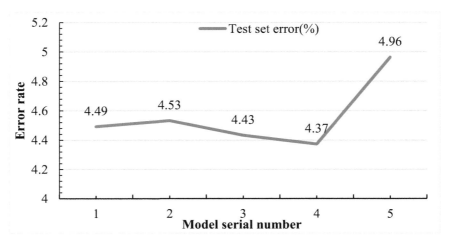

Fig. 3. Identify error rates

As can be seen from Table 2 and Fig. 3, with the increase in the m-scale parameter, although the network depth is decreasing, the model continues to become increasingly complex and more parameters need to be trained, but, as a result. The error rate of classification of the model in the set of tests is becoming ever lower. The scale parameter of model 8 is 5, which is greater than that of model 7. The number of parameters is ten meters longer, but the accuracy decreases. According to our analysis, this may be due to the too complex structure and too many parameters of model 8, resulting in increased training difficulty and slight over fitting. Therefore, in terms of our current experimental results, the network with scaling parameter set to 4 and depth of about 26 has the best effect.

5   Conclusion

This paper mainly describes the research of CNN. In this paper, we propose several methods to recognize the image in EEG mode, and give the flow charts of the three algorithms. Finally, we compare and analyze the two effects to determine which is more suitable for the whole video model and its application scene. Through comparison, it is found that each algorithm has its own advantages and disadvantages, and there is a certain relationship between different types. This paper is also the reason for the expansion of the range of weight selection.

References

1. Li, D., Zhao, X., Yuan, G., Liu, Y., Liu, G.: Robustness comparison between the capsule network and the convolutional network for facial expression recognition. Appl. Intell. **51**(4), 2269–2278 (2020). https://doi.org/10.1007/s10489-020-01895-x

2. Kong, Q., Cao, Y., Iqbal, T., et al.: PANNs: large-scale pretrained audio neural networks for audio pattern recognition. IEEE/ACM Trans. Audio Speech Lang. Process. **28**, 2880–2894 (2020)

3. Dai, Y., Qiu, D., Wang, Y., et al.: Research on computer-aided diagnosis of Alzheimer's disease based on heterogeneous medical data fusion. Int. J. Pattern Recogn. Artif. Intell. **33** (5), 1957001.1–1957001.17 (2019)

4. Zhang, Q., Zhang, M., Chen, T., et al.: Recent advances in Convolutional Neural Network acceleration. Neurocomputing **323**, 37–51 (2019)

5. Ahmed, T., Parvin, M.S., Haque, M.R., et al.: Lung cancer detection using CT image based on 3D CNN. Comput. Commun. (English) **3**, 35–42 (2020)

6. Afridi, M.J., et al.: On automated source selection for transfer learning in CNNs. Pattern Recogn. J. Pattern Recogn. Soc. **73**, 65–75 (2018)

7. Nguyen, H.D., Yeom, S., Lee, G.S., et al.: Facial emotion recognition using an ensemble of multi-level CNNs. Int. J. Pattern Recogn. Artif. Intell. **33**(11), 139–149 (2018)

8. Banharnsakun, A.: Towards improving the Convolutional Neural Networks for deep learning using the distributed artificial bee colony method. Int. J. Mach. Learn. Cybern. **10** (6), 1301–1311 (2019)

9. Liu, W., Wang, Z., Liu, X., et al.: A survey of deep neural network architectures and their applications. Neurocomputing **234**, 11–26 (2017)

10. Zahangir, A.M., Mahmudul, H., Chris, Y., et al.: Improved inception-residual convolutional neural network for object recognition. Neural Comput. Appl. **32**, 1–15 (2017)

11. Wu, Z., Jiang, S., Zhou, X., et al.: Application of image retrieval based on Convolutional Neural Networks and Hu invariant moment algorithm in computer telecommunications. Comput. Commun. **150**, 729–738 (2020)

12. Ashiquzzaman, A., Tushar, A.K.: Handwritten Arabic numeral recognition using deep learning neural networks. In: IEEE 2017 IEEE International Conference on Imaging, Vision & Pattern Recognition (icIVPR) - Dhaka, Bangladesh, 13–14 February 2017, pp. 1–4 (2017)

Ancient Village Culture Based on Big Data Technology

Fugang Li[1(✉)] and Insaf Abdullah[2]

[1] Zibo Normal College, Zibo 55130, Shandong, China
59133995@qq.com
[2] University of Sulaimani, Sulaimaniyah, Kurdistan Region, Iraq

Abstract. With the rapid development of China's economy and the continuous improvement of people's living standards, people's demand for material and spiritual aspects is higher and higher. Under such a background, traditional culture is facing great challenges. In recent years, China's tourism industry has risen rapidly. At present, there is no unified standard for the development and management of ancient villages, and there are even phenomena that no one is responsible in many places. Tourists can understand the ancient village culture through various media platforms, and can also use some social software to communicate online and browse the information of scenic spots and related history on the web. This paper mainly discusses the problems existing in the construction of ancient village culture under big data technology, in order to study the ancient village culture based on big data technology and understand big data technology and its advantages. This paper uses statistical analysis and experimental research methods to study the ancient village culture based on big data technology. The survey results show that most respondents believe that the promotion effect of ancient villages is relatively good, and most foreign respondents obtain ancient village information from TV, mobile phone and the Internet. On the basis of the Internet, more people get information about ancient villages. The research on ancient village culture based on big data technology is a subject of great research value.

Keywords: Big data technology · Ancient village culture · Data collection · Cultural research

1 Introduction

BDT has affected and changed the lifestyle and daily behavior of almost everyone. People are also exploring the connections and relationships between things. The arrival of the big data era is unstoppable. Therefore, we need to pay close attention to the impact and value of BDT, as well as the infinite opportunities and challenges, advantages and risks it brings.

At present, there are many related researches in the field of big data and ancient village culture. For example, A scholar said that big data technology brings new possibilities to the protection of ancient villages, and the rich cultural information carried by ancient villages can be preserved more completely and permanently [1]. A scholar proposed to promote the transformation and innovation of traditional village cultural resources products with big data technology and promote the development of new business forms of traditional village culture in the era of big data [2]. A scholar believes that the construction of characteristic towns from the perspective of traditional ancient village culture protection has become the focus of national attention [3]. Therefore, this paper has great practical value to study the ancient village culture based on big data technology.

The research contents of this paper are as follows: introduce big data technology and its advantages, understand ancient village culture, the composition and value of ancient village cultural resources, understand the collection work and difficulties of ancient village culture big data, investigate and analyze the promotion effect of ancient village culture and the ways and channels for respondents to obtain ancient village information. Finally, the conclusion is drawn: most respondents believe that the promotion effect of ancient villages is relatively good. On the basis of television and other media and the Internet, more people get information about ancient villages.

2 Big Data Technology and Ancient Village Culture

2.1 Big Data Technology

In today's world, the characteristics of globalization, networking, and digitization are becoming more and more obvious, Internet speeds are constantly increasing, and various technical means are emerging one after another. In the era of big data, information technology changes and value is provided, and data is flooded with the Internet. "Big data" has appeared more and more times, has been accepted by all walks of life around the world, and has quickly become a hot topic and search topic of people's attention. For the research of BDT, big data can be summarized into two aspects: one is massive, huge, complex, and multi-dimensional information; the other is traditional data analysis software and tools [4, 5].

They can be effectively recorded, analyzed and processed in a short time. Big data can be described as a huge, fast-growing, complex, and constantly evolving whole, fast and diverse, profound and influential, and unable to effectively use traditional means, methods and technical tools, thinking model storage, extraction, and analysis And a collection of various data processed. BDT is the development, use, and popularization of big data, which analyzes and processes massive amounts of data, and then obtains valuable conclusions, laws, judgments, predictions, etc. [5, 6].

(1) The Value of Big Data Technology
     Through massive data analysis, it puts forward prediction, new theories, new methods, new technologies and new knowledge, and creates great value in various fields such as behavior analysis, market prediction, decision support and real-time monitoring. First, study and predict the future, understand the relationship between things and the laws behind them, and make accurate and reasonable judgments and predictions of events that have not occurred. Second, supporting and improving decision-making ability can help decision-makers contact and learn information, make it fully effective, and make decision-making more efficient, reasonable, timely and objective. Third, it helps to achieve the accuracy of analysis, effectively

capture, operate, integrate and process data and information, help decision makers find more opportunities, reduce risks, subdivide objectives, shorten response time, implement reform and innovation, and finally obtain accurate analysis and evaluation. Fourth, discover and cultivate potential. With the help of the information resources of big data technology, decision makers can conduct a comprehensive review and analysis of their organizations and better understand their environment, so as to fill the gap, reduce risks and opportunities, tap potential and enhance their own value. At the same time, sharing big data technology helps to share and improve more useful resources and create a smoother and more efficient operating system [6, 7].

(2) EigenTrust Algorithm and PeerTrust Algorithm

The EigenTrust algorithm obtains the credibility of the service provider node through multiple iterations of the direct credibility values of two interacting entities. Before the calculation, the problem of standard unification needs to be solved, so all the direct credible values need to be normalized. The calculation formula for the global credibility of all nodes is shown in Equation.

$$r^{(l+3)} = (3 - b)DRr(l) + bo \tag{1}$$

Among them, C represents its matrix, and $r^{(l)}$ represents the result of iterations of the global credibility vector of l.

The PeerTrust algorithm is mainly used in direct trusted value calculation, and the specific formula of PeerTrust is shown in the formula.

$$R(v) = \lambda DR(v) + \mu DE(v) \tag{2}$$

The credibility of node v is denoted by $R(v)$, $CT(u)$ represents the credibility due to the transaction, and $CF(u)$ represents the credibility due to the transaction environment.

2.2 Ancient Village Culture

The social production and skills, living habits, and economic and technological achievements obtained by the residents of the ancient villages through labor are reflected in the cultural production and life of the ancient villages, such as education, science, ethics, customs, beliefs, etc., which form the ancient villages culture.

(1) The Composition of Cultural Resources in Ancient Villages

The cultural resources of ancient villages include all tangible and intangible resources. The cultural resources of ancient villages are mainly divided into: One is the cultural and natural resources of ancient villages. Many ancient villages that have been preserved to this day are located between natural mountains and rivers, and the site selection focuses on the integration of natural mountains and rivers. The second is the cultural and social resources of ancient villages. People, villages and environment are an organic whole with ecological balance. The third is the cultural and architectural resources of the ancient villages. The buildings left by the ancient villages have not only colorful ancient cultural architectural landscapes, but also architectural cultural connotations related to architectural landscapes. Due to the influence of environment, society and culture, the characteristics of architectural landscape have obvious regional characteristics. The fourth is the human resources of the ancient village [6, 8].

(2) The Value of Cultural Resources of Ancient Villages.

The ancient villages in our country have a long history and rich and profound cultural connotations. Analyzing the value function of the cultural resources of ancient villages will help improve people's understanding of the value of cultural resources of ancient villages and provide a benchmark for the better development of ancient villages. The value of cultural resources of ancient villages includes the following aspects: First, historical value. These historical relics more or less recorded the living conditions and social, economic, political and cultural conditions of people in different historical periods. They are the true incarnation of national culture and regional culture. The second is cultural value. The ancient village retains real historical and cultural information. Modern people can experience the humanistic spirit, traditional natural philosophy, and the cultural contrast in the architecture and folk customs of the ancient village. The third is economic value. The rational and effective use of ancient village cultural resources can increase the content of local tourism resources and increase the attraction of foreign tourists, thereby attracting tourists to increase local financial income and improve the lives of local residents. The fourth is scientific value. The study of the composition of cultural resources in ancient Chinese villages is of great significance to the development of cultural industries, urban and rural environmental planning, tourism planning, and cultural attractions design [9, 10].

2.3   Collection of Big Data on Ancient Village Culture

Collection of original data in order to ensure the integrity and integrity of digital information of ancient village culture, a very large amount of data needs to be collected and stored in the form of image, text, audio and video as much as possible. Due to the large workload and complex data collection of ancient village cultural data, it is much larger than the traditional scale [11]. At the same time, ancient villages are geographically dispersed, mostly far away from cities, backward information infrastructure, and there are some difficulties in collecting big data. For example, the historical and cultural materials of ancient villages are huge, the types are complex, the coverage area is large, and the village forms in different regions are very different. Data collection and preprocessing should last for a long time. Leave enough expansion space for the future. With the rapid development of industrialization and urbanization, villages are rapidly disappearing, and the remains of ancient villages are scarce. The current situation is not optimistic, which will consume a lot of manpower and time, and may need to use different standard processing methods for data from different sources. Therefore, we must find ways to collect and preprocess big data at low cost and efficiently, and solve the difficulties in the work to the greatest extent.

3 Experimental Investigation and Research of Ancient Village Culture Based on Big Data Technology

3.1 Questionnaire Design Process

The subjects of the questionnaire survey were local ancient village residents and foreign respondents in ancient villages in G city. Through the distribution of network questionnaire or paper questionnaire, collect and quantitatively analyze the information filled in by users, and draw the conclusion of the questionnaire.

(1) In the preliminary preparation of the questionnaire, the number of topics shall be as simple as possible to avoid fatigue of the interviewees.

(2) The questionnaire was distributed through online questionnaire, on-site questionnaire and inviting friends to help ask and fill in with their friends and students. A total of 200 questionnaires were distributed, including 100 local ancient village residents and 100 foreign respondents. 200 valid questionnaires were recovered, with a questionnaire recovery rate of 100%.

(3) Questionnaire analysis will sort out the collected questionnaire information to obtain the required information data. The results of the questionnaire were analyzed, including the ways and channels for respondents to obtain ancient village information, respondents' views on the promotion effect of ancient village culture, etc. Some results of this questionnaire are as follows.

3.2 Questionnaire Content

The first part is to collect information about the promotion effect of ancient village culture from 200 respondents;

The second part is a survey of 100 foreign respondents about the ways and channels for respondents to obtain ancient village information;

The third part is to sort out the data collected from the questionnaire to understand the respondents' views on the promotion effect of ancient village culture, as well as the ways and channels to obtain ancient village information.

4 Investigation and Analysis of Ancient Village Culture Experiment Based on Big Data Technology

4.1 Analysis of the Promotion Effect of Ancient Village Culture

This time, 200 interviewees were surveyed on the promotion effect of AVC, and related information was collected. The survey results are shown in Table 1.

Table 1. Promotion effect of ancient village culture

Project	Local people	Outsider
Very good	34	42
Good	28	29
General	22	18
Not good	16	11

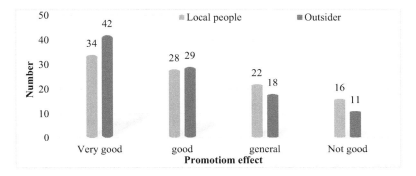

Fig. 1. Promotion effect of ancient village culture

It can be seen from Fig. 1 that among the local residents and foreign respondents of the ancient village, 76 people think that the promotion effect of the ancient village is particularly good, and 27 people think that the promotion effect is poor. It can be seen that most interviewees believe that the promotion of ancient villages is particularly effective.

4.2   Analysis of the Interviewees' Methods and Channels for Obtaining Information on Ancient Villages

This survey of 100 foreign interviewees conducted research and analysis on the ways and channels for interviewees to obtain information on ancient villages. The results of the survey are shown in Table 2.

Table 2. Ways and channels for respondents to obtain information on ancient villages

Ways and Channels	Number
Television	24
Phone	28
Internet	32
Newspaper	10
Else	6

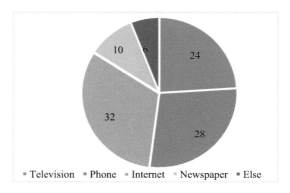

Fig. 2. Ways and channels for respondents to obtain information on ancient villages

As can be seen from Fig. 2, 32 of the 100 external interviewees obtained information on ancient villages from the Internet, and 24 and 28 persons obtained information on ancient villages from TV and mobile phones, respectively. It can be seen that most people get information about ancient villages from TV, mobile phones, and the Internet. On the basis of the Internet, more people get information about ancient villages.

5   Conclusions

The era of big data is inevitable, and it is urgent to understand the practical value of big data technology. By understanding, mastering and applying big data technology, we can more efficiently study ancient village culture and improve the use efficiency of relevant materials, effectively promote digital management and the application of ancient village culture in China, and have important trend leading significance for the protection and inheritance of ancient village culture in China. Therefore, the research on ancient village culture based on big data technology has great practical significance.

References

1. Cook, R.A., Schurr, M.R.: Growth of a village: using fluoride analysis and artifact frequencies to examine early fort ancient/mississippian household and site formation. Am. Antiq. **83**(03), 552–564 (2018)
2. Li, E., Wong, A.M., Tseng, K.C.: Designing system architecture for the catering management system of chang gung health and culture village. In: Antona, M., Stephanidis, C. (eds.) HCII 2020. LNCS, vol. 12189, pp. 584–596. Springer, Cham (2020). https://doi.org/10.1007/978-3-030-49108-6_42
3. Houk, B.A., Bonorden, B.: The "borders" of British Honduras and the San Pedro Maya of Kaxil Uinic village. Anc. Mesoam. **31**(3), 554–565 (2020)
4. Hyun, Y., Kamioka, T., Hosoya, R.: Improving agility using big data analytics: the role of democratization culture. Pac. Asia J. Assoc. Inf. Syst. **12**(2), 2 (2020)

5. Wanderley, C.: Oral cultures and multilingualism in a world of big digital data: the case of Portuguese speaking countries. Educ. Inf. **34**(3), 239–254 (2018)
6. Mari, L., Petri, D.: The metrological culture in the context of big data: managing data-driven decision confidence. IEEE Instrum. Meas. Mag. **20**(5), 4–20 (2017)
7. Obschonka, M.: The quest for the entrepreneurial culture: psychological big data in entrepreneurship research. Curr. Opin. Behav. Sci. **18**, 69–74 (2017)
8. Gil, M., Kozio, P., Wróbel, K., et al.: Know your safety indicator – a determination of merchant vessels bow crossing range based on big data analytics. Reliab. Eng. Syst. Saf. **220**, 108311 (2021)
9. Wood, G., Szmuilowicz, E., Clepp, K.: Creating a village and supporting culture change: facilitating advance care planning by teaching everyone how to have an "introductory conversation" (FR453). J. Pain Symptom Manag. **53**(2), 373 (2017)
10. Sinha, J.N.: Amrita's village. Sci. Cult. **84**(9–10), 333–335 (2018)
11. Fileva-Ruseva, K.G.: Nestinar feast in the village of Stomanovo, the Smolyan region. Int. J. Literat. Arts **5**(5), 37–47 (2017)

Method of Moving Target Detection in 3D Image of Side Leg Skill Movement in Free Combat

Xiaopeng Ji[(✉)]

Physical Education Research Department, Xinjiang University,
Urumqi 830046, Xinjiang, China
Jxpl3565834075@126.com

Abstract. Side leg is an important skilled movement in free combat. In order to improve the training effectiveness of Free combat side leg skill, a 3D moving target detection method is proposed based on adaptive multi-multiple dynamic enhancement skill. Based on multimedia vision, the key technical features of Free combat side leg technology are extracted, and the movement characteristics of Free combat edge leg technology are analyzed. The 3D image reconstruction of Free combat edge leg technology action is carried out by using 3D image tracking recognition method. Using image segmentation method to extract the movement characteristics of Free combat side leg technology, combining with adaptive multiple dynamic enhancement technology to enhance the information of 3D image moving object. The edge information fusion method is used to fuse the feature of the 3D motion image of the free combat leg technology. The proposed method works well in free combat side leg motion 3D image detection, and it has a good ability to extract the action features of Free combat technology.

Keywords: Free combat · Image · Moving object detection · Information enhancement · Edge profile detection

1 Introduction

Free combat is a modern athletics event in which two people use the methods of kicking, hitting, falling and defending in wushu, which is an important part of Chinese Wushu. Free combat in China has a long history. Free combat is not only the skill and tactics of military combat, but also an important sport item. The movement often used in Free combat is the side leg, which needs to be detected by the side leg movement of Free combat. Image acquisition and analysis technology based on computer vision analysis and image processing has been applied to image acquisition and evaluation of moving scenes. The application of high-tech image processing method in 3D target detection of side leg image in Free combat can better analyze the movement and technical characteristics of side leg in Free combat. It provides accurate data support for side leg training and match referee in Free combat. It has great significance to study the three-dimensional image object detection method of Free combat leg technology [1].

In order to realize the three-dimensional image motion target of the technical action in Free combat, the optimization design of image processing algorithm is carried out, the whole and local image information are highlighted pertinently [2]. Combined with adaptive image denoising technology and feature extraction technology, the three-dimension moving target detection method of Free combat leg technology is realized, and good image detection effect is obtained, in which, in reference [3], a new method for detecting the Free combat shape features in the image is proposed. In reference [4], a multi-feature based video shot detection method based on Free combat motion feature template matching detection method is proposed. In this paper, the segmentation and detection of 3D moving object of Free combat edge leg technology in Free combat 3D image is realized under multimedia vision, and the detection effect is high, but the convergence and robustness are not satisfactory. In the complex scene, the extraction accuracy of the athlete's detailed action is not good.

To this end, a method of moving target detection based on adaptive multi-multiple dynamic enhancement skill is proposed in this work.

2   Image Acquisition and Enhancement Processing

2.1   Image Acquisition of Free Combat Leg Motion Under Multimedia Vision

The image acquisition is conducted under the multimedia vision, and image enhancement should be carried out to facilitate visual enhancement of side leg behavior in the technical action of Free combat side leg under complex background [5]. $u^n(x, y; d)$ represents the pixel sequence of the edge leg visual image in the Free combat edge leg movement. The undirected feature distribution graph $G' = (V, E')$ of edge leg is obtained, where E' is a non-empty subset of E. By affine transformation, a plane decomposition model of $M \times N$ bit binary image is constructed to represent the side leg visual image in Free combat edge leg skill. In L bit plane, the state equation of edge leg visual image data acquisition can be expressed:

$$\begin{cases} p_{th}^{(b_{int})} = C_t \sum_{x_i \in w} k(\|x_i\|) \delta(h(x_i) - b_{int}) \\ p_{te}^{(b_{ine})} = C_e \sum_{x_i \in w} k(\|x_i\|^2) his_{x_i} \delta(v_{x_i} - b_{ine}) \end{cases} \quad (1)$$

where $C_t = C_e = \frac{1}{\sum_{x_i \in w} k(\|x_i\|^2)}$ denotes the edge contour marking line set of the edge leg in the Free combat leg skill, and $b_{ine} \in [1, M]$ represents the multi-scale edge contour decomposition coefficient. G represents the output image of human motion morphological analysis under multimedia vision:

$$G_2 = AF_{t-1} + t \quad (2)$$

where, $F_t = [x_t, y_t]^T$ represents the associated pixel value of the t frame of the edge leg visual image. The two-dimensional coordinates are distributed in the bitplane subblock [6]. In the gradient direction, the edge leg action at the pixel (x, y) is filtered by Retinex corner to satisfy:

$$\text{trace}(x, y, \sigma^{(0)}) > \text{trace}(x, y, \sigma^{(l)})l \in \{n - 1, n + 1\} \tag{3}$$

where, trace(.) indicates that the corner of the side leg joint is tracking the track of gray pixel in pixel (x, y). The image acquisition of the Free combat leg action image under the multimedia vision is realized:

$$X_t = AX_{t-1} + t \tag{4}$$

where, $X = [x_t, y_t]^T$ represents the pixel position of the edge leg visual image in the technical action of the t frame Free combat edge leg.

2.2   Image Segmentation and Motion Feature Extraction

Three-Dimensional image tracking recognition method for free combat leg motion 3D image reconstruction is proposed. Firstly, $\dot{\sigma}_i$ is defined as the scale information of the 3D visual information collection of the side leg in the morphological action of Free combat, and the transformation relationship between the two frames of the side leg visual image acquisition in the Free combat edge leg skill can be obtained:

$$\dot{\sigma}_i = \begin{cases} \mu \sin\frac{\pi e}{2\mu}, |e_i| < \mu \\ \mu, |e_i| \geq \mu \\ -\mu, |e_i| \leq -\mu \end{cases} \tag{5}$$

where, $\sigma_x, \sigma_\theta, e_i$, represent the fusion parameters of the shape change of side leg, μ denotes the threshold of 3D moving object segmentation of edge leg technology, and $\mu > 0$, stereoscopic segmentation method is used to obtain the edge leg process in Free combat morphological action [7].

3   Image Target Detection Algorithm Optimization

3.1   Adaptive Multiplex Dynamic Enhancement

This paper extracts the key technical features of the Free combat side leg movement under the multimedia vision, analyzes the movement characteristics of the Free combat edge leg technology, and adopts the image edge information fusion method to fuse the 3D image of the Free combat edge leg action [8]. The statistical features of the 3D moving object of the Free combat edge leg action are obtained as follows:

$$|W_{\Phi^H}\Phi^H(a, \tau)| = \sqrt{2}/2 \tag{6}$$

F represents the image gray level of the edge - leg visual image, and $X(t)$ represents the visual error compensation output obtained by the 2D image, which can be expressed:

$$X(t) = A_0 R_{ux}(\tau - \tau_0) + \sum_{i=1}^{m} A_i R_{ux}(t - \tau_i) + V(t) \tag{7}$$

The image stabilization processing is conducted by using the electronic image stabilization technology [9]. The discrete form of the edge pixel of the image is obtained:

$$\mathbf{r} = [r(0), r(1), \Lambda, r(N-1)]^T \tag{8}$$

The classification attribute weight of 3D visual contour feature points in Free combat leg skill is evaluated. The quadrilateral (E_i, E_j, d, t) denote the trace corresponding to the side leg, and the stable value of the side leg motion object meets the F. then the spatial scale delay is $\min_{c \in (r,g,b)} (\min_{y \in \Omega(x)} (\frac{I^c(y)}{A})) \to 1$, the process of gray level transformation is expressed as:

$$S_{i,j}(t) = \frac{p_{i,j}(t) - sp_{i,j}(t)}{p_{i,j}(t)} \tag{9}$$

$$T_{i,j}(t) = \frac{|p_{i,j}(t) - \Delta p(t)|}{p_{i,j}(t)} \tag{10}$$

$$U_{i,j}(t) = \exp\left[-b\left[Z_i(t) - Z_j(t)\right]^2\right] \tag{11}$$

where, $p_{i,j}(t)$ represents a 2D array of edge and leg action features in t time dispersion, $sp_{i,j}(t)$ is the convolution of multiscale space, $\Delta p(t)$ is the reference value of the position of the sportsman's body, $Z_i(t), Z_j(t)$ is the output of the edge contour of each pixel point, and b is the comparison coefficient of the moving target.

3.2   Corner Detection Skill and Image Moving Target Detection

The edge information fusion method is used to fuse the feature of Free combat leg action 3D image, the feature quantity of moving object is extracted from the image, and the Free combat side leg is adopted [10]. The feature of edge contour is extracted by moving object segmentation method, and the 3D mapping region of edge contour of Free combat leg is constructed.

$$G_x(x, y; t) = \partial u(x, y; t)/\partial x \tag{12}$$

$$G_y(x, y; t) = \partial u(x, y; t)/\partial x \tag{13}$$

Based on the visual sampling value of the edge leg moving target of each pixel point, the image smoothing is carried out, and the edge contour of the 3D moving object is obtained by using the Free combat edge leg technology. The output of the extracted result is obtained:

$$p(z_t \mid x_{t-1}, d_{0,n,t-1}) = p(x_t \mid x_{t-1}, u_{t-1})\mathrm{Bel}(x_{t-1})dx_{t-1} \tag{14}$$

$$p(z_t \mid x_t) = p(x_t \mid x_{t-1}, u_{t-1}) \tag{15}$$

Assume that the point $p(x_t \mid x_{t-1}, u_{t-1})$ in the multi-scale Retinex corner set $p(z_t \mid x_{t-1}, d_{0,n,t-1})$ of the edge leg behavior image is located in the gray pixel space $P(x, y)$. The upper motion parameter Hessian matrix is written as:

$$H = \begin{vmatrix} \frac{\partial^2 f(\sigma)}{\partial x^2} & \frac{\partial^2 f(\sigma)}{\partial x \partial y} \\ \frac{\partial^2 f(\sigma)}{\partial y \partial x} & \frac{\partial^2 f(\sigma)}{\partial y^2} \end{vmatrix} \tag{16}$$

A number of closed regions are distributed in the side leg visual image in the Free combat leg skill, and the parameters of the moving object are estimated.

4   Simulation

Simulation is conducted to validate the feasibility of the proposed algorithm.

Fig. 1. Three-dimensional image of technical action of side leg in Free combat

The image segmentation method is adopted to extract the motion features of the technical movement of the edge and leg of the Free combat (see Fig. 1), and the information enhancement processing of the motion target of the 3D image is conducted by the adaptive multiple dynamic enhancement technology, the moving target of the 3D image is detected, and the output results are obtained, as shown in Fig. 2.

Figure 2 shows that the accuracy of using this method to detect the movement features of Free combat leg skill is high, and the detection performance is tested.

Fig. 2. Output of target detection results

Figure 3 shows the comparison of accurate detection probability. The accuracy of moving object detection in three-dimensional image of Free combat leg skill is high.

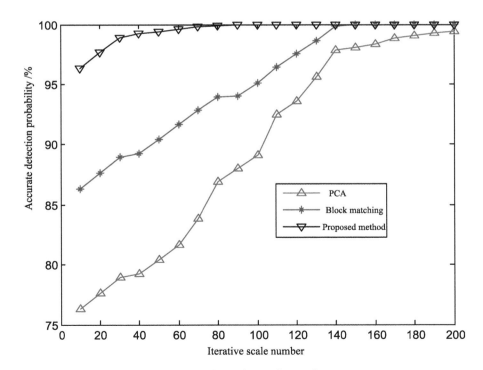

Fig. 3. Comparison of detection performance

5 Conclusions

In this paper, a 3D moving target detection method is proposed based on adaptive multi-multiple dynamic enhancement skill. Based on multimedia vision, the key technical features of Free combat side leg technology are extracted, and the movement characteristics of Free combat edge leg technology are analyzed. The 3D image reconstruction of Free combat edge leg technology action is carried out by using 3D image tracking recognition method. Using image segmentation method to extract the movement characteristics of Free combat side leg technology, combining with adaptive multiple dynamic enhancement technology to enhance the information of 3D image moving object. The edge information fusion method is used to fuse the feature of the 3D motion image of the Free combat leg technology. The feature quantity of the moving object is extracted from the image. The proposed method works well in free combat side leg motion 3D image detection, and it has a good ability to extract the action features of Free combat technology. This method has good application value in Free combat image recognition and detection.

Acknowledgement. This work was supported by "National Social Science Foundation of China (NO. BLA200218)", Research on the construction of physical education curriculum objectives in Xinjiang multi-ethnic areas from the perspective of national identity.

References

1. Abreu, A.D., Frossard, P., Pereira, F.: Optimizing multiview video plus depth prediction structures for interactive multiview video streaming. IEEE J. Sel. Top. Signal Process. **9**(3), 487–500 (2015)
2. Borji, A., Sihite, D.N., Itti, L.: Quantitative analysis of human-model agreement in visual saliency modeling: a comparative study. IEEE Trans. Image Process. **22**(1), 55–69 (2013)
3. Gupta, R., Elamvazuthi, I., Faye, I., Vasant, P., George, J.: Comparative analysis of anisotropic diffusion and non local meanson ultrasound images. J. Mach. Mach. Commun. **1**(1), 51–68 (2014)
4. Ramos-Llorden, G., Vegas-Sanchez-Ferrero, G., Martin-Fernandez, M., Alberola-Lopez, C., Aja-Fernandez, S.: Anisotropic diffusion filter with memory based on speckle statistics for ultrasound images. IEEE Trans. Image Process. **24**(1), 345–358 (2015)
5. Rahimizadeh, N., Hasanzadeh, R.P.R., Janabi-Sharifi, F.: An optimized non-local LMMSE approach for speckle noise reduction of medical ultrasound images. Multimed. Tools Appl. **80**(8), 1–23 (2021)
6. Sudeep, P.V., et al.: Speckle reduction in medical ultrasound images using an unbiased non-local means method. Biomed. Signal Process. Control **28**(6), 1–8 (2016)
7. Roth, M.: Parameterized counting of partially injective homomorphisms. Algorithmica **83**(3), 1829–1860 (2021)

8. Ono, T., Kimura, A., Ushiba, J.: Daily training with realistic visual feedback improves reproducibility of event-related desynchronisation following hand motor imagery. Clin. Neurophysiol. **124**(9), 1779–1786 (2014)

9. Janssen, L., Creemers, R., Bodegraven, A.V., Pierik, M.: P183 A systematic review on outcome measures of long-term efficacy in clinical trials concerning Crohn's disease patients. J. Crohn's Colitis **16**(1), 248–249 (2022)

10. Reddy, G.S., Lakshmi, S.M.: Exploring adversarial attacks against malware classifiers in the backdoor poisoning attack. In: IOP Conference Series Materials Science and Engineering, vol. 1022, no. 1, p. 012037 (2021)

The Implementation Method of Battlefield Situational Awareness System Based on Human Brain Perception Mechanism

Yuqiang Ke[✉]

Units Command Department,
Officers College of Chinese People's Armed Police Force,
Chengdu, Sichuan, China
wwjwrc@163.com

Abstract. In this paper, I propose a computational method for incorporating attentional mechanisms that mimic those of the human brain into a battlefield situational awareness system by dividing attention-related circuits in the human brain to make the attentional mechanisms of the human brain computable and applying them to battlefield multi-source situational analysis. This allows effective fusion of multi-source signals, and reduces the computational effort of battlefield situational awareness and effectively assists commanders in judgment and decision making.

Keywords: The human brain mechanism of attention · Feature extraction · Multi-source information fusion · Battle-field situation analysis

1 The Physiological Basis of Spatial Attention and Knowledge Attention

In the early stage of attention research, people mainly pay attention to the psychological explanation of attention and the phenomenon analysis of neurophysiological methods. In recent years, with the development of artificial intelligence and computer vision, it has become very important to study computable models of attention. Psychological and neurophysiological studies show that attention consists of two different processes, one is bottom-up, and the other is a top-down process. Therefore, attention modeling is generally considered separately from these two processes.

Bottom-up attention has long intrigued researchers. In some contexts or backgrounds, some visual input itself is prominent and eye-catching, and this prominence is independent of the task, and the observer's attention will be unconsciously drawn to these areas. Spatial attention is a visual process driven by spatial features without conscious participation. Some psychologists call this bottom-up process pre-attention. Top-down attention has been slow to be studied because it is not as easily defined as bottom-up attention. Top-down attention models can be divided into three types, visual search model, context model and task-driven model.

In order to better reference the attention mechanism of human brain in battlefield intelligence information fusion, we divide the attention related circuits in human brain.

V. Sugumaran et al. (Eds.): ICMMIA 2022, LNDECT 138, pp. 319–327, 2022.
https://doi.org/10.1007/978-3-031-05484-6_40

Attention mechanism is a part of the visual processing system of human brain. We can consider attention mechanism from the structure of visual system. Anatomical and visual physiological tests have shown that the human brain contains two different visual channels [1]: one is known as the Dorsal Stream, which starts at V1 and passes through V2 to the Dorsal medial and Middle parapet-MT or V5, and then reaches the inferior parapet lobule. The dorsal pathway, often referred to as the Where Pathway, is involved in processing information about an object's spatial location and related motion control. The other channel is called Ventral Stream, which also starts from V1 and enters into the Inferior temporal lobe through V2 and V4 successively, as shown in Fig. 1. This pathway is often referred to as the "What Pathway" and is involved in target recognition, etc. [1, 2]. These two channels are involved in attention production in different ways. The two distinct attention-forming channels in the visual cortex are analogous to the two main components of our proposed model of attention. These two different modes of attention production correspond to characteristics of different levels of complexity.

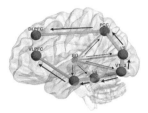

Fig. 1. Les bases physiologies de attention

The dominant mechanism of dorsal attention is spatial attention, and the generation of such attention does not involve the participation of consciousness, nor is it task-driven [3]. Spatial attention is based on underlying features such as brightness, orientation, color, texture, etc. Spatial attention can also be driven by more complex object shape features. All of these characteristics do not involve the knowledge and tasks required for high-level processing. At the front end of the visual processing system arise some underlying features that can drive spatial attention. Therefore, we believe that due to the early appearance of some common features in the visual pathway, some spatial attention can be generated in V1 region, which is consistent with the research conclusions in the literature. According to the previous discussion, spatial attention is generated by the entire dorsal channel, at least involving V1, V2, V5 (MT) and other visual cortex regions.

Ventral attention mechanism is closely related to target information fusion, requiring the invocation of prior knowledge and knowledge required for performing specific tasks [4, 5]. Therefore, in this paper we call it knowledge attention. This kind of attention is based on more complex features. Such complex features can be roughly divided into two categories: One can be called Common knowledge based feature, which comes from the knowledge acquired by human interaction with the environment in the process of evolution. For example, humans can easily recognize what an "object" is; another complex feature can be called Task related feature. For example, when a driver is performing a driving Task, he will quickly allocate his attention to special "objects" such as vehicles and pedestrians in front of him. Mission characteristics can be acquired through training.

A unified attention model is established by considering the top-down process and the bottom-up process. We learn from the attention mechanism in the human brain to study the relationship between different levels of features and attention. Based on the two attention circuits in human brain, we finally draw on the unified Bayesian model of attention. In general, this attention model mainly contains two main modules, one is spatial attention module, and the other is knowledge attention module [6].

2   Establishing the Relationship Between Target Features and Attention

2.1   Analyze the Relationship Between Features at Different Levels and Attention

An automatic battlefield situation perception system based on human brain attention mechanism, including battlefield perception module and feature extraction module [7]. The feature extraction module also includes attention module and fusion module, as shown in Fig. 2.

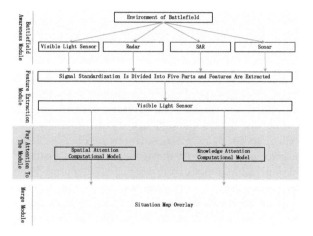

Fig. 2. Note the application framework of mechanism in battlefield intelligence information fusion

The battlefield perception module, feature extraction module, attention module and fusion module are connected in sequence. Attention module includes knowledge attention module and space attention module. Knowledge attention module and space attention module are in parallel relationship, and the working mode of each module is as follows:

The battlefield perception module is used to obtain various signals of battlefield intelligence information.

The feature extraction module is used to standardize and normalize the signals of the same space-time, and obtain the real-time feature set corresponding to the signals [8].

Knowledge attention module is used to process the real-time feature set and obtain the battlefield situation knowledge activation diagram. It includes: feature extension module, which is used to add relevant features of objects; Training module for training classifier; Prediction module for multi-modal feature fusion. The training module includes: feature database, which is used to store data and obtain prior knowledge [9]; Centralization module for strengthening the central area; the classifier training module is used to train the classifier to obtain the classifier for the signal, and the classifier combination is strengthened to obtain a strong classifier; Training module and prediction module are parallel.

The spatial attention module is used to process the real-time feature set and get the spatial activation map of battlefield situation. It includes: feature space generation module, which is used to generate different feature space; Space transformation module, used for space transformation, get the feature set of different Spaces; Activation diagram selection module, used to select the optimal activation diagram; a weighted summation module for a weighted summation of the activation graph. Feature space generation module, space transformation module, activation map selection module and weighted summation module are serial relations.

In the fusion module, the battlefield situation knowledge activation map and the battlefield situation spatial activation map are superimposed on the situation map, and the results are visualized.

2.2   The Characteristics of the Target are Divided According to Different Levels

According to different levels, the characteristics of the target can be divided into four categories:

(1) Underlying characteristics. Simple features can lead to attention. Therefore, in this paper's model, simple features are taken as a factor driving attention. Simple features refer to some simple attributes of pixels or regions in the image and their values, or linear filters that can be convolved with the image and their responses;

(2) Universal target features. In the absence of any visual search task, humans are still very sensitive to objects in images. In this case, we are not defining any specific goal. We call such an unqualified goal a generic goal. The goal of the visible scene is one of the factors leading to attention. The generic target characteristics here are more complex than the underlying characteristics. This is similar to the characteristics of closure and connectivity described by gestalt theory, as well as the edge density and independent dividing line referred to in the study of objective measurement. The characteristics of this level are task independent;

(3) Specific target characteristics. By specific goals, I mean task-related goals. In a confrontation between two armies in a battlefield environment, commanders pay more attention to tanks, aircraft and other entities closely related to operations. This is actually equivalent to object search or target detection. We define

task-related target features as image region description features that can be used for object detection.

(4) Global features of the scene. The global context information of the scene is closely related to the generation of attention. The global feature of the scene plays a role in both saliency detection and target detection.

2.3   Establish a Unified Bayesian Model of Attentional Relationship

In order to systematically analyze the relationship between various features and attention, we used Bayesian formula to express the relationship between them [10]. We represent the underlying feature variable, the pan-object feature variable, the task-related object feature variable and the global context feature variable with v_l, v_o, v_t, v_c respectively. Then, given these features, the conditional probability density of the location o where object x appears is $p(o, x \mid v_l, v_o, v_t, v_c)$. According to Bayes' formula, the conditional probability density is expressed as follows:

$$p(o, x \mid v_l, v_o, v_t, v_c) = \frac{p(o, x, v_l, v_o, v_t, v_c)}{p(v_l, v_o, v_t, v_c)} \tag{1}$$

$$= \frac{p(v_t \mid o, x, v_l, v_o, v_t, v_c)p(v_o \mid o, x, v_l, v_c)p(v_l \mid o, x, v_c)p(o, x, v_c)}{p(v_l, v_o, v_t, v_c)} \tag{2}$$

In order to simplify the problem, given the global feature v_c of the scene, we do not consider the relationship between the underlying feature, object feature and task-related object feature, that is, we assume that the conditions of the three variables v_l, v_o, v_t are independent under the given condition of variables. Therefore, Formula (2) can be written as:

$$p(o, x \mid v_l, v_o, v_t, v_c) = \frac{p(v_t \mid o, x, v_c)p(v_o \mid o, x, v_c)p(v_l \mid o, x, v_c)p(o, x, v_c)}{p(v_l \mid v_c)p(v_o \mid v_c)p(v_t \mid v_c)} \tag{3}$$

In Formula (3) $p(v_t \mid o, x, v_c)$, $p(v_o \mid o, x, v_c)$ and $p(v_l \mid o, x, v_c)$ are the three characteristic likelihood terms.

We define the note as: 2222:

$$S(x) = \underbrace{p(v_l \mid v_c)^{-1}p(v_o \mid v_c)^{-1}p(v_t \mid v_c)^{-1}}_{\text{Spatial attention}} \underbrace{\underbrace{p(v_t \mid o, x, v_c)}_{\text{Task note}} \underbrace{p(v_o \mid o, x, v_c)}_{\text{Pay attention to the object}} \underbrace{p(o, x \mid v_c)}_{\text{Center of preference}}}_{\text{Pay attention to the knowledge}}$$

$$\tag{4}$$

According to Formula (4) attention in our model can be divided into four parts: spatial attention, pan-object attention, task-related attention and central preference effect. Since the last three are related to prior knowledge, we collectively refer to them as knowledge attention. In this paper, $S_s(x)$ is used to represent spatial attention and $S_s(x)$ is used to represent knowledge attention [11].

3 Realization Method of Computational Model Based on Knowledge Attention

(1) The Perception of Specific Objects is Equivalent to Target Recognition in the Traditional Sense

From the perspective of statistical pattern recognition, there are three main steps to complete the perception of task-related specific objects: first, the extraction of underlying features; The second is to construct complex features that can describe objects based on underlying features, which is generally achieved by constructing descriptors (feature vectors). The third is training classifier [12], as shown in Fig. 3.

Fig. 3. Application of knowledge attention model in battlefield intelligence information fusion

(2) Features used for object property measurement include multi-scale saliency (MS), color contrast (CC), edge density (ED), and super pixel crossover (SS). Object attention based on object property measure is defined as:

$$S_o(x) = k_1 CC(p(x)) + k_2 ED(p(x)) + k_3 SS(p(x)) \tag{5}$$

Here k_i is the weight of each feature, which is obtained through learning. p Is the image block.

(3) We can estimate $p(o, x \mid v_c)$ from two aspects:

1) The probability of objects appearing in different positions is estimated according to the internal and external parameters of road model and camera calibration;

2) Use historical information (recorded road scenes) to learn the probability of objects appearing at different locations.

4 A Computational Model for Simulating Spatial Attention is Proposed

The realization method of the calculation model simulating spatial attention is based on battlefield situational awareness model 7 of spatial attention, as shown in Fig. 4.

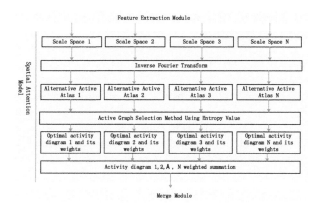

Fig. 4. Application of spatial attention model in battlefield intelligence information fusion

Its calculation process can be roughly divided into four parts:

(1) Pre-processing: registration association and standardization are carried out for various information sources in the same battlefield. Treat each of these different sources as a channel for processing below;

(2) Calculate the activity graph of each channel: (1) for images of any channel or other sources, generate spectral scale space according to Formula (1):

$$\Lambda(u, v; k) = (g(.,.; k) * A)(u, v) \tag{6}$$

where, $\Lambda(u, v; k)$ is the amplitude spectrum of a source, $\Lambda(u, v; k)$ is the scale space derived from $A(u, v)$, which is obtained by the convolution of Gaussian kernel $g(u, v; k)$ with the original amplitude spectrum $A(u, v)$, and the Gaussian kernel is defined as:

$$g(u, v; k) = \frac{e}{\sqrt{2\pi}2^{k-1}t_0} e^{-(u^2 + v^2)/(2^{k-1}t_0^2)} \tag{7}$$

where $t_0 = 0.5$ is the scale parameter $k = 1, 2, \ldots, K$ and K is a natural number determined by the size of the signal:

$$K = [log_2 \, min\{H, W\}] + 1 \tag{8}$$

where, H, W is the length and width of the source (image).

(3) The inverse Fourier transform is performed on each layer of the scale space to generate a series of alternative active graphs; then, according to Formula (8), the optimal one is selected as the final activity map of the channel, and its entropy value is recorded [13]:

$$k_p = arg_k \, min\{H(S_k)\} \tag{9}$$

where, H is the entropy of the activation is graph.

(4) Integrated activation map: according to the feature integration theory, each major feature in the process of visual system processing is processed separately and represented as different active maps. Finally, these activity maps can be integrated in a linearly weighted manner. In the SSS model, we determined the weight of each channel according to the entropy value of the activity graph of the three channels, so as to carry out weighted linear fusion for each channel, and the fusion result was taken as the final significance distribution map. The weight is determined according to the following formula:

$$w_i = \text{entropy}_i / \sum_{\{Avalablaactivationdiagram\}} \text{entropy}_i \qquad (10)$$

5 Conclusions

Compared with the existing technology, the main advantages of the proposed method are as follows:

(1) The application of human attention mechanism in the battlefield intelligence information fusion system can reduce the calculation of battlefield situation awareness and effectively assist commanders in judgment and decision-making.

(2) The traditional target recognition calculation method is used to model task-related attention in knowledge attention, object attention is modeled based on object property measurement, and battlefield situation is constructed for specific targets by using database.

(3) Using the global inhibition mechanism of human brain and spectral scale theory to model spatial attention can effectively suppress repeated patterns, effectively fuse multi-source signals, and effectively respond to emergencies by situation awareness and prediction of unknown situations, without being limited by database storage capacity.

References

1. Leggings, M.E., II., Chong, C.-Y., Kadar, I., Alford, M.G.: Distributed fusion architectures and algorithms for target tracking. In: Proceedings of the IEEE, vol. 85, no. 1, pp. 95–107, January 2020
2. Jayaram, S.: A new fast converging Kalmar filter for sensor fault detection and isolation. Sens. Rev. **30**(3), 219–224 (2020)
3. Karunanithi, N., Whitley, D., Malaiya, Y.K.: Using neural networks in reliability prediction. IEEE Softw. **9**(4), 53–59 (2020)
4. Vvedensky, V.L.: Cimon model of evolution for living cell and central nervous system. Bioelectrochem. Bioenerg. **48**, 343–347 (2021)
5. Afzal, W., Trocar, R., Felds, R.: Prediction of fault count data using genetic programming. In: IEEE International Multitopic Conference 2020, Karachi, Pakistan, pp. 349–356 (2020)

6. Pang, Y., Ganesh, A., Wright, J., et al.: Rash: robust alignment by sparse and low-rank decomposition for linearly correlated images. In: Proceedings of IEEE Conference on 23th Computer Vision and Pattern Recognition, San Francisco, USA, pp. 763–770 (2010)
7. Grille, C., Lucio, L., Vincenzo, L.: Agent-based architecture for designing hybrid control system. Inf. Sci. **176**, 1103–1130 (2021)
8. Koop, G.F.: Dynamics of neuronal circuits in addiction: reward, antireward, and emotional memory. Pharmacopsychiatry **42**(Suppl.), 1S32–41 (2019)
9. Jenkinson, M., Bannister, P., Brady, M., Smith, S.: Improved optimization for the robust and accurate linear registration and motion correction of brain images. Neuroimage **17**, 825–841 (2020)
10. Abernethy, M., Rai, S.M.: Cooperative feature level data fusion for authentication using neural networks. In: Loo, C.K., Yap, K.S., Wong, K.W., Teoh, A., Huang, K. (eds.) ICONIP 2014. LNCS, vol. 8834, pp. 578–585. Springer, Cham (2014). https://doi.org/10.1007/978-3-319-12637-1_72
11. Khaleghi, B., Khamis, A., Karray, F., Razavi, S.: Multi-sensor data fusion: a review of the state-of-the-Art. Inf. Fusion **14**(4), 28–44 (2013)
12. Nitish, S., Geoffrey, H., Alex, K., et al.: Dropout: a simple way to prevent neural networks from over fitting. J. Mach. Learn. Res. **15**(6), 1929–1958 (2014)
13. Curran, E.A., Stokes, M.J.: Learning to control brain activity: a review of the production and control of EEG components for driving brain-computer interface (BCI) systems. Brain Cogn. **51**(3), 326–336 (2021)

English Linguistics Multimodal Emotion Recognition System Based on BOOSTING Framework

Ying Liu and Dandan Zhao[✉]

Foreign Language Teaching Department, Changchun University of Chinese
Medicine, Changchun 130117, Jilin, China
ccucmfld@163.com

Abstract. With the emergence of artificial intelligence, the realization of more humane and intelligent human-computer interaction has always attracted attention, and emotion recognition has become one of the global hot spots. Traditional language translation systems focus on translating voice and text messages into English. However, the way of communication between people is not simply the exchange of textual information, there are also rich emotional exchanges. Therefore, recognizing the emotion of English language has become an indispensable part of realizing natural language translation system. For this reason, this paper proposes to design an English linguistics multimodal emotion recognition system based on the BOOSTING framework. The purpose is to improve the accuracy of the emotion recognition system. This paper mainly uses the methods of comparison and experiment to analyze the single-modal and multi-modal English language emotion recognition technology. Experimental data shows that the accuracy of multi-modal emotion recognition results after fusion for feature extraction can reach more than 47%. And its recognition level basically remains at the same level, with little change.

Keywords: BOOSTING framework · English linguistics · Multimodal recognition · Emotion recognition system

1 Introduction

In today's information society, people have higher and higher requirements for their own quality of life, and work and study are the most important and fundamental link. Therefore, we need a need to be able to communicate with others, express emotions and share emotions. In the current situation of the rapid development of the Internet, people have also produced great changes in the way of communication. How to judge multimodal emotion in English linguistics has become the focus of research.

There are many theoretical results in the research of English linguistics multimodal emotion recognition system based on the BOOSTING framework. For example, some people have proposed two effective methods for detecting visual and sound multimodal fusion emotion based on the boosting framework [1, 2]. Someone also proposed a recommendation system architecture based on the boosting framework, including a strong recommendation system based on multiple core recommendation algorithms [3,

V. Sugumaran et al. (Eds.): ICMMIA 2022, LNDECT 138, pp. 328–335, 2022.
https://doi.org/10.1007/978-3-031-05484-6_41

4]. In addition, some people raised existing problems. They believe that the interaction between humans and machines is inseparable from the recognition of emotions. This has the problems of low detection rate and low robustness [5, 6]. Therefore, based on the purpose of solving the problem, this paper studies the BOOSTING framework and links it with the English linguistics multimodal emotion recognition system to improve the accuracy of the system.

This article first studies some related theories and methods of monomodal emotion recognition. Secondly, the multi-modal emotion recognition method and the AdaBoost framework are studied. Then enumerate the emotional feature extraction of English speech, and conduct detailed analysis. Finally, the multimodal emotion recognition system of English linguistics is tested through experiments, and the data results are obtained.

2 English Linguistics Multimodal Emotion Recognition System Based on BOOSTING Framework

2.1 Monomodal Emotion Recognition

Facial expressions, voices, language, gestures and posture are important indicators of people's emotional expression. They are interdependent and form a basic communication system between people in the social environment. People use the information received by human ears and eyes, even if the signal is very weak, humans can recognize emotions. Database is the cornerstone of emotion recognition research, and a high-quality database helps to further deepen the research. Among various multi-modal emotion databases, the most suitable is the emotion database based on facial expressions and English speech [7, 8].

The most important steps in facial emotion detection include image preprocessing, feature extraction and emotion classification. Image preprocessing includes scaling, normalization, etc., to calculate size, mean, and variance [9, 10].

When a person is in a state of anger, fear or joy, the sympathetic nervous system is activated, leading to increased heart rate and blood pressure, resulting in dry mouth and muscle tremor. The voice becomes higher, the speaking speed is faster, and the high frequency energy is more. Researchers generally believe that there are three types of English phonetic features that are closely related to human emotional expression: prosodic features, spectral features, and tonal features. The prosodic feature describes the phonetic structure of the language, which is a typical feature of human natural language. Prosodic features are mainly reflected in intonation, intonation, accent and rhythm. The frequency spectrum characteristics commonly used in speech sentiment analysis are mainly divided into linear frequency spectrum characteristics and cepstrum characteristics. Sound quality is an indicator of people's subjective evaluation of sound, and it is the natural expression of emotions [11, 12].

Multi-Modal Emotion Recognition

Hidden Markov Model (HMM) can be effectively applied to multimodal (channel) fusion detection technology. In the feature-level fusion method, the features of English language and image channels can be combined to form an observation vector.

Standalone HMM uses dual-component HMM to represent audio and video material. In order to participate in the establishment of the AdaBoost framework, it is necessary to modify the EM learning method of the coupled HMM. In the learning process, it is difficult to identify the weight information of the sample by paying attention to the learning mode.

Based on traditional classification and recognition methods, a large amount of data needs to be preprocessed, including feature extraction such as word frequency, vocabulary, and grammar. But because the process of emotion recognition is a non-linear and complex behavioral process. Therefore, in order to improve the accuracy of model discrimination, reduce the amount of calculation, and reduce the error rate, it is necessary to establish a classification system for the analysis and understanding of the emotional semantic relationship between all types of topics in the text field involved in the BOOSEING framework. Use the system to help us complete the identification of conceptual semantic features, so as to provide basic data support for subsequent program code design.

Create a BoostedHMM multi-layer classifier (MBHMM) for each emotion category to be recognized. Since three eigenmode data are used, the MBHMM classifier has three layers. In each layer of the MBHMM classifier, the component HMM classifier is trained based on the framework of the AdaBoost algorithm to generate a powerful global classifier. The observation probability density function of the model is a mixed Gaussian density function:

$$y_a(x) = \sum_{b=1}^{v_a} q_{a,v}^o A(P_s^o, W_{a,v}^o, R_{a,v}^o) \tag{1}$$

where: $A(P_s^o, W_{a,v}^o, R_{a,v}^o)$ represents the v-th Gaussian probability density function of the hidden state x. Each of the equations $\frac{1}{Q}$ for the sum of the four model parameters of the HMM will be re-evaluated, and the updated equation after replacement is:

$$m(v \mid k) = \frac{\sum_c \frac{q(c)}{Q_c} \sum_v b_{k,v}(s)}{\sum_c \frac{q(c)}{Q_c} \sum_v \sum_i b_{k,v}(s)} \tag{2}$$

where: Q_c refers to the sample observation sequence. It can be proved that the improved Baum-Welch training algorithm is still convergent.

At present, the related research of multi-modal information fusion mainly focuses on the data layer, feature layer and decision-making layer. Among them, the combination of decision-making layers has the advantages of small data volume, strong anti-interference ability, and high fault tolerance, and is widely used. Therefore, this paper is based on a decision-level fusion strategy to perform multi-modal emotion recognition that combines English and facial expressions.

2.2 Emotional Feature Extraction of English Speech

The sound quality, rhythm and frequency spectrum characteristics of English speech signals can reflect most of the information of the speaker's emotional state. In the process of dialogue and communication, English pronunciation plays a vital role in emotional expression. The tone, volume, and speed of speech are all related to the emotional state of the speaker at that time. The methods used include preprocessing of English speech signals, windowing and grating, Fourier transform and inverse Fourier transform, etc., which are analyzed in the time domain or cepstrum. The emotional features extracted in the time domain include short-term energy, duration of speaking English, short-term zero-crossing rate and related statistical parameters.

Perform short-term analysis of English speech signals. The signal flow here is realized by framing. In order to facilitate the transition between fields, framing takes place in the form of overlapping segmentation, that is, there is overlap between adjacent images. The overlapping part is called shift frame. Short-term energy reflects the characteristics of English language energy changes over time, and amplitude changes reflect the short-term energy of English speech signals. It also allows the unvoiced segment to be better distinguished from the voiced segment. When the speaker's emotional state changes, the energy of the speech will also change, so the energy of English speech can distinguish different emotional states.

The emotional features extracted from the cepstrum domain include pitch frequency and related statistical parameters. The pitch frequency and the pitch period are the reciprocal of each other, and most of the commonly used methods for finding the pitch frequency are to first estimate the pitch period to obtain more pitch frequencies. The basic period is the period of vocal cord oscillation, which refers to the time when the vocal cord opens and closes each time a vowel (including consonants) is emitted.

In practical applications, whether you use the cepstrum method or the LPC method to extract the formants, better results may be obtained if the English material is a single frame. But if you detect continuous English language, you will eventually get the wrong Bump peaks and formant fusion. The influence of high-pitched English leads to greater deviation of the recognition results. On this basis, using an improved LPC root search method to identify the trajectory of continuous English language formants can achieve better results. The importance of extracting Mel Frequency Cepstral Coefficient (MFCC) parameters for extracting the characteristics of English speech signals is reflected in the spectrum analysis of English language based on the mechanism of human hearing.

3 Simulation Experiment

3.1 Experimental Platform

The experiment in this article mainly involves the processing of pictures and English speech signals. The experimental hardware environment is a PC, the operating system is Ubuntu14.04.5, equipped with a Linux kernel, and the GPU is used to accelerate deep learning. The configuration is as follows:

Processor: CPU Intel Broadwell E5
Graphics card: NVIDIA 1080 ×4
Memory: 128G
Hard Disk: 5T
Python 2.7.13.

This section briefly introduces the databases used in the experiment, mainly four databases: FER2013, RML, AFEW6.0 and enTERFACE'05. Among them, the FER2013 database is a facial expression database, and the other three are bimodal emotion databases.

3.2   Experimental Program

In order to verify the effectiveness of the different emotion recognition methods proposed in the article, three different bimodal emotion databases eNTRAFACE'05, RML and AFEW6.0 were experimentally analyzed. In the experimental setup stage, the video samples were preprocessed in three databases, mainly including video framing and English language extraction. Among them, eNTRAFACE'05 database intercepted 100 facial expression images and 100 English samples of all videos, RML database intercepted 150 facial expression images and 50 English samples, AFEW6.0 database intercepted 120 facial images. English expression pictures and 30 Language samples. At the same time, the non-sampled eNTRAFACE'0 and RML databases are divided into training set and validation set according to the proportion.

3.3   Experimental Indicators

In the experiment, the fundamental frequency, range, shaping, cepstral coefficient Mel, English strength, English speaking speed and other features and their minimum, maximum, average, median and standard deviation are extracted from the English speech signal. Acquire the emotional attributes of English speech, extract Gabor features from the expression signal, and then use principal component analysis to reduce dimensionality. For posture feature extraction, the EyeWeb platform is used for posture tracking, and the minimum and maximum values, as well as the values, mean, median, and standard deviation of various indicators, are obtained. Finally, the emotional characteristics of the posture are obtained.

4   Analysis of Simulation Results

4.1   Monomodal Emotion Recognition

This article exemplifies the facial expressions of the speech recognition module, and displays the normalized and unnormalized data results. The specific facial expression recognition situation is shown in Table 1:

Table 1. Speech emotion recognition rates under normalized versus unnormalized levels

	Unnormalized	Normalization
Glad	40%	43%
Sad	30%	35%
Fear	26%	29%
Detest	39%	41%
Angry	45%	47%

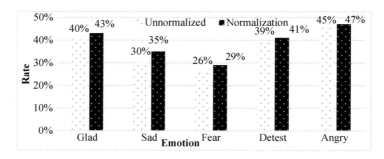

Fig. 1. Speech emotion recognition rates under normalized versus unnormalized levels

As shown in Fig. 1, we can see that the improvement of the recognition rate by normalizing the data is obvious: normalization eliminates the influence of dimensionality on the data, and limits the range of data changes to a small range, which is simple accurate. After the data is selected by the feature, the recognition rate can be improved, and the use of the feature selection technology reduces the dimensionality of the feature data.

4.2   Multi-modal Emotion Recognition

Through principal component analysis and dimensionality reduction normalized speech emotion features, expression emotion features, and posture emotion features, the three-peak emotion features are combined through the method of discriminant multiple canonical correlation analysis, and then the boosting framework is used to conduct experiments, as shown in Table 2:

Table 2. Identification rate of Trmode fusion with feature selection

	Uncharacterized selection	Feature selection
Glad	57%	59%
Sad	46%	47%
Fear	38%	41%
Detest	45%	46%
Angry	55%	56%

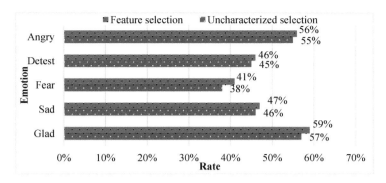

Fig. 2. Identification rate of Trmode fusion with feature selection

As shown in Fig. 2, after all the emotional information of the three modalities are used, the emotional attributes of the three modalities are merged together, which improves the accuracy of emotion recognition to a certain extent, showing that multi-modal analysis The algorithm judges the validity of the association. It confirms the statement that the various modal emotions mentioned above are complementary.

5   Conclusion

In the single modal mode, the recognition rate when using expressions as input is higher than the other two modalities. In a certain emotion category, the recognition rate is higher when using voice or gesture as input, which also confirms that each modal compensation argument is in the process of emotional expression. In the case of bimodal fusion, the recognition rate is higher than that of single modality. In the case of three-modal fusion, the recognition rate is the highest, which proves the effectiveness of the multi-modal fusion process. Today's classifiers cannot achieve good classification results for the time being, so we hope to develop classifiers that automatically optimize classifier parameters or automatically formulate classification strategies in the future.

Acknowledgements. This work was supported by Higher Education Teaching Reform research project of Jilin Province in 2021: College English Information Teaching Design and Application Research with Integrated Mobile Tools (JLJY202128256704);The Scientific Research Project of Higher Education in Jilin Province in 2020: Integrated application and research on the synchronous/ asynchronous online teaching mode under the circumstances of emergencies (JGJX2020D157);Scientific Research Project of Jilin Provincial Department of Education in 2021: Research on the Design of MOOC+SPOC Mixed Advanced English Teaching Model in the Post-epidemic Era (JJKH20211001SK).

References

1. Kumar, M., Jindal, S.R., Jindal, M.K., Lehal, G.S.: Improved recognition results of medieval handwritten gurmukhi manuscripts using boosting and bagging methodologies. Neural Process. Lett. **50**(1), 43–56 (2018). https://doi.org/10.1007/s11063-018-9913-6
2. Kokel, H., Odom, P., Yang, S., et al.: A unified framework for knowledge intensive gradient boosting: leveraging human experts for noisy sparse domains. In: Proceedings of the AAAI Conference on Artificial Intelligence, vol. 34, no. 4, pp. 4460–4468 (2020)
3. Rahdari, F., Rashedi, E., Eftekhari, M.: A multimodal emotion recognition system using facial landmark analysis. Iran. J. Sci. Technol. Trans. Electr. Eng. **43**(JUL.SUPPL.1), S171–S189 (2019)
4. Grosche, S., Regensky, A., Seiler, J., et al.: Boosting compressed sensing using local measurements and sliding window reconstruction. IEEE Trans. Image Process. **PP**(99), 1 (2020)
5. Goulet, N., Ayalew, B.: Distributed maneuver planning with connected and automated vehicles for boosting traffic efficiency. IEEE Trans. Intell. Transp. Syst. **PP**(99), 1–15 (2021)
6. Shao, Z., Chandramouli, R., Subbalakshmi, K.P., et al.: An analytical system for user emotion extraction, mental state modeling, and rating. Expert Syst. Appl. **124**(JUN.), 82–96 (2019)
7. Tissari, H.: Current emotion research in english linguistics: words for emotions in the history of English. Emot. Rev.: J. Int. Soc. Res. Emot. **9**(1), 86–94 (2017)
8. Anjani, S., Ch, S.: An efficient facial emotion recognition system using novel deep learning neural network-regression activation classifier. Multimed. Tools Appl. **80**(6), 1–26 (2021)
9. Shavit, Y., Klein, I.: Boosting inertial-based human activity recognition with transformers. IEEE Access **PP**(99), 1 (2021)
10. Torres-Valencia, C., Álvarez-López, M., Orozco-Gutiérrez, Á.: SVM-based feature selection methods for emotion recognition from multimodal data. J. Multimodal User Interfaces **11**(1), 9–23 (2016). https://doi.org/10.1007/s12193-016-0222-y
11. Andrews, S.C., Staios, M., Howe, J., et al.: Multimodal emotion processing deficits are present in amyotrophic lateral sclerosis. Neuropsychology **31**(3), 304–310 (2017)
12. Latha, G., Priya, M.M.: A novel and enhanced facial electromyogram based human emotion recognition using convolution neural network model with multirate signal processing features. J. Comput. Theor. Nanosci. **14**(3), 1572–1580 (2017)

Application of Intelligent Recommendation Algorithm in Data Mining of Scientific Research Management System

Lijuan Zhao[1]([✉]), Jinhua Yang[2], Haonan Zha[3], and Eric Rosales[4]

[1] Office of Academic Research, Dianchi College of Yunnan University,
Kunming 650228, Yunnan, China
qqzljuan213@126.com
[2] Institute of Science and Technology, Dianchi College of Yunnan University,
Kunming 650228, Yunnan, China
[3] College of Architectural Engineering, Dianchi College of Yunnan University,
Kunming 650228, Yunnan, China
[4] Universidad Autonoma de Zacatacas, Jardin Juarez 147, Centro Historico,
98000 Zacateca, Mexico

Abstract. With the improvement of the status of scientific research and the improvement of scientific research level, traditional management methods and methods are becoming more and more difficult to meet the current increasingly complex scientific research management needs. Under the background of the rapid development of information technology, scientific research management informatization has become inevitable choice for future scientific research management innovation. This article aims to study the application of intelligent recommendation algorithm in scientific research management system data mining. Based on the analysis of recommendation system, intelligent recommendation algorithm and system design principles, a scientific research management system based on intelligent recommendation is designed. When it is 15, the minimum support is 120, and the minimum confidence is 0.8, the test effect is the most ideal.

Keywords: Intelligent recommendation algorithm · Scientific research management system · Data mining · Expert system

1 Introduction

The recommendation system is a very popular technology today with a wide range of uses. Business, news, and entertainment are all very convenient. These systems allow people to choose their preferences from a large amount of data according to their preferences [1, 2]. For users, they can improve user experience, enjoy more personalized services, and reduce unnecessary waste of time [3, 4].

The recommendation system is a unique data screening system that emerged during the development of modern computer technology. The basic idea of the recommendation system is to infer the user's interest based on the user's past activities, access

V. Sugumaran et al. (Eds.): ICMMIA 2022, LNDECT 138, pp. 336–344, 2022.
https://doi.org/10.1007/978-3-031-05484-6_42

files, personal information and other data, so as to give suggestions that the user may be interested in [5, 6]. The biggest difference between the recommendation system and the traditional information retrieval service is that in the traditional search business, when the user enters what they want or the object of interest, the system will match the items in the system according to this item or display it. Related items, although visible to users, because the recommendation system is less passive, it can directly recommend active items to users. In the process of interacting with users, traditional information retrieval services usually do not express, analyze or use the knowledge of users' historical interests [7, 8]. However, the recommendation system will also actively record the historical information of interested users when communicating with users, and put forward service requirements, so as to establish a cognitive model that meets user interests and preferences, and complete the final message push business goal through this model. The research on recommender systems began in the early part of the last century, when recommender systems had summarized most of the results in related fields. Cognitive science, approximation theory, information retrieval, prediction theory, business science, and market models are applied to the research of recommender systems and other knowledge applications [9, 10]. In the recommendation system, the most important thing is the promotion strategy or recommendation algorithm adopted in the promotion system. The algorithm can determine the required input category and the output category generated. Therefore, the algorithm research and development, analysis and implementation process represents the research and development process in the entire promotion system [11, 12]. For the most critical positions, recommendation algorithms usually include content-based filtering algorithms, collaborative filtering algorithms, and hybrid recommendation algorithms.

On the basis of consulting a large number of related references, this paper combines the recommendation system, intelligent recommendation algorithm and system design principles to design a scientific research management system based on the intelligent recommendation algorithm. The system mainly includes 5 modules, namely the system login module and the scientific research project. Management module, scientific research achievement module, expert recommendation module and system management module. Finally, the recommended test of the system is carried out to verify whether the requirements of this article are met.

2   Application of Intelligent Recommendation Algorithm in Data Mining of Scientific Research Management System

2.1   Recommendation System

(1) Collaborative filtering recommendation
    The collaborative filtering system is a recommendation system that has been widely proposed and widely adopted. The basic idea can include the following two parts. One uses the historical information of the target user to count the highest

similarity between target users, and the second uses the highest similarity between the product and the target user and other neighbor product reviews to predict the target user's preference for a certain product. Then the system recommends to target users based on its settings. The starting point of collaborative filtering technology is that no one's interests are isolated. Assuming that a certain user has the same evaluation for a certain item, then the evaluation for other items is the same. The basic idea of this recommendation method is to discover some of the nearest neighbors of the target user (users with similar interests to the target user) by using certain techniques, and make recommendations through the evaluation of the nearest neighbors of the object, and finally select the target and predict the score. The higher items are used as the recommended list of target users.

(2) Content-based recommendation

Information retrieval and information filtering technology are the main theoretical cornerstones for the development of content-based information processing methods. The content-based introduction method is to introduce to the user information content that the user has never contacted based on the user's behavior or historical files. Many methods used in content-based introduction are information retrieval methods, such as tf-idf and text sorting technology. The essence of content-based recommendation is to obtain user attributes based on the user's access to files, calculate the value of the utility function according to the role of the user attributes and the proposed elements, and recommend the results to the user.

2.2   Implementation of Intelligent Recommendation Algorithm

In the process of collating scientific research projects with experts, this article evaluates the technical guidance of experts with the dimensional information of the experts' scientific research information, patent information, theoretical research capabilities, and the actual development skills of the experts. According to the actual needs of the project, four capacity experts are weighted to obtain a perfect match. Therefore, the results recommended by the experts are optimized to produce the best results recommended by the experts. Therefore, the results recommended by the experts are optimized to produce the best results recommended by the experts. This algorithm is suitable for the intelligent recommendation part of scientific research management system. Figure 1 is the realization flow chart of the intelligent recommendation algorithm.

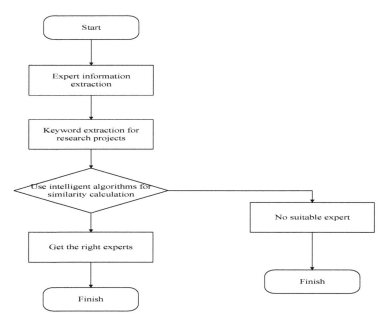

Fig. 1. Intelligent recommendation algorithm implementation flowchart

The recommendation system is the core of the scientific research project management system. Through the analysis of system modules and characteristics, the system can use recommended tag-based algorithms. Experts are interested in the calculation of scientific research projects, combined with the previously introduced TF-IDF algorithm, using the TF-IDF algorithm for authority calculation. The system can create two tables, the project label scientific research table and the label specific information table. Expert predictions select the top N experts who are most interested in a particular scientific research project.

2.3   System Design Principles

(1) Practicability

At present, each stage of scientific research management has corresponding management methods, some are manual processing, and some are small software processing. Although the types of management methods are different, the operation methods and basic requirements of these systems are basically the same, which is conducive to the formation of a single scientific research management system. The scientific research management system starts from meeting the basic needs of scientific research management, integrates data and file formats into business management, and uses computer networks and hardware tools to perform manual and iterative tasks to improve the effectiveness of scientific research.

(2) Reliability

The most basic principle of a software system is reliability. After the software system is officially put into operation, a large amount of data and documents need to be published on the Internet, so the reliability of the software system must be ensured. Ensure timely and consistent data transmission. According to the management characteristics of scientific research projects of various units, the system must do a lot of work for reliability to ensure the data security and stable operation of the system.

(3) Ease of operation

For a software system, convenient operation and concise interface are also important factors to measure its quality. The operating interface of the software system is very complicated, and the logo description is obscure and difficult for users to accept. Therefore, the design of the software system should follow the basic principles of man-machine interface design, and the interface design should conform to the basic business management process specifications. The menu should be divided into a first-level menu and a second-level menu. Messages are added to an impressive location, and function buttons are integrated, making the operation uniform and convenient.

(4) Scalability

Demand is not static. The system must carry out certain backups and scalability fields during the design process to ensure scalability. Large-scale software and hardware systems are easy to upgrade. In addition, the system must submit certain backup and scalability fields so that it can be upgraded at any time to meet new requirements.

(5) Ease of maintenance

Because the user's computer level is uneven, the software system must not only be easy to use, but also easy to maintain. It enables users to operate in the simplest way, including automatic database backups, software upgrades and backups, and a locked system-ready interface.

3   Experiment

3.1   System Login Module

When logging in to the system, you must use the same interface to log in regardless of roles and permissions. Every logged-in user must verify the authenticity of the user ID. When the user enters the login user name and password on the login interface, the management system verifies the authenticity of the identity according to the system user data stored in the database. If the login user name and password are correct, the verification is complete and you can log in to the system. If the system user forgets the login password, he can select the reset button and follow the system prompts to reset the password.

3.2   Scientific Research Project Management Module

Item list: The items are displayed as a table on the page in turn, with 10 items displayed on each page. The list page has add, delete, search, view, print and export buttons in the upper right corner of the page. Click different buttons to respond to different functions.

> Project review: The last cell of the list has two function links, edit and modify. Only the research secretary or supervisor can perform the functions of these two functions. General researchers can only view the information on the page, and there is no license to change it.
>
> New project: Click the new project button to enter the recently added page. After adding a research project, the project has not yet been reviewed and must be managed by a research secretary or super administrator. After the review is completed, the project becomes the actual storage.
>
> Project query: Divided into simple query and complex query. For simple queries, you need to enter the corresponding keywords, and then display the corresponding query items on the list page according to the keywords. For complex queries, it is necessary to combine query conditions that can be combined. It can also be intersected. According to different query conditions, the corresponding query items will also be displayed on the list page.

3.3   Scientific Research Achievement Management Module

The scientific research achievement management module helps scientific research system administrators to add, delete, modify, control and manage various scientific research achievements. In the scientific research achievement management module, click the drop-down menu of scientific research achievement type, select the type of scientific research achievement, and query the corresponding scientific research achievement table. The system administrator can add, delete, modify and check. After the scientific research results are reviewed and approved by the scientific research system administrator, the scientific research personnel can only view but not modify.

3.4   Expert Recommendation Module

First of all, use data mining technology to analyze based on the data of the scientific research system, establish the data model of the scientific research project, and recommend the experts of each scientific research project according to the data model of the scientific research project. In the future, relatively fixed rules will be used as guidance, but this kind of guidance is not the only one. Closed data mining groups can be created continuously and the rules can be modified adaptively. Secondly, in a scientific research project, it is necessary to calculate the data similarity of similar scientific research projects and perform statistics on the calculated data. The basic idea is to recommend scientific research projects with similar data and information, and similar types of scientific research projects to professionals. This part of the technology uses the past behavior data of scientific research projects as initial data, calculates the similarity between different scientific research projects, selects very similar resources from here, and completes the expert recommendation process. Scientific research projects are calculated based on the original data of scientific research projects. You can extract expert information and data from the content and make relevant recommendations. The basic idea is to extract the scientific research direction that the expert is good at from the topics previously considered by the expert, and extract the content features related to the scientific research direction at the same time. The above algorithm not only considers content recommendation, but can also find recommendations based on rules and perform two-way recommendations.

3.5   System Management Module

System management mainly completes the basic configuration of scientific research management system, including system role management, authority management, standard evaluation configuration, basic dictionary data management, etc. The scientific research management system is divided into a variety of management personnel, and the scope of personnel management is quite different. For flexibility, the system can dynamically configure different types of employees, that is, based on system roles.

System role management includes functions such as adding, modifying, deleting, disconnecting, activating, scrolling up and scrolling down. Enter the name of the role to be created and save to add a new system role. You can make changes after selecting the changes in the specified row. Deletion is mainly related to system roles that have not been used since creation. It is possible to deactivate and activate system roles reported since their creation to change their status.

4   Discussion

The data used this time is the Delicious data set developed for the public, which contains a large number of data tags. Some researchers associate education with the Delicious data set for two main reasons. One is that 57% of education-related keywords are used in this tagging system, and the other is that the delicious data set contains a large number of education-related keywords. Therefore, this project will use the Delicious data set as the test data.

The project is based on the association rule test recommendations derived from experiments. Whether to use tags as a template for measuring experiment results. You will receive the recall rate and correct answer rate as evaluation indicators. In this experiment, all users are represented by U, the recall rate calculation formula is shown in formula (1), and the accuracy rate calculation formula is shown in formula (2).

$$Re = \frac{\sum_{u \in U} |R(u) \cap T(u)|}{\sum_{u \in U} |T(u)|} \tag{1}$$

$$Rre = \frac{\sum_{u \in U} |R(u) \cap T(u)|}{\sum_{u \in U} |R(u)|} \tag{2}$$

In this experiment, the arrangement of tags is used to form a recommendation list. The recall rate reflects the completeness of the recommended results and accurately reflects the degree of correspondence between scientific research projects and expert information. Use the recommended Top-N method to set the value of N to {5, 10, 15, 20, 25}, and experiment. The results obtained are shown in Table 1 and Fig. 2.

Table 1. Test results of accuracy and recall

	Accuracy	Recall rate
5	74.3%	67%
10	80%	75.2%
15	88.2%	88.6%
20	82.7%	89.4%
25	86.5%	92.3%

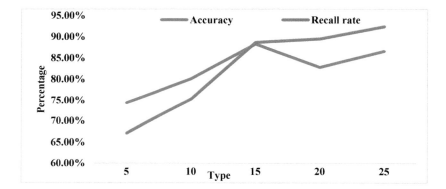

Fig. 2. Test results of accuracy and recall

When the value of N is 15, the accuracy rate is 88.2%, and the recall rate is 88.6%. When N is greater than 15, the recall rate is increasing, but the accuracy rate is decreasing. Therefore, considering the data of these two indicators comprehensively, if N is 15, the minimum support is 120 and the minimum confidence is 0.8, which is the most ideal test result.

5 Conclusions

This paper takes the development and design of scientific research management system as the research background, analyzes the current situation of domestic project application management system, and proposes a scientific research project management system that meets actual needs. Finally, a scientific research project management system based on intelligent recommendation algorithm is implemented to meet the needs of construction projects.

References

1. Biscarri, F., Monedero, I., Larios, D.F., et al.: A data mining environment for management of ground testing of the A400M aircraft. IEEE Aerosp. Electron. Syst. Mag. **36**(6), 56–64 (2021)

2. Ruiz, R., Roubickova, A., Reiser, C., et al.: Data Mining and machine learning for porosity, saturation, and shear velocity prediction: recent experience and results. First Break **39**(7), 71–76 (2021)

3. Papík, M., Papíková, L.: Application of selected data mining techniques in unintentional accounting error detection. Equilibrium **16**(1), 185–201 (2021)

4. Bratsas, C., Chondrokostas, E., Koupidis, K., et al.: The use of national strategic reference framework data in knowledge graphs and data mining to identify red flags. Data **6**(1), 2 (2021)

5. Tripathi, A.K., Mittal, H., Saxena, P., et al.: A new recommendation system using map-reduce-based tournament empowered whale optimization algorithm. Complex Intell. Syst. **7**(1), 297–309 (2021)

6. Al-Rubaie, H.A., Abbas, A.S.: Software effort estimation using data mining techniques based on improved precision. J. Adv. Res. Dyn. Control Syst. **12**(5), 176–185 (2020)

7. Muangprathub, J., Boonjing, V., Chamnongthai, K.: Learning recommendation with formal concept analysis for intelligent tutoring system. Heliyon **6**(10), e05227 (2020)

8. Anand, J.V.: A methodology of atmospheric deterioration forecasting and evaluation through data mining and business intelligence. J. Ubiquit. Comput. Commun. Technol. **2**(2), 79–87 (2020)

9. Fitriati, D., Amiga, B.P.: Implementation of data mining to determine the association between body category factors based on body mass index. J. Riset Inform. **2**(4), 233–240 (2020)

10. Kalathas, I., Papoutsidakis, M., Drosos, C.: Optimization of the procedures for checking the functionality of the Greek railways: data mining and machine learning approach to predict passenger train immobilization. Adv. Sci. Technol. Eng. Syst. J. **5**(4), 287–295 (2020)

11. Sharma, G.: Evaluation of data mining categorization algorithms on aspirates nucleus features for breast cancer prediction and detection. Int. J. Educ. Manag. Eng. **10**(2), 28–37 (2020)

12. Al-Hagery, M.A., Alzaid, M.A., Soud, T., et al.: Data mining methods for detecting the most significant factors affecting students' performance. Int. J. Inf. Technol. Comput. Sci. **12**(5), 1–13 (2020)

Design of Specific Scene Recognition System Based on Artificial Intelligence Vision

Guozhen Gao and Yuhua Li[✉]

Zhujiang College of South China Agricultural University,
Guangzhou, Guangdong, China
bekili@yeah.net

Abstract. Scene recognition technology, as a key research topic in the field of computer vision applications, aims to semantically label a large number of image content through computer technology. Natural scene recognition technology can directly map image information from the pixel level to the semantic level, which has great scientific research significance and application value. This paper aims to study the design of a specific scene recognition system based on artificial intelligence vision. Based on the analysis of the scene recognition technology and the application field of scene recognition, the specific scene recognition system is designed and implemented. Finally, the system in this paper and the other two specific. The scene recognition system is experimentally compared. The experimental results show that the average recognition accuracy of this system for 8 groups of images is always higher than the other two specific scene recognition systems, which proves that this system has strong recognition accuracy.

Keywords: Artificial intelligence vision · Specific scene recognition · System design · Convolutional neural network

1 Introduction

In recent years, due to the vigorous development of artificial intelligence, intelligent systems with functions similar to human cognition have also become a hot topic in the field of artificial intelligence research and development. Intelligent systems mainly use computers and visual equipment as human sensory systems to learn and imitate people's thinking and behavior [1, 2]. In today's world, the main methods of computer science account for about 75% of people's total information intake. Therefore, the development of intelligent information systems is inseparable from the progress of machine vision information technology, and machine vision is also a major research direction in the field of artificial intelligence [3, 4].

Scene recognition methods usually include two methods based on global features and local features based on visual features. By applying this method, people can obtain low- and medium-level characteristics such as color and texture, and divide different types of scenes. Some researchers have constructed a confidence function specifically used to predict the category label of a scene, and used low-level general features to implement the situation classification. During the training process, a classifier was set

© The Author(s), under exclusive license to Springer Nature Switzerland AG 2022
V. Sugumaran et al. (Eds.): ICMMIA 2022, LNDECT 138, pp. 345–351, 2022.
https://doi.org/10.1007/978-3-031-05484-6_43

for the type label of each scene. These classifiers are used in each image test, so as to calculate the probability of each type in the image. But the main disadvantage of this method is the inflexible use of the training set [5, 6]. At the same time, because the local attribute method usually does not consider spatial information, it is easy to transform it into a lower-level general attribute method. For example, you can directly divide the graphics block into several partial graphics blocks, then extract the attributes in turn, and then use each classifier. Finally, by using the overall visual characteristics of the perceptual image block, the vocabulary distribution of the vocabulary was divided into scenes, which achieved good results [7, 8]. Later, some researchers proposed a scene recognition algorithm that combines local and global information. The local information is mostly interested in relatively sparse blocks, while the general information is to divide the entire image into a relatively clean and tight grid. The experimental results also confirm that the sixty-seven indoor situation images in the MIT_indoor data set are at this moment [9, 10]. Although they cannot use specific target items, it is believed that some indoor scenes can be displayed better than the target items they involve, and target recognition can also be directly related to the recognition results of indoor scenes [11, 12]. In addition to the methods introduced above, there are many tasks that use topological spatial positioning methods based on visual symbols, and they are widely used in robot vision processing methods. However, because these identification methods are often used in special environments, they usually do not work well when used in other occasions.

On the basis of consulting a large number of relevant references, this paper designs a specific scene recognition system based on artificial intelligence vision based on scene recognition technology and the application field of scene recognition. The system mainly includes three modules, namely the artificial intelligence visual recognition module. The image preprocessing function module and the scene recognition model construction function module. After implementing these three modules one by one, the system in this paper is compared with the other two systems to verify that the system in this paper is effective in scene recognition.

2 Design of Specific Scene Recognition System Based on Artificial Intelligence Vision

2.1 Scene Recognition Technology

(1) Convolutional neural network
   Convolutional neural network is a type of neural network with a deep structure, derived from the natural visual knowledge mechanism of biology, and is one of the most representative network algorithms for deep learning. Today, CNN is one of the main research hotspots in many fields, especially in the field of standard classification. The Internet is widely used because they can directly use the original image without editing the image at an early stage. Convolutional neural networks can not only extract high-level features, but also do not generate a large

number of model parameters. The main sign is that the structure contains a feature derivator composed of a cohesive layer and a sampling layer, also known as a pooling layer.

(2) AlexNet model

AlexNet is composed of five convolutional layers, five pooling levels, three fully connected levels, and about 50 million selectable parameters. Finally, the completely combined layer is exported to the Softmax1000 dimensional layer, thereby establishing the 1000 classification distribution. The AlexNet model design uses a non-linear activation function, and uses dropout and data enhancement methods to avoid over-adapting to the model.

Dropout: AlexNet has a 0.5 chance to set the output value of each hidden layer neuron to zero. This will help prevent neurons from traveling back and forth. Whenever a new sample is imported, the network will test the latest network structure, but each network structure shares parameters, reducing the complexity between network neurons. Once there is no dropout, the network will be significantly overclocked.

Data enhancement: By increasing training data samples, over-adaptation of the model can be avoided and the accuracy of the network algorithm can be improved. If the training data is limited, you can increase the amount of training data by using traditional simple image conversion methods to generate new data from the existing data set. Common conversion methods include horizontal image inversion, random cropping of the original image, translation conversion, color conversion, and lighting.

(3) Deep learning mobile terminal framework

TensorFlow Lite is a lightweight mobile solution and embedded TensorFlow that can help developers easily develop artificial intelligence applications on mobile devices. It is a mobile framework for predicting low-latency deep learning on small binary devices. TensorlowLite uses a small, fast model to quantify the core and reduce system latency.

(4) MobileNet V1 network

MobileNetV1 is a deeply separable network model based on convolution. Deep convolution is defined as a standard convolution solution consisting of deep convolution and 1×1 point increments. The deep coil mode runs the coil mode, each channel input has the same single coil core, and the point-to-point coil combines the outputs of all deep coils. Compared with the standard convolution, the depth separable convolution processes the two-step operation template in one step, which reduces the complexity and model size.

MobileNet has a small network structure and short delay, but it is often included in specific application requirements. The application requirements of the model are small, the network structure is properly adapted, and the effect is better in the research environment. Therefore, enter the width multiplier parameter α. For a given amplitude multiplier α, its function is to uniformly shrink and compress each layer of the network. As a result, the number of input channels changes from M to αM, and the number of output channels changes from N to αN. As the amplitude multiplier parameter a increases, the complexity of the updated separable depth convolution calculation is:

$$D_K \times D_K \times \alpha M \times D_F \times D_F + \alpha M \times \alpha N \times D_F \times D_F \qquad (1)$$

$$D_K \times D_K \times \alpha M \times \rho D_F \times \rho D_F + \alpha M \times \alpha N \times \rho D_F \times \rho D_F \qquad (2)$$

2.2   Application Areas of Scene Recognition

(1) Image and video retrieval field
     Image and video capture is one of the most direct scene analysis and understanding applications. Through the analysis and understanding of the scene, you can find the pictures you need faster. In recent years, content-based image and video search has become commonplace, attracting people's attention, and research centers have become a hot topic for faster and more accurate information retrieval. At present, various image and video retrieval systems are being developed at home and abroad.
(2) Robot
     In fields such as computer vision, it is even more difficult to provide a pair of eyes for the machine. The machine's ability to observe and understand as an individual is a hot topic in current research.
(3) Biometric identification
     In today's society, national and personal information security and network security have also attracted much attention. Some traditional information protection methods, such as passwords, identity verification and some passwords, are easily forgotten or lost. The use of biometric technology for identity recognition has the advantages of convenience, portability, and no loss. For example, face recognition technology has been applied in many fields.
(4) Computer vision
     Another important aspect of scene recognition is to provide prior knowledge of target recognition and motion detection. If you want to identify the target in the image, the scene type will definitely provide more information about the target and help in future work. If you want to identify whether there is a vehicle in the image, if the scene recognition process recognizes that the vehicle may be under the road scene image at this time, the probability is very high. Preliminary information to improve the performance of related recognition tasks.

Therefore, with the continuous advancement of science and technology and the continuous increase of image data, scene recognition technology will be applied to more and more fields and will become the basic technology of social research.

3   Experiment

3.1   Artificial Intelligence Visual Recognition Module

The artificial intelligence visual recognition module is based on a camera head and an accelerometer. The camera it uses is usually an artificial intelligence visual camera recently developed by a company. The camera adopts artificial intelligence vision design, standard definition is 1080P, camera angle is 150° wide-angle, and it has super-strong object recognition, color perception, voice recognition, and even radar detection function. The basic structure diagram of the artificial intelligence vision camera is shown in Fig. 1.

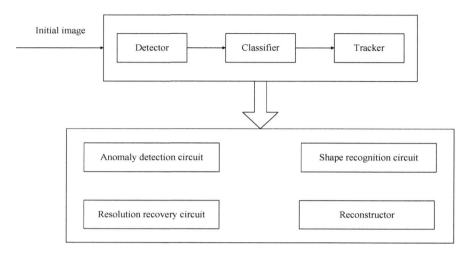

Fig. 1. 360 artificial intelligence vision camera structure diagram

3.2   Image Preprocessing Function Module

In order to facilitate subsequent image processing and improve the accuracy of scene recognition, image preprocessing must be performed after receiving a large number of image signals. The main purpose of picture preprocessing is to eliminate signals that are not related to the content in the picture, so as to simplify the data as much as possible. The image preprocessing scheme mainly includes the training data set and the captured video pictures. First, according to the size of the input layer of the scene recognition model established in the selection of the convergence neural network, linear interpolation is selected in turn and linear interpolation is performed in the x and y directions; then the average is removed, so that each dimension. The value of the large-scale data center has become zero to reduce the amount of calculation.

3.3   Scene Recognition Model Construction Function Module

Scene recognition and classification are the basic functions of the system. With the introduction and improvement of deep learning technology, scene recognition and classification technology has been very successful. Some major data sets are important for extracting good models, and some of them have good generalization. After comparing and analyzing each data set, we selected 11 scene image types from the most complete place205 data set as the scene recognition model training data set. Each scene type has about 5000 images and 50000 image data sets, which are a lot of different. The training data set guarantees the accuracy of the generated scene recognition model.

4   Discussion

In order to test the recognition effectiveness of this specific scene recognition system based on artificial intelligence vision, this article randomly produced eight sets of photos from a certain city, and each set of images has a total of six. Eight groups of images were identified by the system in this paper, specific scene recognition based on multi-modal fusion and specific scene recognition based on local feature matching, and the test results were analyzed. The average recognition accuracy of the three systems in the experiment is shown in Table 1 and Fig. 2.

Table 1. Average recognition accuracy of the system

	Text system	Specific scene recognition system based on multi-modal fusion	Specific scene recognition system based on local feature matching
1	98.46%	95.39%	96.56%
2	97.22%	94.78%	95.12%
3	97.54%	91.15%	95.47%
4	98.28%	91.64%	94.23%
5	98.16%	92.55%	95.82%
6	97.59%	94.64%	96.14%
7	98.73%	95.71%	95.31%
8	97.02%	93.54%	95.44%

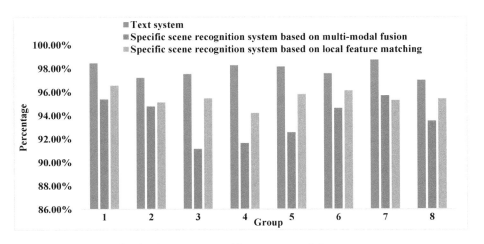

Fig. 2. Average recognition accuracy of the system

It can be seen from Table 1 and Fig. 2 that the average recognition accuracy of this system for 8 groups of images is always higher than that of the other two specific scene recognition systems, which proves that the system has strong recognition accuracy.

5   Conclusions

Specific scene recognition technology is an important field in the field of image processing. It is an image understanding technology that analyzes semantic image information and makes classification decisions. This image semantic understanding technology is widely used in content-based image retrieval, 3D scene reconstruction, robot navigation and other fields. The semantic information in the image is for human use, and the underlying pixel space information is stored by the computer. Solving image recognition can reduce the semantic gap between humans and machines.

Acknowledgment. Exploration on the Teaching Reform of Visual Communication Design under the Background of Artificial Intelligence.

References

1. Qin, N., Hu, X., Dai, H.: Deep fusion of multi-view and multimodal representation of ALS point cloud for 3D terrain scene recognition. Isprs J. Photogram. Remote Sens. **143**(SEP.), 205–212 (2018)
2. Anbarasu, B., Anitha, G.: Indoor scene recognition for micro aerial vehicles navigation using enhanced SIFT-ScSPM descriptors. J. Navigat. **73**(1), 1–19 (2019)
3. Guo, J., Nie, X., Yin, Y.: Mutual complementarity: multi-modal enhancement semantic learning for micro-video scene recognition. IEEE Access **pp**(99), 1 (2020)
4. Xia, J., Ding, Y., Tan, L.: Urban remote sensing scene recognition based on lightweight convolution neural network. IEEE Access **pp**(99), 1 (2021)
5. Xie, L., Lee, F., Liu, L., et al.: Hierarchical coding of convolutional features for scene recognition. IEEE Trans. Multimed. **22**(5), 1182–1192 (2020)
6. Xiong, Z., Yuan, Y., Wang, Q.: ASK: adaptively selecting key local features for RGB-D scene recognition. IEEE Trans. Image Process. **pp**(99), 1 (2021)
7. Seong, H., Hyun, J., Kim, E.: FOSNet: an end-to-end trainable deep neural network for scene recognition. IEEE Access **pp**(99), 1 (2020)
8. Wang, S., Yao, S., Niu, K., et al.: Intelligent scene recognition based on deep learning. IEEE Access **pp**(99), 1 (2021)
9. Khan, N., Chaudhuri, U., Banerjee, B., et al.: Graph convolutional network for multi-label vhr remote sensing scene recognition. Neurocomputing **357**(SEP.10), 36–46 (2019)
10. Morio, Y., Hanada, Y., Sawada, Y., et al.: Field scene recognition for self-localization of autonomous agricultural vehicle. Eng. Agric. Environ. Food **12**(3), 325–340 (2019)
11. Chen, L., Zhan, W., Tian, W., et al.: Deep integration: a multi-label architecture for road scene recognition. IEEE Trans. Image Process. **pp**(99), 1 (2019)
12. Tang, J., Qian, W., Zhao, Z., et al.: Multi-view non-negative matrix factorization for scene recognition. J. Vis. Commun. Image Represent. **59**(FEB.), 9–13 (2018)

The Method of Antibiotic Potency Measurement Based on Image Processing

Dewei Zhu[1,2(✉)]

[1] College of Marine and Bio-engineering, Yancheng Teachers University,
Yancheng 224007, Jiangsu, China
zhudw2021@163.com
[2] Jiangsu Province Engineering Research Center of Agricultural
Breeding Pollution Control and Resource, Yancheng Teachers University,
Yancheng 224007, Jiangsu, China

Abstract. With the development of social economy, antibiotics have an increasing impact on people's lives and agricultural production. The antibiotic dose is usually expressed by the potency. Potency is a relevant standard to measure the quality of antibiotics. Therefore, determining the titer of antibiotics is a very important method in the research, production and application of antibiotics. Therefore, the development of a new type of intelligent antibiotic titration system has important practical significance. This article studies the antibiotic potency measurement method based on image processing, and understands the correlation of the antibiotic potency measurement method on the basis of literature data, and then designs the antibiotic potency measurement system based on image technology, and the designed system the detection results show that the system's image processing frame rate and throughput are more than twice that of other programs, and its program duty cycle is much larger than other programs.

Keywords: Image processing · Antibiotics · Potency measurement · Measurement system

1 Inductions

In biomedicine, anti-infective drugs are divided into antibacterial drugs, antifungal drugs, antiviral drugs, antituberculosis drugs, and antiparasitic drugs. Antibacterial drugs are antibiotics [1, 2]. Antibiotics are widely used medicines in medicine. Commonly used antibiotics can be divided into many types according to their structure. Examples include chloramphenicol, tetracycline, aminoglycosides and macrolides [3, 4]. Antibiotics are widely used in clinical, veterinary and agricultural fields because of their excellent killing and prevention effects on various infections caused by Gram-positive bacteria and Gram-negative bacteria. The hazards of antibiotic residues in human animal food have also received attention, and there are more reports on antibiotics research abroad [5, 6]. In order to comprehensively manage the quality of pharmaceutical products and ensure the safety, scientific rationality and efficiency of

V. Sugumaran et al. (Eds.): ICMMIA 2022, LNDECT 138, pp. 352–359, 2022.
https://doi.org/10.1007/978-3-031-05484-6_44

drug use, strict analysis and testing are required in the process of drug manufacturing, storage, supply, development and clinical application [7, 8].

Regarding the study of antibiotic potency measurement, some researchers said that the commonly used method of testing antibiotics in the Pharmacopoeia is to test the efficacy of microorganisms. In other words, antibacterial activity is used to measure the effectiveness of antibiotic active ingredients. This method can directly reflect the effectiveness of antibiotics, and has the characteristics of high sensitivity and small dose of the test substance. Therefore, this method is widely used to determine the efficacy of antibiotics. The accuracy of antibiotic efficacy test results directly affects the safety and effectiveness of substance use. As a classic detection method, measuring the efficacy of microorganisms has its own advantages. The method of determination is a hot issue in determining the efficacy of antibiotics, and it is not static. A lot of domestic research in this area has been accumulated and some experience has been provided. However, this method also has some disadvantages, including time-consuming, multiple operation steps, and many factors that affect the test. In order to make the verification method more perfect, many verification workers are actively looking for improved methods and looking for alternatives [9]. Researchers have also discovered that the principle of antibiotic microbial analysis is to use a special method to detect the inhibitory effect of antibiotics on microorganisms under specific temperature conditions and to calculate the quantitative response of their efficacy. Since this method measures the efficacy of antibiotics by the degree of inhibition of the antibiotics in the tested microorganisms, it is necessary to compare the test product with a known standard product to determine the efficacy. In addition, the molecular structure of antibiotics is complex, and more exclusive physical and chemical analysis methods have not been developed in the early stages of drug development. The microbiological analysis results of antibiotics are relatively intuitive and the most basic method. Even today, microbial detection is still the preferred analytical method for multi-component antibiotics [10]. In summary, there are many research results on the measurement of antibiotic potency, but there are few researches on the application of image technology in it.

This article studies the antibiotic potency measurement method of image technology, analyzes the antibiotic potency measurement method and image processing technology based on the literature data, then designs the antibiotic potency measurement system based on image processing technology, and design the system performs testing and draws relevant conclusions through the testing results.

2   Research on the Method of Antibiotic Potency Measurement

2.1   Methods and Characteristics of Antibiotic Potency Measurement

(1) Tube sheet method
   The tube plate method uses the diffusion of antibiotics in agar medium to compare the size of the inhibition zone produced by the standard product and the test product inoculated with the test bacteria to determine the efficacy of the test product [11]. There are many methods to determine the internal and external

suspension areas, including adaptive image segmentation methods based on gray histograms, non-contact optical measurement, and measurement based on scanners and computers. These three methods can quickly and accurately measure the size of the zone of inhibition to achieve the purpose of determining the efficacy of antibiotics. Among them, the non-contact optical measurement method is highly subjective, difficult to operate, and cannot achieve very ideal measurement accuracy. Scanners and computer-based measurement systems do not require manual measurement due to their high equipment utilization and low cost. The suspension error is highly reliable.

(2) Routine turbidimetric method antibiotic efficacy determination system
The turbidity method is to measure the turbidity value of the bacteria after culture, and use the antibiotics to inhibit the growth of the tested bacteria in the liquid medium. This is a method to determine the efficacy of the test product by comparing the degree of inhibition of the growth of the tested bacteria by the standard product and the test product [12].

(3) Drilling method
The drilling method is a new method improved on the basis of the tube-coil method. This method is to first cast the double-layer board according to the tube-sheet method, solidify, and then use a sterile perforator to drill a hole in the place where the Oxford cup was originally placed. Punch the hole within 5–6 mm in diameter and lift it up with a sterile toothpick. Take out the agar block and leave a small round hole. Use a dropper to drop a drop of melted agar into the small hole to seal the bottom to prevent the liquid from flowing out of the bottom of the plate. The advantages of this method: the terracotta lid of the Oxford cup and petri dish is omitted, the liquid surface is not tightened during operation, the liquid in the hole does not flow at a low slope, a small amount of ideal circular bacteriostatic agent can be used by adding The sample gets sampler, more accurate quantification.

(4) Chemiluminescence analysis method
Chemiluminescence analysis is an analysis method based on the phenomenon of chemiluminescence. The content of the analyte is related to the intensity of chemiluminescence. Because the chemiluminescence analysis device is simple, it does not need a light source or a scattering device, but it avoids many interferences such as measurement error caused by the unstable light source, interference caused by scattered light or stray light. In short, this method has the advantages of wide linear range, high sensitivity, fast resolution, and simple equipment. Chemiluminescence analysis also has the function of "shrinking the chemiluminescence area", which can generate a chemiluminescence signal before the sample is completely diffused into the bulk solution, realizing fast online detection and recording of peaks, and there is no obvious bandwidth.

(5) Spectrophotometry
Spectrophotometry is a widely used analytical method. The light source emits white light, the spectrometer receives monochromatic light, and the monochromatic light passes through the colored solution. The change in the intensity of the emitted light is measured by the detector to obtain the antibiotic content.

Spectrophotometry is widely used to measure the efficacy of antibiotics because they have groups that absorb ultraviolet and visible light, or because they can react with reagents of specific colors. The Chinese Pharmacopoeia uses ultraviolet spectrophotometry to determine the solid dosage form of chloramphenicol.

2.2   Image Processing Technology

Ten years ago, when the digital technology environment was not very advanced, a lot of image creation work was done using optical analog equipment. Those involving images are stored in digital format, that is, digital image processing. Digital production technology was first introduced in the 1920s, but it was limited to the computer level. As we all know, the digital imaging process begins immediately after 30 years. The digital sign process can be divided into four levels: retrieving image information, resetting the image, testing and calculating the digital image, and post-processing events. The digital imaging process is characterized by rich information. Compared with traditional image processing systems, digital image processing has many advantages.

1) Editing: Without image storage, image duplication, image overlay or other conversion, the image quality is not changed. In other words, during the creation of a digital image, the original image cannot always be edited, which is not possible when creating an analog image.
2) High process accuracy: According to the logarithmic relationship between processor accuracy and instrument complexity in the analog image production process, the accuracy of digital image production can cover all applications, leading to major differences, engine modifications and upgrades.
3) A wide range of fields: The digital imaging process can track images that can only be seen under a microscope, and even large images, such as remote graphic satellite antennas. In addition, the digital image processor runs on a variety of sources, including visible and invisible images as well as spectral images under visible light (such as infrared images).
4) The highest resolution: image editing can be divided into three parts: image enhancement, image resolution and image correction. Generally speaking, traditional imaging processes can only perform simple functions. The digital imaging process can perform various complex mathematical or logical operations, which greatly increases the flexibility of image creation.

3   Design of Antibiotic Potency Measurement System Based on Image Processing

3.1   Basic Principles of the System

Automatic measurement and analysis of microbial areas based on image generation is based on digital computer image generation technology. A new type of automatic measurement system integrating image sensor, calculation, debugging and other technologies.

The starting point of the system is shown in Fig. 1. First, light transmission or transmission, and then light transmission or transmission can be converted into digital images using transducers and photoelectric data collectors, and then the computer creates a processing module for the process to capture the antimicrobial area The size and the effect of antibiotics can be calculated.

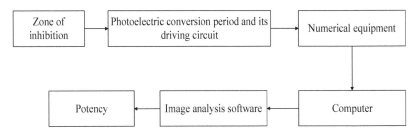

Fig. 1. Basic principles of the system

3.2   Data Acquisition Module

The task of the photography unit is to obtain images from the digital driver (scanner, digital camera, etc.). Based on standard digital media devices such as computers and scanners, that is, standard interfaces, intermediate programs are created to input, store, and distribute data collected from computers.

3.3   Image Processing

(1) Sobel edge detection

The edge of an image is an important part of image information, including internal information such as contour, direction, and sudden change. Edge detection is an important feature often used in image segmentation, sorting and pattern recognition. When using a camera to capture images, due to factors such as ambient light, thermal noise of imaging elements, and electromagnetic interference, the captured images contain a lot of high-frequency noise. Therefore, in order to improve the accuracy of classification or recognition, the original image must be preprocessed before classification or recognition. An efficient pre-processing edge detection algorithm can effectively remove noise and improve the accuracy and stability of post-processing.

(2) Gaussian smoothing filter

Gaussian smoothing, also known as Gaussian blur, is a very effective low-pass filter, which has been widely and effectively applied in the field of digital image processing. Gaussian smoothing filter has many excellent characteristics and details of rotational symmetry, uniqueness, spectral veneer, and smoothness. Calculated as follows:

$$g(x, y) = \frac{1}{2\pi\sigma^2} e^{\frac{-(x^2+y^2)}{2\sigma^2}} \tag{1}$$

Among them, σ is the standard deviation, x and y are Cartesian coordinates, and the size of the standard deviation determines the smoothness of the image processing result. Filter parameters are usually pre-calculated to improve calculation performance. For example, the commonly used 5 * 5 integer standards are:

$$\frac{1}{273} \times \begin{bmatrix} 1 & 4 & 7 & 4 & 1 \\ 4 & 16 & 26 & 16 & 4 \\ 7 & 26 & 41 & 26 & 7 \\ 4 & 16 & 26 & 16 & 4 \\ 1 & 4 & 7 & 4 & 1 \end{bmatrix} \tag{2}$$

(3) Dynamic reconstruction technology

Although FPGAs can be used for image processing to achieve a high degree of parallelism, changing system functions usually requires FPGAs to perform activation reset and configuration operations. In real-time applications, image editing will be suspended due to the above situation. The dynamic reconstruction technology provided by Zynq-7000 can solve the above problems. By using Zynq-7000's dynamic reconfiguration technology to separate the FPGA dynamic and static modules, the dynamic logic modules can be reconfigured in non-reset mode, resulting in changes in the behavior of the hardware configuration, which can be flexibly adjusted according to the situation. Perform different image editing functions by quickly switching between different image editing units and adjusting their sizes. It not only saves chip resources, but also improves response speed.

3.4 Data Processing Unit

This part mainly covers the steps of product production, data display and automatic improvement. Mainly include: 1) Adjust the margins of the antibacterial ellipse image. This method calculates the geometric size of the ellipse based on the image pattern, center position and average length of the ellipse. 2) Measure the error rate of the root scanner and the error correction rate of the scanner, use the standard to obtain the position of the antibacterial part, and automatically measure the performance after the position is corrected.

4 Image Processing Detection

Use software to implement Sobel edge detection and compare performance with FPGA solutions. The experimental results are shown in Table 1.

Table 1. Image processing test results

	FPGA solution	Software solution
Sobel processing time	2709287	70324580
Sobel processing start time	2698745	77279775
Actual frame rate	29.55	12.45
Actual throughput program duty cycle	26.34	11.23

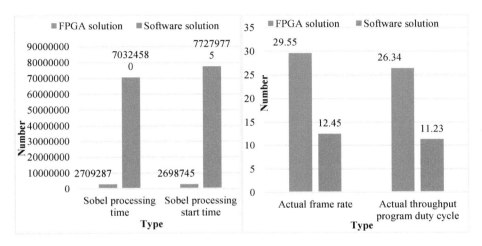

Fig. 2. Image processing test results

It can be seen from Fig. 2 that using FPGA for Sobel processing has obvious advantages. Both calculation and startup time are orders of magnitude shorter than software solutions. The frame rate and performance are more than twice that of software solutions, and the program life cycle is much longer than software solutions.

5   Conclusions

This article studies the antibiotic potency measurement method based on image processing. After analyzing the related theories, designs the antibiotic potency measurement system based on image processing, and tests the designed system. The detection result shows that the FPGA is used Sobel processing has obvious advantages, and its computing time and startup time are both an order of magnitude less than software solutions.

Acknowledgements. Supported by Jiangsu Provincial Research Foundation for Basic Research, China (Grant No. BK20201063); Supported by the Opening Project of Jiangsu Province Engineering Research Center of Agricultural Breeding Pollution Control and Resource (Grant No. 2021ABPCR006).

References

1. Lab, Asia, group. Automated system helps veterinary pharmaceutical company rapidly generate accurate data on antibiotic potency. Lab Asia **24**(2), 32–32 (2017)
2. Ma, A., Gn, A., Fl, B., et al.: Antimicrobial defined daily dose adjusted by weight: a proposal for antibiotic consumption measurement in children - ScienceDirect. Enfermedades Infecciosas Microbiol. Clín. **37**(5), 301–306 (2019)
3. Mydock-Mcgrane, L., Cusumano, Z., Han, Z., et al.: Anti-virulence C-mannoisdes as antibiotic-sparing, oral therapeutics for urinary tract infections. J. Med. Chem. **59**(20), 9390–9408 (2016)
4. Munz, A., Hoefert, S., et al.: Effect of bisphosphonates on macrophagic THP-1 cell survival in bisphosphonate-related osteonecrosis of the jaw (BRONJ). Oral Surg. Oral Med. Oral Pathol. Radiol. **121**(3), 222–232 (2016)
5. Sugo, K., Kawashima, R., Nakasu, M., et al.: Antibiotic elution profile and physical properties of a novel calcium phosphate cement material. J. Ceramic Soc. Jpn. **124**(9), 954–958 (2016)
6. Dronamraju, V., Moffat, G.T., Nakeshbandi, M.: Adherence to Procalcitonin measurement in inpatient care: a guide for antibiotic stewardship and high value care. J. Gener. Intern. Med. **35**(2), 609–610 (2020)
7. Verma, P., Samanta, S.K.: Comparative assessment of antibiotic potency loss with time and its impact on antibiotic resistance. Comp. Clin. Pathol. **25**(6), 1163–1169 (2016). https://doi.org/10.1007/s00580-016-2321-2
8. Patel, M.B., Garrad, E., Meisel, J.W., et al.: Synthetic ionophores as non-resistant antibiotic adjuvants. RSC Adv. **9**(4), 2217–2230 (2019)
9. Odusami, J.A., Ikhile, M.I., Fotsing, M., et al.: Towards eradicating antibiotic-resistant bacteria: synthesis and antibacterial activities of substituted N-(2-Nitrophenyl)pyrrolidine- and piperidine-2-carboxylic acids. J. Chin. Pharmac. Sci. **28**(10), 704–715 (2019)
10. He, Y., Nurul, S., Schmitt, H., et al.: Evaluation of attenuation of pharmaceuticals, toxic potency, and antibiotic resistance genes in constructed wetlands treating wastewater effluents. Sci. Total Environ. **631**(AUG.1), 1572–1581 (2018)
11. Saidykhan, L., Bakar, M., Rukayadi, Y., et al.: Development of nanoantibiotic delivery system using cockle shell-derived aragonite nanoparticles for treatment of osteomyelitis. Int. J. Nanomed. **11**(11), 661–673 (2016)
12. Schwalen, C., Hudson, G.A., Kille, B., et al.: Bioinformatic expansion and discovery of thiopeptide antibiotics. J. Am. Chem. Soc. **140**(30), 9494–9501 (2018)

Acquisition and Application of Internet Medical Big Data Based on Text Mining Technology

Teng Ma[✉], Yan Wang, Yan Li, and Hui Zheng

Zaozhuang Vocational College of Science and Technology,
Zaozhuang 277599, Shandong, China
mdmt1103@163.com

Abstract. With the development of computer and communication technology, the Internet has gradually penetrated into various fields of production and human life, and has become an important source of knowledge for mankind. People can receive information from the Internet to guide their work and life. This paper studies the acquisition and application of Internet medical big data based on text mining technology, and understands the relevant knowledge of Internet medical big data on the basis of literature data, and then designs an Internet medical big data analysis platform based on text mining technology. The applied data analysis model is tested, and the test results show that the TSDDN model is significantly better than all benchmark methods (including TSSDA) in F-measure and accuracy indicators.

Keywords: Text mining · Internet data · Medical big data · Data application

1 Inductions

With the development of computer technology, digital medicine is becoming more and more perfect, and medical data processing is increasing at an unprecedented rate [1, 2]. Biomedical scientific research has evolved into a typical high-data-intensity scientific research, and has produced a data explosion phenomenon called the "big data" era [3, 4]. In the era of big data information, data has become a new type of strategic resource and a key factor in technological innovation, and it is gradually changing the working methods of biomedicine research and development and the daily life and thinking of human beings [5, 6].

Regarding data mining research, with the rapid development of social Internet, sensor technology, mobile network technology, etc., some researchers believe that the total amount of data content of China's Internet legal person has reached the PB level, and it will become a key strategy for the development of Chinese companies and the country resource. At the same time, due to the active promotion of the Chinese government, big data service platforms are also constantly emerging. They will have the potential to capture the required resources from the "data sea" and discover the potential value behind big data, providing the cornerstone for technological progress [7]. However, China's current big data service market is still in its infancy, and the

V. Sugumaran et al. (Eds.): ICMMIA 2022, LNDECT 138, pp. 360–367, 2022.
https://doi.org/10.1007/978-3-031-05484-6_45

overall development has not yet been completed. In meeting the various requirements of big data mining users, individual businesses still have the problem of insufficient service capabilities. To this end, relying on the big data industry chain, make full use of the talent resources and technological resources of the upper, middle and rear companies in the industry chain to further improve the innovation ability and service level of the data mining business, and use the big data to analyze the industry alliance. The data mining services provided by all sectors of society play a great role in further enriching China's big data service market [8]. The business model used by the big data alliance directly affects the quality and level of enterprise data mining business. In meeting the characteristics of enterprise big data service needs, and creating a service process that helps to effectively integrate data resource mining business models, big data resources. Dataization is an urgent problem that needs to be solved urgently in the enterprise/personal data analysis service industry [9]. Some researchers have defined big data from three levels of reality and technology. The first level of "big data" refers to a large amount of data quantification and differentiation, in fact a large amount of data, technically a large amount of data storage; the second level of "big data" refers to big data technology. In practice, it refers to the analysis and use of large amounts of existing or newly acquired data. Technically speaking, it refers to cloud storage and cloud computing. "Data" refers to the big data mindset. In fact, it refers to concepts such as the research method of sampling the entire target and the research of relevance. Technically speaking, it refers to the use of batch data for analysis, processing and use. Big data technology used to assist decision-making or direct mechanical and semimechanical decision-making [10]. In summary, there are many research results on big data, but there are few researches on internet medical big data.

This paper studies the acquisition and application of Internet medical big data based on text mining technology, analyzes the characteristics and applications of Internet medical big data on the basis of literature, and then designs a medical big data analysis platform based on text mining technology. The designed platform is tested, and relevant conclusions can be drawn from the test results.

2 Research on Internet Medicine Big Data

2.1 The Characteristics of Internet Medical Big Data

Feature 1: Polymorphism refers to the doctor's description of the patient. It has the subjective characteristics of physician experience, and it is difficult to achieve standardization [11].

Medical big data includes all kinds of text information (patient complaints, medical consultation, medication, etc.), digital information (physiology, biochemical data, vital signs data, etc.), image information (including color Doppler ultrasound, magnetic resonance imaging, X-ray film wait). This is the most important and essential difference between large-scale medical data and data from other disciplines.

Feature 2: Confidentiality refers to the high degree of privacy of medical and health data. In the analysis of large-scale medical data, it is difficult to avoid the intervention of patients' personal data, and the leakage of personal information brings anxiety and harm to patients. Especially in the development of the Internet health system that combines Internet big data and mobile health monitoring, personal data leakage will cause more serious damage.

Feature 3: Uncertainty refers to the deviation and lack of explanation of the patient's condition in medical analysis [12].

There is no process of collecting and analyzing large amounts of medical data in the market. A large amount of data is taken from manual records, leading to heterogeneous big data in medical records. Many recorded data reflect uncertainties based on physician experience. This is especially evident in medical history records and drug dosages, which introduce uncertainty in large medical records.

Feature 4: Redundancy refers to the fact that medical data contains a large amount of irrelevant data information.

The electronic medical record contains many files with the same or similar information, such as disease diagnosis and treatment records, patient description information, and laboratory information unrelated to pathological characteristics. It also contains many duplicate, irrelevant or inconsistent medical records.

Feature 5: Efficiency means that the data is only useful for a period of time, and the time of onset is sudden.

The continuous development of medical technology has created immediate and effective diagnosis and treatment methods for intractable new diseases that are difficult to overcome. The timeliness of big data mining objects continues to increase, and the effectiveness of diseases is very important. For example, vaccinia treatment was initially incurable, but it can now be avoided with a vaccination.

Feature 6: Sequence refers to a certain time interval between the onset of the disease when the patient seeks medical attention.

Like ECG recording, standard ECG cannot detect signs of paroxysmal heart disease and should be based on long-term real-time monitoring of heart conditions. In other words, there is a time series and a time series period.

2.2   Application of Internet Medical Big Data

(1) Focus on the effective application of big data analysis in the field of public health: strong evidence-based medicine thinking and principle integration capabilities encourage the wide application of big data analysis methods in the field of public health data. Some R&D personnel use social multimedia data sharing technology to remotely monitor children with AIDS. The results are consistent with the recent CDC monitoring results. The estimated cost based on network data analysis is high, but it is up-to-date. Our extensive data creates opportunities for early public health warning. In the past decade, big data has also been successful in diagnosing heart disease and liver cancer classification.

(2) Improve the quality of diagnosis and treatment: There are many differences among individuals in terms of age, gender, body size, metabolic rate, and gene mutation types. Based on the above reasons, it can be proved that standard identification measures are not reasonable for everyone, and the effect of drugs depends on the population. This appeared in the late 1990s, but due to the lack of

comprehensive patient health statistics, doctors were unable to formulate prevention and medical plans for each patient's condition. However, in the era of big data, by using a collection of genomes, wireless sensors, digital imaging, and medical information technology, it is possible to collect and analyze each patient's health data to integrate and analyze the patient's inheritance, medical history, and genes. Doctors can measure the risk of patients, provide early diagnosis and prevention and medical procedures, and provide personalized guidance for patients.

(3) Promoting medical and pharmaceutical research: In the context of traditional clinical research, clinical research by clinical researchers is often separated. If clinical staff want to do research, they usually need to collect data from the beginning. Due to funding and time constraints, scientific research projects often cannot be carried out for a long time, especially for chronic diseases, which may take a long time. This makes it difficult for clinicians to plan and carry out long-term clinical research and hinders clinical research. The rise of large-scale health data provides an opportunity to improve this situation. Through the application of a large amount of data collected over a period of time in clinical research, the integration of clinical and scientific research is realized.

(4) Big data provides decision support. Create a population and patient database and use it to monitor the country's health economy, patient health, and measure the implementation of medical services. The data in this database is combined with high-quality, complete data from other studies, including basic patient demographics, outpatients, hospital expense lists, imaging and laboratory examinations, etc. These data can be used to analyze the entire or multiple the prevalence, overall prevalence and disease burden of chronic diseases in each cohort. A big data platform that fully integrates healthcare and clinical data is an effective tool for healthcare and decision-making systems.

3   Design of Internet Medical Big Data Platform Based on Text Mining Technology

3.1   Big Data Platform Design Analysis

Compared with traditional data warehousing and application methods, the medical data analysis process based on the big data platform adopts big data storage, processing, and analysis technologies to meet the characteristics and application requirements of current medical big data. The data analysis application technology based on the big data platform adopts service-oriented architecture design technology, and divides the entire data analysis application process into four parts: data collection, data integration, data analysis, and data application presentation. The box is shown in Fig. 1.

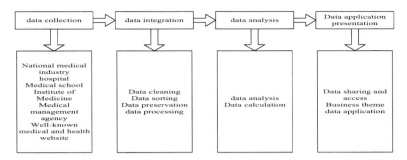

Fig. 1. Big data platform framework

3.2   Data Collection

Medical data retrieval, as the name suggests, is a way to capture specific data on the subject of "medicine". When designing a search, you must first ensure that the subject of the detected information is very relevant and accurate, and at the same time avoid traps in the search as much as possible, comply with the ethics of the search procedure, and minimize the impact. There are two main types of medical search strategies.

The first is a targeted crawling of pharmaceutical-related websites that are known and recognized in the industry. For example, in the national medical industry, hospitals, medical schools, medical institutes, medical management institutions, and well-known medical and health websites have been included in the scope of testing. The content of each page has analyzed the source code structure, extracted the required information and saved it in In the database.

The second way is to first define the subject terms, and then determine whether the received web content is related to the subject terms based on the subject terms. Subject terms can be contained in various places on the page, such as the title and text, and the URL is related to the subject.

3.3   Data Integration

The data needs to be cleaned and sorted before it can be saved and processed. Also known as the ETL process, it mainly performs functions such as extracting, converting, cleaning, and reading data. The big data medical platform designed in this paper uses TongETL software tools for data preprocessing. ETL is based on Java technology and standard database interfaces (JDBC, 0DBC, etc.), supports development on a variety of mainstream operating systems and home operating systems, can access a variety of mainstream databases, open source and national databases, supports various structures, and supports standard/non-standard write files in a structured format and interact with other application systems through multiple protocols. TongETL provides a large number of job data and transformation elements. These elements can be transferred to complete the integration requirements, without manual coding to programmatically control various complex data, and quickly create data integration applications.

3.4   Data Analysis

(1) Data analysis

There are many ways to analyze large medical data. It can be divided into two main types. The first is SQL queries for medical data sets, such as Hive analysis. The other is to perform classification extraction, text extraction and other extraction analysis on medical data sets.

(2) Data calculation

The method of calculating medical data is distributed computing. Due to the large amount of medical data, the use of distributed computers can save the total calculation time and greatly improve the calculation performance, and the distributed lapReduce calculation framework can be used, according to the design of the specific data analysis method of the platform, calculate the data set. The basic idea of this framework is "dismantle tasks and merge results". First, the Hap() function obtains the key-value/input pair (ey.Value), then divides the larger task into several smaller parts (S1ip), and obtains the intermediate result of the key-value pair (Key.Value) to generate a set, and finally, the Reduce() function executes all stages and merges the average netkey value.

3.5   Data Mining Algorithm Implementation

The purpose of this article is to classify disease-related text into appropriate topics. An effective idea is to introduce topic supervision into an unsupervised detailed pre-learning process to make the trained hidden SDA layer specific to the topic and improve the performance classification problem. The idea is similar to SDA for emotional supervision.

The process of the native SDA model adds a fully connected accounting regression layer, and uses the softmax function to predict the data subject of the output layer. Therefore, the output level of each DA is used not only to construct the reconstruction level, but also to predict the subject category of the input data. In this way, the TSDDN model can extract better features for the specific topic of the classification task. In the TSDDN model, W1 is the connection weight between the wear level and the output layer, and W2 is the connection weight between the output layer and the topic category of the input data. The probability that the data at output level 0 belongs to a specific topic t can be expressed as:

$$P(T = t \mid o) = softmax(OW_2 + c) \tag{1}$$

where c is the node deviation of the regression level.

For multi-class regression, the negative record probability is usually used as the loss function. Therefore, this section defines the subject's loss of supervision function as follows:

$$loss(T) = -\log P(T = t \mid 0) \tag{2}$$

4   Platform Test

In order to prove the effectiveness of the algorithm, this paper uses the data analysis model of this paper to conduct a 10-fold cross-validation experiment on the data set. At the same time, the t-test p value of the F value and the accuracy rate between DSSDA, LDADBN and each reference method are also calculated (because it is a complete measure of accuracy and recall, only the F value is considered here). Table 1 shows the performance comparison results of each method.

Table 1. Performance comparison results of each method

	Precision %	Recall%	F-measure%	Accuracy%
DSSDA	71.5	64.9	67.9	78.6
TSSDA	68.5	64.5	66.4	77.5
SDA	67.2	64.3	65.2	77.1
LDADBN	63.1	61.2	62.1	75.2
DB N	62.5	62.6	62.6	74.6
SVM	79.5	51.1	62.1	75.1
GNB	42.1	50.5	45.6	53.2
PE	64.3	59.1	61.3	73.5
DT	51.4	51.3	51.3	67.1

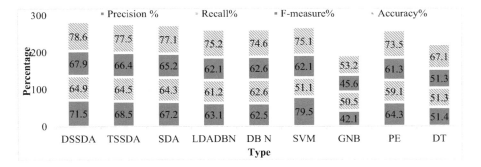

Fig. 2. Performance comparison results of each method

It can be seen from Fig. 2 that the performance of the TSDBN model proposed in this paper is better than the reference method on each scoring index. In particular, the TSDBN model is significantly better than all reference methods (including TSSDA) in F-measure and Accuracy indicators, and the LDADBN model is significantly better than all non-DBN models in F-measurement indicators.

5 Conclusions

This article studies the acquisition and application of Internet medical big data based on text mining technology. After analyzing the relevant theoretical knowledge, designs the Internet medical big data analysis platform based on text mining technology, and tests the TSDDN model applied in this article. The test results show that the performance of the TSDDN model proposed in this paper is better than the benchmark method in each evaluation index.

References

1. Leo, J., Leiva, V., Saulo, H., et al.: Birnbaum-Saunders frailty regression models: diagnostics and application to medical data. Biometr. J. Biometrische Zeitschrift **59**(2), 291–314 (2017)
2. Siuly, S., Zhang, Y.: Medical big data: neurological diseases diagnosis through medical data analysis. Data Sci. Eng. **1**(2), 54–64 (2016). https://doi.org/10.1007/s41019-016-0011-3
3. Gottlieb, L., Tobey, R., Cantor, J., et al.: Integrating social and medical data to improve population health: opportunities and barriers. Health Aff. **35**(11), 2116–2123 (2016)
4. Lomotey, R.K., Nilson, J.A., Mulder, K., et al.: Mobile medical data synchronization on cloud-powered middleware platform. IEEE Trans. Serv. Comput. **9**(5), 1 (2016)
5. Karimi, N., Samavi, S., et al.: Toward practical guideline for design of image compression algorithms for biomedical applications. Expert Syst. Appl. **56**(C), 360–367 (2016)
6. Kalantari, A., Kamsin, A., Shamshirband, S., et al.: Computational intelligence approaches for classification of medical data: state-of-the-art, future challenges and research directions. Neurocomputing **276**(FEB.7), 2–22 (2017)
7. Reichel, J.: Oversight of EU medical data transfers – an administrative law perspective on cross-border biomedical research administration. Heal. Technol. **7**(4), 389–400 (2017). https://doi.org/10.1007/s12553-017-0182-6
8. Pliss, I., Perova, I.: Diagnostic neuro-fuzzy system and its learning in medical data mining tasks in conditions of uncertainty about numbers of attributes and diagnoses. Autom. Control. Comput. Sci. **51**(6), 391–398 (2017). https://doi.org/10.3103/S0146411617060062
9. Robinson, K.L., Bryan, M.E., Atkinson, E.S., et al.: Neutering is associated with developing hemangiosarcoma in dogs in the veterinary medical database: an age and time-period matched case-control study (1964–2003). Can. Vet. J. Revue Veterinaire Canadienne **61**(5), 499–504 (2020)
10. Iida, S., Ishida, T., Horimoto, K., et al.: Medical database analysis of Japanese multiple myeloma patients with planned stem cell transplantation (MEDALIST) – a focus on healthcare resource utilization and cost. Int. J. Hematol. **113**(2), 271–278 (2021)
11. Lima, M., Lima, J., Barbosa, R.M.: Medical data set classification using a new feature selection algorithm combined with twin-bounded support vector machine. Med. Biol. Eng. Compu. **58**(3), 519–528 (2020). https://doi.org/10.1007/s11517-019-02100-z
12. Singh, A.K., Ghalib, M.R.: KM-LA: knowledge-based mining for linear analysis of inconsistent medical data for healthcare applications. Pers. Ubiquit. Comput. (2), 1–15 (2021)

Underwater Image Processing and Target Detection Algorithm Based on Deep Learning

Yangning Zheng[1](✉) and Yang Yang[2]

[1] School of Medicine, Nanchang Institute of Technology,
Nanchang, Jiangxi, China
zyn@nut.edu.cn
[2] School of Electronics and Information, Nanchang Institute of Technology,
Nanchang, Jiangxi, China

Abstract. With the growth of the earth's population, people's demand for space and resources continues to grow, making marine environmental research and exploration a new field that human development needs. The composition of the marine environment is unique to humans, and the application of intelligent underwater vehicle equipment with the ability to autonomously detect and recognize the environment can expand the scope of human research in the ocean. This article focuses on the research of underwater image processing and target detection algorithms based on DL, and understands the relevant theories of underwater image processing and target detection based on the literature, and then the underwater image processing and target detection algorithms based on DL In the design, in order to improve the accuracy of target detection, the DL model is optimized, and the optimized model is tested. The test result shows that if the Atrous convolution operation is added, the calculation amount of the network model will increase, resulting in detection.The speed drops to 50FPS. But after adding Atrous processing, the detection accuracy will be improved accordingly.

Keywords: Deep learning · Underwater images · Image processing · Target detection

1 Inductions

In recent years, with the dynamic development of ocean information processing technology, research and application of underwater mechanical vision is rapidly progressing, and the application of underwater vision imaging technology in ocean exploration and development is also deepening [1, 2]. Compared to underwater sonar technology, underwater optical imaging technology has the advantages of wide field of view, high resolution, fast information retrieval speed, and low interference. Suitable for detecting targets at close range and accurate positioning. Therefore, it occupies a very important position in the underwater application of mechanical optical imaging technology, scientific development, and even the military field [3, 4]. However, due to the extreme complexity of the marine environment and the inherent characteristics of the underwater visual environment (mainly reflected in the absorption and scattering of light by the body of water and the non-linear interference in underwater imaging), underwater imaging techniques are ideal. The useful information contained is also relatively rare [5, 6]. Videos and images shot underwater have many problems, including low contrast, blurred textures, severe color distortion, and noise interference. How to achieve effective and consistent target detection in such complex scenes is an open issue so far [7, 8].

According to the research on image processing, some researchers have adopted multi-scale pyramid as the main tool for image classification. The multi-scale pyramid is a hierarchical data structure, which is composed of a variety of images with different resolutions. Multiple images with different resolutions can be formed through mathematical algorithms of different dimensions. The base of the image pyramid is the original image, and the definition of each pixel layer is gradually reduced from low to high. These structures make full use of global and local signals, spatial and gray-scale image signals, and they are all hierarchical [9]. Some researchers must determine the camera parameters underwater to accurately identify the target. They can also construct a mapping matrix through the parameters to more accurately convert the two-D coordinates of the plane into three-D. Related camera calibration methods are divided into traditional calibration methods, self-calibration-based methods, and active vision system-based methods. At the calculation level, there are mainly linear calibration methods, nonlinear calibration methods and their combination [10]. In summary, there are more researches on traditional image processing methods, but less research on underwater image processing methods.

This paper studies the underwater image processing and target detection algorithms based on DL. It analyzes the characteristics of underwater images and the application of DL based on the literature, and then conducts the underwater image processing and target detection algorithms based on DL. The designed algorithm is design and tested, and verifies the effectiveness of the algorithm through the test results.

2 Research on Underwater Image Processing and Target Detection

2.1 Analysis of Characteristics of Underwater Images

Before talking about underwater image preprocessing technology, this article first analyzes the physical characteristics of light when it travels in water. These physical properties are different from those when light propagates in a normal environment (such as air) [11]. Underwater images usually give a blurry, low-contrast impression because the light continues to dim as it passes through the water body. The attenuation of light limits the visible distance of the underwater environment. In clear water, the visible distance can reach about 20 m. In deep water, only a range of less than 5 m can be seen [12]. The process of underwater light attenuation is caused by two main factors. Absorbing and scattering light will reduce the total energy of underwater light, while light scattering will change the path and direction of light propagation. The scattering effect can be further subdivided into forward scattering and back scattering. When the light reflected from the target and returned is scattered at a small angle on the way to the camera, forward scattering occurs, which blurs the characteristic target in the image.

Backscattering is caused by some light reflected by the water body to the camera before reaching the target of interest, which reduces the contrast of the entire image, like a layer of "fog" covering the image. Adjusting the artificial light source in the water is an effective way to improve visibility, but it can also cause uneven lighting conditions, bright spots in the center of the image, and dark corners around the image. As the depth increases, the diversity of colors in the underwater environment gradually decreases, because the degree of light attenuation depends on the color (wavelength). The shorter the wavelength of light (green light, blue light, etc.), the deeper the water body can penetrate. Therefore, in the deep sea environment, green and blue are the main bright colors. In summary, images and videos taken in underwater environments generally have the following problems: low visibility, low contrast, uneven lighting conditions, severe blurring, color distortion, and large noise interference.

Underwater image processing algorithms can generally be divided into two categories: image restoration technology and image enhancement technology. Image restoration refers to the use of prior knowledge of the degradation process to estimate the underwater image degradation model and restore the degraded image to its original appearance, which is essentially a reverse problem. This kind of method is more rigorous in theory, but many unknown parameters in the model, such as damping coefficient, diffusion coefficient, etc., need to be defined to explain the nature of the water system. Image enhancement is a specific qualitative way of image quality enhancement that adapts to the human visual experience. Such methods are generally simpler, easier to understand, more convenient, and faster than image restoration techniques because they are not based on physical models of underwater image imaging.

2.2 Application of DL

Each commonly used target recognition method has its limitations. DL is based on a powerful learning ability, independently extracting suitable and efficient functions from a large amount of changing physical data, without monitoring, reducing the burden of people in target recognition, and using these features to solve practical problems. In some complex situations, it will get very good results as well as good generalization ability and robustness. At the same time, with the continuous development of computer technology that provides effective computing power and the continuous development of Internet technology that provides a large amount of learning data, DL has grown exponentially. However, image target recognition based on DL still faces many challenges.

(1) Small sample problem. DL applications are usually based on large amounts of data. However, in practical applications, it is difficult to collect images, and it is difficult to meet the large number of samples required for learning. Therefore, how to create a DL model suitable for small problem research samples is a major challenge.

(2) Network structure and parameter design issues. The DL model has many parameters. How to obtain the parameter values with good performance is a time-consuming and laborious process. The difference in network structure will also affect the training results. Therefore, how to solve the problem intelligently is the basic parameter of network structure optimization is also the challenge of DL for target recognition.

(3) Theoretical issues of DL. The theoretical basis of DL does not have the complete theoretical basis of support vector machines. It is often unclear which features are being learned and why they are effective.

3 Underwater Image Processing and Target Detection Algorithms Based on DL

3.1 Image Target Recognition and Classification Process

The key to image object recognition and classification technology is to allow the computer to recognize the type and location of objects in the scene. The accuracy of target recognition and classification is closely related to the way the target is described. The results of image segmentation and target boundary extraction provide material for target recognition and classification. Figure 1 shows the basic algorithm flow of a typical image target recognition and classification system.

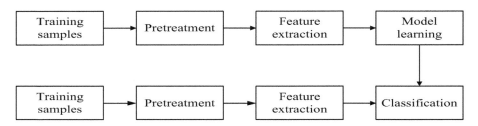

Fig. 1. Image target recognition and classification process

3.2 Feature Extraction

(1) Perform image preprocessing, image denoising, contrast enhancement and other processing.
(2) Segment the threshold area in the image, use the peripheral connection to extract the target area, or use the edge detection algorithm to get the contour of the object and fill it to get the target area.

This method is simple, but not a panacea. Underwater images are severely disturbed by noise, which will cause large errors when segmenting regions. This error affects the accuracy of the output. In view of the existence of some typical geometric objects in the experimental images, this study uses spatial geometry methods to extract objects.

In this paper, a centrifugal target extraction geometric algorithm based on Hough transform is used. The number of endpoints detected by the Hough transform is used to classify the target, determine whether it is a linear target or an arc-shaped target, and use various algorithms to center the number according to the detected straight line to obtain the detected target.

3.3   Learning Model

Since the RPN network is used in the original converged SSD network, the pixels in the center of each window can create different anchor frames with different proportions. However, in a multi-scale image, the target scale varies greatly, so if a single RPN is used to obtain the anchor frame, small targets in the image may be ignored, target detection may be lost, and the required failure check rate is low. In addition, this article needs to identify multiple types of targets to expand the self-generated data set, while also taking into account the differentiation sensitivity of the network. Therefore, this paper adopts multi-scale feature maps and difficult sample mining strategies, as well as gradual training to improve the performance of the algorithm model.

(1) Multi-scale feature map
    After the feature is derived and retrieved, the original feature map is subjected to two convergence changes of different scales, and a feature map convergence group is added to the sum of the original feature map size and the last three different size maps, and then the RPN network is used to edit these feature maps, to receive multiple scales of different anchor point contours at different scales. Experiments show that multi-scale feature maps can be more sensitive to multi-scale targets.

(2) Difficult sample mining strategy
    Since this article aims to detect multiple targets, the self-generated data set defines 1000 images for each target category, for a total of 4000 positive sample images. In addition, difficult sampling strategies can improve the accuracy of the network, adding 1200 negatively sampled images. Experiments show that a reasonable increase in the number of negative samples can improve the accuracy of detection.

(3) Step by step
    Experiments show that the distributed batch training network has better classification accuracy than the network that puts all the training data in the training process of the aggregated network at the same time, but the network training time is longer. This is because as each batch of data enters the network, the entire network becomes more sensitive, avoiding over-adaptation. Experiments show that stage training can improve the robustness of the network.

3.4   SSD Optimization

Generally speaking, the higher the resolution of the image, the clearer the characteristic information contained in the image. Object detection makes it easy to detect these features. There are two commonly used methods to increase the resolution of an image. One is to use scrolling to create top-down feature map information to improve resolution, and the other is to use Atrous filter to merge, use to enlarge image size, and use decoupling method in the context of SSD algorithm. This article attempts to improve the detection of small objects by using the Atrous filter method to increase the resolution of the image and highlight the small objects in the image.

Atrous mode is divided into one-dimensional signal mode and two-dimensional image mode. If the input is a one-dimensional signal, that is:

$$y[i] = \sum_{k=1}^{k} x[i + r \cdot k]w[k] \tag{1}$$

Among them, w[k] is the Atrous filter, K is the length of the Atrous filter, x[i] is the one-dimensional input signal to be processed, and y[i] is the output signal after the Atrous convolution mode, which represents the r of the input signal rate or sample size.

Therefore, assuming that the filter size is kxk and the rate is r, after entering a zero value r − 1 between the filters, the length of the filter will be expanded as follows:

$$k_e = k + (k - 1)(r - 1) \tag{2}$$

4   Target Detection Effect Detection

In order to verify the effectiveness of the Atrous filter and test the improved detection effect of SSD on small targets, this article will design an Atrous experiment. Atrous experiments have confirmed that the functions of the previous layer of the SSD network structure can be effective. Table 1 shows its experimental data.

Table 1.   The detection result of the target

Feature extraction	FPS	mAP (%)
conv2_ 2	60	70.1
conv3_ 3	60	72.2
conv4_ 3	60	74.3
conv5_ 3	60	68.1
conv3_ 3, conv4_ 3	60	74.6
conv3_ 3, conv4_ 3, conv5_ 3	60	73.4
Conv3_ 3, (conv4_ 3, Atrous)	51	75.4

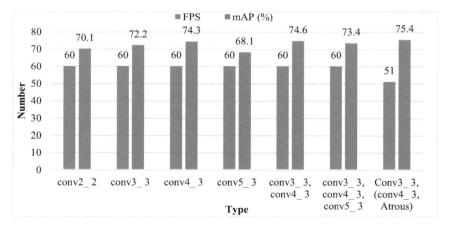

Fig. 2.   The detection result of the target

It can be seen from Fig. 2 that there is no change in the detection speed, whether it is a single layer or multiple layers to provide feature information for small targets, unless each layer adds Atrous spool features. Adding the Atrous convolution function increases the complexity of the network model and reduces the detection speed to 50 FPS. However, adding Atrous processing will improve the detection accuracy accordingly.

5   Conclusions

This paper studies the underwater image and target detection algorithms of DL. After analyzing the phase theory, the underwater image processing and target detection algorithms based on DL are designed. In order to improve the detection accuracy, the learning model is optimized, and use experiments to verify the effectiveness of the optimization. The experimental results show that after adding Atrous processing, the detection accuracy will be improved accordingly.

Acknowledgment. 2021 Science and Technology Research Project of Jiangxi Provincial Department of Education-Research on Real-time Recognition of Sea Cucumber Based on Underwater Machine Vision.

References

1. Kamel, I.R., Georgiades, C., Fishman, E.K.: Incremental value of advanced image processing of multislice computed tomography data in the evaluation of hypervascular liver lesions. J. Comput. Assist. Tomogr. **27**(4), 652–656 (2017)
2. Membarth, R., Reiche, O., Hannig, F., et al.: HIPAcc: a domain-specific language and compiler for image processing. IEEE Trans. Parallel Distrib. Syst. **27**(1), 210–224 (2016)
3. Vijayan, S., Geethalakshmi, S.N.: A survey on crack detection using image processing techniques and deep learning algorithms. Int. J. Pure Appl. Math. **118**(8), 215–219 (2018)
4. Sowmiya, E.: A noval method for detecting plant leaf disease using image processing and deep learning. Turk. J. Comput. Math. Educ. (TURCOMAT) **12**(2), 337–340 (2021)
5. Aoki, K.: Server for implementing image processing functions requested by a printing device. Environ. Pollut. **152**(3), 543–552 (2018)
6. Vieira, A., Ribeiro, B.: Introduction to deep learning business applications for developers. Image Process. (Chapter 4), 77–109 (2018). https://doi.org/10.1007/978-1-4842-3453-2
7. Kim, J., Kwon, O.S., Hai, N.L.D., et al.: Study on the design of an underwater chain trencher via a genetic algorithm. J. Marine Sci. Eng. **7**(12), 429 (2019)
8. Matsuyama, T.: Image processing device, image processing method and computer readable storage medium. Biophys. J . **89**(4), 2443–2457 (2016)
9. Erden, F., Velipasalar, S., Alkar, A.Z., et al.: Sensors in assisted living: a survey of signal and image processing methods. IEEE Signal Process. Mag. **33**(2), 36–44 (2016)
10. Kervrann, C., Sorzano, C., Acton, S.T., et al.: A guided tour of selected image processing and analysis methods for fluorescence and electron microscopy. IEEE J. Sel. Top. Signal Process. **10**(1), 6–30 (2016)
11. Geetharamani, R., Balasubramanian, L.: Retinal blood vessel segmentation employing image processing and data mining techniques for computerized retinal image analysis - ScienceDirect. Biocybern. Biomed. Eng. **36**(1), 102–118 (2016)
12. Yamane, T., Chun, P.J.: Crack detection from a concrete surface image based on semantic segmentation using deep learning. J. Adv. Concr. Technol. **18**(9), 493–504 (2020)

Application of Convolutional Neural Network in Image Processing

Yan Sun[✉], Wenxi Zheng, and Zhenyun Ren

College of Information and Communication Engineering University,
Harbin 150001, Heilongjiang, China
aidenby@163.com

Abstract. With the continuous development of computer technology, the wave of machine learning has become popular in various fields. Convolutional neural networks are based on the development of artificial neural networks and are widely used. How to optimize the algorithms of convolutional neural networks in image processing has become a research hotspot. There are many optimization algorithms for convolutional neural networks, but the results have not been ideal. In order to overcome the over-fitting phenomenon of convolutional neural network in image processing, and improve the accuracy of image classification and description conformity in complex scenes, this paper makes improvements based on the VGG model.First of all, this study chooses the VGG16 convolutional neural network model as the basic model, and introduces the Batch Normalization method to standardize the data. Then the Relu activation function with faster convergence speed is used to activate the neurons. Then use the improved dropout regularization method to optimize the model parameters. Finally, the improved model is tested by simulation experiments.The experimental results show that the error rate of the algorithm is as low as about 7% when iterates 300 times.Keywords: Convolutional Neural Network, Image Processing, Deep Learning, Image Classification, Dropout Regularization Method.

Keywords: Convolutional neural network · Image processing · Deep learning · Image classification · Dropout regularization method

1 Introduction

1.1 Background Significance

Convolution neural network model algorithm in deep learning has a high recognition rate, which is a research hotspot in various fields, especially in digital image recognition, processing and target tracking. However, the traditional convolution neural network often has the problem of over fitting, and the accuracy of image classification in complex scenes has not reached the ideal state. Therefore, it is necessary to further improve the weight parameter optimization algorithm of convolutional neural network model to speed up the convergence speed so as to improve the optimization ability, which can improve the image recognition and classification ability of the model, and provide a more scientific and effective method for the field of image processing.

1.2   Related Work

There are many research results on convolutional neural network algorithms. Mishkin D studied the impact of convolutional neural network structure and learning methods on object classification problems, and evaluated and tested factors such as nonlinearity, pool variables, network width, classifier design, image preprocessing, and learning parameters [1]. Min C studied non-invasive load monitoring based on power measurement, constructed a one-dimensional convolutional neural network based on GoogLeNet structure with two-dimensional convolution, and proposed linear programming boosting (ALPBoost) based on adaptive weights and thresholds [2]. Although the algorithm they proposed has high accuracy, it has a large amount of calculation and a very complicated process. At the same time, there are many research results on image processing. Osher S provides an algorithm that can automatically generate a high-performance plan for the Halide program, and only needs to provide an advanced strategy for mapping the image processing pipeline to a parallel computer to generate high-performance image processing code [3]. Bergounioux processed a second-order image decomposition model for denoising and texture extraction. He regarded the decomposition of the image as the sum of three different order terms, and used the dual problem and the inf convolution formula to give the qualitative properties of understanding [4]. Although their image processing algorithms can eventually produce clearer and more accurate images, once the scene is complicated, the results of image processing will have larger errors.

1.3   Innovative Points in This Paper

In order to overcome the over fitting phenomenon of convolutional neural network model algorithm in image processing as far as possible, improve the accuracy of image classification in complex scenes and the consistency of image description, so as to improve the ability of convolutional neural network model algorithm in image processing. The innovation points are as follows: (1) based on VGG16 convolutional neural network model, batch normalization method is used to process data, Relu activation function is used to activate neurons. eed after 300 iterations, and overcome the over fitting phenomenon to a certain extent.

2   Convolution Neural Network and Image Processing Method

2.1   Deep Learning Theory

Deep learning does not require human participation in the process of image extraction. Unsupervised learning algorithm is used to automatically learn the low-level features of the image from the original input image, and then these features are abstracted, mapped and combined layer by layer to obtain high-level semantic features [5].

Automatic encoder is one of the deep learning models of unsupervised learning, which uses unlabeled data sets for learning. The automatic encoder establishes an approximate equation which represents the same setting of output and input. When the image passes through the automatic encoder, the sparse feature vector is obtained to reconstruct the image. Moreover, comparing the difference between the target output and the actual output, the network parameters can be adjusted to better reconstruct the image. The automatic encoder consists of an encoder and a decoder [6]. After the data is input, it enters the encoder to re encode. The input decoder of the new code is reconstructed to judge whether the new code is consistent with the original input data. If it is consistent, the new code is output. If not, the encoder parameters are adjusted again, and the encoding and decoding operation is carried out again until the error is reduced to the minimum.

2.2 Basics of Convolutional Neural Networks

(1) Convolutional neural network structure

The structure of convolutional neural network generally includes convolutional layer, pooling layer and fully connected layer [7]. The output process of the convolutional layer is to convolve the input feature map of the previous layer through the filter, and apply the bias term to the nonlinear activation function, and its output value is the feature map of this layer. Each filter produces an output feature map, which is expressed in Formula 1:

$$T_i^k = n(M_i^k \otimes T_{i-1}^k + c_i^k) \tag{1}$$

Among them, T_i^k represents the k-th feature map of the i-th layer of the network; M_i^k, c_i^k respectively represents the weight matrix and bias term of the k-th filter of the i-th layer; n is the number of filters.

In order to further reduce the number of parameters and reduce the complexity and overfitting of the model, a pooling layer will be set after the convolutional layer. The pooling layer can integrate the output of adjacent neurons in the same feature map, which is a down-sampling process. The pooling layer can retain the effective information of the image on the basis of effectively reducing the feature dimension, eliminate redundant data, and accelerate the training speed of the network. The process of the pooling layer is shown in Formula 2:

$$T_i^k = n(down(T_{i-1}^k) + c_i^k) \tag{2}$$

Among them, $down()$ is the downsampling function, T_i^k is the k-th feature map of the i-th layer of the network; c_i^k is the bias term; n is the number of filters.

(2) Parameter optimization of convolutional neural network

The optimization training methods of convolutional neural networks are generally divided into three categories: using Bayesian optimization to find the best parameters, regularization calculations and seeking a better convolutional neural network model structure [8]. This article mainly explains the regularization calculation method.

Dropout regularization method suppresses a part of the activated units during training, and the same data is trained by different sub-models in each iteration. In the test, the method of predictive averaging is used to convert different sub-models into a predictive value integrated by all models. Although some neurons will be suppressed during each iteration of training, they will still be retained because they may be activated in the next iteration of training. Dropout regularization method is used in supervised learning. Each half of the network can obtain the recognition result through learning, but the accuracy of the result cannot be guaranteed. Generally, the more iterations, the higher the accuracy. The advantage of this method is that the update of the weight does not depend on the nodes with fixed affiliation, and it can learn more robust features.

2.3   Image Processing Algorithm

(1) Super resolution reconstruction algorithm

Image super-resolution reconstruction can obtain high-resolution images from one or more low-resolution images with poor imaging quality. The ideal high-resolution image is blurred by motion blur, optical blur and sensor after rotation and translation deformation. After sampling and adding noise, it can be transformed into the image of real scene, which is also the working principle of image imaging equipment.

Super resolution algorithms based on reconstruction include iterative back projection method and maximum a posteriori estimation method. Iterative back projection algorithm takes the low resolution image after bicubic interpolation as the initial high-resolution solution a, according to the degradation model H, the low-resolution image K predicted by the algorithm is obtained. The process is shown in Formula 3:

$$K = Ha + n \tag{3}$$

Then compare the calculated low-resolution image K with the actual low-resolution image Q. if the difference is too large, the error between the two is projected back to a to correct the high-resolution solution. The process is shown in formula 4:

$$a = a + H_{BP}(K - Q) \tag{4}$$

The two processes are repeated until the error between them meets the requirements. Iterative back projection algorithm is simple, but there is no unique solution, so it is difficult to select the degradation model.

In the maximum a posteriori probability estimation method, the high and low resolution images are regarded as different random processes. On the basis of statistics, the maximum a posteriori probability estimation method is used to calculate. The priori knowledge is fully utilized. The unique solution of the problem can be obtained. The quality of the reconstructed image is also very good, but the amount of calculation is large and the operation is complicated [9].

(2) Skew correction algorithm for text image

Hough transform algorithm considers that the curve can be transformed into a corresponding point in the parameter space from the original image space after passing the appropriate expression operation [10]. Therefore, the problem of curve detection in the original image space can be transformed into finding the peak value in the parameter space. The algorithm only needs to find the points with the highest cumulative peak value in the parameter space to determine the approximate lines that may exist in the original image. However, this algorithm is only suitable for processing pure text images. If the background with the same color as the text appears in the image, it will get wrong results, and the calculation of this algorithm is very large.

3   Experiments on Intelligent Algorithm Model of Ecological Water Environment Management

3.1   Image Processing Model Based on Convolutional Neural Network

(1) Model selection

The VGG convolutional neural network has clearer post-holiday and good migration application capabilities, so this research chooses the VGG16 convolutional neural network model as the basic model. The model contains 16 layers of convolutional layers and fully connected layers, and only accepts 224×224 image input, 3×3 convolution filters and 2×2 pooling areas.

(2) Data standardization

Data standardization is the first step of data preprocessing, which can reduce the difficulty of training. This study introduces the Batch Normalization method to standardize the data.

First, insert the Batch Norm layer in the convolutional layer or the fully connected layer, and then calculate the mean and variance of the Batch Norm layer input when each batch of data is trained, and finally standardize and de-normalize the input and output respectively. Assuming that there is an n-dimensional batch input data $f_n = (f_1, f_2, \ldots, f_n)$, the n-dimensional input is standardized according to Formula 5:

$$\hat{f}_n = (f_n - \phi)/\sqrt{\alpha^2 + \varpi} \tag{5}$$

Among them, ϕ, α^2, is the mean and variance of each batch of training data, ϖ is a constant, and then output according to Formula 6:

$$g_n = BN_{\eta,\chi}(f_n) = \eta \hat{f}_n + \delta \tag{6}$$

Among them, η, δ are all trainable parameters.

3.2   Improved Dropout Regularization Method

The optimization of the convolutional neural network parameters in this study adopts an improved dropout regularization method. In the training phase, the maximum pooling dropout can suppress some activation values in the modified input feature map, and the pooling starts to be random.

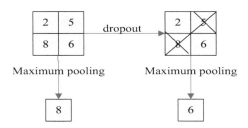

Fig. 1. Schematic diagram of maximum pooling dropout

As shown in Fig. 1, if the activation value of a pooled area is 2, 5, 6, 8 respectively. When dropout is not used, the largest 8 is output. After dropout is used, if 5 and 8 are suppressed, the result of pooling is to select the maximum 6 from the remaining 2 and 6.

4   Discussion on Image Processing Effect Based on Convolution Neural Network

4.1   Algorithm Performance Test

In order to verify the performance of the algorithm, the error rate of the fully connected layer using dropout is benchmarked, and the algorithm is calculated for 300 iterations. The error rate curve is as follows:

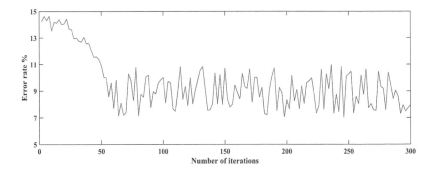

Fig. 2. Error rate curve for 300 iterations

As shown in Fig. 2, although the number of iterations continues to increase, the error rate after convergence begins to fluctuate around 9%, without further decline. However, the convergence speed of the fully connected layer using the dropout method is very fast, and the error rate of about 9% is actually relatively low, and the error rate can be as low as about 7%. This shows that the algorithm has good performance and can overcome the occurrence of over-fitting to a certain extent.

4.2   Accuracy of Image Classification of the Model

The VGG model optimized by the improved dropout regularization method is compared with the original model without improvement. The results are as follows:

Table 1. Comparison of training error rate

Type	Number of sample sets	Training error rate
Improved model	2776	8.11
Original model	2776	14.37

As shown in Table 1, the sample size of both models is 2776. When training, the error rate of the improved model is only 8.11%, while the error rate of the original model is 14.37%, which is 6.26% higher than the original model.

In order to further evaluate the image classification ability of the improved VGG model in complex scenarios. Record the numerical changes of the loss function of the training set and the validation set of the model, and calculate the accuracy of the model, if it is as follows:

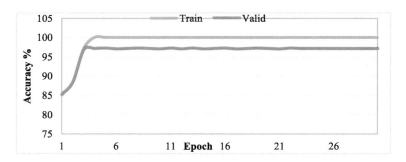

Fig. 3. Accuracy of the model

As shown in Fig. 3, the accuracy of the improved VGG model on the training set is higher than that of the validation set, that is, the convergence value on the training set is higher than the validation set. This shows that the model tends to be stable, so training time can be effectively shortened, performance fluctuations are also reduced, and the convergence speed is faster.

4.3   Manual Evaluation Test Set Image Description Results

The 398 test set images were scored according to the rules of manual evaluation of the description results automatically generated by the model. There are four types: the description is perfect, there are some errors in the description, the description is partially related to the image, and the description has nothing to do with the image. The results are as follows:

As shown in Fig. 4, all four cases occur, with different proportions. Among them, the best description is about 48%, accounting for almost half. Secondly, 26% of the descriptions were partially wrong, and 22% of the descriptions were partially related to the images. Description has nothing to do with image only 4%. This proves that the model has a good ability in image description.

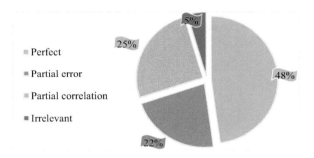

Fig. 4. Image result of manual evaluation test set

5   Conclusions

This research is based on the improved convolutional neural network model based on the VGG16 model, which has good performance in image processing. It can achieve a faster convergence speed, overcome overfitting, and better classify and describe images in images with complex scenes. It can be said that it has strong image processing capabilities.

Acknowledgements. This paper is funded by National Natural Science Foundation of China under Grant No. 51679057.

References

1. Apraites, L., Atlas, A., Karami, F., et al.: Some class of parabolic systems applied to image processing. Discrete Continuous Dyn. Syst. – Ser. B (DCDS-B) **21**(6), 1671–1687 (2017)
2. Mishkin, D., Sergievskiy, N., Matas, J.: Systematic evaluation of convolution neural network advances on the Imagenet. Comput. Vis. Image Underst. **161**(aug.), 11–19 (2017)
3. Min, C., Wen, G., Yang, Z., et al.: Non-intrusive load monitoring system based on convolution neural network and adaptive linear programming boosting. Energies **12**(15), 2882 (2019)

4. Osher, S., Shi, Z., Zhu, W.: Low dimensional manifold model for image processing. SIAM J. Imaging Sci. **10**(4), 1669–1690 (2017)
5. Bergounioux, M.: Mathematical analysis of a inf-convolution model for image processing. J. Optim. Theory Appl. **168**(1), 1–21 (2015). https://doi.org/10.1007/s10957-015-0734-8
6. Oshea, T., Hoydis, J.: An introduction to deep learning for the physical layer. IEEE Trans. Cogn. Commun. Netw. **3**(4), 563–575 (2017)
7. Sirinukunwattana, K., Raza, S.E.A., Tsang, Y.W., et al.: Locality sensitive deep learning for detection and classification of nuclei in routine colon cancer histology images. IEEE Trans. Med. Imaging **35**(5), 1196–1206 (2016)
8. Feng, J., Cai, S., Ma, X.: Enhanced sentiment labeling and implicit aspect identification by integration of deep convolution neural network and sequential algorithm. Clust. Comput. **22**(6), 1–19 (2018)
9. Aoki, K.: Server for implementing image processing functions requested by a printing device. Environ. Pollut. **152**(3), 543–552 (2018)
10. Lim, S.C., Kim, S.H., Kim, Y.H., et al.: Training network design based on convolution neural network for object classification in few class problem. J. Korea Inst. Inf. Commun. Eng. **21**(1), 144–150 (2017)

The Shape Model of Ceramic Products Based on 3D Image Reproduction Technology

Wenyu Zhang[✉]

Dalian University of the Arts, Dalian 116600, Liaoning, China
zl3694461333@163.com

Abstract. With the development of computer technology, three-dimensional spatial data has become a mainstream trend. In industrial design, the use of two-dimensional or four-dimensional projection to simulate and reconstruct the surface of ceramic products can reduce a lot of unnecessary troubles. At the same time, the ceramic industry is China's pillar industry. With economic globalization and the acceleration of the international trade process, ceramic products play an increasingly important role in the international market. Due to factors such as technical level, production scale, and product quality, most domestic enterprises are still stuck in a simple and extensive business model. In order to increase market share and enhance competitiveness and reduce costs, so that the company can obtain better benefits, it is necessary to continuously innovate and optimize product design and manufacturing processes to meet customer needs. This article uses experimental analysis and data analysis to better understand the performance of the model and the degree of deviation of the angle of the ceramic thin-walled parts through experiments. According to the experimental results, as the diameter of the extrusion head and the height of the extrusion layer change, the angle of the ceramic thin-walled part will also have a certain deviation. When the extruder diameter is 0.6 mm, the formability of the layer height under 0.4 mm is the best, the deviation value is the smallest, and the ceramic product has the highest precision.

Keywords: 3D image reproduction · Ceramic products · Feature extraction · 3D modeling

1 Introduction

The ceramic industry is an important industrial pillar for the development of China's national economy, and it also occupies a pivotal position in the national economic strategy. Therefore, it is necessary to improve the comprehensive strength of the ceramic industry to adapt to the new changes brought about by the international competitive environment. At the same time, the 3D image reproduction technology has the characteristics of visualization and high precision, and has become one of the indispensable key technologies in the modern industrial production process. It can digitally convert the two-dimensional space, establish graphical models, and construct continuous surfaces, etc. operate. Ceramic product modeling design is a very intuitive form of visual expression. In computer digital image processing technology, three-dimensional space analysis methods are used to model objects.

At present, many scholars have conducted research on ceramic production, molding and materials, and have obtained quite a wealth of research results. For example, Yu

V. Sugumaran et al. (Eds.): ICMMIA 2022, LNDECT 138, pp. 384–390, 2022.
https://doi.org/10.1007/978-3-031-05484-6_48

Fei pointed out that 3D ceramic printing has application characteristics such as flexibility, efficiency and precision in the design and manufacture of customized ceramic tiles, and has important application value to the development mode and development direction of ceramic tile design and production [1]. Liu Chuan believes that extrusion-freezing 3D printing is a suitable molding process for ceramic parts. However, the space structure of thin-walled ceramic parts is weakly supported and is easily deformed during the printing and molding process [2]. Chen Limin proposed that ZrB2-based ultra-high temperature ceramics have the advantages of high melting point, high strength, high thermal conductivity, etc., and are a high-temperature structural material with excellent performance [3]. Therefore, this article starts from a new perspective, combined with three-dimensional image reproduction technology, to study the construction of ceramic product shape models, which to a certain extent has important research significance and reference value.

This article mainly discusses these aspects. First of all, the three-dimensional image reproduction technology and related research are explained. Then, the ceramic product molding technology and related research are introduced. In addition, the application of 3D image reproduction technology in the construction of ceramic product shape models is also discussed. Finally, the research is carried out around the construction of the ceramic product shape model, and the corresponding experimental results and analysis conclusions are drawn.

2 Related Theoretical Overview and Research

2.1 3D Image Reproduction Technology and Related Research

In recent years, with the improvement of computer processing and storage capabilities, three-dimensional image reproduction technology has also made significant progress. At the same time, it is increasingly used in all aspects of life, including virtual reality, 3D video games, industrial automation production, historical buildings and cultural relics protection, and the film and television industry (3D movies) and many other fields.

Three-dimensional image reproduction technology enables industrial robots to use product photos to reconstruct product models, and to automatically process and compare products, which also plays an important role in the final quality inspection. This greatly improves the level of intelligence and mechanization in the product manufacturing process, and greatly improves the production efficiency and precision of the product.

Three-dimensional information has brought great benefits to people's lives, making people's lives more colorful, efficient and convenient. There are many ways to obtain three-dimensional information. First, use the existing 3D modeling application software to complete the construction of the 3D model. This method needs to know the specific information of things in the scene in advance, such as objects. Size, location, shape, etc. Second, use a complex 3D scanning system (such as a depth scanner) to obtain 3D information. This method uses laser or infrared to measure the distance between the object and the instrument and stores the data. It is suitable for applications that require high precision, but it also has some disadvantages, for example, more expensive equipment and poor portability [4, 5].

The real-time performance of 3D image reproduction has always been the bottleneck of this technology. The rendering process includes various aspects such as the complexity of modeling, the deletion and calculation of brightness. These problems will undoubtedly increase the complexity of modeling and slow down the speed of graphics rendering. The classification of 3D image reproduction technology is known in Fig. 1.

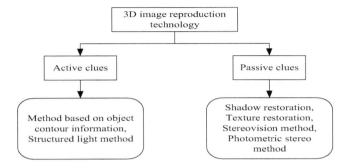

Fig. 1. Classification of 3D image reproduction technology

The objectively existing world is embodied in three dimensions in space, while existing image acquisition equipment can only display and record three-dimensional objects as two-dimensional information. Although this two-dimensional image information always contains a certain amount of three-dimensional spatial information, after the scene is restored and processed by three-dimensional reconstruction technology, the photo can be used to directly reproduce and restore the scene, and then perform digital processing to obtain the processed three-dimensional information In order to use the reconstructed digital model.

When the target is in a natural environment, it will be affected by inevitable factors such as uneven illumination and complex target background. At the same time, changes in the image sensor parameters of the image acquisition device will also affect the quality of the image, which in turn affects the detection and extraction of the contour and corner points of the target contour.

In order to make the final results corresponding to the feature points accurate, the image needs to be preprocessed before performing these operations. Preprocessing can reduce or eliminate the influence of various external factors on the image mapping information to achieve the desired effect. To extract the contour of the target contour in image preprocessing, it is necessary to select a filtering method that can better retain the contour information of the target [6, 7].

In order to subsequently extract the effective feature points of the target in the image for better feature point correspondence and obtain more effective feature correspondence points, the target area in the image must be separated from the image background.

The edge can well retain the structural information of the target in the picture, and express the most important feature of the target in the most intuitive way. There are many commonly used edge detection algorithms. Among them, the most widely used algorithm with better detection performance is the Canny operator. The goal of Canny edge detection is to find an optimal target edge, that is, to ensure that only one pixel plane is detected at the target edge. The specific calculation method is shown in formula (1) (2).

$$H = \sqrt{H_a^2 + H_b^2} \tag{1}$$

$$\alpha = arctan\left(\frac{H_b}{H_a}\right) \tag{2}$$

Among them, H is the gradient amplitude, Ha and Hb are a pair of convolution arrays in the a direction and the b direction, respectively, and α are the gradient directions.

Due to the error of the edge detection algorithm itself, the detected target edge is not ideal, which affects the result of subsequent processing steps. The contour following algorithm is generally used to follow a closed contour to obtain a closed contour curve.

The contour tracking algorithm finds the endpoints that follow the boundary and records the information about the boundary at the same time. The purpose is to preserve the attributes of the outer contour of the target area and prepare for shape analysis and target recognition.

2.2   Application of 3D Image Reproduction Technology in the Process of Constructing Ceramic Product Shape Models

As a material widely used in medicine, industry and other fields, ceramics have unique and excellent high temperature resistance, high strength and wear resistance.

The strength and density of the ceramic body depends on the influence of the applied force field and does not require any additives. However, the main disadvantage of these two methods is that the ceramic ligands cannot be completely uniform, and it is difficult to achieve mass production of high-performance ceramics. Since then, the research on the ceramic molding process has focused on improving the uniformity and reliability of the material body. Now, more and more new molding processes are gradually introduced and successfully applied and developed.

Filter press molding is a rapidly developing ceramic molding method in recent decades. The main method is to inject ceramic slurry with good dispersion and stability into the cavity of the porous mold through the slurry pipeline. At this time, under the action of external force, part of the liquid medium in the slurry is filtered out of the porous filter layer, so that the alumina particles are arranged firmly and solidified into a ceramic body. During this process, the green body is directly dried without powder agglomeration. Secondly, the filter press molding adopts liquid slurry for direct molding, and the agglomerates, inclusions and large particles in the ceramic powder can be removed by methods such as elimination and separation, thereby greatly improving the uniformity of the green body structure [8, 9].

When modeling with 3D image reproduction technology, we must first have a complete understanding of the generated image data and the corresponding target points, line segments and other parameters. Secondly, according to the information, design the corresponding surface map and volume map to form a physical modeling model; finally, transform it into a physical structure model and establish a database for storage and management, and finally perform the 3D reconstruction process of the solid object. At present, there is a relatively successful three-dimensional surface reconstruction method, which is to use a digital model to construct a two-dimensional plane figure on an object to obtain the desired image.

3D image reproduction technology includes extracting the texture and shape attributes of ceramic parts, then calculating the appropriate size, thickness and other attributes according to the parameters of the scanned data, and finally multiplying the obtained digital map by a predefined curve model to obtain the final result. Construct the shape model of the ceramic product.

The method based on 3D model reconstruction is a relatively common, universal, practical, efficient and feasible method for extracting information such as the shape characteristics and intrinsic quality parameters of ceramic products, and analyzing and identifying them to determine the required products size or structure to get. This process can obtain accurate information about the contour and internal structure, shape characteristics and internal quality parameters of ceramic parts in real time [10].

The three-dimensional surface reconstruction of ceramic products is basically divided into the following three steps. First, after the model is established, the network is rebuilt and the parameters are determined according to the required requirements. First, the local minimum variance method is used to solve the corresponding relationship between the positions of points, lines or surfaces in the two-dimensional image. Secondly, use the built-in texture mapping function to generate solid contours and shape curves in the target area to obtain graphics that meet the conditions. Then the target surface is imported into the constructed geometric database, the model parameters are calculated and the results are analyzed and processed, and finally the shape information required for the reconstructed ceramic product is obtained.

First, perform three-dimensional modeling of the product, make two-dimensional graphics, and extract the necessary elements. Then import these items to the computer. Finally, the relevant dimensions are calculated according to the obtained model parameters. The second is to digitize the surface of all parts after the solid shape is generated, obtain the feature point set of the surface, define its geometric information and other attribute point set; then use the contour curve of the workpiece and the created surface data to perform fitting analysis and contour cutting and get the three-dimensional graphics of the corresponding size.

3   Experiment and Research

3.1   Experiment Preparation

Since the molding process of ceramic products is more complicated, there are many aspects to consider when using three-dimensional modeling. A clear understanding and extraction of the relationship between the required information and data can ensure the quality requirements of the model, improve work efficiency, and reduce costs. Based on the 3D image reproduction technology, this paper constructs a ceramic product shape model to simulate the changes in the contour or spatial position of the object in the two-dimensional or multi-dimensional digital image. In this experiment, the model size is

H = 12 mm, D = 25 mm, and the extruder diameter is 0.6 mm, 0.8 mm, 1.2 mm, 1.6 mm, and the extrusion environment temperature is −10 °C.

3.2   Experimental Process

In this experiment, in order to better understand the performance of the model, the degree of deviation of the angle of the ceramic thin-walled parts was tested. The test items include the diameter of the extrusion head and the height of the extrusion layer. The diameter of the extrusion head is 0.6, 0.8, 1.2, 1.6 mm, and the extrusion layer height is 0.4, 0.6, 0.8, 1 mm. The experimental results are shown below.

4   Analysis and Discussion

In this experiment, in order to test the performance of the model, the degree of deviation of the angle of the ceramic thin-walled parts was tested. The test items include the diameter of the extrusion head and the height of the extrusion layer. The test results are shown in Table 1.

Table 1. Test results of the model

Storey height (mm)	0.6	0.8	1.2	1.6
0.4	0.1	0.3	0.2	0.4
0.6	0.3	0.5	0.4	0.2
0.8	0.2	0.7	0.5	0.7
1.0	0.5	0.6	0.6	0.6

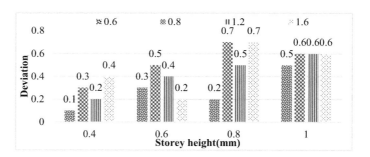

Fig. 2. Test results of the model

It can be seen from Fig. 2 that as the diameter of the extrusion head and the height of the extrusion layer change, the angle of the ceramic thin-walled piece will also have a certain deviation. When the diameter of the extrusion head is 0.6 mm and the height of the extrusion layer is 0.4 mm, the resulting deviation is the smallest, with a deviation of 0.1 mm. Therefore, when the extruder diameter is 0.6 mm, the layer height under 0.4 mm has the best formability, the smallest deviation value, and the highest precision of ceramic products.

5   Conclusion

After years of development, the ceramic industry has played an important role in China's national economy and is a symbol of the comprehensive national strength of a country and a region's economic strength. At the same time, as the process of industrialization accelerates, the automotive industry and the transportation industry have higher and higher requirements for the expression of 3D spatial images. Three-dimensional image reproduction technology can use a computer to perform two-dimensional or three-dimensional modeling, which can realize functions such as face recognition, target object contour extraction and texture analysis. Therefore, based on the three-dimensional image reproduction technology, this article conducts research on the construction of ceramic product shape models, which has important practical significance and research value to a certain extent.

References

1. Kherraz, N., Chikh-Bled, F.H., Sainidou, R., et al.: Tunable phononic structures using Lamb waves in a piezoceramic plate. Phys. Rev. B Condensed Matter Mater. Phys. **99**(9), 094302.1–094302.12 (2019)
2. Corporation, N.: Electronic device, imaging device, image reproduction method, image reproduction program, recording medium with image reproduction program recorded thereupon, and image reproduction device. Bond Univ. **182**(9), 793–802 (2018)
3. Taimori, A., Marvasti, F.: Adaptive sparse image sampling and recovery. IEEE Trans. Comput. Imaging **4**(3), 311–325 (2018)
4. Aksoys, F., Altinmakas, E., Candaa, E., et al.: 3D real time image reproduction of the prostate: can it be used on virtual reality (VR) headsets and/or tilepro of Da Vinci surgical system as a guide during robotic radical prostatectomy? Eur. Urol. Suppl. **18**(6), e2660–e2661 (2019)
5. Agi, A., Jurei, D., Agi, D., et al.: Importance of extended separation in NIR technology for secure secondary image identification, a survey. ACTA GRAPHICA J. Print. Sci. Graph. Commun. **29**(2), 19–26 (2018)
6. Snow, J., Zink, M., Schnuelle, D., et al.: Reports from the SMPTE technology committees. SMPTE Motion Imaging J. **127**(8), 16–28 (2018)
7. Tanaka, M., Takanashi, T., Horiuchi, T.: Glossiness-aware image coding in JPEG framework. J. Imaging Sci. Technol. **64**(5):50409-1–50409-15 (2020)
8. Shouman, M.: Use the 3D printer technology to obtain molds for ceramic products. J. Design Sci. Appl. Arts **2**(1), 259–268 (2021)
9. Yongxiang, W., Peng, W.: Three-dimensional image data reconstruction method under cloud computing environment. Mod. Electron. Technol. **040**(020), 108–110 (2017)
10. Saadi, E., Azbouche, A., Benrachi, F.: Spatial distribution of natural radioactivity and statistical analyses of radioactive variables in Algerian ceramic products. Int. J. Mater. Prod. Technol. **10**(02), 47–54 (2020)

Application of Computer Image Technology in Traditional 3D Movie Animation Production

Ziwei Xu[✉]

Wuchang University of Technology, Wuhan, Hubei, China
xuyoyo2021@126.com

Abstract. In recent years, the development of animation industry has been accelerated and good results have been achieved in creating economic and social benefits. Traditional animation must be deeply innovated. Adapt to the trend of the times, make a good transformation, combine with computer image technology and solve different problems according to a scientific and reasonable model. In this paper, a simulation experiment is established. The human body is modeled using a "rod" model. The motion of the simulated model is obtained using computer imaging technology and the processed image is output. The processed image retains its original quality. In this paper, low-cost MEMS sensors are used for node sensing acquisition, and calibration is performed before acquisition to reduce the error of individual sensors and make the output value closer to the real value.

Keywords: Computer technology · Image processing · 3D animation · Traditional animation

1 Introduction

The development of Chinese animation has experienced several ups and downs. In terms of animation originality, the animation style was lost in the early stage, the production style was missing, and the content positioning was deviated. The whole animation industry chain cannot maintain a virtuous cycle [1, 2]. Currently, after a long period of adjustment, learning and integration, the Chinese animation market is in a state of poise. Enterprises and production teams must learn and utilize advanced foreign experience in production management, production quality, project operation, etc. The development and other aspects have laid a good foundation, and the rise of Chinese animation industry is just around the corner [3, 4]. In the 1980s, the development of computer graphics and imaging technology led to a revolution in film production with the strong support of the government and the unremitting efforts of university research and production companies [5, 6]. With the support of computer graphics imaging technology, many unprecedented cinematic art films were born and showed great commercial value. Since the 1990s, the computer graphics and imaging technology industry has gained great attention and flourished in various countries [7, 8].

With the continuous innovation of the production mode and skill characteristics of animation, 3D animation jumped out of the limelight at this time [9, 10]. The design mode of 3D animation not only overturns the traditional animation design, but also makes a bold attempt for the innovation of animation creation. Film and television special effects technology as a breakthrough of modern computer graphics application technology. It provides a broader fantasy space for the audience [11, 12].

Digital preservation is one of the ways in which ethnic cultures can survive. The role of information and communication technology is a key factor in preserving this cultural and natural tourism. prayoonrat C proposes a framework for a national animated film that follows cultural heritage and natural history. prayoonrat C uses storytelling techniques to build the film storyboard and uses adobe photo-shop, blender, audacity and avoided as tools to create this race-based 3D animated film [13, 14]. To simplify and speed up the process of drawing animation sequences, CHANG developed an automated CEL drawing system.CHANG proposed a method to reduce production costs and improve the overall quality of animation using computer-aided animation, which marks a significant improvement in the working environment for producing 2D animation.CHANG also proposed a new integrated annotation algorithm for constructing statistical regions and topographic map features. And a partial matching algorithm for attribute maps was proposed [15, 16]. Mo C proposed a new method for automatic control of stereo camera parameters. By optimizing the camera parameters, stereo content with better depth perception is generated to ensure visual comfort. mo C describes how to achieve this on the MAYA plug-in and test the stereo effect using a professional stereo 3D animation scene. Experimental results showed that Mo C's method improved the stereoscopic effect [17, 18].

In this paper, simulation experiments were established. A "rod-like" model was used. Computer imaging techniques are used to capture the motion of the simulated model. The processed images still maintain the original quality. In this paper, low-cost MEMS sensors are used for node sensing acquisition, and calibration is performed before acquisition to reduce the errors caused by individual sensors, so the output values are closer to the real values.

2   Proposed Method

2.1   Computer Image Processing Technology

1) Basic meaning of computer image technology
   Computer image technology is the use of computers to process images to obtain the desired results and target images through the analysis and processing of images.
2) Type of image
   According to whether the image can be processed on the computer, the categories of images can be divided into the following two main categories. First, analog images. The advantage of analog images is that the output is generally faster and more convenient, and the disadvantage is that the accuracy is not high and the application is not flexible. Second, digitize the image. Digital images have many advantages such as high precision, easy processing and flexibility, which meet the actual needs of modern development.

3) Features of computer image technology
   (1) Computer image processing technology has high precision.
   Using computer image processing technology, images can be digitized to obtain a two-dimensional array. This two-dimensional array can be digitized by certain equipment images to achieve any size variation of this two-dimensional array. With modern scanning instruments, the grayscale of each pixel can be anesthetized to 16 bits or higher, which makes the accuracy of computer image processing technology improved to meet more practical needs.
   (2) Good reproducibility of computer image technology.
   The traditional analog image processing technology degrades the image quality after image processing. Computer image processing technology can maintain the original quality of images in the process of image storage, reproduction and transmission, and has good reproducibility.
   (3) Wide application of computer image technology.
   Computer image technology can handle images from different information sources. As long as different image information acquisition measures are taken for different information sources, these images can be stored digitally in the computer and processed by computer technology to meet the needs of people's actual work and life.

2.2   3D Animation Technology

1) 3D filming in the film
   (1) Live-action system
   The 3D live shooting system is an important guarantee for shooting scenes with 3D visual effects. Before the improvement of the technology, the shooting system was mainly constructed by two cameras working side by side. As the improved live-action filming system is built from two high-definition digital lenses and a camera, which can be manually adjusted, plus the former has the advantages of easy movement, small size, different lens split filters, etc.. The scene can reflect and refract two images separately, and then enter two cameras for recording, thus forming a more realistic three-dimensional vision. Its operation is simpler and faster than the original technology.
   (2) Virtual shooting system
   The so-called virtual shooting system, is the use of virtual cameras in the virtual CG scene for three-dimensional shooting work, and then use three-dimensional tracking software and model software to shoot the scene in the computer converted into a three-dimensional scene, and then the virtual camera will be added to the virtual three-dimensional scene.
2) The application of 3D capture technology in movies
   Motion capture technology can express the characters' expressions in a delicate and realistic way. Through this technology, the 3D movie Avatar realistically reproduces every action of the actors. During the filming process, the actor needs to wear a straitjacket with capture points when doing the required action, and then walk through the studio. The captured data is transmitted to the system through the

near-infrared spectral reflection tracking of the live ED lights, and the actors' movements and spectral reflections are analyzed as a whole to obtain the corresponding 3D models, presenting a more realistic action effect.

2.3   3D Animation Film Production Key Technology

1) 3D modeling, materials, textures and lighting

   3D modeling is the basis of 3D animation film production. Material is a property that visualizes the surface of an object, reflecting a concrete or abstract perception of the surrounding environment. Lighting is the soul of virtual scene production in 3D animation film. Highlight the visual effect and attractiveness of the cartoon. Rendering refers to the construction of scene models, materials, textures, lighting presets and program screens, or animations that must be rendered to produce static or dynamic renderings.

2) Animation

   Animation is the basis of 3D animation film production. Each animated film needs to have a variety of characters and different traits, and the extrapolation of character and traits can be achieved through the movement of the characters, so in the production of animated films, the combination of the laws of motion and 3D software technology is the key to give life and uniqueness to the characters. The 3D software can produce animation effects close to the laws of human movement through the binding of skeleton and the adjustment of skin.

3) Virtual reality technology

   Its biggest feature is that it uses real characters instead of virtual characters, so it can use various programs involving virtual characters, such as modeling, texturing, motion capture, etc. It greatly improves the progress of animation production and saves cost.

4) Cloud Rendering

   Cloud rendering is a kind of rendering service platform based on network sharing. It has many common points with cloud computing, such as contextualization technology, massive data storage and processing technology, workflow technology, etc. Cloud rendering has three functions in 3D animation movie production: firstly, it increases efficiency and improves the efficiency of users' rendering tasks; secondly, it optimizes users' rendering tasks through resource sharing; and thirdly, it saves costs and reduces hardware configuration costs.

5) Post-synthesis

   The post-synthesis of 3D animation movie production is similar to the post-synthesis of film and TV drama. Through linear editing software to complete the material grouping, and finally in accordance with the film release requirements of the format to generate an animated film and television.

3 Experiments

3.1 Experimental Implementation

(1) Establish a simulation model
    Establish a simulation experiment object. Motion capture technology is to set up a tracker in the key parts of moving objects, record the movement of the object, and then obtain the three-dimensional data through computer processing. This technology is widely used in digital protection, games, animation, ergonomics research, simulation training and other fields. Such as virtual reality, motion capture and other research fields can not only capture all kinds of movements, but also quantify the characteristics of movements, effectively realizing the scientific and digital teaching. In the human body modeling, the "rod-like" model is adopted, and each limb belongs to a contiguous relationship. Through the mutual transformation of the human coordinate system and the absolute coordinate system, the motion relationship between various parts of the human body can be obtained; thereby the human body model is established to simulate the movement of human body and capture.

(2) Image processing
    The human body simulation model is captured using a motion capture system, and the resulting image is processed using a computer image processing technique to output an image.

3.2 Sensor Calibration

In this paper, low-cost MEMS sensors are used for node sensing acquisition. Due to the low-cost MEMS manufacturing process problems, there are some manufacturing variations in the factory. The data collected directly in the sensor is not suitable in practice. Application, calibration and compensation are required to compare the measured changes in the real response.

The main calibration tasks include:

(1) Gyro zero offset calibration
(2) Acceleration zero offset calibration
(3) Geomagnetic sensor hard calibration and soft calibration.

4 Discussion

4.1 Simulation Model Experimental Analysis

In this paper, a simulation model is established to simulate the experiment, and a simulated human body model is adopted. In the human body modeling, a "rod" model is selected, as shown in Fig. 1.

Fig. 1. Simplified structure of the human body model

From Fig. 1, it can be seen that there is a linkage relationship between the limbs of the "stick" model. The interconversion of the human coordinate system and the absolute coordinate system allows for simulated motion capture.

The motion capture system is used to capture the human body simulation model, and the obtained image is processed by computer image processing technology to output the image. The results are shown in Fig. 2.

Fig. 2. Image after computer image processing

As can be seen from Fig. 2, the computer image processing generated a 3D stereoscopic image from the "stick" model image, and the processed image is beautiful and three-dimensional.

4.2   Measurement Accuracy Experimental Data and Analysis

The testing principle of pitch and roll angle is similar to that of heading angle, except that the fixation of the sensor needs attention to the lateral fixation with the platform. The pitch and roll angle range is ±90°, where the measurement range is locked. ±60° in increments of 5°, because the experimental data is carried out by observing the debugging results printed on the serial port, and the raw data of the sensor is digital, which is constantly jumping data, so it has to be solved by complementary filters. The error curves of the experiments are shown in Fig. 3, Fig. 4 and Fig. 5 (x-axis is the measured angle and y-axis is the error value in ° for all coordinate systems in the following figure).

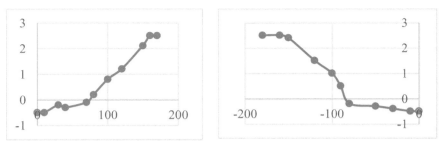

(a) Yaw measurement error in the range of 0 to 180° (b) Yaw measurement error in the range of -180 to 0 degrees

Fig. 3. Yaw measurement error

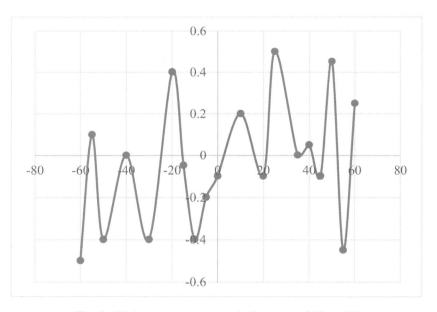

Fig. 4. Pitch measurement error in the range of 60 to 60°

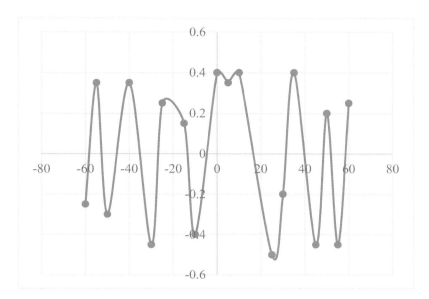

Fig. 5. Roll measurement error in the range of −60–60°

Figures 3, 4 and 5 can reflect the magnitude of the error variation more visually. From Fig. 3, Fig. 4 and Fig. 5, the measurement deviation of heading angle is relatively large. In the range of −0.5–2.6°, the error fluctuation range is 3.1°; the accuracy of pitch angle and roll angle is relatively high, in the range of −0.4–0.5° and −0.5–0.4° respectively, and the error fluctuation range is basically 1°. The data are fused by complementary filters to reduce the impact of errors caused by a single sensor, so the output value is closer to the actual value.

5   Conclusions

This paper introduces the historical development process of Chinese animation production, introduces the 3D animation technology, combines computer image processing technology with 3D film animation production through experiments, and studies the application of computer image processing technology in traditional 3D film animation production.

A human body simulation model experiment was established, and a "rod" model was used. A computer motion capture system was used to drive the human 3D model for real-time motion display, so as to achieve gesture recognition and scene reproduction. After the experiment, it was found that the processed image could still maintain the original quality, indicating that the computer image processing system could promote the development of 3D film animation.

In this paper, the sensor is calibrated for initial distortion calibration. Three sets of experiments were then conducted to test the heading angle, pitch angle and roll angle based on the sensor measurements after entering the data fusion algorithm. It is concluded that the error value of heading angle is larger (about 3° range) compared to the pitch and roll angles (0.) ° range), which basically meet the requirements of the motion capture system.

References

1. Braga, P.H.C., Silveira, I.F.: SLAP: storyboard language for animation programming. IEEE Lat. Am. Trans. **14**(12), 4821–4826 (2017)
2. Taylor, S., Kim, T., Yue, Y., et al.: A deep learning approach for generalized speech animation. ACM Trans. Graph. **36**(4), 1–11 (2017)
3. Gu, X., Song, H., Chen, J., et al.: A review of research on pig behavior recognition based on image processing. Int. Core J. Eng. **6**(1), 249–254 (2020)
4. Fang, J., Guo, Z., Wu, T.: Research on computer image processing technology and application of wavelet lifting algorithm. J. Comput. Theor. Nanosci. **13**(7), 4465–4469 (2016)
5. Jie, Z., Jie, Z., Wang, Q., et al.: An overview on passive image forensics technology for automatic computer forgery. Int. J. Digit. Crime Forensics **8**(4), 14–25 (2016)
6. Zhang, J., Chen, D., Liao, J., et al.: Model watermarking for image processing networks. In: Proceedings of the AAAI Conference on Artificial Intelligence, vol. 34, no. 7, pp. 12805–12812 (2020)
7. Jean-Pierre, A.: Two-dimensional directional wavelets and image processing. Int. J. Imaging Syst. Technol. **7**(3), 152–165 (2015)
8. Ong, F., Milanfar, P., Getreuer, P.: Local kernels that approximate Bayesian regularization and proximal operators. IEEE Trans. Image Process. **28**(6), 3007–3019 (2019)
9. Dorado, A., Martinez-Corral, M., Saavedra, G., et al.: Computation and display of 3D movie from a single integral photography. J. Displ. Technol. **12**(7), 695–700 (2017)
10. Wang, D., Yang, X., Hu, H., et al.: Visual fatigue during continuous viewing the 3D movie. Electron. Imaging **2016**(5), 1–6 (2016)
11. Lee, S., Jin, Y.: The potential of a text-based interface as a design medium: an experiment in a computer animation environment. Interact. Comput. **28**(1), 85–101 (2016)
12. Boutekkouk, F., Sahel, N.: Color image processing under uncertainty. Int. J. Technol. Diffus. **12**(2), 46–67 (2021)
13. Prayoonrat, C.: Raising energy awareness through 3D computer animation. Appl. Mech. Mater. **752–753**, 1116–1120 (2015)
14. Bluff, A., Johnston, A., Clarkson, D.: Interaction, narrative and animation in live theatre. IEEE Comput. Graph. Appl. **38**(2), 8–14 (2018)
15. Chang, C.-W., Lee, S.-Y.: Automatic cel painting in computer-assisted cartoon production using similarity recognition. Comput. Anim. Virtual Worlds **8**(3), 165–185 (2015)
16. Aoki, K.: Server for implementing image processing functions requested by a printing device. Environ. Pollut. **152**(3), 543–552 (2018)
17. Mo, C.: Application of computer virtual simulation technology in 3D animation production. IOP Conf. Ser. Earth Environ. Sci. **94**(1), 012003 (2017)
18. Avril, Q., Ribet, S., Ghafourzadeh, D., et al.: Animation setup transfer for 3D characters. Comput. Graph. Forum **35**(2), 115–126 (2016)

Prediction and Early Warning Model of Substation Project Cost Based on Data Mining

Shili Liu[1], Liming Chen[2], Xiaohu Zhu[1], Fan Yang[1], Jianqing Li[1(✉)], and Mahamat Ali Diallo[3]

[1] State Grid Anhui Electric Power Co., Ltd. Economic Research Institute, Hefei, Anhui, China
Li_Jian_QQing@163.com
[2] State Grid Anhui Electric Power Co., Ltd., Hefei, Anhui, China
[3] School of Economics and Management, North China Electric Power University, Beijing 102206, China

Abstract. Under the dual pressure of the reform of the power system and the increasing demand for electricity in society, power grid companies urgently need to improve the level of lean management of power engineering costs. Substation project is an important part of power engineering construction. Accurate cost forecasting can effectively control project budgets, improve corporate efficiency, and standardize capital management and control. Traditional cost prediction methods are difficult to solve the high-dimensional and non-linear problems of cost data. This paper builds a machine learning model based on data mining techniques to predict the cost of substation projects and improve the cost prediction ability. First, the importance of random forest is used to extract key influencing factors, and the support vector machine regression (SVR) cost prediction model based on Grid Search (GS) optimization is constructed. Then, it is compared with the prediction results of the Random Forest (RF) model to verify the effectiveness of the model proposed in this paper for the cost prediction of substation projects. Finally, the probability density function is constructed according to the distribution of historical data, and the threshold of cost early warning is determined based on the "3 sigma principle" to realize the early warning control of the cost of substation projects and provide better guidance for cost management decisions.

Keywords: Substation project · Cost forecast and early warning · Data mining · Support vector machine · Random forest

1 Introduction

The rapid growth of electricity demand brought about by the sustained and rapid economic development has put forward higher requirements for the construction and management of power grid companies. However, many power grid companies are facing tremendous operating pressure because they cannot accurately grasp the level of modern power engineering construction costs, and their capital utilization rate is low.

© The Author(s), under exclusive license to Springer Nature Switzerland AG 2022
V. Sugumaran et al. (Eds.): ICMMIA 2022, LNDECT 138, pp. 400–407, 2022.
https://doi.org/10.1007/978-3-031-05484-6_50

As an important part of power engineering construction, substation project directly affects the operating level and benefits of power grid companies. Accurate forecasting and early warning are of great significance to improving the level of lean management of power companies.

In substation project cost prediction, the choice of influencing factors has an important impact on the model prediction results. Wang et al. [1] believed that the voltage level of the substation and the capacity of the main transformer are the important factors affecting the cost prediction of the substation project. Yabo, S et al. [2] took the number of tower materials, steel price, terrain distribution and other indicators as the key influencing factors for the cost prediction of power transmission and transformation projects.

The results of substation project cost prediction not only depend on the selection of influencing factors but also the appropriate prediction method is also very important to it [3]. Early engineering cost forecasting methods mainly include qualitative or quantitative forecasting methods such as Building Cost Information Service (BCIS) [4], Monte Carlo estimation method [5], the quota set usage, and the inventory valuation method. However, traditional cost forecasting methods often have a certain lag. At the same time, they cannot fit non-linear, high-dimensional modern engineering cost data, which has certain limitations. With the gradual maturity of computer technology, more intelligent forecasting technology has been introduced into the field of cost forecasting. Hegazy T. and Ayed A. [6] optimized the BCIS model to improve the accuracy of the prediction model. Guo Hehong [7] improved the accuracy of project cost forecasting by using improved project cost estimate indicators. Kim Sooin et al. [8] constructed a pipeline construction cost prediction model based on multivariate time series, which improved the efficiency and accuracy of cost prediction. Miao Fan et al. [9] proposed the Least Squares Support Vector Machines (LS-SVM) model based on the support vector machine model (Support Vector Machine support, SVM). The support vector machine model optimized by the swarm algorithm improves the prediction accuracy.

Risks run through the construction of substation projects, and risk early warning after project cost prediction is of great significance for improving the level of cost management. In terms of pre-warning of project engineering, Dileep et al. [10] used the three-sigma principle to control and define the value deviation warning range of specific categories of age and gender, and achieved the improvement of prediction accuracy.

In summary, based on the preprocessing of the original data, this paper uses the importance of random forest features to screen the key influencing factors and establishes a grid search optimized support vector machine model for project cost prediction. Finally, the scale factor is screened among the key influencing factors, and the probability density function is established based on historical cost data to divide the early warning level. Through quantitative and qualitative analysis of forecast and early warning results, it provides a reference for the lean management and decision-making of power grid enterprises.

2 Methodology

2.1 Support Vector Machine

Support Vector Machine theory is developed based on the statistical theory of scholar Vapnik and his collaborators. The basic principle is to find an optimal hyperplane, separate the two types of samples with the largest interval, that is, all training samples are away from the optimal hyperplane. The error is the smallest. The algorithm assumes that there is a non-linear mapping relationship F so that $y = F(x)$. The regression equation $f(x) = \sum_{i=1}^{n}(a_i^* - a_i)K(x_i, x) + b^*$ is solved by a support vector machine and y_i represents the i predicted value. Transform the problem into a quadratic convex optimization problem, effectively minimize the scope of predictive structural risk, solve or avoid problems such as "over-learning", and have a good regression prediction in processing nonlinear, high-dimensional data and small sample learning effect [11–14].

The estimation function of the SVM algorithm is as follows.

$$y = f(x) = \omega^{T}\phi(x) + b \tag{1}$$

Among them, y represents the linear regression prediction value, ω^{T} represents the weight vector, $\phi(x)$ represents the non-linear function, and b represents the degree of offset. The radial basis kernel function is as follows.

$$K(x_i, x_j) = \exp\left(-\frac{\| x_i - x_j \|^2}{\sigma^2}\right) \tag{2}$$

2.2 Grid Search

Grid Search (GS) [15] is an exhaustive algorithm. Its basic principle is to divide the parameter space into several grids and traverse all grid intersections in the input space to obtain the optimal solution. The algorithm firstly determines the value range of each parameter, and then interpolates the value range of each parameter according to a certain rule to obtain several sets of parameter combinations; calculates each set of parameter combinations once, and applies cross-validation to calculate the prediction error; The parameter combination with the smallest prediction error is the optimal parameter value. At the same time, each group of parameters is decoupled from each other in the grid search optimization process, which can effectively avoid the complexity of the solution caused by the mutual influence of the parameters.

3 Prediction and Early Warning Model of Substation Project Cost

3.1 Predictive Model Construction

The process of the substation project cost prediction model is constructed in this paper is shown in Fig. 1. The model uses the random forest model to perform feature selections and uses a grid search algorithm to optimize the parameters of the support vector regression model. The specific steps are as follows.

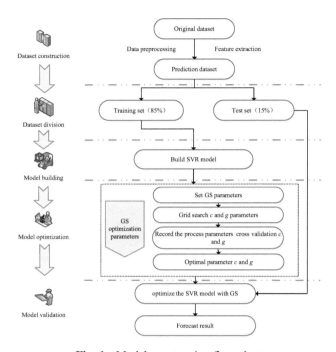

Fig. 1. Model construction flow chart

(1) Build a data set. Perform data preprocessing on the sample set and use the importance of random forest features to select key influencing factors; (2) Divide the data set. Divide the sample data into the training set and the test set according to the ratio of 8.5:1.5; (3) Construct a substation project cost prediction model based on GS-SVR. Input the data of the training set into the support vector regression machine model for training, adjust the model parameters, obtain the optimal training model, and verify its generalization ability through the verification set, and output the prediction results of the substation project cost.

3.2 Early Warning Model Construction

Early warning of substation project cost risk refers to issuing early warning signals based on the nature and degree of deviation between the predicted value of substation project cost and the early warning threshold, and taking corresponding measures to avoid risks. The basic process is to collect and process cost information, determine the early warning threshold and dynamic early warning. The specific steps are as follows.

(1) Screen scale factors through the importance of random forest features to construct a static investment early warning indicator system.

(2) Construct the normal distribution function $N(\mu, \sigma^2)$ as shown below.

$$f(x) = \frac{1}{\sqrt{2\pi}\sigma} e^{-\frac{(x-\mu)^2}{2\sigma^2}} \tag{3}$$

Among them, μ and σ are the expected value and standard deviation of the historical data of early warning attributes.

(3) According to the "Sigma Principle", "Small Probability Event" and hypothesis testing, early warning is given for the cost falling outside the warning interval.

4 Empirical Analysis

4.1 Original Data and Preprocessing

In this paper, through the power grid companies' substation cost analysis datasheet, the data for the new substation project from 2015 to 2021 is obtained for 7 years. Among them, there are a total of 224 data records for 110 kV and 220 kV new substation projects.

(1) Data preprocessing

After sorting out the data set, it was found that there were data incompleteness, duplication of indicators, and serious affiliation in the data.

Fill in missing values. For discrete (sub-type) variables, the mode of 7-year data is used for filling, and for continuous variables, the mean value is used for filling.

Outlier handling. The 3σ principle is used to identify the outliers in the numerical data, that is, the samples with an average distance greater than 3σ are judged as outliers. Then, the outliers are eliminated, and the above-mentioned outliers eliminated data are filled by the numerical data missing value filling method.

(2) Feature coding

To simplify the data set, reduce the complexity of the model and improve the prediction performance, the discrete data is encoded and quantized. Taking into account the advantages of one-hot encoding in dealing with disordered discrete variables, optimizing feature calculation distance, and expanding features, this paper selects the one-hot encoding to deal with disordered discrete variables. At the same time, use label coding to quantify the characteristics of ordered discrete variables.

(3) Random forest feature importance selection

This data set has many feature variables. After data preprocessing, the importance of random forest features is used to further screen the key factors, and the feature variable with the smallest out-of-bag error rate is selected to construct the final influencing factor index set.

4.2 Evaluation Index

In this paper, the prediction accuracy is used to evaluate the prediction effect of the model, and the calculation formula is

$$P = (1 - MAPE) \times 100\% \tag{4}$$

Among them, P is the prediction accuracy, MAPE is the Mean Absolute Percentage Error (MAPE), and the MAPE calculation formula is

$$MAPE = \frac{1}{n} \sum\nolimits_{t=1}^{n} \left| \frac{y_i - \hat{y}_i}{y_i} \right| \tag{5}$$

Among them, n represents the number of samples, y_i represents the true value of the sample, and $\hat{y}_i$ represents the i predicted value of the sample.

4.3 Forecast Model Comparison and Result Analysis

The Support Vector Machine model can make the result of data analysis and prediction tend to the result obtained when the data is infinite when the sample data is small. The Random Forest model is predicted by multiple decision trees at the same time, and the results of each tree are simply averaged to get the final value. Use "collective wisdom" to make final decision-making judgments, avoid over-fitting, and strengthen predictive capabilities. Considering a large number of feature variables in this project, to verify the prediction performance of the model proposed in this article, this project selects the Support Vector Machine Regression model and the random forest for comparison. After multiple experiments, the prediction accuracy and fitting diagrams of different models are shown in Table 1 and Fig. 2.

Table 1. Evaluation results of substation project prediction model

Evaluation index	Random forest	Support vector machine
Precision	88.94%	91.03%

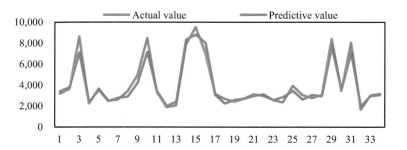

Fig. 2. Fitting diagram of the predicted value of substation project

It can be seen from Fig. 2 that the prediction accuracy of the Support Vector Machine is 91.03%, which is significantly better than the Random Forest model. Prove its effectiveness in processing small samples and non-linear data. At the same time, from the fitting effect diagram of the predicted value and the true value, it can be seen that the support vector machine has a better fitting effect and trend.

4.4 Early Warning Result Analysis

Take a 110 kV substation (the single capacity of the main transformer is 50MVA (three-phase), the number of main transformers in the current period is 1 (three-phase), and the high-voltage side power distribution device type is AIS column type) to realize the static investment deviation level early warning.

(1) Construct the probability density function. Calculate the expected value and standard deviation of 33 historical data.

$$f(x) = 379.55e^{-\frac{(x-2142.65)^2}{2 \times 379.55}}$$
$$(\mu = 2142.68, \sigma = 379.55)$$

(6)

Among them, μ and σ respectively represent the expected value and standard deviation of the 33 historical data of the calculation example.

(2) Construct the early warning grade interval according to the "Sigma Principle", as shown in Table 2.

Table 2. 110 kV-50MWA-1 set-AIS column transformer static investment warning zone

Warning level	Warning zone
Low	$(814.25, 1004.03) \cup (3281.33, 3471.11)$
Middle	$(624.48, 814.25) \cup (3471.11, 3660.88)$
High	$(0, 624.48) \cup (3660.88, +\infty)$

(3) Early warning of substation project cost risk. Through the early warning of the cost of the substation project, a three-level early warning interval of the high, medium, and low is constructed. In the future, when static investment volatility is in the low and mid-level range, managers should do a good job of inspecting and screening errors and omissions. When the fluctuation range increases to a high level, the management personnel must analyze and correct deviations promptly manner to ensure the reasonable construction of the project.

5 Conclusion

This paper constructs a substation project cost prediction model based on GS-SVM and a substation project cost early warning model based on probability density function, aiming to provide managers with a basis for cost management, improve the level of lean management, thereby alleviating the pressure on the operation and construction of power grid enterprises. First, the original data is preprocessed, and the importance of random forest features is used to extract key influencing factors. Then, build a prediction model based on GS-SVM, and use the training set and test set to train and verify the model. Then, the scale factors are screened among the key influencing factors, the probability density function warning model is established based on the historical data of scale factors, the low, medium, and high-cost warning intervals are determined, and corresponding management measures are proposed.

References

1. Wang, H., et al.: Cost forecasting model of transformer substation projects based on data inconsistency rate and modified deep convolutional neural network. Energies **12**(16), 3043 (2019)
2. Yabo, S., et al.: Research on overhead line engineering cost prediction based on PCA-LSSVM model. In: E3S Web of Conferences, vol. 185, p. 02023 (2020)
3. Zhou, M., Lu, S., Wang, S., et al.: Analysis of the influence factors of transformer substation project cost based on the random forest model. In: Analysis of the Influence Factors of Transformer Substation Project Cost Based on the Random Forest Model, Wuhan, Hubei, China (2020)
4. Liu, F., Huang, Z.: Comparison and analysis of fixed asset investment estimates at home and abroad. Infrastr. Optim. (04), 26–29 (2003)
5. Tummala, V.M.R., Burchett, J.F.: Applying a risk management process (RMP) to manage cost risk for an EHV transmission line project. Int. J. Proj. Manag. **17**, 223–235 (1999)
6. Hegazy, T., Ayed, A.: Neural network model for parametric cost estimation of highway projects. J. Constr. Eng. Manag. **124**, 210–218 (1998)
7. Guo, H.: Research on transmission project cost forecast based on regression analysis. Smart Power (03), 75–77 (2008)
8. Kim, S., Abediniangerabi, B., Shahandashti, M.: Pipeline construction cost forecasting using multivariate time series methods. J. Pipeline Syst. Eng. Pract. **12**(3), 04021026 (2021)
9. Miao, F., Ashutosh, S.: Design and implementation of construction cost prediction model based on SVM and LSSVM in industries 4.0. Int. J. Intell. Comput. Cybern. **14**(2), 145–157 (2021)
10. Dileep, M.R., Danti, A.: Human age and gender prediction based on neural networks and three sigma control limits. Appl. Artif. Intell. **32**(3), 281–292 (2018)
11. Fayed, H.A., Atiya, A.F.: Decision boundary clustering for efficient local SVM. Appl. Soft Comput. **110**, 107628 (2021)
12. Zouhri, W., Homri, L., Dantan, J.: Handling the impact of feature uncertainties on SVM: a robust approach based on Sobol sensitivity analysis. Expert Syst. Appl. **189**, 115691 (2022)
13. Albashish, D., et al.: Binary biogeography-based optimization based SVM-RFE for feature selection. Appl. Soft Comput. **101**, 107026 (2021)
14. Baldomero-Naranjo, M., Martínez-Merino, L.I., Rodríguez-Chía, A.M.: A robust SVM-based approach with feature selection and outliers detection for classification problems. Expert Syst. Appl. **178**, 115017 (2021)
15. Erdogan Erten, G., Bozkurt, K.S., Yavuz, M.: Grid search optimised artificial neural network for open stope stability prediction. Int. J. Min. Reclam. Environ. **35**(8), 600–617 (2021)

Multi-feature Fusion Image Recognition System Based on Machine Learning Algorithm

Qiakai Hailati[✉], Tao Wang, Jiangtao Guo, Liu Yang, and Hong Li

State Grid Xinjiang Information and Telecommunication Company,
Urumqi, Xinjiang, China
kayrat23@163.com

Abstract. The branch BP neural network of machine learning can recognize image information, so it can effectively improve the application effect of artificial intelligence. Compared with humans, machine learning has faster data processing speed, storage information and the ability to extract features. Machine learning models that deal with image recognition and cognitive abilities have a profound impact on applications such as massive image classification, image recognition positioning, image-text conversion, and image segmentation. This paper uses BP neural network to design a multi-feature fusion image recognition system. After using this system to extract the color and texture feature vector values of the image, the image recognition accuracy is calculated. Research shows that the system has the highest recognition accuracy of the image in the forest, reaching 93.68. %, the recognition accuracy of grass is the lowest, only 9.52%.

Keywords: Machine learning algorithm · BP neural network · Multi-feature fusion · Image recognition system

1 Introduction

After decades of development of machine learning, there are still many problems that have not been solved, such as image recognition, positioning and detection, and image and text generation. In the field of machine learning research, scientific research on image recognition has always been the most basic, traditional and most urgent research direction. Machine learning algorithms can recognize images from multiple features such as image color, texture, geometric shape, etc. Therefore, this article uses this as an entry point to study the application of multi-feature image fusion recognition system.

Many scholars at home and abroad have conducted in-depth research on the multi-feature fusion image recognition system based on machine learning algorithms, and have achieved good research results. For example, a research team developed a computer vision image system based on a deep-condensed neural network, which surpassed human cognition and greatly reduced the error rate in image classification [1]. Aiming at the problems of long-distance oblique imaging and angle changes, geometric distortion, and radial distortion overlap, some scholars have proposed a new point-based method, which creates a non-linear space across domains for feature extraction and simulates the original image for sampling [2]. A scholar studied image recognition based on density grouping and multi-feature fusion, analyzed the image features such as grayscale, texture, and edge of the image, and extracted these features of the recognized object as the preferred antibody, completing the medical adjustment data association fusion [3]. Although the research on the multi-feature fusion image recognition system is progressing smoothly, in practical applications, the recognition accuracy of the image recognition system needs to be improved.

This article introduces several machine learning algorithms, among which the BP neural network can be used to construct an image fusion recognition system. Combining the multiple features of the image, the image can be recognized through image preprocessing, feature extraction, and classification. This article takes a natural landscape image as an example to color the forest, grass, shrubs and other elements on the image And texture feature extraction to analyze the accuracy of image recognition.

2 Machine Learning Algorithm and Multi-feature Image Fusion Theory

2.1 Machine Learning Algorithm

The machine learning algorithm first needs to preprocess the data, then extract the selected image attributes from the preprocessed data according to the artificially defined pattern, and finally develop a complex function through the image recognition modeling process to complete the image correction [4]. Machine learning algorithms can be roughly divided into three categories. One is the supervised learning algorithm. Based on the supervised learning model, you can learn or test the model from a set of training data, and then infer new cases from the model. The main algorithms are neural networks, support vector machines, and naive Bayes methods, decision tree [5]. The second category is unsupervised learning algorithms. This type of algorithm has no specific target output. The algorithm divides the data set into different groups. The third category is reinforcement learning algorithms. Decision-based reinforcement learning weighting training is common. The algorithm is trained based on whether the decision is successful or not. After extensive experience training, the algorithm can make better predictions. The algorithm is in the environment. The rewards and punishments given have gradually raised expectations for stimulus [6].

(1) Decision tree
    The decision tree is based on the structure of the tree and uses the processing methods that people use to make decisions. Moreover, the decision tree is very interpretable. From the structure of the tree, we can see which characteristics determine the decision behavior. Usually, a decision tree will have a root node, an intermediate node and a leaf node. For example, the leaf node corresponding to the decision result and the root node corresponding to the decision test sequence, the goal of training is to create a decision tree with strong generalization ability [7].
(2) BP neural network
    BP neural network is used for image classification, which is a supervised classification method. Based on the progress feedback mechanism, the BP neural network needs to set the target error of the system before training, and the weight limit of each threshold of each input level is continuously adjusted to minimize

the sum of the square errors. After training, the neural network that receives the input and output, the neural network can classify the target [8]. Image classification through neural network can get a high classification accuracy, but because it is trained by controlling the error, the recognition effect cannot reach 100%.

(3) SVM classifier

The SVM classifier can also be regarded as a neural network algorithm. Compared with the traditional neural network algorithm, the SVM classifier method appears later and the method is updated, but it is not as mature as the neural network. Support vector machine classification is a supervised classification method. The SVM method can only solve two classifications at first, but with the continuous improvement of the SVM method, it can be applied to the multi-object classification method, but the effect is not good. The SVM classification algorithm has a good recognition advantage when there are few target samples, non-linearity, and high feature dimension [9]. The SVM classification method is widely used, from the initial application in mathematics, language recognition, face recognition, to the current three-dimensional object recognition.

2.2   Multi-feature Fusion

In image recognition, a single feature is used to recognize images, which often fails to achieve good recognition accuracy. Therefore, combining multiple features of the image to recognize the image can improve the accuracy of recognition [10]. There are many feature fusion methods. The three main fusion methods will be introduced below. Each of these three methods has its own advantages and disadvantages. For different identification objects and requirements, different fusion methods are adopted.

Data-level fusion refers to the direct fusion of the image data obtained by the collector, and feature extraction and analysis of the images obtained after the fusion; the data-level fusion can retain the collected original information to the greatest extent, and can achieve high recognition accuracy, but The disadvantage is that the amount of data is too large, the processing is time-consuming and laborious, and cannot meet the real-time requirements. Data-level fusion can only fuse data obtained by similar collectors. For data or images obtained from different sources, only feature-level fusion or decision-level fusion can be used.

Feature-level fusion has a stronger purpose, which refers to combining different features into a set of features, and then using a classifier for classification.

Decision-level fusion is to use different features to train and classify respectively to obtain their respective matching values. Since each feature is independent of each other, the matching values of various features can be combined using addition rules or multiplication rules to obtain the final matching result.

2.3   Image Recognition Preparation

(1) Image preprocessing

Image preprocessing should prepare in advance for image recognition. Its purpose is to eliminate obstacles and noise in the image, increase the accuracy of effective information extraction from the image, and improve the visibility of the recognized object. Due to the existence of objective factors such as equipment, weather, lighting, and light painting techniques and methods, it will affect the quality of pixels when acquiring images. To have the best recognition effect, we must try our best to avoid the influence of irrelevant information on image pixels, restore and improve useful image information, and at the same time, detect useful information to determine image attribute data, in order to improve image processing speed. Reduce the later workload [11].

(2) Image feature extraction

Deriving image features is a key step in image recognition. Through the feature extraction, the feature classification and matching are carried out, so as to achieve the purpose of image recognition and classification. The characteristics of the image are mainly color, texture, shape characteristics and so on. Color is usually described by three relatively independent quantities. The three variables are the three components of the color, and the three variables constitute a space. Texture usually refers to a pattern arranged regularly in a certain area of an image. The image texture feature also shows the surface properties of the object. The dimension of the extracted texture feature is not large, which reduces the computational complexity of subsequent applications, has strong feature identification ability, good stability, and fast extraction and calculation speed. Compared with the color and texture characteristics, the shape feature of the image is more complicated for the expression of the meaning of the image, and the description and expression of the shape of the image lack a precise definition. The image can be extracted with the same shape after transformation. According to the image range parameters, it can be divided into methods based on external parameters and internal parameters [12].

(3) Image classification

Image classification is an operation to be performed after image processing and feature extraction, which is classified according to the characteristics of the image target. Image classification mainly refers to the design of classifiers and the realization of their functions, such as SVM classifiers.

2.4   Multi-feature Extraction

(1) Gray features

Generally speaking, two images with larger differences will have a greater difference in the mean gray value of the two images, so the mean gray value can be used as an image matching feature. The average gray value is calculated using formula (1).

$$u = \frac{\sum_{i=0}^{m} \sum_{j=0}^{n} I(i,j) \cdot II(M(i,j))}{\sum_{i=0}^{m} \sum_{j=0}^{n} II(M(i,j))} \tag{1}$$

Among them, I is a grayscale image. M is the mask image of I. I() is an indicator function. When the parameter value is 0, the function value is 0; when the parameter value is greater than 0, the function value is 1.

(2) Texture characteristics

The texture feature can describe the uneven grooves on the surface of the object, which appear as bright and dark changes in the pixel gray value on the image. Therefore, texture features can effectively describe the spatial position and intensity information of image pixels. The study of texture features is also a very important research direction in the field of image processing.

$$C(x,y) = \sum_{i=-k}^{k} \sum_{j=-k}^{k} I_1^{'}(x+i,y+j) \cdot I_2^{'}(x+i,y+j) \qquad (2)$$

Among them, $C(x,y)$ represents the correlation between two image pixels, and I_k is a grayscale image.

3   Research on Image Recognition System

3.1   Research Content

This paper first uses BP neural network to establish a multi-feature fusion image recognition system, and then shoots a natural landscape image. The image has six types of land including forest, grassland, high shrub forest, low shrub forest, bare land, and construction land. The system is used The color and texture features of these land types on the image are extracted, and finally the extracted feature values are integrated to calculate the image recognition accuracy.

3.2   Research Innovation

The innovation of this paper is that the image recognition system introduces the BP neural network model of the machine learning algorithm, which can improve the recognition accuracy by training the image data information for multiple times.

4   Analysis of Multi-feature Fusion Image Recognition System Based on BP Neural Network Model

4.1   System Construction

As shown in Fig. 1, this system accepts the pictures uploaded by users on the web, and after the pictures are resized and formatted, they are sent to the server running the BP neural network model through socket communication. The server is based on Multiview. The target object in the image is predicted, and then the prediction result is combined with the relevant knowledge information about the target object we obtained from wikipedia, and the result is returned to the web server through the socket, and the web server obtains real-time monitoring whether there is a new return The result is returned to the user. In subsequent system updates, we will add user feedback information, that is, users can judge whether the returned results are accurate and label them, and then add the images that have been labeled by users to the training data to retrain the BP neural network model.

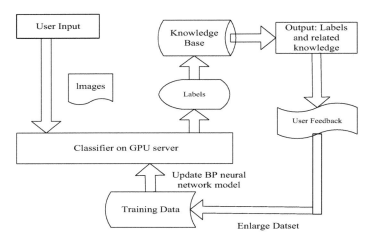

Fig. 1. Multi-feature fusion image recognition system architecture based on BP neural network

4.2   System Application

(1) Color feature fusion recognition

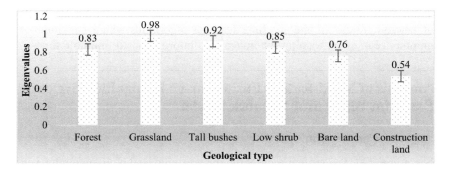

Fig. 2. Color feature vector value

Extract the color features of the image area, and the color feature values represented by the six types of features are shown in Fig. 2. The color characteristic value of grassland and high shrub forest is above 0.9, and the color characteristic value of forest and low shrub forest is above 0.8.

(2) Texture feature fusion recognition

Table 1. Texture feature extraction parameter values

	Energy	Entropy	Moment of inertia	Correlation
Forest	0.0876	2.5342	0.6867	0.5358
Grassland	0.2813	1.4770	0.2139	1.7912
Tall bushes	0.0845	2.5109	0.7214	0.5223
Low shrub	0.3042	1.4964	0.1853	1.6847
Bare land	0.3589	2.1775	0.5267	2.3586
Construction land	0.1286	0.0948	1.2764	0.0689

The four statistics of texture feature energy, entropy, moment of inertia, and correlation are used to identify and classify aerial images by taking the average of these four statistics as feature vectors. The texture feature values of various typical training samples are shown in Table 1. It can be seen from Table 1 that the various texture feature values of forest and high-shrub forest are similar, and the difference in texture feature values of low-shrub forest and grassland is relatively small, while bare land and construction land are not related to any type of texture.

(3) Recognition rate of land image based on multi-feature fusion

Table 2. Comparison of land recognition accuracy

	Recognition rate (%)
Forest	93.68
Grassland	9.52
Tall bushes	65.74
Low shrub	47.39
Bare land	82.62
Construction land	50.13

Combine the color and texture feature values of the image to recognize the image and calculate the recognition accuracy. From the recognition results of various ground objects in Table 2, we can see that the forest has the highest recognition rate and the best recognition effect, with a recognition accuracy of 93.68%; while the grassland's recognition rate is the worst, only 9.52%. Recognized as bare land. The grass is sparse and the color of the bare land is close, which seriously affects the recognition accuracy; the color of high shrub forest is in the middle of the various object categories, so the recognition results are scattered, the recognition rate is not high, and many construction land are also recognized as the recognition accuracy is not high in bare ground.

5 Conclusion

Image is a way of information expression. Recognizing image information helps people clearly recognize the content of the image. Traditional image recognition systems cannot decompose image pixels, but the image recognition system based on machine learning algorithms in this paper can analyze and process image data, and then combine the characteristics of multi-feature fusion to perform feature extraction and matching on the image to complete the image recognition process. This paper uses the system to extract the color and texture characteristics of a natural landscape image, and obtains the image recognition accuracy through neural network training. I hope that the system can be applied in more fields to help people complete image recognition.

References

1. Ogata, T., Ueyama, T., et al.: Automated field-of-view, illumination, and recognition algorithm design of a vision system for pick-and-place considering colour information in illumination and images. Sensors **18**(5), 55–59 (2018)
2. Lim, J., Baek, Y.: Low-power communication method using on-device deep neural network for low-power image recognition system. J. Korean Inst. Commun. Inf. Sci. **44**(8), 1588–1596 (2019)
3. Ergn, C., Norozpour, S.: Farsi document image recognition system using word layout signature. Turk. J. Electr. Eng. Comput. Sci. **27**(2), 1477–1488 (2019)
4. Ashaduzzaman, M., Hasan, S.M.: Recognition method for aggressive behavior of group pigs based on deep learning. Nongye Jixie Xuebao/Trans. Chin. Soc. Agric. Mach. **35**(23), 192–200 (2019)
5. Demir, H.S., Christen, J.B., Ozev, S.: Energy-efficient image recognition system for marine life. IEEE Trans. Comput. Aided Des. Integr. Circ. Syst. **39**(11), 3458–3466 (2020)
6. Et, A.: Automated bird species recognition system based on image processing and svm classifier. Turk. J. Comput. Math. Educ. (TURCOMAT) **12**(2), 351–356 (2021)
7. Maggay, J.G.: Mobile-based eggplant diseases recognition system using image processing techniques. Int. J. Adv. Trends Comput. Sci. Eng. **9**(1.1 S I), 182–190 (2020)
8. Sl, B.: Multidomain feature fusion for varying speed bearing diagnosis using broad learning system. Shock. Vib. **2021**(5), 1–8 (2021)
9. Beyaz, A., Ozlu, S., Gerdan, D.: Experimental recognition system for dirty eggshell by using image analysis technique. Turk. J. Agric. - Food Sci. Technol. **8**(5), 1122–1126 (2020)
10. Karimi, D., Akbarizadeh, G., Rangzan, K., et al.: Effective supervised multiple-feature learning for fused radar and optical data classification. IET Radar Sonar Navig. **11**(5), 768–777 (2017)
11. Majumder, A., Behera, L., Subramanian, V.K.: Automatic facial expression recognition system using deep network-based data fusion. IEEE Trans. Cybern. **48**(99), 103–114 (2017)
12. Fakhrurroja, H., Machbub, C., Prihatmanto, A.S., et al.: Multimodal fusion algorithm and reinforcement learning-based dialog system in human-machine interaction. Int. J. Electr. Eng. Inform. **12**(4), 1016–1046 (2020)

Prediction of the Path Level of Carbon Emission Reduction Based on the IPAT Model from the Computer Vision

Yaru Shen[1(✉)], Shuai Hu[1], Haizhong Ma[2], Peng Huang[2],
and Junxian Ma[1]

[1] State Grid Ningxia Electric Power Corporation Limited Economic and
Technical Research Institute, Yinchuan 750004, China
a2233177650@126.com
[2] State Grid Ningxia Electric Power Company, Yinchuan 750001, China

Abstract. Currently, global warming (GW) problems have become a hot spot that humans have to pay attention. Carbon dioxide is the most important greenhouse gas and the main cause of GW. In response to the huge environmental pressure, countries around the world have begun to explore the path of carbon emission reduction (CER) in order to control the excessive emission of carbon dioxide gas. IPAT model plays an important role in the research of CER field. This article starts from the computer perspective to study the CER path level prediction based on the IPAT model. This article introduces the expression and meaning of the classic IPAT model, selects the random model STIRPAT model of the IPAT model, and constructs the CER path level prediction model. This article takes the CER path of H province as the research object, predicts the CER effect of the province in different situations, and derives the CER path policy enlightenment. The survey data shows that under the balanced emission reduction scenario, the total carbon dioxide emissions of H province will reach a peak before 2029, with a peak of 525.23 million tons. This shows that H province should attach importance to the suppression of carbon dioxide emissions by technological innovation, energy structure adjustment, and industrial structure optimization to improve energy efficiency.

Keywords: IPAT model · Carbon emission reduction · Carbon emission forecast · Carbon emission influencing factors

1 Introduction

At present, the excessive emission of carbon dioxide, mainly caused by industrialization construction, has caused problems such as GW, and is highly valued by countries all over the world [1, 2]. In response to this problem, countries have formulated and implemented corresponding CER goals and measures [3]. China is the world's largest developing country and a major climate negotiation party, and shoulders a huge responsibility for emissions reduction [4]. The IPAT model can reflect the impact of social and economic development on the environment, and is the most commonly used and recognized model for studying carbon emission (CE) and CER

© The Author(s), under exclusive license to Springer Nature Switzerland AG 2022
V. Sugumaran et al. (Eds.): ICMMIA 2022, LNDECT 138, pp. 416–423, 2022.
https://doi.org/10.1007/978-3-031-05484-6_52

path issues [5, 6]. From the computer perspective, the research on CER path level prediction based on the IPAT model has attracted more and more attention.

Regarding the research on CER, many scholars have conducted multi-angle explorations. For example, Wang B took China as the research object and studied the influencing factors of the country's CE [7]; Tian X took the Shanghai area as an example to study the competitiveness of China's CER in the industry [8]; Starting from international environmental protection, Xia X proposed a CER strategy to strengthen the integration of upstream and downstream in the industrial chain [9]. It can be seen that the research on CER has always attracted much attention. From the perspective of computer, this article has carried out research on the prediction of CER path level based on the IPAT model, and proposed a carbon emission prediction model and CER strategy, which are of practical significance.

This paper studies the prediction of CER path level based on the IPAT model under the computer vision. This article first introduces the expression and meaning of the classic IPAT model, selects the random model STIRPAT model of the IPAT model, and constructs the CER path level prediction model. Then, this article takes the CER path of H province as the research object, and predicts the CER effect of this province in different situations. Finally, this article draws the CER path policy enlightenment.

2 CER Path Level Prediction Based on IPAT Model Under Computer Vision

2.1 IPAT Model

The calculation method of environmental pressure specified by the classic IPAT model is shown in formula (1):

$$I = P * A * T \tag{1}$$

In the formula, I represents environmental pressure, usually expressed in terms of CE; P represents population size; A represents wealth per capita; T represents technical level, usually expressed in terms of energy consumption intensity.

This paper selects the random model STIRPAT model of the IPAT model, and its specific form is shown in formula (2):

$$I = aP^b * A^c * T^d e \tag{2}$$

Among them, a is the coefficient of the model, b, c, and d are the indexes of humanistic driving force such as population, affluence, technology, etc., and e is the model error. The IPAT model is a special form of the STRIPAT model. The introduction of the index makes the model can be used to analyze the non-proportional effects of multiple environmental pressures of various driving factors.

2.2 Carbon Emission Prediction Model Based on IPAT Model

The five factors of population, GDP per capita, urbanization rate, energy intensity, and CE per unit of energy have a very large impact on CE. Therefore, this paper selects these 5 factors as the influencing factors of CE to construct an extended STIRPAT model, whose expression is shown in formula (3):

$$\ln I = \ln a + b(\ln P) + c(\ln A) + d(\ln T) + e(\ln U) + f(\ln E) + \ln g \qquad (3)$$

In the formula, I means CE, P means population, A means GDP per capita, T means energy intensity, U means urbanization rate, E means CE per unit of energy, and a means model coefficient. b, c, d, e, f correspond to the elasticity coefficients of population, GDP per capita, energy intensity, urbanization rate, and CE per unit of energy, respectively, and g is a random error term.

2.3 CER Path Policy Enlightenment

(1) In-depth understanding of the current carbon emission stage of the country, and formulate targeted carbon emission measures. At different development stages, the effects of the influencing factors of CE are also different. Therefore, in order to achieve effective control of CE at a certain stage, it is necessary to clarify the CER stage that the country is currently in, and then formulate measures to control CE based on the key factors affecting CER at this stage to achieve effective control of CE.

(2) Control the population. In all stages of CER, population is the main influencing factor. In the early stage, CE were affected mainly through labor supply and other channels, and the population size needed to be controlled [10, 11]. However, considering the development demands of various countries in the initial stage and the relatively small energy consumption of residents in the initial stage, the feasibility and practical significance of controlling the population are limited. In the later stage of CE, the amount of CE generated by domestic household consumption is relatively high, and the CE effect of an aging population is obvious. Therefore, during this period, it is more necessary to control the population size and structure in advance to alleviate the impact of CE caused by residential consumption and population aging.

(3) Strengthen international technical support and cooperation. Regardless of whether it is a developed or a developing country, the carbon emission problem brought about by economic construction is very obvious, but it is impossible in practice to achieve CER by reducing economic growth. Therefore, in the early stage of CER, countries need to reduce CE through international technical support and cooperation [12]. In the later stage of CER, the CER effect produced by technological progress will be more obvious. Therefore, international cooperation and support, through the introduction of low-carbon and environmentally friendly technologies, can improve the CER effect of countries in the process of economic development as much as possible, and enable CER to develop together with the economy, society, and resources. However, the proposal requires the mutual cooperation of

countries around the world, especially the support of developed countries in environmental protection technology of developing countries, so it is necessary to encourage mutual cooperation and assistance between countries from a global perspective.

(4) Trade globalization has formed a carbon transfer effect, which is manifested in the fact that countries with low economic growth and high CE are under the pressure of "carbon pollution". To solve this type of problem, we can proceed from the following two aspects:

From a global perspective, the carbon transfer effect will not only reduce global CE, but will increase CE due to the backward energy use efficiency of the transferred countries, resulting in unnecessary CE. This phenomenon is caused by the inevitable law of economic development and requires some artificial adjustments. From an economic point of view, increasing the cost of carbon transfer can help alleviate the problem, but it also has a certain impact on the economic development of the transferred country, which is not in line with the concept of sustainable development. Therefore, the formation mechanism of this problem is complicated, and the measures to solve this phenomenon need more detailed argumentation.

(5) The CER effect brought by the adjustment of the industrial structure during the rapid growth and decline of CE is limited, and the adjustment of the industrial structure as soon as possible can maximize the CER effect. Therefore, all countries must accelerate the adjustment of industrial structure and develop the service industry and high-tech as the main tertiary industry as soon as possible in order to effectively tap the CER role of the industrial structure, compress the carbon emission space of the industrial structure, and effectively realize the CER effect of the industrial structure.

3   CER Path Level Prediction Experiment Based on IPAT Model

3.1   Research Objects and Tools

This article takes the carbon emission forecast of H province as the research object, uses 2017 as the base year, and is based on the data provided by the H province statistical yearbook. The ridge regression method is selected to fit the carbon dioxide emissions of H province, and the carbon emission regression model of H province is constructed, and finally the STIRPAT model expression of H province is obtained as shown in formula (4):

$$\ln I = 2.184\ln b + 1.011\ln c + 0.449\ln d + 1.052\ln e + 1.03\ln f + 9.174 \qquad (4)$$

In the formula, b, c, d, e, and f respectively represent the elasticity coefficient of population, GDP per capita, energy intensity, urbanization rate and CE per unit of energy.

3.2   Experiment 1

From 2008 to 2020, the energy-related carbon dioxide emissions of H province continued to rise, especially from 2011 to 2015, the carbon dioxide emissions of H province increased rapidly. After 2016, the growth rate of carbon dioxide emissions in province H began to gradually slow down. Therefore, based on the historical trend of CE in H province from 20058 to 2020, the GM (1, 1) model is used to predict the CE in H province from 2022 to 2030.

3.3   Experiment 2

(1) Planning scenario assumption
   This section uses the extended STIRPAT model to predict the carbon dioxide emissions of each province from 2022 to 2030. Among them, the average annual growth rate of each influencing factor is calculated from the indicators of the corresponding policies of H province. Since the data of unit energy CE and per capita GDP in 2022–2030 have not yet been obtained, this paper uses the grey forecast method to fill in the data.

(2) Scenario assumptions for energy conservation and emission reduction
   This scenario does not adjust the economic, urbanization, and population-related policies of H province, but only adjusts energy-related policies. For example, H province will increase the implementation of energy policies, actively adjust the energy structure, and promote technological progress.

(3) Assumptions of balanced emission reduction scenarios
   This scenario refers to the annual average growth rate of population, per capita GDP, and urbanization rate by adjusting the population, economy, urbanization and energy-related policies of province H on the basis of the planned scenario, and the annual unit energy CE and energy intensity. The increase in the average reduction rate will enable the overall CE of H province to reach its peak in 2029. In this context, H province focuses on green development and can fully consider future social, economic and environmental development needs, implement CER, optimize industrial structure and other measures to promote green development.

4   Analysis of CER Path Level Prediction Results

4.1   H Province Carbon Dioxide Emission Analysis

The collected CE data from 2008 to 2020 in H Province was sorted and a model was established. The original data and accumulated data were used for error analysis. The results are shown in Table 1:

Table 1. Comparison of carbon emission forecasts (ten thousand tons)

Years	Actual value	Predictive value	Relative error (%)
2008	23931	23931	0
2009	24923	24951	0.11
2010	26747	26762	0.06
2011	31788	31802	0.04
2012	35274	35292	0.05
2013	40272	40304	0.08
2014	44530	44571	0.09
2015	47820	47860	0.08
2016	48744	48798	0.11
2017	49833	49879	0.09
2018	50790	50818	0.06
2019	51902	51931	0.06
2020	52172	52204	0.06

As shown in Fig. 1 that in the carbon emission prediction, the GM(1,1) model has a good predictive ability, and the relative error range of the prediction is 0.04%–0.11%, indicating that the GM(1,1) model can be used the carbon dioxide emissions forecast for H province is in progress.

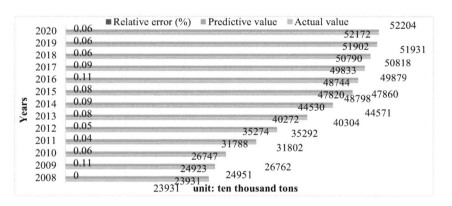

Fig. 1. Comparison of carbon emission forecasts (ten thousand tons)

4.2 Prediction of Carbon Dioxide Emissions in Province H Under Various Scenarios

The prediction results of the carbon dioxide emissions of H province under various scenarios from 2022 to 2030 are shown in Table 2: Under the planning scenario, the CE of H province in 2022 and 2030 will reach 525.84 million tons and 535.49 million tons respectively; under the control scenario, In 2022 and 2030, H province's carbon dioxide emissions will reach 524.7 million tons and 538.23 million tons, respectively. Under the balanced emission reduction scenario, the carbon dioxide emissions of H province in 2022 and 2030 will reach 522.68 million tons and 524.19 million tons, respectively.

Table 2. Projections of carbon dioxide emissions in province H under each scenario

Years	Planning scenario	Control scenario	Balanced emission reduction
2022	52584	52470	52368
2023	52640	52543	52416
2024	53692	52581	52431
2025	53737	53614	52462
2026	53789	53664	52497
2027	53834	53721	52519
2028	53880	53750	52520
2029	53927	53792	52523
2030	53949	53823	52419

As shown in Fig. 2, the growth rate of CE in province H will slow down in the future. Compared with the baseline scenario, the slowdown in the growth rate of CE under the planning scenario is more pronounced. And under the balanced emission reduction scenario, the total carbon dioxide emissions of H province will reach a peak before 2029, with a peak of 525.23 million tons. Therefore, H province should pay attention to the suppression of carbon dioxide emissions by technological innovation, energy structure adjustment, and industrial structure optimization to improve energy efficiency.

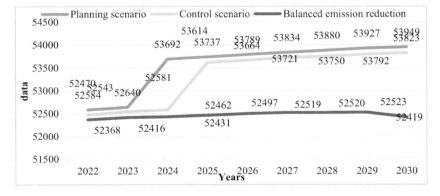

Fig. 2. Projections of carbon dioxide emissions in province H under each scenario

5   Conclusions

There is no doubt about climate change. Human activities consume a lot of energy, leading to a rapid increase in CE, which is not conducive to environmental protection. This paper mainly establishes a carbon emission prediction model based on the IPAT model, and uses the STIRPAT model to predict the carbon emission of H province from 2008 to 2020. Based on the future social, economic, energy and other development plans of H province, this paper conducts a scenario analysis of the peak situation in 2029 and discusses the CER path of H province.

References

1. Chontanawat, J.: Decomposition analysis of CO2 emission in ASEAN: an extended IPAT model. Energy Proc. **153**(Complete), 186–190 (2018)
2. Hwang, Y.S., Um, J.S., Schlüter, S.: Evaluating the mutual relationship between IPAT/Kaya identity index and ODIAC-based GOSAT fossil-fuel CO2 flux: potential and constraints in utilizing decomposed variables. Int. J. Environ. Res. Public Health **17**(16), 5976 (2020)
3. Elsayed, G.F., Shankar, S., Cheung, B., et al.: Adversarial examples that fool both human and computer vision. **7**(1), 1 (2018)
4. Baghdadi, A., Cavuoto, L., Hussein, A.A., et al.: PD58-04 modeling automated assessment of surgical performance utilizing computer vision: proof of concept. J. Urol. **199**(4), e1134–e1135 (2018)
5. Khan, S., Rahmani, H., Shah, S., et al.: A guide to convolutional neural networks for computer vision. Synthesis Lect. Comput. Vis. **8**(1), 1–207 (2018)
6. Minu, S., Shetty, A.: Prediction accuracy of soil organic carbon from ground based visible near-infrared reflectance spectroscopy. J. Indian Soc. Remote Sens. **46**(5), 697–703 (2017). https://doi.org/10.1007/s12524-017-0744-0
7. Figuero, A., Rodriguez, A., Sande, J., et al.: Dynamical study of a moored vessel using computer vision. J. Mar. Sci. Technol. **26**(2), 240–250 (2018)
8. Maggipinto, M., Terzi, M., Masiero, C., et al.: A computer vision-inspired deep learning architecture for virtual metrology modeling with 2-dimensional Data. IEEE Trans. Semicond. Manuf. **PP**(99), 1 (2018)
9. Brunetti, A., Buongiorno, D., Trotta, G.F., et al.: Computer vision and deep learning techniques for pedestrian detection and tracking: a survey. Neurocomputing **300**(jul.26), 17–33 (2018)
10. Mcroberts, R.E., Næsset, E., Gobakken, T.: Comparing the stock-change and gain-loss approaches for estimating forest carbon emissions for the aboveground biomass pool. Can. J. For. Res. **48**(12), 1535–1542 (2018)
11. Setiawan, F.B., Wijaya, O., Pratomo, L.H., et al.: Sistem navigasi automated guided vehicle berbasis computer vision dan implementasi pada Raspberry Pi. J. Rekayasa Elek. **17**(1), 7–14 (2021)
12. Bhattacharjee, S.: Prediction of satellite-based column CO concentration by combining emission inventory and LULC information. IEEE Trans. Geosci. Remote Sens. (sep.), 1–16 (2020)

Design of Mobile App Interactive Interface Based on 3D Image Vision

Li Ma[(⊠)]

College of Internet and Communication, Anhui Technical College of Mechanical
and Electrical Engineering, Wuhu 241002, Anhui, China
maives@126.com

Abstract. With the development of science and technology and the advent of
the 5G era, smartphone apps are being integrated into our daily lives, and even
most of the daily activities need to be implemented by mobile apps. The
importance of human-computer interaction in the era of big data is increasing
day by day. Faced with this situation, the design of interactive interfaces needs
to get rid of solidified thinking and transform to emotional and personalized.
The user's innovative demand for APP interface is also the main driving force
for the development and innovation of interactive interface design. This paper
analyzes the design method of mobile APP interactive interface based on the
visual perspective of 3D images. First, through the combing of related concepts,
the scope of the 3D image and the level of coverage of the mobile APP interactive interface are clarified; then, the factors affecting the user's satisfaction
with the APP interactive interface are analyzed to analyze the priority of the
design of the APP interactive interface to the involved elements; finally The
literature analysis method and the case analysis method focus on the analysis of
examples of 3D image vision in mobile APP interactive interface design. It not
only mentions the widely used 3D vision methods, but also elaborates the
cutting-edge 3D technology methods that are still in the research stage. This
article aims to explore the new style and future development trend of interactive
interface design, in order to pave the way for future research.

Keywords: Three-dimensional image · Interactive interface · Visual update ·
Smart phone application design

1 Sorting Out Related Concepts

1.1 Three-Dimensional Image Vision

The development and growth of 3D vision technology is not only a major achievement
in academia, but also has been widely used in real life. Recently seen unmanned control
systems, robots, AR (augmented reality), VR (virtual reality) and other intelligent
unmanned systems have also benefited from the development of three-dimensional
vision. On the one hand, these applications are rooted in 3D vision technology, and on
the other hand, they also accelerate the development of 3D vision technology. For
example, the three-dimensional data that needs to be collected in three-dimensional
vision technology comes from the popularization of a wide range of sensors in life.

V. Sugumaran et al. (Eds.): ICMMIA 2022, LNDECT 138, pp. 424–432, 2022.
https://doi.org/10.1007/978-3-031-05484-6_53

At first, ordinary users could not own these devices. The earlier the device is produced, the larger the body size, and the average person cannot use it in life. Sensing equipment has ranged from lidars used in airplanes to small three-dimensional scanners in cars to mobile phones held by everyone today. With the development of the times, the size of the sensor is gradually shrinking. Just like the depth camera of an Apple mobile phone, ordinary people can also have a sensor with three-dimensional data. As a discipline, 3D vision is the intersection and fusion of several disciplines, including computer vision, graphics and artificial intelligence. Many elements of artificial intelligence are built into three-dimensional vision. Therefore, artificial intelligence will play an increasingly important role in it. As shown in Fig. 1, the research content related to 3D vision mainly includes 3D perception, posture perception, 3D modeling, and even 3D cognition [1].

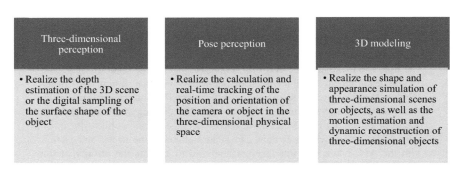

Fig. 1. Three-dimensional image research scope

1.2 Smart Phone App Interactive Interface

Advances in science and technology have gradually improved the functions of smart phones, making them ubiquitous and becoming an indispensable part of people's daily lives. With the popularity of smart phones, the man-machine interface of applications has become more complex, and the interaction between users and APP has become more difficult. According to iiMedia Consulting's statistics on the Chinese application market in 2020, many applications have very little user retention, and there is a development trend of "throw away when used up" [2]. Therefore, how to design the APP interface and how to improve the APP interface evaluation system is particularly important. The appeal of APP to users lies in the ergonomic design of the APP interactive interface, that is, the technical design of human factors and the unique functional design. If the APP interface can be carefully designed from the perspective of human factors engineering, the human factors are highly compatible and suitable for users (see Table 1). The corresponding user score will be very high, so that such an APP can quickly occupy the market. Otherwise, it will be replaced by apps with similar functions, and then eliminated by the market.

Table 1. Influencing factors of human suitability of APP interactive interface

Factor	Index
Interface design	Reasonable interface layout Operating habits Element design Color matching
Ease of use	Operation convenience Prompt clarity Help practicality Recommendation accuracy Rapid feedback Shortcuts Sharing information
Feedback	Error prompt Illegal operation prompt Unrecoverable operation prompt
Functionality	Function meets the needs The guiding role of the elements The rationality of the function
Reliability	Account security Privacy protection User permission settings and changes Reliability of links Reliability of elements/controls reliability of interface switching
Satisfaction	Operation feedback form Content presentation interaction Prompt information form Recommendation desire

2   Analysis of Influencing Factors of Smart Phone App Interactive Interface Design

2.1   User Habits and Operational Thinking

Human memory is limited and varies from person to person. If the purpose is to reduce the burden on users and improve work efficiency, then the consistent design of the interactive interface is currently the most effective method. For example, you can use commands and dialog boxes of the same structure to delete folders, text, and graphics. The training process and methods will affect the efficiency and usage. Therefore, reducing the user's memory burden can enhance the use efficiency and form an unconscious use habit. First, smart phone applications are usually pre-configured to predict user behavior, reduce user operational thinking and memory, and make operational behavior unconscious and intuitive [3, 4]. Secondly, the operation of smart phones is greatly affected by environmental factors. Users usually use mobile phones during fragmented time. Therefore, designers need to follow the user's habits when designing the interface, simplify the amount of reading, and reduce the user's reading time. The design also requires the design of graphics instead of text in the interactive interface, including memo buttons, fields, navigation bars, list boxes, icons, and even fields and controls, so as to solve the problem of use.

2.2 IOS and Android Operating Systems

The design of smart phone app interactive pages is greatly influenced by the operating system. Today, iOS and Android systems are the two systems with the highest market share, providing users with a variety of services, and each has its own advantages and characteristics.

For the design of app interactive interfaces of different systems, it is necessary to compare the characteristics of the two systems from the perspective of interactive behavior analysis, and summarize the general characteristics and methods. Users of the Android system can perform tasks in a variety of ways. For example, Huawei phones can support downloading apps from the Internet or download apps through the app market. The function settings of the Android system are more open. However, it can also cause problems such as system speed lag, excessive running steps, and secondary pages. Apple iOS system is generally based on guided thinking for functional configuration, users usually perform tasks based on limited methods and operating methods, which limits the user's choices to a certain extent. Therefore, the iOS system provides a clear operation flow. Compared with the Android system, the iOS system reduces a lot of sub-pages and operation steps, and avoids the cognitive interference and selection burden caused by humans. However, the interaction design model is too fixed and conservative, and it is far inferior to the Android system in terms of customization, freedom of choice and flexibility [5].

2.3 Multi-channel Interaction

Psychological channel refers to the way that human senses transmit and retrieve information. Common channels include visual, auditory and sensory channels. The channel in human-computer interaction refers to all the information exchange methods used between humans and intelligent machines. The multi-channel interactive mode allows mobile smartphone applications to recognize the information they represent and provide feedback to users in various media formats. There are two types of input channels and output channels [6].

(1) Visual hierarchy
    The important function of the eye is to obtain information and occupy an important part of interface interaction, and its importance is self-evident. The types of external information that people can visually perceive include images, text, graphics, symbols, and codes, and the content includes shapes and colors. Vision can be used to search, detect, identify, identify, and confirm. It is the process of visual thinking behavior that processes external information. Based on this visual thinking process, the smart phone interface design methods and precautions include visual feedback such as how the information levels of the created app interface are distributed, whether the graphic symbols are concise and easy to understand, and

whether the interface color is clean and tidy. The more organized and concise the interface information is displayed on the smartphone, the faster and easier the user can obtain the information. The scenes in which people use mobile phones are mostly fragmented, and information is easily overlooked by vision. Therefore, visual design needs to refine the interface. A structured visual hierarchy design mainly requires the processing, placement and integration of visual elements, including text, icons, graphics, grids, navigation tools and other interface elements [7, 8].

(2) Fingertip touch

Gestures are at the core of the mobile experience. The fingertips have a variety of tactile peripheral organs, and the ability of tactile perception is strong. In addition to communication, the main function of the hand can also receive tactile signals and execute brain commands. With the introduction of smart phone multi-touch technology, 3D pressure detection and other technologies, the tactility of mobile phone users continues to improve. When users perform gesture operations, designers need to consider the visual design of the interface, choose appropriate UI controls and effective visual feedback. Many smartphone applications are trying to incorporate gestures into their interface design. APP Clear is a typical example. Its interface does not contain UI elements dedicated to navigation and navigation functions. All operations must be performed with gestures such as gesture shooting and multi-point zooming [9].

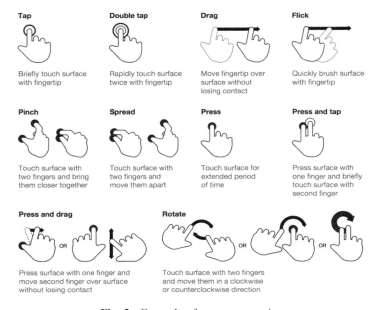

Fig. 2. Example of gesture operation

(3) Sound feedback

The so-called natural sound is not the sound produced by the device itself, on the contrary, it is the sound that occurs in the natural environment. This means that any collision, pinch, click, etc. between the user and the phone will naturally produce different sounds. You can refer to the working principle of a smart phone. These voices allow users to have a complete understanding of this information. Users can perceive materials (metal, leaves, glass) and space through sound. The iOS operating system consists of a specialized sound system, such as telephone ringtones. Send and receive messages from the keyboard and other voice input, allowing Apple mobile phone users to receive voice and pre-allocated information (Fig. 2). This is a new way of mobile artificial intelligence. For example, the Hello Siri function of the iOS system is aimed at the user's functional requirements (call or send a message to a specific contact). Through human-computer interaction and voice recognition system, although sound has signal transmission function, it may also cause confusion to users. For example, noise interference caused by public places is likely to affect the user's voice input. So the challenge for designers is how to use sound perception levels reasonably.

3   Interactive Page Application for 3D Image Visual Design

3.1   AR Design of APP Interactive Interface

In recent years, in the development trend of APP interactive interface design, AR graphics occupy a very important position. However, how to integrate various AR graphics into mobile and web interfaces is a considerable challenge, requiring both skills and artistic vision. The advantage of AR graphics is that it can attract users, and the realistic AR rendering images look gorgeous and real, which will enhance the sense of reality of APP use. When the content required by the user is not available or the APP charges are too expensive, this kind of graphic design may prolong the user's attention and increase the download rate and retention rate. This type of technology is most widely used in digital game app interactive interfaces. It not only combines real-time and interactivity, but also uses realistic AR images to restore the background of the story, enhance the reality of user interaction, have more visual appeal and affinity, and have a higher sense of user immersion [10]. Take the mobile game "Talking Tom Cat" developed by Outfit7 in July 2010 as an example. The game is available in IOS and Android versions. This AR interactive game application uses AR images as a selling point, and uses voice and gesture interaction to give users a new experience, Which has been downloaded more than 80 million times. In the game, players can interact with AR-designed kittens through different three-dimensional navigation buttons, which is very interesting.

Fig. 3. AR design application example

AR interactive design has been well received by users since its inception (see Fig. 3). It is also more effective than traditional icon graphics when using 3D graphics and charts to capture visual information. And there are new discoveries in social, scientific and professional applications. However, as shown in Table 2, AR images have the best distance for viewing, and if the distance is not appropriate, people will feel dizzy and dizzy. In order to further explore the optimization and upgrade of users' visual effects, the design must correct and eliminate 3D visual defects, and present users with a perfect app interactive interface design.

Table 2. The best distance when viewing AR images

Distance	Effect
0.5 m	When the text is too close to you it will make it difficult for your eyes to focus, especially when moving between the near focal plane and the far focal plane
1 m	This is the closest distance to maintain good readability of the interface, but over time, such close text can still cause eye fatigue
1.5 m	The text can be read comfortably, but switching the focus between near and far may still cause eye fatigue
2 m	When the text is farther away, the three-dimensional effect will be reduced, but this helps reduce eye fatigue. Starting from 2 m, the subject is easier to focus (the final reading effect depends on which VR lens is used)
3 m	This is considered a better interface display distance. It is not only clear and comfortable to read, but also does not interfere with most scenes
6 m	It is possible to maintain the readability of the interface at a longer distance, but the occlusion of the front object to the interface will reduce the readability of the text. If it is not occluded, it may make the user feel a little strange

3.2   Holographic Gesture Control

From multi-touch to gestures, from two-dimensional (2D) to three-dimensional (AR), human-computer interaction is evolving in a more natural and intuitive way. The combination of the holographic gesture control interface and the APP interactive interface allows users to directly control the APP from the air, making the operation easier and more enjoyable [11]. Interactive holographic displays can provide vivid, accurate views and a natural interactive experience. At present, gesture control can be divided into three main types: wearable device-based, visual recognition and AR touch-based interaction [11]. Hand-held device-based gesture recognition is very accurate, but the limitations of hand-held devices make it difficult to perform comfortable operations. Vision-based and AR-based touch interaction methods allow users to control apps naturally without wearing a device. Early gesture recognition systems were inseparable from handheld devices. The oldest touch interaction method appeared in 1997. However, due to the limitations of computer capabilities and algorithms at the time, the dialogue process was slow. Later, data gloves equipped with accelerometers or geomagnetometers appeared, and people used them to collect user gesture signals, and then used Hidden Markov Model (HMM) algorithm for gesture recognition [12].

4   Summary

Science and design are not a hostile relationship. On the contrary, from the perspective of the APP interface, science and design have always been in a state of mutual influence and common progress. In recent years, smart phone apps have completed the transition from complex and redundant design to three-dimensional simplicity, which has also promoted the development of mobile devices. With the research and development of interactive user experience, the scope of APP interface designers' considerations has also expanded to humans. Research on visual and physiological and psychological mechanisms. In this paper, the research on the design of the APP interactive interface for 3D image vision is based on the human biological system, extending to real examples, and finally attributed to the prospect of the future development trend of interactive interface design. The design of the interactive interface has changed from two-dimensional, 2.5-dimensional and finally to three-dimensional. The development of the design concept shows that the design of the interactive interface should conform to the development of science and technology, and maintain high sensitivity and consistency to the development of contemporary science and technology. Ultimately maintain the relative market advantage of APP.

Acknowledgments. This work was supported by 2017 Anhui Province University Humanities and Social Sciences Research Key Project "Wuhu Iron Painting Digital Protection and Inheritance Research" (SK2017A0809).

References

1. Song, Z., Jin, S.: Research on the Design of mobile app product—taking 'wei weather' APP as an example. In: Proceedings of 4th International Conference on Wireless Communications and Applications (ICWCA 2020), Part1, pp. 60–67. Springer, Cham (2020)
2. Ren, Q.: Design of mobile APP user behavior analysis engine based on cloud computing. J. Phys. Conf. Ser. **1533**(2), 022092 (2020)
3. Porta, M.: Vision-based user interfaces: methods and applications. Int. J. Hum. Comput. Stud. **57**(1), 27–73 (2002)
4. Chen, Y., Zou, W., Sharma, A.: Graphic design method based on 3D virtual vision technology. Recent Adv. Electr. Electron. Eng. (Formerly Recent Patents Electr. Electron. Eng.) **14**(6), 627–637 (2021)
5. Wei, Y., Qian, C., Li, J.: Modular design of mobile APP interface based on the visual flow. Autom. Control. Comput. Sci. **53**(1), 56–62 (2019)
6. Anonymous. New year...New App! Download the ELFA mobile app and win! Equip. Leas. Financ. **37**(1), 9–19 (2021)
7. Prakash, M., Aparna, S., Srivastava, Y.: Enhanced image vision and resolution during low light conditions using GANs. Int. J. Recent Technol. Eng. (IJRTE) **9**(1), 1544–1547 (2020)
8. Argyros, A.A., Lourakis, M.I.A.: Vision-based interpretation of hand gestures for remote control of a computer mouse. In: European Conference on Computer Vision, pp. 30–39. Springer, Heidelberg (2019)
9. Samsung Electronics Co. Ltd.: Image vision processing method, device and equipment. In: Patent Application Approval Process (USPTO 20180137639). Electronics Newsweekly, pp. 682–689 (2018)
10. Murthy, G.R.S., Jadon, R.S.: A review of vision based hand gestures recognition. Int. J. Inf. Technol. Knowl. Manag. **2**(2), 405–410 (2009)
11. Anonymous. DISNEY+ was most-downloaded U.S. Mobile app in Q4. Broadcast. Cable **150**(1) (2020)
12. Shu, Y., Xiong, C., Fan, S.: Interactive design of intelligent machine vision based on human–computer interaction mode. Microprocess. Microsyst. **75**, 30–59 (2020)

Sports Performance Prediction Model Based on Stochastic Simulation Algorithm

Zhou Yang[1(✉)] and Cheryl[2]

[1] Physical Education Department, Dalian Polytechnic University,
Dalian 116034, Liaoning, China
yzl8900997788@163.com
[2] Physical Education Department, University of Toronto,
Toronto M5S 2E8, Canada

Abstract. With the continuous in-depth study of random theory, the description of various events in life is becoming more and more scientific. The methods provided by stochastic process theory are very important to mathematics. The purpose of this article is to study sports performance prediction models based on random simulation algorithms. Introduce the stochastic simulation based on Markov process, firstly give a general overview of the problems involved in the Markov process, and then construct the prediction model of the stochastic simulation algorithm in detail. Track and field performance is the essential test of athletic ability. With the help of the complementarity between the advantages of Markov and the stochastic simulation algorithm, the improved Markov stochastic simulation algorithm prediction model analyzes and predicts the annual best results of the women's 1000 m in M college. The experimental results show that it is similar to the gray GM prediction model. In comparison, the accuracy is improved to 98.20% .

Keywords: Stochastic simulation · Simulation algorithm · Sports performance · Predictive model

1 Introduction

Sports performance prediction refers to the prediction and conclusion of many future or unknown factors in the sports field [1, 2]. Sports performance plays a very important role in the development of the entire sports industry or understanding the development culture of the sports industry. It provides an important decision point for decision makers. The field of sports performance prediction is very wide. Macro strategic forecasts include Olympic strategic forecasts, sports system development forecasts, and sports industry forecasts [3, 4]. Unsurprisingly, it includes the prediction of the scientific management of sports teams, the prediction of the results of major events, and the prediction of the management of the sports industry [5, 6].

With the deepening of people's understanding of sports science prediction, the system structure of sports prediction methods will be further improved. Cao P proposed a solution for the transmission and utilization of athletes' sports data based on decentralized blockchain. In addition, the available athlete motion data generated in past exercises or test activities is usually sparse and time-related, which requires powerful and time-conscious data fusion and processing solutions. In this case, they use the HMM model to cope with the sparseness and dynamics of the data, so as to further make accurate predictions of sports performance for athletes. Finally, they designed a set of experiments on real-world data sets to verify the feasibility of their proposal in terms of effectiveness and efficiency [7]. Deo RC uses a computationally efficient artificial intelligence (AI) model called Extreme Learning Machine (ELM) to analyze patterns embedded in continuous assessment to simulate weighted scores (WS) and exams (EX) Score. Advocate the first to implement robust artificial intelligence methods to reveal the relationship between students' learning variables, and develop teaching interventions and curriculum health checks to solve problems related to the results of graduate students and the learning attributes of students in higher education departments [8]. In short, the perfection of the sports forecasting system will surely bring unprecedented development prospects to sports forecasting.

This article first classifies the initial data of the annual best women's 1000-m sprint performance of the M Academy, and uses Markov process theory to develop a random simulation algorithm prediction model. Secondly, the accuracy test of the prediction model of the standard random simulation algorithm is carried out, and the test method adopts the test method of average relative error in this article. Third, through the introduction of the annual best performance gray GM model (1.1) of women's 1000 m M college women's track and field events for gray prediction analysis, and finally established an improved random simulation algorithm prediction model. Scientific prediction and analysis of future development of sports performance. It can provide reference materials for the scientific training and sustainable development of the track.

2   Research on Sports Performance Prediction Model Based on Random Simulation Algorithm

2.1   Stochastic Simulation Based on Markov Process

In the real world, many surprises will change and develop over time [10]. The random events discussed in basic probability processes and mathematical calculations are usually described by one or final number of random variables, and the test results are generally described by one or final number of real numbers [11]. But in fact, when writing these events, since each event changes randomly, one or a limited number of random variables cannot explain all the mathematical laws of these random phenomena. In other words, when studying natural or socioeconomic events, the things that need to be investigated are not only random, but also a process of change. For an infinite family of random variables, at this time, in-depth study of the application of the principle of randomness is essential [12].

In practice, there is a random process: the state of the process or system at t0 is known, and the conditional distribution of the process state at t > t0 has nothing to do with the state of the process before time t0. Simply put, the "future" state of the process has nothing to do with the "past" state. Markov property is a very important concept in probability theory. When a random state and all previous states are given in a random process, the conditional probability distribution of its future state depends only on the current state.

2.2   Stochastic Simulation Algorithm Prediction Model

The random simulation algorithm prediction model can be divided into the following three steps:

R0 can be assumed to be the best estimate of sports performance at t0, which is not a random variable here. Assuming that R0 is obtained by the common algorithm M, which can be simulated repeatedly, the future results can be simulated by the bootstrap method and the Bayesian method. For sports test scores, we can find the progress factor through the usual results, and use the regression model to smooth and expand the progress factor to get the progress factor that exceeds the observation year. Then the generalized CapeCod method is used to stabilize the sports performance data of the most recent years, and the sports performance data of the earlier years can be adjusted based on the loss data. In order to ensure repeatable simulation, subjective judgments generally cannot be added to this algorithm.

The second step is to simulate the changes of various sports performances in this year based on D0. Let Cij represent the results of various sports i, the progress year j, and n represent the final progress year. All tests have been completed. What we need to estimate is the new diagonal value in the traffic triangle. The value on the new diagonal is added up to get C1, which is the sum of the simulated increments of each sports item in this year. The estimated value of the new diagonal can be simulated from a normal distribution or a lognormal distribution.

The third step is to estimate the score R1 at t1. The method is to calculate R1 based on the data simulated in step 2, that is, the new diagonal value in the flow triangle, using the same algorithm M when calculating R0. After calculating the N simulation values of R0, C1 and R1 according to the above steps, the N simulation values of CDR can be obtained, and then the empirical distribution of CDR can be obtained as our estimation of the probability distribution of CDR. With the probability distribution of CDR, it is easy to find the standard deviation of the performance prediction. The N-time simulation of CDR can also be used by the performance prediction model of multiple sports items, and then the overall performance of sports can be obtained.

In the above calculation, we did not consider the accident factors in the sports test. We can adjust the upper triangle data and then simulate, and then adjust the performance R0 of the current calendar year according to the current price. Considering the details of accident situations in sports testing is not the focus of this article.

3   Investigation and Research on Sports Performance Prediction Model Based on Random Simulation Algorithm

3.1   Mathematical Statistics

This article collects the data of the best performance of the women's 1000 m track and field of M Academy from 2018 to 2021. Organize and analyze the data, analyze and analyze the original data using EXCEL. Then use the statistical analysis process and the probability process to further analyze and infer the statistical data, and present the statistical results in the form of graphs, and then present the research results openly and intuitively. Secondly, using the algorithm of this paper, analyze and predict the development of the annual best performance of the women's 1000-m track and field event of M College.

3.2 Algorithm Improvement

For the improvement of the stochastic simulation algorithm, it is necessary to combine the method of fluid approximation to reduce the warm-up time from the initial state to the equilibrium state during the simulation process, that is, let the system directly start from the equilibrium state instead of starting from the given initial state for simulation This equilibrium state is CTMC's expectation of steady-state probability distribution.

For any given model, each state in its state space can be represented by a numeric vector. Each position of the numeric vector is a non-negative integer, and the transition between states passes through the corresponding position of the corresponding numeric vector. Add 1 or subtract 1 to achieve. When the number of processes in the model is large, the change in the number of each position of the numeric vector during state transition is relatively small, so it can be approximately considered that the process of transition between states is continuous Rather than jumping, this method will derive a system of ordinary differential equations. The action matrix and the transition delay function put forward a more effective method: the fluid approximation method.

Denote the state at time t as x(t), then after Δt time has elapsed, as shown in formula 1:

$$x(\cdot, t + \Delta t) - x(., t = F(x\cdot, t))\Delta t = \Delta t \sum_{l \in A | \text{abe}|} lf(x(., t), l) \qquad (1)$$

Therefore, the ODEs of the model can also be quickly obtained, as shown in Eq. 2:

$$\begin{cases} \frac{dx(P_1)}{dt} = -min(ax(P_1), cx(P_1)) + bx(P_2) \\ \frac{dx(P_2)}{dt} = min(ax(P_1), cx(P_1)) + bx(P_2) \\ \frac{dx(P_1)}{dt} = -min(ax(P_1), cx(P_1)) + dx(P_2) \\ \frac{dx(P_2)}{dt} = min(ax(P_1), cx(P_1)) + dx(P_2) \end{cases} \qquad (2)$$

Because the numerical solution of the differential equation derived from the fluid approximation is not affected by the explosion of the state space, x^* can be calculated quickly and easily, so that CTMCxn can be directly simulated from the state close to $n \cdot x^*$ rather than from the initial state.

3.3 Quantitative Testing of the Model

As we all know, no predictive model can be completely accurate. Then whether the prediction of the Markov stochastic simulation algorithm model is true and reliable, the accuracy of the prediction model of the Markov stochastic simulation algorithm needs to be quantitatively tested. In this paper, residual test, posterior test, grey system theory correlation analysis method is used to quantitatively test the Markov stochastic simulation algorithm prediction model constructed in this paper.

4   Investigation and Analysis of Sports Performance Prediction Model Based on Random Simulation Algorithm

4.1   Quantitative Test of the Model

The overall trend of the annual best performance of the women's 1000 m items shows a gradual increase and decrease; the specific data has increased and decreased in each year. As a traditional gray GM (1.1) prediction model, the original data sequence is required to show an exponential and regular change, and its predicted value shows a monotonous rise or a monotonous decline. This state of prediction is obviously related to the tortuous changes in sports performance. Inconsistent. Therefore, the accuracy of long-term performance prediction is not very high. In view of this, this paper uses Markov stochastic simulation algorithm prediction model to predict and analyze the annual best results of the women's 1000 m project of M College, and then effectively solve this problem.

The 4 years closest to 2021 are selected as the basis for the prediction of the state of the 1000 m women in M College, as shown in Table 1. According to the status and the number of transition steps in each year, the cumulative state transition probability of 1000 m women in M college is obtained, and the state 1 with the largest cumulative state transition probability is taken as the possible state in 2021. Therefore, the state in 2021 is determined as state 1, that is, In 2021, the women's 1000 m annual best performance of M Academy is underestimated, as shown in Fig. 1.

Table 1. The model's prediction of women's 1000 m in M college in different years (unit: second)

Years	Actual value	Markov stochastic simulation algorithm prediction model Correction value
2018 year	198.67	197.39
2019 year	199.73	196.56
2020 year	204.15	200.02
2021 year	207.59	208.18

4.2   Model Comparison

This paper uses residual test and posterior test to carry out residual test and post-test on the women's 1000 m race of M College from 2018 to 2021. The relative error of the gray GM (1.1) model has 1 at the first level and 3 at the second level; the relative error of the Markov stochastic simulation algorithm prediction model has 3 at the first level, and only 1 is at the second level. The mean square error ratio of the gray GM (1.1)

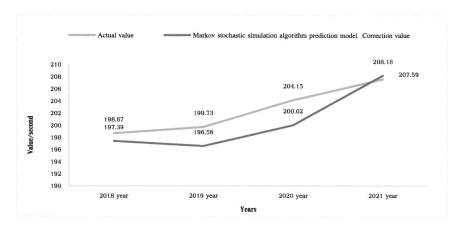

Fig. 1. The model's prediction of women's 1000 m in M College in different years (unit: second)

model is below the fourth level (0.77 > 0.65), and the mean square error ratio of the Markov stochastic simulation algorithm prediction model is above the first level (0.18 < 0.35); gray GM (1.1) The small error probability of the model is below the fourth level (0.55 < 0.60), and the Markov stochastic simulation algorithm predicts that the mean square error ratio of the model is above the first level (1.20 > 0.95); at the same time, it is predicted by the Markov stochastic simulation algorithm The model has higher accuracy in predicting women's 1000 m performance. From the above description, it can be seen that the Markov random simulation algorithm prediction value of the M college women's 1000 m annual best performance in 2021: 186.22 s, as shown in Fig. 2; compared with the gray GM (1.1) prediction model, the accuracy Increase it to 98.20%, and in the same way, the state and predicted value of the future years can be determined.

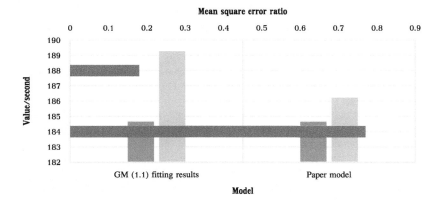

Fig. 2. The predicted state and predicted value of sports performance (unit: second)

5   Conclusions

In recent years, with the strong penetration of science and technology in the field of sports and the frequent sports exchanges between countries, the gap in training conditions, methods, and methods between countries has been narrowing, and the level of exercise has become increasingly close. Therefore, "the urgent need Explore and grasp the new situation of future sports development. This article comprehensively analyzes the current characteristics of sports prediction research, and on this basis, puts forward the scientific prediction of sports performance from the perspective of methodology and the application of scientific prediction methodology to guide. Make formulation feasible Effective sports decision-making and sports planning are possible. This development has greatly enriched the theoretical system of sports prediction methods.

References

1. Vögele, C., van Hoorn, A., Schulz, E., et al.: WESSBAS: extraction of probabilistic workload specifications for load testing and performance prediction—a model-driven approach for session-based application systems. Softw. Syst. Model. **17**(2), 443–477 (2016). https://doi.org/10.1007/s10270-016-0566-5
2. Ahn, S.H., Xiao, Y., Wang, Z., et al.: Performance prediction of a prototype tidal power turbine by using a suitable numerical model. Renew. Energy **113**(dec.), 293–302 (2017)
3. Saneie, H., Nasiri-Gheidari, Z., Tootoonchian, F.: An analytical model for performance prediction of linear resolver. IET Electr. Power Appl. **11**(8), 1457–1465 (2017)
4. Muneera, C.P., Karuppanagounder, K.: Performance prediction model for urban dual carriageway using travel time-based indices. Transp. Dev. Econ. **6**(1), 2.1–2.10 (2020)
5. Tirupathinaidu, C., Jain, N., Renganathan, T., et al.: Unified thermodynamic model for performance prediction of adiabatic feedstock gasifiers. Industr. Eng. Chem. Res. **59**(44), 19751–19769 (2020)
6. Luini, L., Emiliani, L., Boulanger, X., et al.: Rainfall rate prediction for propagation applications: model performance at regional level over Ireland. IEEE Trans. Antennas Propag. **PP**(11), 1 (2017)
7. Cao, P., Zhu, G., Zhang, Q., et al.: Blockchain-enabled HMM model for sports performance prediction. IEEE Access **PP**(99), 1 (2021)
8. Deo, R.C., Yaseen, Z.M., Al-Ansari, N., et al.: Modern artificial intelligence model development for undergraduate student performance prediction: an investigation on engineering mathematics courses. IEEE Access **PP**(99), 1 (2020)
9. Li, G., Wang, Z.: Prediction model and shrinkage character of high performance concrete plate component at early age. Tumu Jianzhu yu Huanjing Gongcheng/J. Civ. Archit. Environ. Eng. **39**(1), 93–100 (2017)
10. Guimares, B., Coelho, C., Woolnough, S.J., et al.: An inter-comparison performance assessment of a Brazilian global sub-seasonal prediction model against four sub-seasonal to seasonal (S2S) prediction project models. Clim. Dyn. **56**(6), 1–17 (2021)
11. Zeng, Y., Liu, J., Sun, K., et al.: Machine learning based system performance prediction model for reactor control. Ann. Nuclear Energy **113**(MAR.), 270–278 (2018)
12. Ghosh, D., De, R.K.: Block search stochastic simulation algorithm (BlSSSA): a fast stochastic simulation algorithm for modeling large biochemical networks. IEEE/ACM Trans. Comput. Biol. Bioinform. **PP**(99), 1 (2021)

Interactive Garden Landscape Digital Reconstruction System Based on Particle Swarm Algorithm

Wenda Ou[✉]

Nanchang Institute of Technology, Nanchang, Jiangxi, China
yanfei81213@126.com

Abstract. Garden art is a living three-dimensional art, which comes from nature but is higher than nature. Therefore, imitating the real natural environment is the basic element of garden design, including the simulation of terrain, roads, buildings, climate and other factors. To reconstruct and design real scenes, the digital reconstruction system has been continuously exploring in this field. The purpose of this paper is to study the design of interactive garden landscape digital reconstruction system based on particle swarm algorithm. From the analysis of the current limitations of the garden and the requirements of the times, digitization and particle swarm algorithms are proposed to solve them. Among many professional technologies, GIS is selected as a technical tool, and the virtual technology function of VRGIS is selected in GIS to break through, and finally, an example is used to demonstrate its reconstruction. The function of the system. When the slope is above 4°, the terrain undulating height difference is large, which can promote the digitalization and scientificization of garden reconstruction.

Keywords: Particle swarm algorithm · Interactive design · Garden landscape · Digital reconstruction

1 Introduction

According to the information's direction and information structure to make it practical, the effect of imitation can be observed from multiple angles and multiple directions. The data of three-dimensional objects is only scattered, without any structural information [1]. The general method is to modify the standard model to have a specific model [2]. Gardening is a comprehensive subject that studies structural design. It uses plants, water, rocks, stainless steel, light and other materials to remove cultural features such as light and history, adapting to the environment, and creating beautiful colors and different shapes. It requires a high level of thinking and time and time to observe it, so the digital reconstruction system plays an important role in the garden landscape [3, 4].

Although 3D modeling is so difficult, many researchers are still exploring this field [5]. Some scholars have studied military medical equipment maintenance schedule planning based on interactive particle swarm algorithm. Transform the itinerary planning into a traveling salesman problem, and then use the interactive particle swarm algorithm to solve it. In the evaluation process, the method of relative distance and quantifiable attributes effectively relieves user fatigue. And according to the actual maintenance situation, a solution was formulated, which proved the feasibility and usability of the model. Results The route planning model can formulate work plans and traffic plans according to actual maintenance tasks and requirements, and at the same time meet the needs of users in terms of time cost and expense [6]. Some scholars consider the interaction between all particles and replace the personal best value of each particle with their average value. The simulation and F-test results of several typical test functions show that compared with the standard PSO and other improved PSO algorithms, this algorithm not only has a great advantage in convergence, but also effectively avoids falling into the local optimum. The problem of solution [7]. It can be concluded that it is feasible to study the design of interactive garden landscape digital reconstruction system based on particle swarm algorithm.

This article combines the virtual part of GIS software with outstanding functions and traditional garden design software, CAD gives full play to the advantages of graphic design and GIS technology highlights the characteristics of analysis and real simulation. It can be analyzed and produced, planning and design under complex terrain, using graphic design software and the characteristics of DEM in GIS, combined planning, so that this field is easier for the public to understand. Make particle swarm algorithm play a certain guiding role for the digital promotion and construction of garden planning. This technology will surely bring opportunities for the development of garden digital reconstruction system with the development of computer software and hardware.

2 Research on Design of Interactive Garden Landscape Digital Reconstruction System Based on Particle Swarm Algorithm

2.1 Interactive Evolutionary Computing

Interactive evolutionary computing is developed from evolutionary computing [8]. It is an evolutionary calculation method based on human subjective evaluation to obtain the suitability of evolutionary individuals. It organically combines human intelligence assessment with evolutionary computing, and goes beyond creating an optimized system. Clear performance indicators have significantly expanded the application range of evolutionary computing. Interactive evolution computers are widely used in graphics and animation, face image creation, optimization control, virtual reality and other fields [9, 10].

The particle swarm optimization (PSO) algorithm is an emerging evolutionary algorithm. It is an evolutionary algorithm based on swarm intelligence. Optimize the group intelligence search generated by intra-group collaboration and anti-particle competition, and maintain a global search strategy based on the use of a simple speed transfer model to avoid complex genetic functions. At the same time, its unique memory enables it to dynamically monitor the current search status to customize its search strategy. It has strong global convergence and robustness, and does not require the help of typical information about the problem. Therefore, as the most effective parallel search algorithm, particle swarm optimization is very suitable for solving optimization problems in complex environments. Its theory and application research have important academic significance and mechanical value [11, 12].

Interactive evolutionary computing is based on the application of evolutionary computing in system optimization. The addition of human subjective judgment is simply to replace the fitness function in evolutionary computing with the user's subjective judgment. We know that general system optimization is to quantitatively analyze the system by defining measurement functions. This method is widely used in automatic control, pattern recognition and mechanical design. However, obtaining graphics or music from an interactive system requires human subjective judgment and analysis. In this case, it is generally difficult or impossible to define a measurement function. Generally speaking, the output of image, sound, virtual reality and other systems is better than the user's impression, emotion, preference and understanding. The parameters and structure of the system optimization must be optimized through human subjective judgment, so we need another method different from the traditional method.Therefore, on the basis of evolutionary computing, interactive evolutionary computing is proposed to meet this requirement.

Interactive evolutionary computing is divided into narrow and general definitions. The first definition is based on traditional functional functions instead of user subjective judgments, using Ec to optimize the target system; the broad definition is to optimize the target system through Ec in the interface during human-computer interaction.

2.2   The Goal of the System

The goal of this work system is people-oriented, focusing on the real experience of people, and providing potential tourist customers with a realistic interactive garden landscape experience. The immateriality of the product determines that the public cannot be sure whether the attraction meets their wishes before arriving at the attraction. The interactive digital garden landscape system is designed for this. The mood of the potential audience is the most important. Therefore, the fastest real-time interaction speed and user-friendly interface are the principles that the system must follow. Faster interaction speed can better satisfy potential audience groups' psychology of searching for novelty and beauty.

3   Investigation and Research on the Design of Interactive Garden Landscape Digital Reconstruction System Based on Particle Swarm Algorithm

3.1   The Basic Conditions and Landscape Resources of the Study Area

M Garden is the link between the two main geological building blocks p and k. In the long history of geological development, complex structural schemes have been formed. The mountains are steep and the valleys are deep, mostly in the shape of a "V"; the tops of the mountains are relatively open and partly flat. The whole land is high in the east and low in the west. The highest peak in the east is a garden, with an elevation of 1,023 m and the lowest elevation of 854 m. It is located in a rough stone valley. The climate is warm and humid, with numerous mountains and rivers, dense forests, lush rivers and abundant natural resources. Colorful gardens, fascinating and steep scenery, beautiful clear water views, long-standing cultural landscapes and cool and pleasant climate resources are unique and fascinating.

3.2   Interactive Settings

Flying camera is the flight mode. According to the pre-defined outline, you can visit the scenic spots in a panoramic manner and show the most characteristic scenery of the scenic spots to the public. In this mode, the public can access the graphic points without any intervention. Flying camera mode is more like watching a short landscape film. Walkthrough camera is the roaming mode. If the audience is not satisfied with the angle of the flying camera mode above, they can choose the position and angle they want, and use the keyboard and mouse to control the walking direction to reach the desired position. Like the real world, Xiao must stick to the perspective of the producer and walk at will.

3.3   Mapping Coding of Particle Swarm Algorithm Solution Space

The method used in this document is not sampling (if there are multiple points in the model, sampling can be unified), but directly encoding each feature point of the fabric model according to spatial correlation mapping and execution. The coding rule is to assign values according to spatial sectors. The position of the mass point of the object model to the index value and code of the stored two-dimensional array.

When two objects are coded according to this mapping coding method, the codes of each point of the two objects can be freely combined. The original solution space is a two-dimensional solution space composed of collision pairs, as shown in Eq. 1, but the current solution space is a Na * Nb solution space R* encoded according to space space mapping, as shown in Eq. 2. At present, in this new solution space, the degree of adaptation is almost continuous.

$$R = (P_A, P_b) \quad (1 \leq P_A \leq N_A, 1 \leq P_B \leq N_B) \tag{1}$$

$$R^* = (A_1, A_2, B_1, B_2) \quad (1 \leq A_1 \leq A_r, 1 \leq A_2 \leq A_c, 1 \leq B_1 \leq B_r, 1 \leq B_2 \leq B_c) \tag{2}$$

In formula (1), PA and PB are the arbitrary mass points in object model A and B, respectively, and NA and NB are the total mass points of object A and B respectively; in formula (2), A1 and A2 are arbitrary mass points in model A The point mapping code of B1, B2 is the mapping code of any point in model B·Ar, Ac is the number of rows and columns of the mass point in model A, and Br, Bc are the number of rows and columns of the mass point in model A.

4   Investigation and Analysis of the Design of Interactive Garden Landscape Digital Reconstruction System Based on Particle Swarm Algorithm

4.1   The Composition of the System

The program interface is face to face with the user, so visual design is very important. The entire program view is divided into 5 parts, the main view area, navigation map, tool area, communication area and graphics selection area. The main viewing area is used to define objects in the virtual space and is the most important part of the entire system; the navigation map is used to track the current user location, as well as nearby driving routes, bus stops, etc. Toolbox can choose browsing mode, including scheduled route, free walking, flight mode, etc. The communication area is used to exchange ideas with other users. Users can exchange ideas with other users by entering text. Select this area for unused sites. For example, the user can go directly to any other viewpoint from Leifeng Tower. In this way, multiple scenes can be combined into one, making it easier for the audience to change the speed of the virtual tourist spot and saving production costs. In order to facilitate users to browse the site, the icons in the tool area and field selection area are automatically hidden by semi-transparent etc. When the cursor is moved to the toolbar, the options on the screen will appear, but they are usually hidden. In this way, business needs can be met without disturbing the user interface. The visual design diagram is shown in Fig. 1.

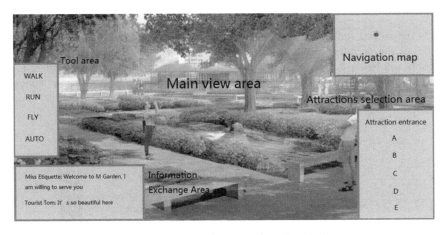

Fig. 1. Schematic diagram of interface design

When a user logs in to the system for the first time, he must first register an account with a password, and then enter the user name and password for authentication before logging in to the system. This system can be considered as a virtual community, and the uniqueness of the user name ensures that each logged-in user follows certain rules when browsing the system and interacting with other users.

4.2 Terrain Simulation

Terrain is the skeleton of landscape garden design. The establishment and expression of terrain model determines the way of landscape garden design to some extent. Traditional landscape design uses Sketchup to build a platform-shaped terrain model, which is time-consuming and labor-intensive, and has poor accuracy. The corresponding approach is an inaccurate design method based on experience. The new digital model building method and its corresponding data analysis and design methods can conveniently query site information, perform relevant data analysis, and present data before and after design in real time. It has a great role in promoting the scientificization of landscape architecture design. At present, professional software such as GIS and Civil3D can easily establish digital terrain model. Since different types of land need to be built on the basis of terrain with different slopes, our first consideration is to analyze the slope of the region when planning land use in mountainous landscapes. The relationship between M garden slope value and degree is shown in Table 1.

Table 1. Relationship between slope value and degree

Category	Slope value (%)	Degree
Flat slope	3	1
Gentle slope	5	3
Middle slope	13	8
Steep slope	30	15

On flat slopes and gentle slopes with small slopes, the terrain is relatively flat and open, and the architectural layout is relatively free, which can obtain the sunshine direction and better landscape vision. The road layout can also adopt the ideal layout form, and the amount of earthwork is generally not large. When the slope is above 4°, the terrain undulation height difference is large, as shown in Fig. 2.

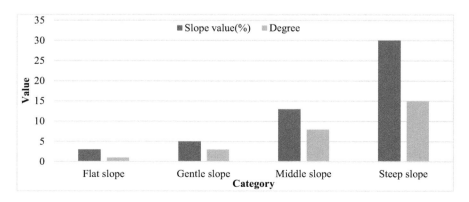

Fig. 2. Relationship between slope value and degree

In order to simulate the real world as much as possible and make the audience feel more real, it is necessary to perform logical and physical calculations on the objects in Changrong. That is, the camera that represents the human eye cannot penetrate the ground in the virtual field, nor can it enter through the wall. When the audience controls the camera line to walk on the wall, adjust the -U collision so that the camera will not cross the wall on the other side. Reasonable collision settings will make the audience feel more real.

5   Conclusions

With the progress of society and the improvement of people's living standards, two-dimensional visual effects can no longer meet people's needs. Because of its strong sense of reality, three-dimensional display technology is more and more popular and accepted by people gradually. Users only need to click the mouse at will to truly understand the surrounding three-dimensional world. This article discusses the research direction and trend of garden digitization, using GIS technology to complete and continue the incompetent analysis and prediction functions in traditional garden planning, and use the simplest GIS virtual technology to achieve the purpose of digital landscape reconstruction. It makes the garden reconstruction have a new field of play and planning tools, making the reconstruction more scientific and effective, and get rid of the limitations of limitations. Provide an effective and feasible technology for the digitization of digital gardens.

References

1. Choi, S., Lee, S., Lee, H., et al.: Development of a prototype chest digital tomosynthesis (Cdt) R/F system with fast image reconstruction using graphics processing unit (GPU) programming. Nucl. Instr. Methods Phys. Res. **848**(MAR.11), 174–181 (2017)
2. Soyoung, P., Guna, K., Chulkyu, P., et al.: Iterative interior digital tomosynthesis reconstruction using a dual-resolution voxellation method. J. - Korean Phys. Soc. **73**(3), 355–360 (2018)
3. Polavaram, S., Ascoli, G.A.: An ontology-based search engine for digital reconstructions of neuronal morphology. Brain Inform. **4**(2), 123–134 (2017)
4. Korolev, A.N., Lukin, A.Y., Filatov, Y.V., Venediktov, V.Y.: Reconstruction of the image metric of periodic structures in an opto-digital angle measurement system. Sensors **21**(13), 4411 (2021)
5. Soltani, S., Seno, S.: A formal model for event reconstruction in digital forensic investigation. Digit. Invest. **30**(Sep.), 148–160 (2019)
6. Adem, P., Isa, Y.: An iterative reconstruction algorithm for digital breast tomosynthesis imaging using real data at three radiation doses. J. X Ray Sci. Technol. **26**(3), 347–360 (2018)
7. Studiawan, H., Sohel, F., Payne, C.: A survey on forensic investigation of operating system logs. Digit. Invest. **29**(JUN.), 1–20 (2019)

8. Zeng, R., Badano, A., Myers, K.J.: Optimization of digital breast tomosynthesis (DBT) acquisition parameters for human observers: effect of reconstruction algorithms. Phys. Med. Biol. **62**(7), 2598–2611 (2017)
9. Boudraa, S., Melouah, A., Merouani, H.F.: Improving mass discrimination in mammogram-CAD system using texture information and super-resolution reconstruction. Evol. Syst. **11**(4), 697–706 (2020). https://doi.org/10.1007/s12530-019-09322-4
10. Banerjee, D., Yu, K., Aggarwal, G.: Robotic arm based 3D reconstruction test automation. IEEE Access **6**(1), 7206–7213 (2018)
11. Zellner, M.B., Champley, K.: Development of a computed tomography system capable of tracking high-velocity unbounded material through a reconstruction volume. Int. J. Impact Eng. **129**(JUL.1), 26–35 (2019)
12. Kolomvakis, N., Eriksson, T., Coldrey, M., et al.: Reconstruction of clipped signals in quantized uplink massive MIMO systems. IEEE Trans. Commun. **68**(5), 2891–2905 (2020)

Analysis of the Influence of Computer Imaging Technology on Chinese Film Special Effects

ChunLiang Wang[✉]

School of Film and Television Media, Wuchang University of Technology,
Wuhan 430223, Hubei, China
chunliang20150284@163.com

Abstract. Since entering the international market in the 1990s, Chinese films have faced fierce competition and challenges from all over the world. In the whole process of the application of computer technology, the traditional concept of film special effects production has been restored and has become one of the main technical means of film production. The vigorous development of computer technology is also an important factor for Chinese films to effectively prevent the influence of foreign films, increase the income of commercial films and understand the internationalization of the film industry. This paper studies the influence of computer image technology on Chinese film special effects and how to apply computer image technology to the field of film special effects. Combined with technology and resources, Chinese films can effectively prevent the invasion of foreign commercial film culture and market, and obtain rich economy and art in the process of international development.

Keywords: Computer imaging technology · Movie special effects · Chinese movie · Film industry

1 Introduction

The film was originally invented as a technological innovation, so the film history over the past 100 years is, from one side, the development history of special effect technology. Therefore, under the unremitting pursuit of many filmmakers and the audience's expectation of novel visual effects, film special effects have always promoted the evolution of human audio-visual aesthetics, thus promoting the formation of the film industry chain, making the film go beyond the field of technology and export to the world with multiple properties such as culture, art, commerce and politics [1]. If film is a comprehensive commercial art integrating a variety of art categories.

With the breakthrough development of computer technology, the innovative competition of world film production gradually focuses on computer imaging technology (CGI), "Computer director" Determines the viewing quality of a film. CGI technology mainly includes two parts: animation and synthesis, which is the focus of the first chapter of this paper. The global wave of 3D film and Max film in 2009 is a complete revolution in film shooting, performance, post synthesis and viewing mode [2]. This technology expands film creation to a boundless realm, and it is making great efforts With the power to transform the virtual visual effect into an immersive reality, the film is no longer limited to the illusion of vision and hearing, but continues to develop in the direction of innovation including touch and smell [3]. Modern films weaken the importance of the plot, enhance the amazing visual impact, and make the film surpass the market restrictions of race, culture and region, and spread to any place in the world. Therefore, the film not only achieves great commercial success, but also develops its own more far-reaching influence. Chinese films have developed by leaps and bounds in the new century. In the process of connecting with international standards, films have rapidly transformed from traditional ideological tools to industrialization [4]. Facing the increasingly young and market-oriented domestic film viewers, digital special effects technology has become the first symbol to determine the success or failure of commercial films. The core purpose of this paper is to vigorously develop computer imaging technology and fully absorb excellent international experience.

2   Overview of Computer Imaging Technology

2.1   Definition of Computer Imaging Technology

Traditional film special effects, also known as film stunts, include artificial illusions and illusions such as makeup, models and digital control. Its purpose is to prevent actors from being in a dangerous situation when performing special performances, reduce the production cost of the film, expand and enrich the theme of the film, and achieve the angle or content that can not be achieved by the real shot, so as to make the creators give more free and perfect play to their creativity.

Computer imaging technology is a new form of film special effects, namely computer generated imaging (cgfb), which is also commonly known as computer special effects. It refers to the creation of characters, scenes and special animation effects required by the film in the form of virtual digital, and the shooting and synthesis pictures. It does not require the participation of cameras and actors, but is completed by software. CGI not only shows the complete hand drawing feeling of traditional animation, but also can shape real and natural photos into images, and perfectly combine virtual animated characters with real photos [5]. These effects can be said to go beyond the special effects of previous films, making the film creation more powerful and the performance of the theme omnipotent. The application of computer special effects is a major innovation in the process of photography, printing, editing and synthesis.

The maturity of computer technology has pushed the production of special effects to the peak of free creation. The application of computer makes the editing materials can be copied without limitation, and the order can be reversed to form a story. The editor can choose a variety of narrative ways to complete the conception [6]; Accordingly, 3D animation and matting synthesis technology make the combination of computer special effects and real shooting characters more comfortable. This post editing method that is not restricted by the process is called non-linear editing, that is, the production method of digital film special effects. The advent and maturity of digital technology has narrowed the technical platform between designers, which has completely liberated the imagination of film creators.

2.2   The Marvelous Charm of Computer Imaging Technology

(1) Smooth super long lens

In the process of film shooting, the adjustment of position is the starting point to obtain a good lens, but the previous shooting was mainly limited to long lens or special angle of view. But in animation software, producers can build countless virtual cameras [7]. As long as the camera is provided with an appropriate path, you can take photos freely according to the settings of the camera. The development of computer technology gives full play to the functions of pushing, pulling, shaking, moving and bending. A long shot is a continuous execution of a shot event or paragraph. Through the natural space-time movement, it shows the reality on the screen and creates a unique reality show style. In addition, a screenshot also contains complex information that needs to be explained by multiple screenshots. As shown in Fig. 1 below, an imaging lens is taken.

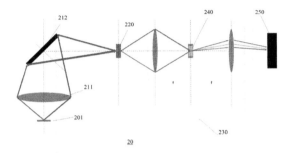

Fig. 1. Shooting imaging lens

(2) Aesthetic film color

The most intuitive and strong visual impact of the film comes from the rich color and aesthetic feeling of the film. Color is an important tool for rendering and explaining the atmosphere. The correct use of color plays an important role in the narration of the plot, the interpretation of the environment, the improvement of visual image, the activity of film atmosphere, the expression of characters' emotion and so on. In the traditional shooting, the color of the film mainly depends on the lighting of the lighting division and the color of the characters' clothes. In the later stage, the color correction can only be properly controlled. This coloring method basically completely depends on the previous records, which is difficult to regenerate or compensate in the future. Computer special effects are free to use digital color. It can not only simulate the actual light source, but also supplement the light of the later text, reduce the image noise, change the local color, and create a more exquisite and unique color style for the film, as shown in Fig. 2.

Fig. 2. Special effects

3 Scientific and Technological Renewal Stage of Special Effects Production in China

3.1 Exploration and Development of Science and Technology from 1940s to 1950s

The 1940s to 1950s was the period when Chinese films began to take shape. The production of film special effects was also separated separately and began to try to produce film special effects by means of modern science and technology.

In 1953, Shanghai film set up the stunt department, which made the film stunt an important technical department for special research and production. The stunt section includes a design unit, a photography unit, a model unit and a fireworks unit. Special technical consultants have been hired to expand personnel and equip the organization with additional facilities and equipment [8]. Members of the Department worked together to solve the problem of high-speed cameras. According to the actual needs of operation, they also designed, manufactured and refitted a number of practical and reliable grid by grid motors, synthetic photography platforms, lathe type synthetic platforms that can print simple optical skills, grid by grid projectors, synthetic photography studios, etc. The completion of these stunt facilities and equipment provides a reliable material guarantee for the substantial development of Shanghai film stunt.

In 1955, the drama myth film Tianxianpei was independently developed by the special effects Department of Shanghai film and used the advanced "color separation and synthesis" special effect, also known as "transparent painting" to complete the special effects production of the film. This technology is a special effect synthesis that has been close to the current "through painting method". It is a technology that combines the photographed characters and the background in the later stage. Why can it mix the effects such as gods falling from the sky, changing character images, or flying through clouds [9]. The film also used a large amount of liquid carbon dioxide to make a sea of clouds for the first time, and its experienced model technology made the film an unprecedented success. The audience in mainland China alone exceeded 100 million, becoming the film with the highest box office at that time.

3.2   Rapid Development in the 21st Century

After 2000, the development of Chinese special effect films has attracted more and more attention from all walks of life. The state has made greater efforts to support the animation education industry and its chain industry, promoted more employees with excellent quality to join, and greatly promoted the development of special effect film industry. From 2000 to 2002 Shanghai Film Studio took the lead in setting up a computer stunt production center and shooting "emergency landing" and "polar rescue". The film replaced the traditional post production with computer special effects, with more than 60% of the special effects lens. However, it tried its best to imitate the production mode and film structure of Hollywood blockbusters, so that the two films that tried computer stunts for the first time were not very successful, but they played a positive leading role, However, the output of Chinese special effects films has increased year by year [10]. As shown in Fig. 3 below, the special effect film processing simulation software.

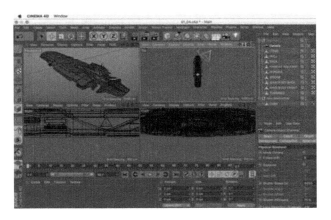

Fig. 3. Special effect film processing simulation software

Even though the exploration road of special effects production is rugged and slow, Chinese special effects films still break through with an unstoppable attitude and develop very rapidly. Around 2005, commercial blockbusters such as "cloud water ballad", "love madness saint", "Seven Swords", "myth", "head character D", "the secret of treasure gourd" and "Crazy Stone" produced and distributed by China Film Group achieved very good box office results, attracting excellent production talents at home and abroad and many well-known companies to join and cooperate one after another [11]. In 2008, the digital production base of China Film Group was listed as a major cultural project in the outline of the national "Eleventh Five Year Plan" for cultural development. The State Administration of radio, film and television has invested tens of millions of funds for special research on digital intermediate films suitable for China's national conditions and spare no effort to develop China's special effect film industry!

4   The Influence of Computer Imaging Technology on Chinese Film Special Effects

4.1   The Rapid Growth of Chinese Film Market

China has become the most promising film market in the world, with an annual growth rate of nearly 50%. In 2009, the annual output of Chinese films reached 456. Film director of the State Administration of radio, film and television, said at the 2010 news briefing of the film bureau held on the 8th that this figure does not include 27 cartoons, 19 documentaries, 52 scientific and educational films and 110 digital films for television broadcast in the program center of the film channel. At the end of the year, only the domestic film box office accumulated about 6.4 billion yuan [12]. If you add the box office and copyright revenue of films outside Shanghai and the revenue of 1.69 billion yuan of national television stations, especially film channels, the revenue of Chinese films will exceed 10 billion yuan in 2009! From this objective income, we can not only see the huge commercial value of the film industry, but also prove that China's film market will become the most attractive cake in the future international market, but also has been expanding. Han Sanping, chairman of China Film Group, said confidently that the box office of China's film market and domestic population and economy should be about 35 billion, and the market share of Chinese films will not be less than 55% in the future, which will also attract more foreign films to make films in China or cooperate with domestic films, The growing film industry will also bring more young people and film lovers broad employment space [13]. The huge audience has led to the rapid development of China's film consumption market. However, in addition to the proud consumer market, we need strong Chinese national film creation. Otherwise, China's surging cinemas, screens and audience groups will pave the way for the entry of foreign films.

4.2   Growth of Excellent Local Production Team

In the past decade, digital special effects technology has developed vigorously in China. The government attaches great importance to animation and film, gives strong support in policy and economy, integrates market forces, focuses on the construction of chain industry, and has achieved good development results in the past decade. China's CG industry has poured in more professional employees at home and abroad. Through mutual reference and learning, China's special effects team has accumulated a lot of valuable production experience and has the ability to independently complete a large number of film special effects. In those years, the special effects of the hero film still need to be completed overseas, In recent years, many directors have also turned their special effects production to local special effects teams, and even some special effects shots of American blockbusters will be processed in China [14]. All this proves the growth and maturity of the ability of Chinese special effects production teams. The practitioners in this industry are generally composed of young people, which has also added vitality to the creation and R & D of film special effects.

5   Conclusion

China has the largest film viewing population in the world. With China's urbanization and the development of per capita GDP, China will become the largest film market in the world. However, it goes without saying that China is not a strong cause of film, which is the mission of our generation of filmmakers, and film special effects production with digital technology as the core will undoubtedly be the core technology of film industrialization in the future. The rapid development of Chinese films not only brings opportunities but also challenges.

References

1. Shi, Q., Qin, B.: Film-forming property and oxygen barrier characteristic of gel-stabilized foam used for controlling spontaneous combustion of coal. Energy Fuels **35**, 12083–12090 (2021)
2. Jiang, W., Yeting, M.A.: The effects of Chinese policy uncertainty on the global economic output and trade markets. Singapore Econ. Rev. (2021)
3. Li, Y., Wang, Y., Liu, H., et al.: Effects of different crop starches on the cooking quality of Chinese dried noodles. Int. J. Food Sci. Technol. (2021)
4. China-Europe Film Fund: Make Chinese Films Visible. (7) (2022)
5. Gao, Q.: Sino-foreign film co-production boosts Chinese films to go global. Comput. Inf. (12) (2022)
6. Zhang, Y., Ou, D., Kang, S.: The effects of masking sound and signal-to-noise ratio on work performance in Chinese open-plan offices. Appl. Acoust. **172**, 107657 (2021)
7. Modwel, G., Mehra, A., Rakesh, N., et al.: Advanced object detection in bio-medical X-ray images for anomaly detection and recognition. Int. J. E-Health Med. Commun. (IJEHMC) **12**, 93–110 (2021)
8. Hufnagl, B., Stibi, M., Martirosyan, H., et al.: Computer-assisted analysis of microplastics in environmental samples based on μFTIR imaging in combination with machine learning. Environ. Sci. Technol. Lett. **9**(1), 90–95 (2021)
9. Dai, Y., Gao, Y., Liu, F., et al.: Mutual attention-based hybrid dimensional network for multimodal imaging computer-aided diagnosis (2022)
10. Gbs, A., Jla, A., Mlt, A., et al.: It's clearly the heart! Optical transparency, cardiac tissue imaging, and computer modelling. Progr. Biophys. Mol. Biol. (2021)
11. Metaxas, D., Qu, H., Riedlinger, G., et al.: Deep learning-based nuclei segmentation and classification in histopathology images with application to imaging genomics. Comput. Vis. Microsc. Image Anal. (2021)
12. Luo, Y., Zhao, Y., Li, J., et al.: Computational imaging without a computer: seeing through random diffusers at the speed of light. eLight **2**(1), 1–16 (2021)

13. Forte, V., Cavallo, A., Bertolo, R.G., et al.: Computer tomography texture analysis: a promising tool aiding in suspecting clear cell renal carcinoma at pre-treatment imaging (2021)
14. Velazco-Garcia, J.D., Shah, D.J., Leiss, E.L., et al.: A modular and scalable computational framework for interactive immersion into imaging data with a holographic augmented reality interface. Comput. Methods Program. Biomed. **198**, 105779 (2021)

Image Processing of Gas-Liquid Two-Phase Flow in High-Speed Photography Based on Reconstruction Algorithm

Ming Liu[✉]

Huanghuai University, Zhumadian 463000, Henan, China
616409324@qq.com

Abstract. In recent decades, with the development of science and technology, gas-liquid flow and gas flow are widely used in petroleum, medicine, manufacturing, aerospace and other fields, biological and other fields. Put great emphasis on the LPG two-phase flow measurement method and gradually switch from two-dimensional measurement to three-dimensional measurement. By studying LPG two-phase flow, a more comprehensive research platform is provided. Image processing and reconstruction as well as visual measurement have become one of the most important scientific research instruments and continue to maintain theoretical and technical achievements. Based on image processing of high-speed two-phase flows, this work investigates the image reconstruction algorithm of liquid vapor phase flows.

Keywords: High speed photography · Reconstruction algorithm · Image processing · Gas liquid two phase flow

1 Introduction

With the development of scientific theory and advanced technology, gas-liquid two-phase flow plays an important role in various fields at home and abroad. With the development of science and technology and image processing technology, industrial production puts forward higher requirements for efficiency and quality. Ocean currents are widely used in chemical industry, petroleum, human activities, biology, aerospace and other fields, which promotes the research in this field. For example, the process of transporting oil, natural gas and low boiling point liquids, the chemical and physical processes in electric heating equipment such as boilers, boiling tubes and condensers, gas mixtures and gas separators [1]. The meteorology and liquid interface of gas-liquid two-phase flow are variable, and the meteorology can be compressed. Therefore, gas-liquid two-phase flow is considered to be the most complex two-phase flow.

Gas-liquid two-phase flow has many characteristics than unidirectional flow. Physical and chemical properties of gas-liquid two-phase medium. The relative velocity between them is an important factor affecting the two upstream flows [2]. In addition, due to the existence of upstream, heat and mass, the flow can be transmitted through the interface, and the shape of the gas-liquid interface will change at any time, resulting in a certain degree of phase transition. For example, bound bubbles in high-density bubbles are combined with bubbles in relatively large bubbles [3]. Because the flow process of the two upstream is very complex and the flow state and phenomenon are diverse, it is necessary to understand the phenomenon, master its flow law, establish a mathematical model, and deeply analyze and study its development. Based on high-speed images, the image processing algorithm can extract image features with high precision by measuring the total flow field and instantaneous parameters of gas-liquid two-phase flow, and effectively realize image matching and three-dimensional reconstruction. Three-dimensional image reconstruction and shape parameter extraction are the focus of this paper. This is the core of research and optimization [4].

2   Overview of Gas-Liquid Two-Phase Flow

Gas-liquid two-phase flow is a complex problem often encountered in nature and industry. It is widespread in strategic production processes such as water saving, energy saving, environmental protection, machinery, petrochemical and metallurgy, and occupies a more and more important position in the national economy and national life.

Gas-liquid two-phase flow is a process of gas-liquid two-phase coexistence and interaction. According to the distribution, it can be divided into single component gas-liquid two-phase flow and anisotropic gas-liquid two-phase flow. Forging gas-liquid two-phase flow, such as gas-liquid two-phase flow composed of steam and water and gas-liquid two-phase flow composed of air and water. In gas-liquid two-phase flow, the shape can flow in the same direction. That is, it can flow in the same direction. Countercurrent is also possible [5].

There are meteorology, gas-liquid interface and liquid in ocean current system. Meteorology can be compressed, and there is shape exchange of energy, quantity and mass between interfaces. The interface is easy to deform and can form various interface combinations, i.e. types. Therefore, in gas-liquid two-phase flow, the type is closely related to the flow and electrothermal characteristics. Based on the relative images of the two upper reaches and the concepts of laminar flow and water potential, the flow types in many cases are studied more deeply. If the gas and liquid are different or the flow conditions are different, different fluids will appear. Flow determines the flow characteristic parameters, which vary under different conditions [6]. The typical types of vertical pipelines include century mammals, snails, massive upstream, annular upstream, etc., while the typical types of horizontal pipelines include century mammals, pistons, annular upstream, etc. Laminar flow, waves, snails, circulation, etc.

Gas-liquid two-phase flow systems and processes involve many fields, such as nature, human living environment and industrial development. For example, the ocean current phenomenon is closely related to human life such as lightning caused by typhoon, soil sand in the river, yellow leaves in late autumn and snow in winter. There are two stages of blood circulation and metabolism in the biological world. Super-cooling treatment is an effective heat exchange method in chemical production. During the heating process, the liquid produces bubbles on the boiling surface, and the bubbles grow and rise to form a gas-liquid flow. In the process of cold compress, studying the formation and rising process of bubbles is helpful to study the electrothermal mechanism of cold compress. Gas ocean currents include satellites, space stations, aerospace rays, oil and gas exploration, development and transportation in the oil industry, boilers and heat exchangers for energy and energy technology. Therefore, the in-depth study of gas-liquid two-phase flow will play an important role in the development of human society and industry.

The study of gas-liquid two-phase flow is closely related to human life and involves many scientific and technological fields [7]. The measurement of fluid flow characteristics is the basis of understanding and controlling fluid development. In order to further study the flow mechanism of gas-liquid two-phase flow and better understand the static and dynamic characteristics of gas-liquid two-phase flow, several characteristic parameters of gas-liquid two-phase flow system are introduced and their flow characteristics are explained. The details are as follows.

(1) Dress up

Cross phase separation is the proportion of gas-liquid phase in phase flow system. Gas liquid two phase flow:

$$\alpha = \frac{p_1}{p} = \frac{p_1}{p_1 + p_2} \tag{1}$$

(2) Flow

The flow is similar to that of single-phase flow. It can be classified according to the parameter characteristics, which can be divided into volume flow and mass flow. Flow refers to the ratio of the mass or volume of the medium passing through the pipeline to the time required for the medium to pass through the pipeline section. In addition, the flow can also be divided into gas phase flow, liquid phase flow and overall flow according to the composition.

3   Image Processing Algorithm

In the gas-liquid two-phase flow measurement experiment based on high-speed photography, the complexity of the experimental environment and the electronic noise and optical noise inside the camera inevitably reduce the quality of the obtained image. For example, the influence of fluorescent lamp light source in the environment increases the image noise, the uneven irradiation of laser and the reflection and refraction of laser in the pipeline, resulting in uneven brightness of bubble image, image noise, poor characteristics of bubble edge and other defects in the processing process.

Based on the above analysis, the image processing system needs to design a set of image preprocessing algorithm model suitable for experimental subjects before image 3D reconstruction. At the same time, it mainly focuses on the influence of image noise on the accuracy of image feature parameters and the edge characteristics of bubbles [8]. The image preprocessing system of image graying, subtraction detection, anti color, median filter and threshold segmentation is designed. Based on image noise analysis and the adaptability of threshold selection in threshold segmentation theory, an adaptive threshold wavelet denoising method is proposed to optimize the image processing algorithm.

3.1   Image Processing Algorithm Design

For the gas-liquid two-phase flow image obtained from the experiment, the image algorithm preparation is as follows:

(1) Graying the original image obtained by the high-speed camera is shown in Fig. 1 (a). According to the image grayscale formula, the original image is transformed into a grayscale image, as shown in Fig. 1(b). Grayscale is convenient for computer processing, speeds up the operation speed, and lays the foundation for related image algorithm processing.

(a) Original image   (b) Gray image

Fig. 1. Image gray processing

(2) The experimental difference gray image is shown in Fig. 2(a), the original background gray image is shown in Fig. 2(b), and the difference result is shown in Fig. 2(c):

(a) Gray image   (b) Original background gray image   (c) Difference result

Fig. 2. Image subtraction processing

3.2   Sparse Bubble Image Threshold Segmentation

Threshold segmentation has a relatively mature technical level in this field. According to its algorithm characteristics, it is classified in order. Iterative method, dynamic threshold segmentation method, maximum interclass variance method (Otsu method), bimodal method and so on are commonly used binarization methods.

In the dynamic measurement experiment of characteristic parameters of gas-liquid two-phase flow sparse bubble flow, Weinberger high-speed camera obtained multiple groups of experimental data at 500 fps. Using the general image denoising method, about 5% of the image threshold segmentation in each group of experiments will produce errors, resulting in a large number of unevenly distributed continuous black areas, which can not be recognized in subsequent target detection, As a result, the precursor data can not be associated with the subsequent data, resulting in faults in the matching, and it is difficult to accurately extract the bubble features in the image, which greatly affects the measurement results and measurement accuracy such as gas holdup and bubble velocity [9]. The position of the above error image in the image sequence is random, discontinuous and has no obvious law. Through analysis, the noise largely comes from the photoelectric noise and environmental noise of the high-speed camera. Due to the interference, the gray difference between the background and the target is not obvious. This kind of noise is difficult to be eliminated by experimental methods, and can only be processed by effective image denoising methods.

4 Slice Image Motion Matching Algorithm for Gas-Liquid Two-Phase Flow

4.1 Overview of Matching Algorithm

Image matching is to find the target image in the subgraph to be tested or the image to be tested through a specific algorithm. The former is the search image and the latter is the template image. Matching algorithm plays an important role in image applications in target matching, medicine, manufacturing, aerospace and other fields. The key links are summarized as follows:

(1) Similarity is the similarity measure, which mainly uses the methods of correlation function degree and minimum distance to measure the similarity between the target image and the search image. Correlation function mainly includes covariance function, product correlation function, etc. The minimum measurement methods mainly include hadorff distance, MSD (mean square difference), SD (mean square deviation), ad (absolute difference), mad (mean absolute difference), etc.

(2) The search strategy, that is, which search method is used to match the search image based on the target image, directly affects the image matching accuracy and the matching speed. Generally, the more common methods are traversal, genetic algorithm and so on.

(3) The search space is the sum of the search space before and after the transformation through some transformation methods such as Fourier transform, wavelet transform and polar coordinate transform. Therefore, in order to achieve faster algorithm speed, it is necessary to reduce the search range as much as possible.

(4) In the feature space, the search image is matched by comparing the feature parameters based on the feature parameters of the target image, such as edge contour, graphic centroid, etc. This method uses some features to replace the local

image for matching, which can reduce the matching complexity, reduce the matching time and improve the matching accuracy.

4.2   Typical Image Matching Algorithm

Image matching is to find whether there are concerned targets in the image. Generally, there are three practical algorithms: global matching, feature matching and transform domain based matching. Among them, based on the transform domain, a simpler matching method is selected according to the image graphics after the transform such as Fourier and wavelet. The following introduces several methods and analyzes their characteristics.

(1)   Global matching
   Find the expected target in a specific area based on the image. It needs to be trimmed as the referenced template. Finally, find whether such template exists based on the second image. The above matching method is represented by an image, as shown in Fig. 3. Let the correlation measure be r (x, y), and use a formula to represent the correlation between the template and the search image, that is:

$$\begin{cases} E(t)\dot{x}_d(t) = f(t, x_d(t)) + B(t)u_d(t) + d_a(t) \\ \quad\quad y_d(t) = C(t)x_d(t) \end{cases} \tag{2}$$

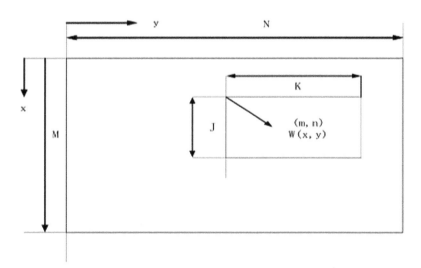

Fig. 3. Global sample correlation at point (m, n)

(2) Feature based matching

Based on image feature matching, it is necessary to select the feature space of the image at first, then detect the feature parameters such as the edge or centroid of the figure in the image by some method, and finally realize image matching.

Feature space is the premise of processing related image processing algorithms and matching algorithms. It is necessary to select the feature space at the beginning. The determination of general feature space needs to comprehensively consider various conditions and factors, which can be summarized into three points. Similarity, the template image needs to have some type of characteristics when matching with the search image, and it is invariant in many cases. Uniqueness, based on the feature parameters detected in the search image, they need to correspond to the template image, otherwise the matching rate will decrease. Stability. If the template image and the search image are obtained in different environments, or at different temperatures and from different detection systems, the characteristic parameters will not be seriously changed [10].

Feature extraction, after the feature space is determined, it is necessary to extract the feature parameters of the image, which can be divided into two categories: point feature and line feature. The feature of point feature is that the algorithm is simple to implement, easy to extract, and has good characterization characteristics when the image matching accuracy is not high. The commonly used point features include line intersection, corner, centroid of closed curve, maximum curvature point on contour, local maximum point of wavelet transform and so on. In addition to the above methods for point feature detection, point feature detection can also be carried out based on detection operators, such as Forstner operator and Moravec operator. Line features are used to characterize the overall structural features in the image, such as edge features, and are also the priority of feature space.

5   Conclusion

Research on image processing of gas-liquid two-phase flow in high-speed photography based on reconstruction algorithm, experimental research is carried out on image preprocessing and image three-dimensional reconstruction. According to the flow characteristics and experimental environment of gas-liquid two-phase flow, the measurement experimental system based on high-speed photography mainly faces the problems of image noise interference and poor bubble edge characteristics in image processing. Based on wavelet theory and the characteristics of wavelet transform at high frequency and low frequency, adaptive wavelet denoising is realized, and the image processing algorithm model of this course is optimized, which lays a foundation for improving the accuracy and visualization of slice image 3D reconstruction.

References

1. C Li, L Sun, J Liu, 等: Improvement of signal processing in Coriolis mass flowmeters for gas-liquid two-phase flow (2021)
2. Fang, L., Zeng, Q., Wang, F., et al.: Identification of two-phase flow regime using ultrasonic phased array. Flow Meas. Instrum. **72**, 101726 (2020)
3. Shen, X., Yamamoto, T., Han, X., et al.: Interfacial area concentration in gas-liquid metal two-phase flow. Exp. Comput. Multiphase Flow 1–15 (2021)
4. Wu, X., Gao, Z., Meng, H., et al.: Experimental study on the uniform distribution of gas-liquid two-phase flow in a variable-aperture deflector in a parallel flow heat exchanger. Int. J. Heat Mass Transf. **150**, 119353 (2020)
5. Zahedi, R., Rad, A.B.: Numerical and experimental simulation of gas-liquid two-phase flow in 90-degree elbow. AEJ – Alex. Eng. J. (2) (2021)
6. Grühn, J., Vogel, M., Kockmann, N.: Digital image processing of gas-liquid reactions in coiled capillaries. Chem. Ingenieur Tech. **93**(5), 825–829 (2021)
7. Sukamta, S.: Correlation between void fraction and two-phase flow pattern air-water with low viscosity in mini channel with slope 30 degrees. Key Eng. Mater. **846**, 289–295 (2020)
8. Birjukovs, M., Dzelme, V., Jakovics, A., et al.: Dynamic neutron imaging of argon bubble flow in liquid gallium in external magnetic field. Int. J. Appl. Electromagn. Mech. (2), 1–7 (2020)
9. Zhai, L., Wang, Y., Yang, J., et al.: Cost-based recurrence analysis of conductance time series for gas–liquid two-phase flow system. Int. J. Mod. Phys. C **32**, 2150161 (2021)
10. Guo, W., Wang, L., Liu, C.: Prediction of gas–liquid two-phase flow rates through a vertical pipe based on thermal diffusion. Industr. Eng. Chem. Res. **60**, 2686–2697 (2021)

Application of Two-Dimensional Code Encryption Algorithm Under Asymmetric Cipher System

Guofang Huang and Xiping Liu[✉]

Changsha Nanfang Vocational College, Changsha 410208, China
21991077@qq.com

Abstract. In order to understand the application mode of TWO-DIMENSIO-NAL code encryption algorithm under asymmetric password system, this paper will carry out relevant research, mainly discusses the basic concept of asymmetric password system, and then introduces the principle and common algorithm of asymmetric TWO-DIMENSIONAL code encryption algorithm under the system, and finally launches the design of two-dimensional code security application system based on relevant algorithm. The two-dimensional code security application system designed in this paper can encrypt the asymmetric password of the two-dimensional code through the algorithm. After the encryption, the security of the two-dimensional code will be significantly improved and can be put into practical application.

Keywords: Asymmetric cryptosystem · QR code · The encryption algorithm

1 Introduction

At present, with the smart phones and information technology, the expansion of the network environment, qr code has become an important tool in people's life and work, people began to frequent using qr code for payment, identity authentication and other activities, but under the background of these activities continue, qr code has exposed the security problems, if someone's private QR Code. There may be others who use QR Code to seek personal gains and harm the parties concerned. Therefore, how to ensure the safety of the application of QR Code has become a concern of relevant fields. Focusing on this problem, the relevant fields have proposed the asymmetric password system and the TWO-DIMENSIONAL code encryption algorithm under the system (hereinafter collectively referred to as the asymmetric two-dimensional code encryption algorithm), these algorithms can encrypt the two-dimensional code, improve the two-dimensional code application security. Because the asymmetric TWO-DIMENSIONAL code encryption algorithm has been proposed for a relatively short time, many people do not know enough about it and cannot be well put into practical application. Therefore, in order to promote the asymmetric two-dimensional code encryption algorithm and improve the two-dimensional code application security as soon as possible, it is necessary to carry out research on the algorithm application.

2 Basic Concepts of Asymmetric Cryptography

Asymmetric cryptosystem is a cryptosystem based on asymmetric private key and public key. The emergence of asymmetric cryptosystem breaks through the situation that traditional symmetric cryptosystem is the only cryptosystem. The main characteristic of traditional symmetric cryptosystem is key is symmetrical, common network login, for example, usually need to users through the registration window to get a first account/password, the account/password will be preserved by the server at the same time, login user needs to enter account/password correctly, and then through the server retrieval to confirm whether there is the same in the store account/password, If so, the user can log in successfully; otherwise, the user cannot log in [1–3]. It can be seen that in the traditional symmetric password system, the keys used for encryption and decryption are completely symmetric. Therefore, as long as the key is exposed, the user will be harmed or otherwise affected. On the contrary, asymmetric cryptography puts forward two new concepts of private key and public key, in which the private key is mastered by the user and used for encryption, while the public key is used for decryption. In the process, the private key, public key and other parameters need to be combined to achieve the purpose. If the private key is not known, it is impossible to calculate by relying only on public key and other parameters. In asymmetric cryptography, the private key will not be used in practice (unless the user wants to modify personal key information, which is rare). Therefore, the probability of disclosure of the private key is very small, while the public key is used in practice, but the public key is different every time. Therefore, leakage is unlikely to affect users. From this point of view, the security protection ability of asymmetric cryptosystem is higher than that of symmetric cryptosystem [4–6].

3 Asymmetric TWO-DIMENSIONAL Code Encryption Algorithm Principle and Common Types

3.1 Algorithm Principle

Figure 1 are asymmetric encryption principle of QR Code encryption algorithm, based on the content, the algorithm based on the information interaction both of us have different keys, and keys will interact via public communication channels, the information sender and the receiver each has a public key and secret key, this case need only in key exchange of information transmission at the same time, The sender can encrypt the information through the public key of the other party, and the receiver can decrypt the public key through its own private key to restore the encrypted digital information. In this way, encrypted communication can be realized [7].

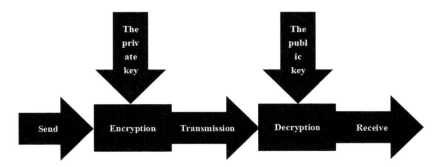

Fig. 1. Encryption principle of asymmetric TWO-DIMENSIONAL code encryption algorithm

In principle, asymmetric QR Code encryption algorithm was able to complete through the different key encryption, decryption, mainly because of the use of a series of inverse algorithm, namely the encryption and decryption process is a typical calculation process, but if to perform calculations to get the answer will bring huge amount of calculation, also makes the computational complexity greatly, Current field also cannot crack the password in this case, so the forward calculation of encryption and decryption is not viable, and encryption and decryption reverse but a small amount of calculation, at the same time also can guarantee the accuracy of the calculated results, so the reverse calculation of feasibility, makes all kinds of reverse calculation method for asymmetric QR Code encryption algorithm. Asymmetric QR Code encryption algorithm against defects in traditional symmetric cryptosystem is to put forward, on the performance is much better than the traditional system, but because the technology is shorter, asymmetric QR Code encryption algorithm application is relatively narrow, the intelligent application in digital signature, and the QR Code is a typical digital signature, Therefore, the current asymmetric TWO-DIMENSIONAL code encryption algorithm is widely used in two-dimensional code encryption. In addition, the asymmetric QR Code encryption algorithm is not without flaws, although its using inverse operation principle to reduce the amount of calculation, operation steps, but cannot reduce index computation complexity is higher in practical application, the computing efficiency is slow, in the face of the defects in related fields has not a good solution, only through a high-performance CPU as much as possible to improve the computing speed [8–10].

3.2   Common Types of Algorithms

Asymmetric TWO-DIMENSIONAL code encryption algorithm research time is relatively short (relatively speaking), there are not many algorithm types, the current more common is RSA algorithm. RSA algorithm is by Ronald lee west, Artie Mr Moore, Leonard DE man three o, was first used in military communication, special force protection of confidential information, and then the algorithm is known, in a short period of time is related areas regarded as one of the most successful public key cryptosystem, especially on the encryption security, There are few alternatives better than this algorithm.

At the same time, compared with other asymmetric TWO-DIMENSIONAL code encryption algorithms, RSA algorithm also has the advantage of simple steps [11]. Figure 2 shows the step flow of RSA algorithm.

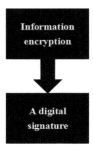

Fig. 2. Step flow of RSA algorithm

According to Fig. 2, first of all, in the middle of the data encryption step to plaintext and ciphertext as A and B, respectively, the private key for n1, key for n2, at this moment the transmission of information, including information sender through information interaction for each other's public key to encrypt information, and then sends the information early, each other can decrypt the private key, public key and other parameters. Meanwhile, when sending data, the method of data encryption by the sender is shown in Formula (1), so that A can be converted into B, and the receiver can decrypt the data through Formula (2), and then convert B into A. Secondly, in the digital signature step, when the information sender encrypts the information with its private key and publishes the public key at the same time, the digital signature is completed, that is, the public key is the digital signature of the information sender. In this way, the receiver can ensure that the information is sent by the sender and avoid information transmission chaos [12].

$$B = A^{n2} \bmod n \qquad (1)$$

$$A = B^{n1} \bmod n \qquad (2)$$

4   Design of Two-Dimensional Code Security Application System

Combined with RSA algorithm, the following will rely on asymmetric password system for two-dimensional code security application system design, the system around RSA algorithm, can provide security guarantee for two-dimensional code application.

4.1   Overall Framework

The two-dimensional code security application system in this paper relies on the PKI public key system, which can carry out closed management of the two-dimensional code to ensure security. In order to better understand the system framework, this paper takes the e-commerce business mode as the application scenario, and generates unified public and private keys through RAS algorithm. The main feature of the system is that the private key and public key are stored in the key management layer at the same time. In this way, the digital signature function of two-dimensional code can be fully used to ensure the legitimacy of two-dimensional code and ensure the security of multiple parties. Figure 3 is the general framework of the TWO-DIMENSIONAL code security application system in this paper.

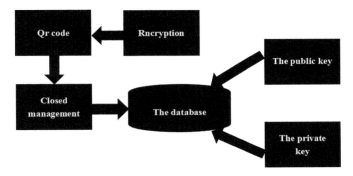

Fig. 3. Overall framework of two-dimensional code security application system

4.2   Design and Implementation Method

The system design is divided into two steps, namely key pair generation module design, data analysis and coding. The specific contents of each step are as follows.

Design of Key Pair Generation Module
According to the application scenario, the design of the key pair generation module firstly takes the TWO-DIMENSIONAL code service management platform as the basis, and is responsible for generating public and private keys and managing the keys. The basic principle is as follows: When the user needs to generate a new TWO-DIMENSIONAL code, he can submit an application to the platform, and then the platform will generate the corresponding private key and public key, and send the public key to the user. In this case, the user can use the public key to transmit information to the platform. It should be noted that although the operation mechanism of the key pair generation module is simple, there is an internal difficulty, that is, how to confirm the module. RSA algorithm while in the current theory is considered to have extremely high security, but no one has been able to verify this, the reason is that mathematics is difficult to decompose factor of RSA algorithm, also because of this related to the security of RSA algorithm is put forward a question, namely the RSA

factor are hard to break down, but does not mean it cannot decompose, It is not impossible for the whole asymmetric cryptosystem to collapse if RSA factors are decomposed in algorithmic encryption applications. Faced with this situation, some researchers proposed a feasible method, which is to reasonably confirm the modulus of RSA algorithm. In this way, although RSA factors cannot be decomposed, the probability of decomposition can be greatly reduced. The confirmation of modulus of RSA algorithm should follow four key points: First, the gap between the double values of modulus should not be too small, so that even after the public key is published, the RSA factor in it is difficult to be decomposed, otherwise, the possibility of factor decomposition will be greatly reduced; Second, the maximum common factor of modulus must be guaranteed to be minimum. Thirdly, modulo binaries must meet the criterion of strong primes. Fourth, the absolute values of the module double values should not be too small, at least to reach more than 8 digits.

Data Analysis and Coding

Firstly, the design work is carried out for data analysis. In the process, the input data stream is analyzed and the character types used for encoding are selected according to the applicable scope of the database. According to this logic, the coded character types supported in the applicable scope of the two-dimensional code data flow are digits and char respectively. Byte, Chinese characters, each character type has unique application value, the specific choice depends on the actual situation. After the character selection is completed, the corresponding mode should be used to efficiently transform the character, and the encoding version should be selected synchronously according to the user's error correction level and data capacity. In the design of this paper, the relevant information of two-dimensional code only contains two coding types, numbers and Chinese characters. Therefore, it is directly classified in the design process, and then the corresponding coding is selected, as shown in Table 1.

Table 1. Coding situation under the code type of TWO-DIMENSIONAL code in this paper

The encoding type	Coding
Digital	0001
Chinese characters	1101

According to Table 1, can effectively improve the efficiency of the system to identify two-dimensional code, and through coding can quickly identify data information according to coding.

Secondly, through data analysis, the basic requirements of data coding are confirmed, and then data coding work will be formally launched. Data coding is to convert real information into information that can be read by terminal devices such as computers. Therefore, according to the recognition range of modern computers, it can only recognize data of high and low levels, where high level is generally represented by 0 and low level is 1. According to the basic requirements, coding for coding, this paper main goal is to embed the character data type information format, such as in Numbers,

Chinese character coding type, both for 0001 and 1101 respectively, and the coded character and Numbers used to represent the number of electricity product, Chinese characters are used to represent the name of the goods, Then the encoding result of the corresponding data is XXXX/1101/XXX/0001 or XXXX/0001/XXX/1101.

5   Conclusion

In conclusion, because of traditional symmetric cryptosystem hole more obvious, so people begin to pay close attention to asymmetric cryptosystem, the asymmetric encryption algorithm is widely used in QR Code scenarios, and of the algorithm is indeed improve the safety of the QR Code application show that the algorithm has a high application value and can give modern people's life and work more convenient. Therefore, relevant fields should have a thorough understanding of asymmetric cryptography and follow the relevant points to promote corresponding algorithms.

References

1. Mihalkovich, A., Levinskas, M.: Investigation of matrix power asymmetric cipher resistant to linear algebra attack. In: Damaševičius, R., Vasiljevienė, G. (eds.) ICIST 2019. CCIS, vol. 1078, pp. 197–208. Springer, Cham (2019). https://doi.org/10.1007/978-3-030-30275-7_16
2. Gorbenko, I.D., Alekseychuk, A.N., Kachko, O.G., et al.: Calculation of general parameters for NTRU Prime Ukraine of 6–7 levels of stability. Telecommun. Radio Eng. **78**(4), 327–340 (2019)
3. Varfolomeev, A.A., Makarov, A.: About asymmetric execution of the asymmetric elgamal cipher. In: 2020 IEEE Conference of Russian Young Researchers in Electrical and Electronic Engineering (EIConRus). IEEE (2020)
4. Almajed, H., Almogren, A.: A secure and efficient ECC-based scheme for edge computing and internet of things. Sensors **20**(21), 6158 (2020)
5. Gorbenko, I.D., Kachko, O.G., Gorbenko, Y.I., et al.: Methods of building general parameters and keys for NTRU prime Ukraine of 5th–7th levels of stability. Product form. Telecommun. Radio Eng. **78**(7), 579–594 (2019)
6. Abed, S., Waleed, L., Aldamkhi, G., et al.: Enhancement in data security and integrity using minhash technique. Indon. J. Electr. Eng. Comput. Sci. **21**(3), 1739–1750 (2021)
7. Aravind Vishnu, S.S., Praveen, I., Sethumadhavan, M.: An IND-CCA2 secure certificateless hybrid signcryption. Wirel. Pers. Commun. **119**(4), 3589–3608 (2021). https://doi.org/10.1007/s11277-021-08422-2
8. Weissbart, L., Chmielewski, Ł, Picek, S., Batina, L.: Systematic side-channel analysis of Curve25519 with machine learning. J. Hardw. Syst. Secur. **4**(4), 314–328 (2020). https://doi.org/10.1007/s41635-020-00106-w
9. Mitin, S.V.: Amount of key information contained in plain and encrypted text sets of the symmetric randomized mceliece cryptosystem. Cybern. Syst. Anal. **56**(5), 726–730 (2020). https://doi.org/10.1007/s10559-020-00288-9
10. Ahamed, B.B., Krishnamoorthy, M.: SMS encryption and decryption using modified vigenere cipher algorithm. J. Oper. Res. Soc. China 1–14 (2020)

11. Osamor, V.C., Edosomwan, I.B.: Employing scrambled alpha-numberic randomization and RSA algorithm to ensure enhanced encryption in electronic medical record. Inform. Med. Unlock. **25**(5), 100672 (2021)
12. Shin, S.H., Yoo, W.S., Choi, H.: Development of modified RSA algorithm using fixed mersenne prime numbers for medical ultrasound imaging instrumentation. Comput. Assist. Surg. **24**(sup2), 1–6 (2019)

Intelligent Translation Recognition Model Supported by Improved GLR Algorithm

Mei Deng and Lijiang Yang[✉]

Jiangxi University of Engineering, Xinyu 338000, China
258172156@qq.com

Abstract. In order to improve the translation accuracy of intelligent recognition English translation model, this paper will focus on the improvement of GLR algorithm. The research mainly starts from GLR algorithm, gradually constructs the intelligent recognition English translation model, and then improves the algorithm for the defects of the model to obtain the improved GLR algorithm. Finally, the improved algorithm model system is designed and verified to confirm whether the improved algorithm model has advantages. The research proves that the intelligent recognition English translation model based on the classic GLR algorithm is indeed effective, but it still has defects. In this paper, the performance of the improved algorithm model is significantly improved, which has advantages over other algorithms and can improve the accuracy of translation results.

Keywords: GLR algorithm · Algorithm improvement · Intelligent recognition of English translation models

1 Introduction

Because English is a universal language, in the context of international development, people often come into contact with English in their work and life. However, not everyone is familiar with English, so it is inevitable that there will be language barrier. In the face of this phenomenon, people need to get help from English translation tools. The translation result of the tool directly determines whether people can cross the language barrier and communicate correctly. From this perspective, the people of translation tools translation results put forward the higher quality requirements, especially the precision of the results, at least security means the right words, but in the past intelligent identification model of English translation influenced by classical GLR algorithm of defects, translation results slightly lacking in accuracy, is difficult to meet the quality requirements of the current translation result, This shows that the classical GLR algorithm needs to be improved, and how to improve the translation accuracy has become a problem worth thinking about, and it is necessary to carry out relevant research.

2   Construction of Intelligent English Translation Recognition Model and Improvement of GLR Algorithm

2.1   Model Construction

Although the classical GLR algorithm has defects, the defects are not many and do not affect its support for intelligent recognition of English translation model. Therefore, translation model can be built around the classical GLR algorithm before the algorithm is improved. The construction of the model is mainly divided into two steps, namely, corpus creation and part-of-speech recognition. The specific contents of each step are as follows. Figure 1 shows the basic process of model translation [1, 2].

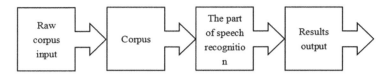

Fig. 1. Shows the basic framework of the model

Combined with Fig. 1, the model first accepts external raw corpus input through the network, and then transfers the original corpus to the corpus. Secondly, through the part of speech recognition analysis in the corpus, the corresponding relationship between the original corpus and the corpus stored in the corpus can be known. After finding the corresponding relationship, the translation result can be obtained and the output can be.

Corpus Creation

As the basic core of intelligent identification of English translation model, corpus must store sufficient corpus. Take English and Chinese phrase translation as an example, corpus should store enough bilingual phrase data, and each corpus should be tagged with part of speech after completion. The annotation results must be completely accurate (the so-called part-of-speech standard is to regulate the function and meaning of each phrase corpus), so that the timeliness and accuracy of translation results of translation models can be basically guaranteed [1, 3, 4]. According to this requirement, the corpus of intelligent identification of English translation model corpus can be realized mainly through information collection tools, and the annotation of each corpus should be expanded according to the specification requirements, that is, the annotation of corpus should be expanded according to the specification content, as shown in Table 1. In addition, the annotation methods of stored corpus should be paid attention to. This paper mainly chooses data, hierarchy and processing as three annotation methods, among which data is mainly to convert corpus into text format, hierarchy is mainly responsible for the part of speech, and processing is to better realize human-computer interaction.

Table 1. Corpus annotation should be based on the content of the specification

Specification content	The size of the
	Application field
	System
	Tense

Part-of-Speech Recognition

Part-of-speech recognition is the main activity of intelligent recognition English translation model in the process of translation. Recognition results directly affect the quality of translation results, so it is very important. Good part-of-speech recognition enables the model to deal with the grammatical ambiguities of phrasal corpus, making the results more accurate, otherwise it will cause errors. The realization of part-of-speech recognition is mainly based on corpus. The basic logic is as follows: First, each word in the segmentation is separated to make the word independent from each other, so that the word can be recognized separately. Then, the accurate meaning of the sentence can be obtained by summarizing the part-of-speech recognition results of all words. Secondly, using relation tree logic, we can get a syntax tree (a kind of relation tree) by analyzing the dependency between syntax and short sentences. The accuracy of translation results can be further improved according to the syntax tree. It is worth noting that the classical GLR algorithm needs to be used in the realization of part-of-speech recognition, which mainly plays a role in judging the context of words and sentences, and generates unconditional transfer statements based on the dynamic recognition table. Classic GLR algorithm does not have detection of grammatical ambiguity function, so the translation results in accuracy, somewhat in the past people face the problem is often through repeated to heavy and calibration method to solve the problem, such as the discovery after grammatical ambiguity, phrases content to make use of syntactic analysis of linear geometry identification, according to local optimal theory to improve the quality of the translation results, This method can indeed improve the accuracy of results to a certain extent, but it still fails to meet the current requirements, which is the main defect of GLR algorithm [5–7].

2.2   Algorithm Improvement

Through the above analysis classic GLR algorithm has defects, the defects in the current is called "contingency" larger problem, namely the results obtained from the classic algorithms is a higher probability of coincidence, this kind of phenomenon is, indeed, by accident, but high probability to cause a decline in quality of translation result, and can't meet the accuracy requirement. From this point of view, the classical GLR algorithm must be improved. As for the method of algorithm improvement, this paper mainly chooses the logic of phrase structure analysis based on phrase center, and then improves the classical algorithm by using quaternion cluster. The improved algorithm will calculate the likelihood of the anteclence of phrase through quaternion cluster, as shown in Formula (1).

$$G_E = (V_N, V_T, S, \alpha) \tag{1}$$

where S is the start symbol cluster, V_N is the cycle symbol cluster, V_T is the end symbol cluster, a is the phrase action cluster.

GLR algorithm identification symbols and actions of the linear table top has been on the right side of symbols, some numerical is the true value, center of the sign for the value (0), on the basis of this rely on improving the GLR algorithm, part of speech of each word recognition will cycle, only get three of the same, and the result of the right standard will be included in the final translation as a result, "Storm", for example, under the GLR algorithm model will continuously circulation is used to identify the part of speech of the word, and generate the corresponding results, as long as when the part of speech translation correct result accumulate to three, part of speech translation results would translate into official statements of the queue, so in this way a sharp rise in the model translation results [8–10].

2.3 The Translation Result Correction Process of the Improved Algorithm

The defects of the intelligent recognition English translation model based on the classical GLR algorithm will lead to inaccurate translation results, while other algorithms are generally unable to relate the context. Therefore, no matter which model is under the applicable algorithm, there are problems. On this basis, the improved GLR algorithm solves relevant problems. Therefore, the improved algorithm should be used to construct a translation model, which has the function of correcting the translation results according to the context and can make the translation results more accurate. The basic principle is as follows (Fig. 2): The pos recognition of phrases is carried out by means of parsing linear tables. After the results are generated, errors will be found in the tagged points in relation to the context. For example, there will be grammatical mismatch and asymmetry between the tagged points of words above or below and the tagged points of pos in the recognized results, so correction will be started.

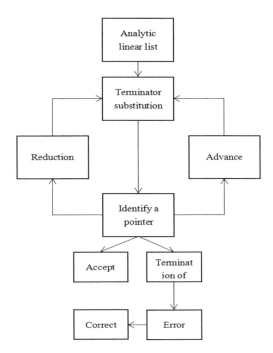

Fig. 2. Translation result correction process of the improved algorithm

3   System Design and Verification

3.1   System Design

In order to verify whether the model translation results of the improved algorithm are accurate and whether the model itself has advantages, this paper will carry out system design first and then verify the system. System design can be divided into three steps: flow design, signal processing and feature extraction. The details of each step are as follows.

Process Design

Based on the improved GLR algorithm intelligent recognition English translation model, the design is carried out, mainly through the hardware equipment to achieve The Acquisition of English signals, and the English signal processing, and then can extract the internal characteristics of the signal, score calculation, and finally complete the recognition. It can be seen that the model has three functions of data collection, output and processing. All three functions are automatic, so the system can run automatically with high convenience. In addition, the improved GLR algorithm is mainly used in the scoring calculation process, and the improved algorithm can play a role in combination with the model framework.

Signal Processing

Because English signal is obtained by hardware acquisition directly from the outside world, so the signal itself is possible interference by outside factors, and the signal itself is also likely to lead to signal quality, for example, in speech recognition is likely to happen in phonetic pronunciation is not standard, or signal interference of peripheral equipment, noise effects, etc., In this case, if the English signal is directly recognized, it is easy to lead to inaccurate translation results or unable to get the translation results. From this perspective, this system requires the signal processing functions, this paper chose the first frame, aggravating, endpoint monitoring, and window of four kinds of processing functions, set up a digital filter to realize these functions, the second to increase processing as an example, combining with digital filter to the vowel spectral characteristics of speech signal, which can identify the first and the second formant Then, neural network is used to measure vowel pitch accuracy and finally optimize English signal. Figure 3 shows the basic flow of signal processing.

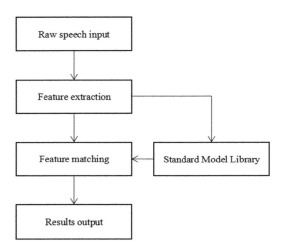

Fig. 3. Basic flow of signal processing

Feature Extraction

In order to realize the above signal processing process and further improve the overall operation efficiency of the system, the feature extraction function needs to be implemented in the system design. First of all, this paper uses artificial intelligence tools to organize all speech signal data in a centralized manner, so as to unify the data, and then find parameterized features from them for subsequent calculation. Secondly, the discrete sampling of the continuous signal can be obtained by Fourier transform, which can be used to calculate the continuous spectrum of the signal. It is important to note that this article chose the Fourier transform method is FFT, this is a kind of based on discrete Fourier transform the odd-even the actual properties, derived from classical discrete Fourier transform algorithm to improve the calculation method, is used in speech signal spectrum calculation, capable of supporting processing functions such as aggravating, add window, framing, also can rapidly in each short-time analysis window spectrum signal, Through these methods, features can be extracted smoothly and the system design can be completed.

3.2   System Verification

In order to verify the effectiveness and advantages of this system and the improved GLR algorithm model, relevant verification work will be carried out in this paper.

Verification scheme: With statistical algorithms, the algorithm of dynamic memory, classic GLR algorithm, improve the GLR algorithm to verify the object, using conventional method to construct the intelligent identification model of English translation (GLR algorithm modeling method is same as above), using the same method to construct the English translation system, and then according to the translation accuracy and speed to test indicators, The same 50 English phrase sounds are input into each system as the original corpus, and the performance of each system is observed and counted synchronously. The results can be obtained by combining the two indexes. In addition, in order to ensure the fairness of verification as much as possible, the corpus contents and annotation results of the four algorithms are completely consistent, containing 780,000 words, which is expected to meet the translation requirements of 22,000 sentences and 12,000 phrases. The 50 English phrase sounds selected in this paper are completely within the scope of the corpus.

Verification results: First, the statistical algorithm model performs poorly in translation accuracy, with 91% translation accuracy. Common errors are grammatical errors and translation speed is 1 s–2 s. Second, the dynamic memory algorithm model performs poorly in translation accuracy, with a translation accuracy of 93%. The common error is incorrect meaning. The reason is that the algorithm needs to accumulate experience, but this is a long process, so this step is not implemented in this verification, and the translation speed is 1 s–1.5 s. Thirdly, the performance of the classic GLR algorithm model in translation accuracy is mediocre, with a translation accuracy of 97%. The common error is word repetition, with a probability of about 31%, and the translation speed is 0.4 s–1 s. Fourth, the improved GLR algorithm model performs well in translation accuracy, with a translation accuracy of 99% and no obvious errors, and a translation speed of 0.4 s–1 s. The comparison shows that the improved algorithm is effective and has advantages, and the system design is successful.

4   Conclusion

To sum up, in order to provide better English translation services for modern people, the machine translation field should pay attention to the defects of the classical GLR algorithm and improve it. The improved GLR algorithm has more advantages. System design based on the algorithm can better improve the accuracy of translation results and avoid the impact of accidental repetition. Meanwhile, the algorithm also inherits the simplification characteristics of the classical algorithm and can improve the translation speed.

References

1. Borsotti, A., Breveglieri, L., Reghizzi, S.C., et al.: Fast GLR parsers for extended BNF grammars and transition networks. J. Comput. Lang. **64**(8), 101035 (2021)
2. Nikiforov, I.V.: Fault detection and isolation based on the constrained GLR Test. IFAC Proc. Vol. **42**(4), 709–714 (2009)
3. Madakyaru, M., Harrou, F., Sun, Y.: Improved data-based fault detection strategy and application to distillation columns. Process Saf. Environ. Prot. **107**, 22–34 (2017)
4. Lundin, G.O., Mouyon, P., Manecy, A.: A GLR algorithm for multiple consecutive measurement bias estimation. In: 2017 Workshop on Research, Education and Development of Unmanned Aerial Systems (RED UAS) (2017)
5. Urbano, S., Chaumette, E., Goupil, P., et al.: Aircraft vibration detection and diagnosis for predictive maintenance using a GLR test. In: IFAC SafeProcess 2018 (2018)
6. Madakyaru, M., Harrou, F., Sun, Y.: Improved anomaly detection using multi-scale PLS and generalized likelihood ratio test. In: Computational Intelligence. IEEE (2017)
7. Bang, T.S., Virach, S.: Sentiment classification for hotel booking review based on sentence dependency structure and sub-opinion analysis. IEICE Trans. Inf. Syst. **101**(4), 909–916 (2018)
8. Khan, Z., Razali, R.B., Daud, H., et al.: Bad data detection in power system state estimation based on generalized likelihood ratio test. Int. J. Energy Stat. **4**(4), 1650016 (2017)
9. Botre, C., Mansouri, M., Nounou, M., et al.: Process monitoring using a PCA-based exponentially weighted generalized likelihood ratio chart. In: AIChE Spring Meeting and Global Congress on Process Safety (2017)
10. Okafor, S.I., Amadi, A.H., Abegunde, M.A.: The choke as a brainbox for smart wellhead control. Eur. J. Eng. Res. Sci. **6**(1), 114–118 (2021)

Construction of Financial Risk Early Warning Model Based on Ant Colony Algorithm

Jing Hu[✉]

Yiwu Industrial and Commercial College, Yiwu 322000, China
569141956@qq.com

Abstract. In order to construct financial risk early warning model through ant colony algorithm and help enterprises to do a good job in financial risk early warning, this paper will carry out relevant research. This paper firstly introduces the basic concept of ant colony algorithm and the importance of financial risk early warning, and then carries out the construction of financial risk early warning model based on ant colony algorithm. Through this study, the ant colony algorithm to build the financial risk early warning model is effective, can improve the financial security of enterprises, ensure the stability of enterprise financial operation.

Keywords: Ant colony algorithm · Financial risk · Risk warning model

1 Introduction

In the long-term business activities of enterprises, it is inevitable to encounter some kinds of financial hidden dangers, which will cause financial risks, which is inevitable. Do it right on the basis of the enterprise financial risk prevention, as far as possible to reduce the outbreak of the financial risk probability, processing a variety of hidden dangers in advance, at least to minimize the influence of the financial risk, but the effectiveness of the modern enterprise financial risk prevention work worrying, so how to improve the quality of the financial risk prevention work is worth people to think about. Inspired by problem research field, puts forward the financial risk early warning model of technical means, in theory to construct such a model can better, faster to find financial risks, the risk of early outbreak or not remind enterprise for processing, and this model is the basis of various algorithms, different algorithms on the accuracy of the early warning model results and related performance has a great influence, Therefore, in order to popularize this model, many attempts have been made in related fields, and some high-performance models have been obtained, among which the financial risk warning model based on ant colony algorithm is more representative, which will be studied in this paper.

2 The Basic Concept of Ant Colony Algorithm and the Importance of Financial Risk Warning

2.1 Algorithm Concept

In essence, Ant Colony algorithm is an algorithm aimed at finding the optimal path to achieve the goal. It was first proposed by Italian scholars Dorigo and Maniezzo, that is, they were inspired by observing the behavior of ants foraging out and put forward the concept of Ant Colony System (ACS) [1, 2]. According to their paper, ant the behavior of the individual in foraging is simpler, but the whole ant colony on macro action has a certain intelligent system characteristics, namely each ant colony's range is limited, assuming that exist within the scope of a piece of food, each ant doesn't know the location of the food, this time will be scattered, ants And leave pheromone on the path to looking for, when an individual ants find food, near the ants will be notified to move food, will pass to find food in the process of the ants walk path, make this path pheromone concentration increased, thus the distance of the ants are influenced by pheromones, postpone the highest path pheromone concentration to the food, So the colony knows the shortest path to the food. On this basis, Dorigo and Maniezzo believed that there was a specific way of information transmission in the ant colony system, and there was a positive feedback mechanism in the information transmission. Grasping these two points, they created the ant colony algorithm. Figure 1 shows the system topology of ant colony algorithm [3].

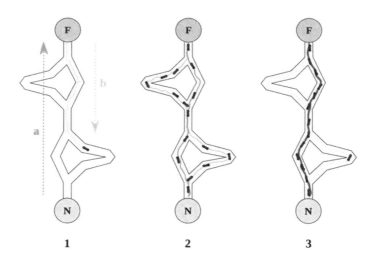

Fig. 1. System topology of ant colony algorithm

In the Fig. 1, N is the starting point, F is the feasible solution of the problem, 1, 2 and 3 are the three stages of path change of ant colony foraging respectively, so it can be seen that the ant colony will finally find the shortest path, namely the optimal solution of the problem.

2.2   Importance of Financial Risk Warning

The emergence of financial risk indicates that the enterprise will face a certain economic loss, and if the impact of financial risk is large, it is possible to create more risks and hidden dangers, so financial risk is endless for the enterprise. At the same time, part of the financial risk also involves the law, indicating that the financial risk may be upgraded to legal risk. From this perspective, any company don't want to own a financial risk, and therefore must to prevention and control of financial risk, but the enterprise continued reform of China's financial operation and potential financial risks will continue to form, so the financial risk may not be fully control, in this case companies will hope oneself can the comprehensive supervision on the financial risk, The purpose is to understand the existing financial risks for the first time, and then deal with the hidden risks before they become risks, in a disguised way to avoid their own financial risks, the concept of financial risk early warning is recognized in this context. As can be seen, financial risk early warning can maximum limit to avoid suffering the influence of the financial risk of enterprise, its main function is to prevent financial risks, secondary development function is to even control the financial risk, financial risk to the enterprise's financial aspects of the influence of minimum, also avoid new risk retention, so the financial risk early warning is very important [4].

3   Financial Risk Early Warning Model Construction

3.1   Setting of Basic Conditions for Model Construction

Because of financial risk early warning model construction of fundamental purpose is to predict the risk, so that the enterprise timely prevention and control, so the models must be completely against financial risk, and the actual situation of the type of financial risk has a lot of, involved in the index is extremely complicated, therefore, for the sake of convenience, this paper mainly based on the model construction of the common investment risk [5].

Investment risk refers to the enterprise for a certain amount of money spent on a project, after waiting for the return of project affected by market demand changes appeared in the process of the phenomenon of lower returns, leading to lower than expected, eventually returns the actual value and may even lower than the cost of risk, the risk not only can reduce enterprise's economic benefits and may also make the enterprise financial losses, Many businesses fail because of this risk. In addition, there are many specific types of investment risks, including interest rate risk, reinvestment risk, exchange rate risk, inflation risk, financial derivatives risk, moral hazard, default risk and so on.

With reference to investment risks, the main indicators for assessing the outbreak probability and influence are shown in Table 1.

Table 1. Evaluation indexes of outbreak probability and influence of investment risks

The index name	Paraphrase
Total amount of capital available to the enterprise	Refers to the enterprise has the specific amount of available capital, represents the enterprise to meet the risk of the tolerable range and unbearable boundaries
Proportion of investment cost	Refers to the proportion of the investment cost in the total available funds of the enterprise in the project. A large proportion represents a high risk incidence and influence, whereas a small risk incidence and influence
Stability of project rate of return	Refers to the amplitude of the fluctuation of the project rate of return. The larger the amplitude is, the more unpredictable the project rate of return is, and the greater the probability of the occurrence of financial risks will be to some extent
Project risk rate	It refers to the risk of the project itself. The higher the risk rate is, the more likely the project will lead to financial risks, so the higher the risk rate is, and the lower the risk rate is
Corporate debt	Refers to the amount of debt of the enterprise after the investment project, the larger the amount is, the higher the risk of investment behavior, the smaller the risk bearing capacity of the enterprise itself, otherwise the opposite
Project market stability	It refers to the recent fluctuation range of the market where the investment project is located, such as the fluctuation range of exchange rate. The larger the fluctuation is, the greater the influence of the market on the stability of the project is, and the greater the incidence and influence of the financial risk of the enterprise is; otherwise, the smaller it is
Cash flow status of the enterprise	The size of the cash flow after business investment so far that, according to the size of cash flow enterprise is currently in a profitable trend (at the same time investment share if cash flow is greater than the same period, means that the enterprise did not profit, profit or conversely is profitable trend), if in earnings trend, is the enterprise interior has erupted investment financial risk

According to the related indexes of Table 1, as long as the information comprehensive analysis can determine the current outbreak of investment enterprise financial risk probability, risk after the outbreak of the early influence of size, investment, financial risk whether had broken out in the enterprise, according to the results of the judgment, can help companies as a financial risk early warning work, Therefore, these indicators will become the basic conditions of the financial risk warning model under ant colony algorithm [6–8].

3.2   Model Construction

The basic framework of financial risk warning model under ant colony algorithm is shown in Fig. 2 [9].

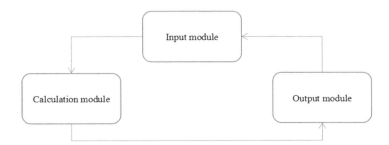

Fig. 2. Financial risk warning model under ant colony algorithm

Combined with Fig. 2, it can be seen that the financial risk warning model under ant colony algorithm in this paper is composed of three modules, namely, input module, calculation module and output module. The specific contents are described below.

Input Module

Input module is mainly to all the indicators in the Table 1 information input to the calculation module, so the input module operation index of the first step is to collect relevant information, this paper except through the enterprise related indicators of a historical record of the collection of data, also with big data technology + smart technology solutions for the project, the market index information, For example, by historical records for the concrete numerical value of the total available funds of the enterprise, through the big data technologies + smart technology solutions for the project risk, project information related to the stability of the market, so in this paper, the data collection scheme for complete coverage of all indexes was carried out in Table 1, can ensure the information is complete, quality for calculation. Calculate the financial risk early warning, because of the need to consider all the indicators, and so on information collection after the completion of the input link, in this paper, the related information of all indicators to quantify process, namely the total funds available to enterprises index vector as the foundation, the remaining indexes changes as a vector, and then different vector generation into the ant computing system of calculation module [10].

Compute Module

The core of the computing module of this model is the ant colony computing system, which carries the basic logic of ant colony algorithm: On basis of the vector, each change vector calculation of the influence of the basic vector, gets some path, on the basis of the intervention of all paths through the observation vector, if according to the changes of basic vector by after intervention than in the past, means that financial risk increases higher incidence, influence, reduce, influence, reduce the incidence of the opposite, Through this kind of performance can carry on the early warning to the enterprise financial crisis.

Output Module

The output module is to output the calculation results of the calculation module, and the output target is the input module. In this way, the ant colony algorithm can keep iterating in the model and finally find the optimal solution. It is worth noting that although the ant colony algorithm needs to obtain the optimal solution through a large number of iterations, this process cannot be endless and must be limited by termination conditions. For this reason, this paper proposes corresponding termination conditions based on enterprise requirements, as shown in Table 2.

Table 2. Termination conditions for output modules

Termination conditions	Stop when the calculation results show that the incidence of enterprise financial risk is less than 10%
	Stop when the calculation results show that the influence of corporate financial risk is less than 10%
	Stop when the calculation results show that the incidence of enterprise financial risk is higher than 30%
	Stop when the calculation results show that the influence of corporate financial risk is higher than 30%

Combined with Table 2, the iteration cycle of ant colony algorithm will be terminated when any condition is triggered, and the calculation result is the optimal solution. Compute the optimal solution was divided into good and bad two kinds, the first, second, the result of the termination condition is good, the third, fourth, the result of the termination conditions, the bad, depending on the type of the result, can help enterprises to accurately forecast financial risk, and at the same time, according to the calculation process each vector index changes in the vector index on the basis of performance, If a certain change vector index in the vector index on the basis of performance, the bad, means that the index is lead to risk looming, enhance its influence factor, enterprise shall timely treatment, such as base vector set enterprises index "enterprise" total funds available for 100%, enterprise change vector index "investment costs accounted for" in the basic vector accounts for more than 70%, This is on behalf of the enterprise risk range of affordable, convenient near unsustainable, then in accordance with the termination conditions index, and the influence of the relationship between each other, accounted for more than 70% above the base vector, on behalf of the enterprise financial risk, as well as the incidence of influence more than 30%, that the incidence of outbreak of enterprise financial risk investment is high, the influence is high, Enterprises should reduce the proportion of investment in time, and it is best to return funds to avoid risk outbreak and excessive influence.

4   Conclusion

In conclusion, the logic of financial risk warning model under ant colony algorithm is relatively simple, so it is convenient to apply. At the same time, the accuracy of warning results is good, indicating that the model has certain practical value, which can help enterprises to break the current predicament that they cannot accurately predict financial risks. On this basis, enterprises should actively introduce the model to improve their financial risk prevention and control ability.

References

1. Anagnostopoulos, I., Rizeq, A.: Confining value from neural networks: a sectoral study prediction of takeover targets in the US technology sector. Manag. Financ. **45**(10/11), 1433–1457 (2019)
2. Bhadani, S., Verma, I., Dey, L.: Mining financial risk events from news and assessing their impact on stocks. In: Bitetta, V., Bordino, Il., Ferretti, A., Gullo, F., Pascolutti, S., Ponti, G. (eds.) MIDAS 2019. LNCS (LNAI), vol. 11985, pp. 85–100. Springer, Cham (2020). https://doi.org/10.1007/978-3-030-37720-5_7
3. Uj, A., Nmb, E., Ks, C., et al.: Financial crisis prediction model using ant colony optimization-ScienceDirect. Int. J. Inf. Manage. **50**, 538–556 (2020)
4. Chi, G., Uddin, M.S., Abedin, M.Z., et al.: Hybrid model for credit risk prediction: an application of neural network approaches. Int. J. Artif. Intell. Tools **28**(05), 1950017 (2019)
5. Eggermont, F., Verdonschot, N., Linden, Y., et al.: Calibration with or without phantom for fracture risk prediction in cancer patients with femoral bone metastases using CT-based finite element models. PLoS ONE **14**(7), e0220564 (2019)
6. Moscato, V., Picariello, A., Sperlí, G.: A benchmark of machine learning approaches for credit score prediction. Expert Syst. Appl. **165**(9), 113986 (2021)
7. Cui, Z., An, F., Zhang, W.: Internet financial risk assessment based on web embedded system and data mining algorithm. Microprocess. Microsyst. **82**(3), 103898 (2021)
8. Mousavi, M.M., Lin, J.: The application of PROMETHEE multi-criteria decision aid in financial decision making: case of distress prediction models evaluation. Expert Syst. Appl. **159**, 113438 (2020)
9. Dreo, J., Siarry, P.: Continuous interacting ant colony algorithm based on dense heterarchy. Futur. Gener. Comput. Syst. **20**(5), 841–856 (2004)
10. Kran, M.S., Zceylan, E., Gündüz, M., et al.: A novel hybrid approach based on particle swarm optimization and ant colony algorithm to forecast energy demand of Turkey. Energy Convers. Manag. **53**(1), 75–83 (2012)

Analysis of Traditional Quyi Emotion Classification Algorithm Driven by Deep Belief Network

Qi Fu[✉]

Southwest University of Science and Technology, Mianyang 621000, China
55607754@qq.com

Abstract. In order to classify traditional Quyi based on emotions through algorithms driven by deep belief network, this paper will carry out relevant research, mainly discussing the basic concepts of deep belief network, and then proposing the algorithm model and matters needing attention of traditional Quyi emotion classification. Through the research of this paper, based on the deep belief network, the traditional folk art can be classified through the algorithm, and the algorithm is effective in the process of application, and the classification results are high accuracy.

Keywords: Deep belief network · Traditional folk art · Emotion classification algorithm

1 Introduction

Under the new era, China's increasing need of people live entertainment, music has become a major "spiritual food", but most users want according to their own emotional state changes to choose music, and the music data is very huge in the network environment, users may be difficult to rely on oneself to choose the right music, Software operators cannot rely on humans to help users. This background, people began to think about whether can through technical means to solve problems, and therefore started a study in related fields, finally proposed the deep belief networks and music emotion classification algorithm, using the former work, driving the latter to effectively understand the mood of different music theme, with corresponding to the user's emotional themes, The problem can be solved by automatically selecting the music that matches the user's mood from the vast data set. The proposal of this method has been widely concerned by people, but because it has been proposed for a short time, there are still many people who do not know enough about it and it is difficult to implement. Therefore, this paper will take traditional Quyi music as an example to carry out relevant research.

2   Basic Concepts of Deep Belief Networks

Deep belief network was first established in 2006 by Hinton, was defined as a top-down automatic learning multilayer neural network, and then with cognitive deepening, its definition changes is a kind of by multilayer neural networks of neurons, could be conducted on the basis of probabilistic graphical just expression, inference network model, This model is formed by a number of restricted Boltzmann units, and the formation mechanism is stacking [1, 2]. Figure 1 is its basic structure.

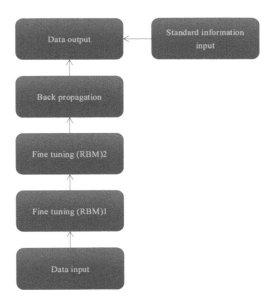

Fig. 1. Basic structure of deep belief network

It can be seen from Fig. 1 that the deep belief network contains at least two restricted Boltzmann machines, which are ordered from top to bottom. At the same time, there is some overlap between each restricted Boltzmann machine, indicating that the former restricted Boltzmann machine is the explicit layer of the latter. It also indicates that the output of the former constrained Boltzmann machine will be input to the latter constrained Boltzmann machine, which is equivalent to the input of the latter constrained Boltzmann machine. According to this process, the deep belief network can carry out independent learning and training until the training link reaches the last link. After the completion of learning and training, the deep belief network will distinguish the link weight and node bias of each link in the training and realize network initialization. Initialization of the network is not yet practical application ability, through the relevant algorithm is needed to adjust, the current widely used algorithm is reverse conduction algorithm, this algorithm can top-down supervise deep belief network according to the supervision of the performance of the network simulation operation adjustment results, every time adjustment is equivalent to a try, until you find the best adjustment as a result, This method can effectively solve the problem that other similar network models are prone to fall into local optimum, and can also effectively improve the efficiency of learning and training and shorten the time. Belief network is important to note that although the depth has great advantage in training, learning ability is very strong, then the training must be large sample data as support, can appear otherwise parameter weights is too small, lead to inadequate training problems, in this case the results of the analysis is not has the application value, Therefore, attention should be paid to this item in the actual operation [3–5].

3   Traditional Quyi Emotion Classification Algorithm Model

Deep belief network application scope is very broad, including the traditional folk art sentiment classification, in the process of network can be used as the driver of the classification algorithm, and the traditional vocal emotion classification algorithms must be based on the deep belief networks to good results, therefore, in order to achieve the target in depth below related algorithm model based on belief network.

3.1   Model Framework

Based on the deep belief network, the model structure of the traditional Quyi emotion classification algorithm is not complicated and only consists of two blocks, as shown in Fig. 2.

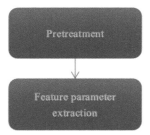

Fig. 2. Model structure of traditional Quyi emotion classification algorithm based on deep belief network

Combined with Fig. 2, it can be seen that the model structure of traditional Quyi emotion classification algorithm based on deep belief network is composed of two blocks: pre-processing and feature parameter extraction. The relationship between the two blocks is as follows: Pretreatment is mainly aimed at the initial training sample, can remove the poor quality of the data in the initial sample, the initial data "purification" samples, to avoid affect the accuracy of the result of poor data quality, and after the "purification" sample data can be input to the next section, the algorithm operation to extract the accurate characteristic parameters [6–8].

3.2   Algorithm Analysis

The following paragraphs will analyze the algorithm around the two major blocks of the model framework.

Pretreatment

The data that the traditional Quyi emotion classification algorithm faces are audio data, not data in conventional format. Therefore, in the preprocessing method, it is impossible to use common keywords retrieval and other methods to process, and other methods must be selected [9–11]. This paper makes a choice for this, as shown in Fig. 3.

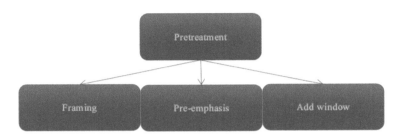

Fig. 3. Pretreatment method

The first is frame splitting, because traditional Quyi has a specific musical signal, which has an obvious feature that all the performance characteristics of the signal will not change in a relatively short period of time, which is theoretically called short-term stability. To grasp the short-term stability characteristics, frame segmentation is to split each frame of traditional folk art audio, confirm the specific situation of the short-term stability characteristics of each frame, such as the audio of a frame is relatively bright, or low, according to this result to classify each frame, so as to complete the frame segmentation processing. It should be noted that in the frame splitting process, the duration of each frame should generally be more than 10 ms, and the longest duration is not recommended to exceed 30 ms. The reason for doing this is to ensure smooth transition between adjacent frames. At the same time, in case of special circumstances, it is allowed to keep the overlapping relationship of 1/3 frame length between adjacent frames during frame splitting. This is also to better achieve the smooth transition of adjacent frames.

The second is pre-weighting, because the high frequency part of the traditional Quyi music signal generally has the characteristics of low energy, so when preprocessing the high frequency signal part, special methods should be taken to enhance the signal energy of this part, this processing method is pre-weighting. The pre-weighting method is mainly realized by filtering, and the specific content is shown in Formula (1) [12].

$$y(n) = x(n) - \mu x(n-1) \tag{1}$$

where, $y(n)$ is the output signal after pre-weighting, and $x(n)$ is the input signal and the pre-weighting factor, whose value is infinitely close to 1.

Finally, add window, its purpose is to increase the continuity between adjacent frames, can greatly weaken the edge effect, control shielding leakage, loss and other problems. There are many windowing processing methods, this paper mainly chose the Hamming window function for processing, the specific process is to multiply the hamming window function and the windowing signal value.

Extraction of Characteristic Parameters

In fact, the method of feature parameter extraction in music emotion classification has been proposed by some researchers for a long time, and this method is indeed feasible. However, in the actual operation process, this method is complicated and the process is very complicated, which requires the extraction of various parameters and comprehensive calculation. However, based on the deep belief network, the feature parameter extraction method can be simplified and only the feature vector of music emotion can be extracted. Therefore, this method will also be adopted in this paper. Firstly, according to music theory, emotion can be regarded as a major attribute of traditional Quyi music, which is obviously representative, so it can be regarded as a characteristic of traditional Quyi music. Secondly according to the traditional vocal music emotion classification, this paper adopted by the mood of the classification model for the AV model, the model of emotion can be divided into positive and negative two plates, according to the characteristic parameters of emotions directly determine the traditional vocal music emotion type and trend, namely if the traditional vocal music emotion characteristic parameter is more close to the positive plate, Then the mood of this piece of Quyi music is more positive. According to the distance between the two, the emotional degree of this piece of Quyi music can be judged. Generally, the emotional degree can be divided into five grades: "very positive, relatively positive, general (neither positive nor negative), relatively negative and very negative". Finally, correlation algorithm can be used to extract feature parameters, and the process is as follows: First, because in depth based on belief network to the traditional vocal music emotion progress characteristic parameters extraction will encounter some unavoidable problems, such as part of the training sample data is less, leading to incomplete characteristic parameters, so in order to solve this kind of method, will reduce the depth of belief network learning training parameter values, This article mainly USES the method to deal with the weights of sharing, this method is derived from the convolution neural network, is mainly used in the convolution of the network model layer, with conventional neural network connections, all parameters of the do not share, this method can give full play to the local connection of convolution neural network, the sharing characteristic of the weights will be a layer of all nodes are interconnected, And let these nodes share a set of parameters, at least in traditional Quyi music feature parameter extraction is feasible, that is, in Quyi music each frame of the signal is not independent of each other, on the contrary, there is obvious continuity, which meets the application requirements of weight sharing method; Second, on the basis of deep belief network, the original time domain characteristics of Quyi music are taken as input data, and then layer by layer training is carried out in the network model, prompting the model to extract the characteristics of the input signal, the main features include short-term average amplitude difference, short-term energy, short-term autocorrelation coefficient, etc. For convenience, this paper takes short-time energy as an example and assumes it as the original input feature of deep neural network, so the energy of this feature in a short period of time can be used to understand the process of acoustic signal energy changing with time. Figure 4 shows the extraction process of characteristic parameters.

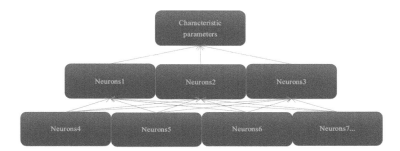

Fig. 4.   Extraction process of feature parameters

3.3   Precautions

Through the above method can obtain the traditional vocal music characteristic parameters, then the mood music are classified, but need to pay attention to because of vocal music emotion of the original time domain feature dimension is higher, inside there are a lot of redundancy and noise, it may affect the accuracy of the classification results, therefore suggested to inspect after extracting feature parameters, If redundancy and noise are found, it is necessary to conduct a learning training again, and before this learning training, dimension reduction of the input training template is required. In terms of dimension reduction processing methods, it is generally recommended to choose principal component analysis. In this case, a multivariate statistical method that examines the correlation between multiple variables can judge according to a few principal components in the sample data, so it has a good effect of dimension reduction. In addition, in the sample data dimension reduction after processing, can be deep belief network training, through training can learn music mood data characteristics, the final output characteristics of the data, for the relevant personnel to the traditional vocal music, and in the process of classification recommended Softmax classification function method, this method is more suitable for deep belief network, treatment effect is good.

4   Conclusion

In conclusion, the depth of the belief network is essentially a deep learning process, mainly by neural network to realize, in the structure of itself on its relatively simple, but the content is more complex, so different about its functionality, use correct methods to use deep belief network, can effectively to the traditional vocal music emotion classification, To meet the personalized needs of users to choose music according to their own mood. At the same time, there are different forms of deep belief network. Although the traditional form can achieve the goal of music emotion classification, its applicability is poor. However, the deep belief network highlighted in this paper belongs to the convolutional network, which is more applicable and therefore has higher application value and practicability.

References

1. Paquette, S., Takerkart, S., Saget, S., et al.: Cross-classification of musical and vocal emotions in the auditory cortex: cross-classification of musical and vocal emotions. Ann. N. Y. Acad. Sci. **1423**, 329–337 (2018)
2. Abeyratne, K., Jayaratne, K.L.: Classification of Sinhala songs based on emotions. In: 2019 19th International Conference on Advances in ICT for Emerging Regions (ICTer). IEEE (2019)
3. Catharin, L.G., Ribeiro, R.P., Silla, C.N., et al.: Multimodal classification of emotions in latin music. In: 2020 IEEE International Symposium on Multimedia (ISM). IEEE (2020)
4. Nascimben, M., Ramsoy, T.Z., Bruni, L.E.: User-independent classification of emotions in a mixed arousal-valence model. In: 2019 IEEE 19th International Conference on Bioinformatics and Bioengineering (BIBE). IEEE (2019)
5. Pandeya, Y.R., Lee, J.: Deep learning-based late fusion of multimodal information for emotion classification of music video. Multimed. Tools Appl. **80**(2), 2887–2905 (2020). https://doi.org/10.1007/s11042-020-08836-3
6. Durahim, A.O., Setirek, A.C., Özel, B.B., et al.: Music emotion classification for Turkish songs using lyrics. Pamukkale Univ. J. Eng. Sci. **24**(2), 292–301 (2018)
7. Phu, V.N., et al.: A valence-totaling model for vietnamese sentiment classification. Evol. Syst. **10**(3), 453–499 (2019)
8. Al-Kharboush, F.M., Al-Hager, Y.M.A.: Features extraction effect on the accuracy of sentiment classification using ensemble models. Int. J. Sci. Res. (IJSR) **10**(3), 228–231 (2021)
9. Rajmohan, A., Ravi, A., Aakash, K.O., et al.: CoV2eX:A COVID-19 website with region-wise sentiment classification using the top trending social media keywords. In: 2021 Sixth International Conference on Wireless Communications, Signal Processing and Networking (WiSPNET) (2021)
10. Dritsas, E., et al.: Pre-processing framework for twitter sentiment classification. In: MacIntyre, J., Maglogiannis, I., Iliadis, L., Pimenidis, E. (eds.) AIAI 2019. IAICT, vol. 560, pp. 138–149. Springer, Cham (2019). https://doi.org/10.1007/978-3-030-19909-8_12
11. Seal, A., Reddy, P., Chaithanya, P., et al.: An EEG database and its initial benchmark emotion classification performance. Comput. Math. Methods Med. **2020**, 1–14 (2020)
12. Sang, H.L., Kim, J.Y.: Classification of the era emotion reflected on the image using characteristics of color and color-based classification method. Int. J. Softw. Eng. Knowl. Eng. **29**(8), 1103–1123 (2019)

Application of Cloud Desktop Technology in Computer Laboratory Management

Yanru Wang[✉]

School of Economics, Jilin University, Changchun 130012, China
wangyanru@jlu.edu.cn

Abstract. The development of computer network technology to promote the development of all walks of life, and the rapid development of cloud computing industry is to let a lot of industry see the advantages of the development of network technology, even in COVID - 19 outbreak of the moment, still did not stop the development of cloud computing industry, whether it is a traditional industry, or for the computer industry, Cloud management, cloud services and cloud technology are the main directions of future development. With the rapid development of Internet and information technology, cloud desktop technology based on cloud service technology is becoming more and more mature, and provides strong support for modern mobile office. For the information teaching of colleges and universities, the computer laboratory as the main means at any time should pay attention to the improvement of the degree of information teaching, and the emergence and application of cloud desktop technology can effectively solve the drawbacks of traditional computer laboratory management. This paper focuses on the application of cloud desktop technology in the scientific management of university computer laboratories.

Keywords: Cloud desktop · Computer laboratory · Management

In recent years, "cloud" has become a new word in modern life, cloud computing, cloud payment, cloud services and other cloud technologies emerge in an endless stream, in addition, Baidu Cloud, micro-cloud and other "cloud concepts" are also more and more, and with the continuous development of cloud technology, cloud desktop began to appear in the public field of vision. In fact, cloud desktop is a concept similar to desktop virtualization. Virtual desktops have many architectures, such as VDI, VOI, and IDV. On the basis of desktop virtualization technology, cloud desktop has advanced in terms of corresponding technology modes and application scenarios [1]. Since 2017, China's cloud computing the size of the market is more and more big, ali, according to relevant data cloud is expected by the end of 2025 the overall size of the cloud computing market in China will be more than 2 trillion, but in 2017 was 34.32% on the basis of the scale of cloud computing market in China has not yet reached 70 billion, only from the perspective of the overall performance of the external market, The development of cloud computing provides profound technical support for the emergence, progress and wide application of relevant technologies [2]. Compared with the traditional PC terminal, the cost of cloud desktop is greatly reduced and the service life is greatly improved. The technical integration of cloud desktop and university computer laboratory management can not only meet the learning needs of all

© The Author(s), under exclusive license to Springer Nature Switzerland AG 2022
V. Sugumaran et al. (Eds.): ICMMIA 2022, LNDECT 138, pp. 494–500, 2022.
https://doi.org/10.1007/978-3-031-05484-6_62

students, but also reduce the cost and management efficiency of university computer laboratory management.

1   Cloud Desktop Technology

1.1   Cloud Desktop Concepts

Cloud desktop (Cloud Deskto) can also be called desktop virtualization or Cloud computer, which is a new technology mode that replaces traditional PC functions more intelligently through Cloud computing related technologies. Cloud desktop technology is similar to desktop virtualization technology, but also an "upgraded version" of desktop virtualization technology. This technology is mainly based on distributed computing and storage technology, and uses highly encrypted technical algorithms to gather all the essence of the Internet applications [3]. Through the integration of technologies and encryption algorithms to provide users at different levels with more secure, convenient and rich and diverse services.

In general, the cloud desktop technology improves the utilization efficiency of network resources by virtualizing different physical devices on the basis of virtual technologies. With the support of virtualization technology, all hardware devices and software applications become more flexible, and the connection between software and hardware is also more flexible. Both software and hardware devices greatly improve their scalability. Essentially, cloud desktop technology is based on virtualization technology, the traditional PC end user information for effective storage and unified management, and then use the most simple network connected to terminals, while users can be realized with the aid of all kinds of terminal equipment software such as the centralized management of information, it is based on the advantage, Therefore, cloud desktop has been widely used in teaching practice in colleges and universities [4].

1.2   Features and Advantages of the Cloud Desktop

The cloud desktop can also be regarded as a collection of cloud terminals. Compared with traditional computer clients, traditional PCS require all computers to install graphics cards independently, and some software updates even require the replacement of graphics cards [5]. However, the emergence of cloud desktop is a good solution to this problem, because the cost of cloud terminal is lower than that of traditional PC, and the service life is 4 times that of traditional computer, and the use cost is greatly reduced [6]. The technical architecture of cloud desktop and cloud terminal is divided into software layer and hardware layer, as shown in Fig. 1:

As can be seen from the figure above, the software layer is mainly some commonly used auxiliary software, while the hardware layer refers to cloud terminals, servers, etc. Supported by desktop virtualization technology, the advantages of cloud desktop technology are mainly reflected in four aspects, as shown in Fig. 2:

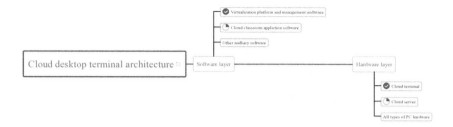

Fig. 1. Cloud desktop terminal architecture

Fig. 2. Advantages of cloud desktop technology

Compared with traditional computer desktop management, cloud desktop technology has the biggest advantages in two aspects: self-developed protocols and convenient maintenance [7]. Cloud desktop technology under the desktop management system includes server, cloud and cloud desktop software, and other important terminal software, but cloud terminal that they cannot be related to operation and information storage, the realization of these functions must have the aid of certain agreements, and be realized by connecting with system server, it is because of the cloud desktop USES is the independent research and development agreement, Therefore, the system based on cloud desktop technology does not have software compatibility problems; Secondly, system maintenance becomes more convenient [8]. This is the most prominent advantage of the cloud desktop. With the cloud desktop technology, users only need to manage and maintain servers, while cloud terminals do not need to perform maintenance. All maintenance operations are fixed to the system, which is very convenient.

Of course, cloud desktop technology also has different technical architectures, which use different scenarios and have different advantages and features, as shown in Fig. 3:

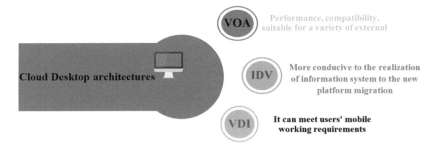

Fig. 3. Advantages of different cloud desktop architectures

2   Practical Application of Cloud Desktop Technology in Computer Laboratory Management

2.1   Centralized Management of Computer Laboratories

The cloud desktop system can realize the daily system maintenance and software update of all computer terminal devices with the help of the central server of the system, and the management of university computer laboratories can realize the unified maintenance and software update of all computer devices with the help of the cloud computing management platform [9]. When students use computers to learn, administrators only need to install the system and related software required by students on the cloud desktop system according to the practical requirements of the specific course before the course starts. Then desktop VMS can be used to implement unified deployment planning. In particular, the administrator can through the cloud desktop system terminal to access and custom computer desktop content, students do not need to install the software on the computer one by one, and managers also do not need on every computer system installation and maintenance, so can maximum increase work efficiency of college computer laboratory management, At the same time, it can also reduce a lot of manpower investment in equipment management and computer maintenance.

Generally speaking, the management of computer laboratories in colleges and universities is scattered [10]. For this reason, the management of relevant data and information in computer laboratories is also scattered, and the management efficiency of computer laboratories is relatively low. In order to improve the overall efficiency of computer room management, can, by using the method of building equipment terminal address pool through cloud desktop technology, to all the computer's network card MAC address for centralized management, all on the cloud desktop server management terminal, which can facilitate unified management of computer information, at the same time can also undertake unity of all the equipment analysis and management. See Fig. 4:

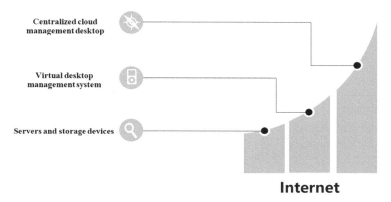

Centralized cloud
management desktop

Virtual desktop
management system

Servers and storage devices

Internet

Fig. 4. Network structure of cloud desktop computer lab

2.2   Dynamic Management of Computer Laboratory Resources

Cloud desktop and desktop virtualization technology both can carry on the dynamic management of computer laboratory resources, in the process of college computer laboratory management, can according to the actual course of students practice demand flow desktop image module, upgrade at the same time, delete, and so on measures to ensure that meet the demand of students at different stages of experimental courses.

With the help of cloud desktop technology, teachers can timely release the subject resources related to current course teaching, which can facilitate the storage and sharing of course resources. Students can use the client of cloud desktop technology to remotely access the operating system of cloud desktop anytime and anywhere, and use the terminal application of cloud desktop anytime. For example, according to the needs of the course, download courseware uploaded by teachers, upload coursework to class modules, carry out programming teaching application tests, program design experiment simulation, database experiment simulation and so on can be completed by remote operation of cloud desktop system.

Teachers and students can remotely access cross-platform and cross-regional systems. With the support of cloud desktop technology, teachers and students can log in to the cloud desktop system by logging in their own accounts through mobile terminals, such as mobile phones, computers connected to local area networks, and ipads. However, if teachers and students need to install specific software in practice, they only need to install the corresponding software in the cloud server and back up their own operation data. Even if the cloud desktop is shut down and restarted, they can find their own data content.

On the other hand, under the coordination of cloud desktop service, cloud storage service, and cloud transport protocol, the cloud desktop system can also support various mobile storage device applications, such as printers, Bluetooth, and wireless network cards, and record users' usage traces at any time. The graphical status monitoring module checks the current system usage, memory status, and system running status to discover problems and bugs in the cloud desktop system and determine whether to restart or reset the system as required.

2.3   Optimize the Desktop Transport Protocol

The efficient management of the computer lab requires comprehensive optimization of the cloud desktop transport protocol, which is also an important step to further improve the management experience of the cloud desktop computer lab. Here we take THE ArHDP secure Desktop Transport protocol as an example. Cloud desktop secure transmission protocol supports scenario-based compression coding (such as H.264 and H.265) technologies in various modes. With the help of ArHDP protocol, computer desktop images can be transmitted through the network. On this basis, local decoding of images can be realized through the processing of intelligent cloud terminals. The decoded image is then displayed on the desktop of the local computer. Desktop transfer protocol can recognize dynamic image and static image in the system with the aid of intelligent monitoring of image. If the network broadband of the university is limited, the system will use the peak bandwidth suppression technology to ensure the best experience of teaching office and student users to the maximum extent.

In most cases, the desktop transport protocol will be compatible with different GPU modes. Especially in the mode of simultaneous high density, the system can timely transfer the image coding and transcoding functions of computer desktop to GPU by virtue of the acceleration technology of GPU, which can greatly reduce the load pressure of computer CPU in the case of high density, especially during the peak use of computer by students. This can further improve the desktop operating experience and system use experience. Meanwhile, in order to meet the work efficiency and system capability of users with high requirements such as 3D software, the pass-through technology in GPU can transfer the physical GPU to the desktop in time. In this case, the GPU will be fully responsible for the encoding and decoding of images, pictures, videos and other contents. In addition, the vGPU technology can be used to better flexibly divide physical Gpus, so that flexible application of network resources can be realized, resource sharing efficiency can be improved, and user experience with more high-end and higher demands can be met.

3   Conclusion

To sum up, the emergence of cloud technology to solve a lot of traditional industry in the aspect of system operation and management efficiency problem, cloud desktop technology and the combination of college computer laboratory management can better play to the cloud management superiority, avoid appear unreasonable resource allocation in the management of computer laboratory, maintenance cost is high, Using desktop virtualization technology, cloud terminal technology can further improve the teaching quality and teaching effect, and with the support of different cloud desktop technology architecture, students can enjoy anytime and anywhere cloud desktop technology experience, further enrich the teaching means, teachers meet the learning needs of students, in the process of differentiation, innovative teaching, The application of cloud desktop technology will certainly provide strong technical support for the innovation and development of parity activities in colleges and universities.

References

1. Tang, Y.: Design of university open laboratory information management system based on cloud desktop technology. J. Ningxia Norm. Univ. **42**(10), 90–97 (2021)
2. Zhang, K.: Application of desktop computing technology based on cloud computing. Int. J. Inf. Technol. Syst. Approach **14**(2), 1–19 (2021)
3. Gu, J.: Application of cloud desktop technology in the construction and management of college computer training room. Inf. Comput. (Theory Edn.) **33**(13), 168–170 (2021)
4. Wang, J., Li, J.: Cloud-based desktop virtualization of laboratory management and application. J. Electron. Test **18**(9), 133–134 (2021)
5. Gómez, C.E., Chavarriaga, J., Tchernykh, A., Castro, H.E.: Improving reliability for provisioning of virtual machines in desktop clouds. In: Schwardmann, U., et al. (eds.) Euro-Par 2019. LNCS, vol. 11997, pp. 669–680. Springer, Cham (2020). https://doi.org/10.1007/978-3-030-48340-1_51
6. Torterolo, L., Ruffino, F.: Desktop cloud visualization: the new technology to remote access 3d interactive applications in the cloud. HealthGrid **175**, 111–112 (2012)
7. Abichandani, P., McIntyre, W., Fligor, W., Lobo, D.: Solar energy education through a cloud-based desktop virtual reality system. IEEE Access **7**, 147081–147093 (2019)
8. Dai, Z.: Application analysis of cloud desktop technology in computer room management. Inf. Comput. (Theory Edn.) **32**(22), 4–6 (2020)
9. Wu, X.: Desktop cloud technology application in the university laboratory management. J. Electron. Test **65**(22), 70–71 (2020)
10. Huang, Y.: Advantages and application analysis of cloud desktop technology in the construction and management of college computer training room. Inf. Comput. (Theory Edn.) **32**(11), 42–44 (2020)

Agent-Based and Multi-Agent Systems for Health and Education Informatics

Construction of a Personalized English Learning System Based on Machine Learning

Yinping Ji[1]([⊠]) and Deepmala Karki[2]

[1] School of Foreign Languages, Shandong Women's University,
Jinan 250000, Shandong, China
JYPjinan@126.com
[2] LBEF Campus (In Academic Collaboration with APU Malaysia),
Kathmandu, Nepal

Abstract. In recent years, the development of countries around the world has gradually globalized, and the communication between countries has increased. English has gradually become important. English learning is a comprehensive learning process, including listening, speaking, reading and writing, but each part is independent and interrelated. Due to the characteristics of learning English, it also puts forward requirements for the English learning system: how to build a multi-functional and integrated English learning environment to make users learn English easily. At the same time, the Internet has been developing continuously, and the number of English learning materials has become very large and rich. How users find suitable resources for learning English from huge materials has also become a problem that has to be considered, which directly affects users' learning time, learning cost and learning interest. In view of the above problems in English learning, this paper studies a personalized English learning system based on machine learning by using the theoretical knowledge of natural language processing, information retrieval and machine learning, so as to meet the needs of personalized organization and search of resources in the construction of English learning system. Combined with the above algorithms, a dynamic English resource retrieval system is constructed in the local retrieval framework, and a system solution of English resource retrieval evaluation, indexing and fast retrieval is proposed. At the same time, in order to meet the needs of word pronunciation and listening in English learning, the functions of reading and listening words are introduced into the system, which is improved on the basis of single model single feature (SMSF) algorithm, and a SMSF combination method is developed. The experimental results show that the accuracy rate can reach 89%. Finally, create a multi-functional and integrated personalized English learning environment for users.

Keywords: Machine learning · English learning system · Personalized system · Learning system construction

© The Author(s), under exclusive license to Springer Nature Switzerland AG 2022
V. Sugumaran et al. (Eds.): ICMMIA 2022, LNDECT 138, pp. 503–510, 2022.
https://doi.org/10.1007/978-3-031-05484-6_63

1 Introduction

The existing software can not meet the personalized needs of users, and can not effectively judge the real English level of users and rearrange and sort the corresponding information. When users use English learning software such as Youdao dictionary and iciba, their main functions are to look up words, see example sentences and listen to example sentences. By analyzing the query results, it is found that for words with different difficulty, the returned sentence results cannot be distinguished, and the difficulty of the word itself and the sentence examples matching the difficulty are not considered. Most importantly, the examples of query statements returned to different users are the same. This situation is not suitable for second language English learners, because users have different levels and return the same results indiscriminately, which is bound to lead users to spend more time looking for their own English teaching materials, so as to reduce users' interest in learning and fail to achieve good learning results. Therefore, it is very necessary to build a personalized learning system of English.

In the research on the construction of personalized learning system of English, many researchers have conducted research and reached many conclusions. Helma C first proposed the use of corpus construction technology for the construction of personalized system based on machine learning, and constructed suitable training corpus and test corpus for different problems [1]. Then, the corresponding text and non text functions are extracted by feature analysis technology. Next, according to the characteristics of feature distribution, an appropriate model algorithm is designed, and k-cross validation or self sampling technology is used to analyze the optimal model parameters. In the research of classification, buczak a adopts Bayesian polynomial model, and the feature set adopts word bag model, and optimizes the smoothing algorithm accordingly. Their experiments show that this method is obviously superior to the traditional language feature analysis method, especially the text with length less than 10 [2]. On the basis of previous experiments, mullainathan s integrates the monolingual model based on sentence length and Gaussian generation model to solve the problem of sparse features and inability to effectively include all features in a single model [3]. These scholars have done a lot of research on the construction of personalized English learning system based on machine learning, which provides a good theoretical basis for this study.

In the first mock exam, we use the first mock exam algorithm and the single model multi feature algorithm to compare the two models. We also study the accuracy of the single feature and multi feature in the single model. The experiment shows that the accuracy of the system is more than that of the single model. Performance has been improved.

2 Research on the Method of Constructing a Personalized English Learning System Based on Machine Learning

2.1 Similarity Calculation Method Based on Space Vector Model

The classification algorithm based on the full-text search method mainly measures the similarity between the search element and the document. The main idea to solve the problem is to decompose the query element into a series of smaller language units (Token for short), and use a similar Boolean model search box to find the corresponding text element [4, 5]. At present, it is necessary to use similarity calculation methods to evaluate the similarity between each document and the query, so as to classify it from large to small [6].

In this process, this article makes full use of the research results of predecessors in the field of information retrieval, mainly based on the specific design methods and practical strategies of the Boolean retrieval model. When the related query results are returned, their similarity is calculated, and a similarity calculation method based on the vector space model is proposed to query the similarity between the element and the document.

2.2   Full-Text Search Sorting Method with User-Defined Information Added

Another issue that needs to be considered when designing a learning system is to add user-defined information to it. This information is mainly used to determine the importance of each domain according to actual needs, as well as the influence weight when MUST query is used in the Boolean recovery model [6]. After considering these two factors, we need to analyze how to incorporate this factor into the previous basic formula.

First, consider calculating the weights of different domains. Let explain the concept of domain first: a document can be composed of different parts [7, 8]. For example, in our research, it can be considered that a document is composed of English text and Chinese translation, where the English text corresponds to a domain, and the Chinese translation corresponds to a domain.

After clarifying the meaning of the department, we need to set the weights related to different domains in our application [9, 10]. This importance is related to our needs. According to the corresponding rules, for example, in our application, the importance of English text is higher than that of Chinese translated text [11]. A document has a higher similarity score when the English domain contains query terms than when the Chinese domain contains query terms.

3   Experimental Research on the Construction of a Personalized English Learning System Based on Machine Learning

3.1   Constructing a Single Model and Single Feature Algorithm Experiment

The experimental results are evaluated according to the performance of different classifiers in actual work as evaluation criteria [12].

The algorithm experiment steps are as follows:

Output: the correct rate of test results.

(1) Stemming each word in Tokens.
(2) Calculate TF-IDE and get the characteristics of the vocabulary.
(3) Perform part-of-speech tagging on Tokens to get the tagging results and proba-bilistic syntactic analysis.
(4) Calculate the syntactic features according to the result of 3, and obtain the syntactic features.

3.2 Single Model and Multiple Feature Algorithm Experiment

Since the performance from a single model with a single feature to a single model with multiple features has improved to a certain extent, whether the single model with multiple features can optimize the effect, this experiment will focus on the fusion of multiple models. Assuming that there is no correlation between different features, we can use these three features to train the six models mentioned above.

Since each feature has its advantages and disadvantages, we combine them and learn from each other to achieve the goal of improving the overall performance of the model. This is also a reference to the idea of integrated learning. We merge these different models in a specific way to improve the generalization error of the model, improve the stability of the model, and achieve the purpose of solving the sensitive problems caused by the continuous changes of training data. The experimental steps of the single-model model multi-feature algorithm are as follows:

Output: the correct rate of test results.

(1) For a single classifier in C, use an algorithm to train it to obtain a set M of models.
(2) For each sentence in T: use the fusion operation on SVM and LR, and predict each sentence to get the result A1. Perform a fusion operation on RF and GBDT, and predict each sentence again to get the result A2.
(3) Perform weighted fusion on the two results obtained in step 2 to obtain the result A3.
(4) Add the results obtained to the final result set.
(5) Count the correct rate of test results and return the result.

3.3 Query Item Analysis Experiment

For the user query analysis experiment, the idea is to conduct modeling research based on the user's English proficiency. The English text is mainly divided into three levels: simple, medium, and high difficulty. In the study of user modeling, these three levels are given different difficulty values, the higher the value, the greater the difficulty. Match the three difficulty levels to 2, 4, and 6. When modeling users, the end user category is taken as the average of multiple categories, thus forming a continuous value between 2–6, because it is between the same text difficulty. There will also be dif-ferences in difficulty, and it makes sense to limit this difference to specific areas.

After the experiment, we estimated the probability of users at different levels of difficulty. Treat each query element generated by the user as a multidimensional variable, and conditionally analyze the probability of each level after the user generates the query element. This conditional probability can be expressed by the following formula (1):

$$p(r = c \mid \text{query}) = \frac{p(\text{query} \mid r = c)p(r = c)}{p(\text{query})} \tag{1}$$

Due to the lack of user information, it is difficult to assess the previous possibility of the user's difficulty level at the beginning. Therefore, at first it was thought that the probability of users in each category was equal. We need to be especially careful when estimating the conditional probability of generating this query item given the known difficulty level of the user. Because in a general search engine like Baidu, the chance of users finding the same search item within a period of time is very low. This is especially true for our application scenarios. When estimating this conditional probability, many features may be the user's problem and the information is not accurate when estimating the query. Probability estimation carried out with insufficient information is prone to over-adjustment, thereby increasing the generalization error of the model (in our model, the estimation bias).

In order to estimate this probability, we use the following alternative method. Specifically expressed as the probability of creating the query item in the set of difficulty level documents c. Since the probabilities before this query item are equal, the denominator of the above type can be omitted. At the same time, the search terms of the query terms are considered independent. Finally, when the query elements generated by the user are known, the conditional probability that the user is proficient in English is as follows:

$$p(r = c \mid \text{query}) = \frac{p(\text{query} \mid D = c)p(r = c)}{\sum_{c=1}^{3} p(\text{query} \mid D = c)p(r = c)} \tag{2}$$

In the above formula, it should be noted here that when processing interrupted words, because interrupted words can be keywords requested by a small number of primary English users, they cannot be filtered. In order to reduce the impact of the interrupted word, the weight of the word is added to the category (this value is the result of the normalization of the value of the chi-square test).

4 Experimental Analysis of the Construction of a Personalized English Learning System Based on Machine Learning

4.1 Experimental Analysis of Single Model Single Feature Algorithm

Based on the algorithm experiment of a single model, the lexical, syntactic and feature test results of the six classifiers are shown in Table 1, where the number after the LSA represents the feature dimension.

Table 1. Analysis of test results of single model and single feature accuracy rate

Feature type	Lexical characteristics	Syntactic features
SVM	79.5	78.8
LR	86.4	84.3
NB	75.7	75.2
KNN	75.8	75.9
RF	75.4	82.4
GBDT	80	86.6

Figure 1 shows that the correct rate of feature words in LR is the highest, achieveing 86.4%, which is much higher than that of other classifiers. The precision of the synthetic characteristics below gbdt is the highest, reaching 86.6%. Generally speaking, these two characteristics have a positive effect on the results and can basically maintain the right percentage about 75%.

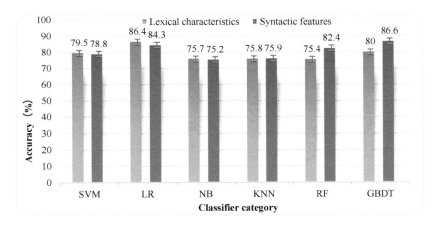

Fig. 1. Experimental analysis diagram of single model single feature algorithm

4.2   Experimental Analysis of Single-Model Multi-feature Algorithm

The effect on the six classifiers (LSA represents the feature dimension) after the feature pairwise combination and the three sets of feature combinations are shown in Table 2:

It can be seen from Table 2 that the syntactic feature and the grammatical feature in the pairwise feature combination can achieve 89.3% accuracy under LR. When these two characteristics are combined with LSA, the accuracy within RF can reach 88.6%, which is obviously better than the result of using each feature on its own.

Table 2. Experimental analysis table of single model multi-feature algorithm

Feature type	Lexical+grammar	Lexical+LSA400	Syntax+LSA400
SVM	81.6	83.9	81.9
LR	89.3	84.6	87.5
NB	83.5	87.6	85.6
KNN	79.6	83.7	86.8
RF	78.1	88.6	82.3
GBDT	85.3	81.5	83.7

It can be seen from Fig. 2 that as the number of feature combinations increases, the accuracy of the final result of difficulty evaluation also increases, and the accuracy of multiple feature combinations can reach 89.3%. By comparing the performance of classifiers from single use to multiple sets of feature combinations, it can be found that as the number of feature combinations increases, the overall performance score of each classifier continues to increase, which also proves that the method is effective to some extent of.

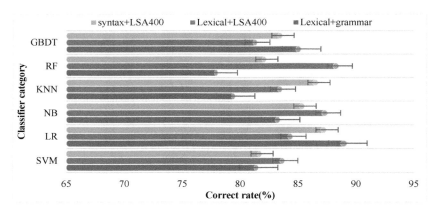

Fig. 2. Experimental analysis diagram of single model and multiple feature algorithm

5   Conclusions

In the research process of this article, there are still some problems, such as verifying the effectiveness of the personalized algorithm, because of the lack of user data, it is impossible to draw very definite conclusions. In response to this problem, an iterative update mechanism of the algorithm is designed to ensure the adaptability of the algorithm. In the research process of this article, only a few problems in English learning are selected for research. In order for users to consolidate the words they have learned in the form of practice questions, this system provides a module for generating practice questions. It can also generate training questions that meet the requirements and are targeted. However, some candidates are not enough to confuse users and cannot connect with some new information now, and then it will be further improved through networking.

References

1. Helma, C., Cramer, T., Kramer, S., et al.: Data mining and machine learning techniques for the identification of mutagenicity inducing substructures and structure activity relationships of noncongeneric compounds. J. Chem. Inf. Comput. **35**(4), 1402–1411 (2018)
2. Buczak, A., Guven, E.: A survey of data mining and machine learning methods for cyber security intrusion detection. IEEE Commun. Surv. Tutor. **18**(2), 1153–1176 (2017)
3. Mullainathan, S., Spiess, J.: Machine learning: an applied econometric approach. J. Econ. Perspect. **31**(2), 87–106 (2017)
4. Raissi, M., Karniadakis, G.E.: Hidden physics models: machine learning of nonlinear partial differential equations. J. Comput. Phys. **357**, 125–141 (2018)
5. Ho, S.C., Hsieh, S.W., Sun, P.C., et al.: To activate English learning: listen and speak in real life context with an AR featured U-learning system. Educ. Technol. Soc. **20**(2), 176–187 (2017)
6. Xu, J., Du, W., Guo, H.: A probe into the establishment and analysis of the mobile English learning system model. Rev. Fac. Ingen. **32**(6), 270–276 (2017)
7. Zhang, Y., Li, W.: A study on the mechanism of the L2 motivational self system of non-English majors in English learning. Int. J. Soc. Sci. Educ. Res. **2**(10), 1–8 (2019)
8. Dadvand, B., Behzadpoor, F.: Pedagogical knowledge in English language teaching: a lifelong-learning, complex-system perspective. Lond. Rev. Educ. **18**(1), 107–126 (2020)
9. Seo, Y.D., Kim, Y.G., Lee, E., et al.: Personalized recommender system based on friendship strength in social network services. Expert Syst. Appl. **69**(mar.), 135–148 (2017)
10. Yi, D., Su, J., Liu, C., et al.: A machine learning based personalized system for driving state recognition. Transp. Res. Part C: Emerg. Technol. **105**(AUG.), 241–261 (2019)
11. Zhang, G.: Study on the construction method of the adult distance learning support service system. Boletin Tecnico/Techn. Bull. **55**(20), 516–521 (2017)
12. Youhei, M., Keisuke, F., Takehiko, M.: Construction of an e-learning system for information processing education. Joho Chishiki Gakkaishi **27**(2), 155–160 (2017)

Computer Technology-Based Surveying and Mapping of Mechanical Products and CAD Training

Lihua Jiang[✉]

Shandong Huayu University of Technology, Dezhou 253034, Shandong, China
jlhjiang_123456@163.com

Abstract. With the continuous deepening of China's industrialization process, the shortage of professional talents has become more and more serious. In particular, the shortage of highly skilled and innovative talents with both theoretical knowledge and strong engineering practice capabilities has become a hot issue of general concern to the whole society. As higher education institutions are the bases for cultivating talents, how to quickly and efficiently train high-skilled talents with innovative spirit and practical ability that meets the needs of the society is an important topic for colleges and universities.

Keywords: High-skilled innovative talents · Mechanical products · CAD training

Practical training is the most basic link for colleges and universities to cultivate students' practical ability and innovation ability, and it is an important way to cultivate talents in applied undergraduate colleges. The course "Mechanical Product Surveying and CAD Training" is a compulsory professional intensive practice course for undergraduates majoring in mechanical design and manufacturing and automation. CAD drawing is an integrated training course for engineering drawing ability, which is an important link for students to comprehensively apply the basic theoretical knowledge and basic drawing skills learned in the classroom.

1 Analysis of the Current Situation of Course Teaching

The course "Mechanical Product Surveying and CAD Training" is a basic practical compulsory course for the major of mechanical design and manufacturing and its automation. It is also an important practical training link for mechanical majors to learn mechanical drawing. It is a combination of theory and practice and in practice. The best way to develop the ability to solve practical engineering problems. Surveying and mapping is essential for improving students' drawing ability, drawing ability, access to data ability, and cultivating students' engineering awareness and professional qualities [1]. However, relevant surveys show that students have various skills in actual employment There is still a certain gap between the project ability and the needs of enterprise talents. The reasons are as follows: ① The teaching content of surveying and mapping is incomplete. At present, the practical teaching of surveying and mapping uses the first-level spur gear reducer as the carrier, which only involves three types of parts: bushing, cover, and box. The structure type of parts cannot cover all the types of parts in practical engineering applications. There is a certain disconnect between engineering practice. ② Single measurement method. In the current teaching surveying and mapping methods, measuring tools such as vernier calipers, internal and external calipers and steel rulers are mainly used to measure machine parts. In this process, due to human subjective factors and differences in technical level, large measurement errors will occur, and the machine parts will be obtained. The size data is not accurate and the work efficiency is low; and for the complex irregular features, it is impossible to directly measure it. Approximate methods are often used to determine the structure size, which is difficult to work and has low accuracy, and it is difficult to meet the actual needs of the project. ③ Students' understanding of the importance of the course of surveying and mapping is basically not enough. Students generally think that in the future manual drawing in enterprises has been completely replaced by computer drawing, and it has no great practical value [2].

Therefore, some students resist the manual drawing method psychologically. Just obey the teacher's order to copy the painting as it is, and did not seriously think about its view expression plan, even at the end of the surveying and mapping course, did not really understand the learning purpose of this course, so there was no correct attitude towards the practical course, which led to the hasty end of the training. The quality of teaching is unsatisfactory. Therefore, how to match the surveying and mapping training teaching with the actual social production, keep up with the pace of the times, connect the surveying and mapping methods with the actual mapping methods used in the society, reform traditional methods, optimize the surveying and mapping teaching, and further expand its extension and deepen its connotation It is very necessary to make it achieve greater educational and teaching benefits.

2   Implementation Basis of Curriculum Teaching Reform

2.1   Equipped with Infrastructure Such as Surveying and Mapping Room, Computer Room, Precision Measurement Technology Laboratory, Etc.

In order to ensure the quality of teaching and smoothly carry out this course, the parts surveying and mapping stage is completed in the surveying and mapping room during the whole teaching process, the CAD drawing is completed in the computer room, and the irregular surface size and shape tolerance measurement are passed the three-coordinate measurement in the precision measurement laboratory [3]. The machine is complete. The surveying and mapping room, microcomputer room, precision measurement technology laboratory and other infrastructure facilities are shown in Fig. 1.

Fig. 1. Surveying and mapping room, computer room, precision measurement technology laboratory

2.2 Equipped with a Team of Teachers with Dual Qualifications and High-Quality Online Resources

The teaching faculty of this training course is composed of professional teachers with long-term teaching experience and three years of teaching this course, as well as teachers with professional training related to three-axis precision testing [4]. Teachers conduct regular teaching seminars to integrate school network teaching The curriculum construction of the platform and the three-coordinate detection resources are the goals, and the development and construction of the teaching resources of this training course are carried out to adapt to the implementation of the teaching reform of this course.

With the help of "Chinese College Student MOOC" and "Superstar" online course platforms, standardized online resource construction is completed, traditional classroom teaching resources and teaching content are enriched, and resource reserves are provided for students to learn independently.

3 The Reform of Curriculum Teaching Implementation

In order to achieve greater educational benefits in surveying and mapping teaching, and to further improve students' engineering awareness, design concepts, and teamwork awareness, this thesis intends to start from the perspective of cultivating innovative talents based on engineering applications, and evaluate the teaching content, teaching methods, and assessment of surveying and mapping practice. Carry out in-depth research in other aspects.

3.1 Reform of Curriculum Teaching Content

The design ideas of teaching content reform are shown in Fig. 2.

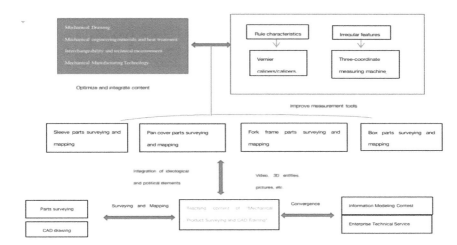

Fig. 2. Ideas for the reform and design of the teaching content of "Mechanical Drawing Surveying and CAD Training"

(1) Optimize and integrate teaching content

Surveying and mapping is an important practical teaching link in the course of mechanical drawing. The purpose of surveying and mapping is to hope that students can cultivate their comprehensive expression ability in mechanical drawing through engineering training in the whole process of surveying and mapping. Therefore, in the process of learning mechanical drawing surveying and mapping, students should not only have relatively rich professional technical knowledge, but also certain knowledge of parts and parts design and some skills in the process of parts and parts surveying and production [5]. However, the current surveying and mapping week teaching is usually arranged in the first semester of the sophomore year. At this time, students have not yet studied professional basic courses such as "Interchangeability and Technical Measurement", "Metal Materials and Heat Treatment" and other professional basic courses, such as the process structure knowledge of parts, Reasonable dimensioning, dimensional tolerances and fits, surface roughness, metal materials and heat treatment, and other knowledge are difficult to understand and master, and common sense errors are prone to occur when drawing [6]. In response to the lack of relevance of relevant basic disciplines exposed by students in the actual surveying and mapping process, after the implementation of the curriculum reform, teachers' selection of learning project carriers not only considers the ability and knowledge requirements of the professional post group, but also considers students In the gradual learning process, the selection of the carrier gradually increases the difficulty from simple to complex. The teaching content of part drawing surveying and mapping covers four types of parts: bushings, plate covers, forks, and boxes. Compilation of supporting teaching guides, The knowledge points related to this course in "Mechanical Drawing", "Mechanical Engineering Materials and Heat Treatment", "Interchangeability and Technical Measurement", "Mechanical Manufacturing Technology" and other courses are fragmented and then optimized and integrated to make the teaching content It is more targeted, encourages students to integrate the breadth and depth of knowledge, and promotes the continuous improvement of students' ability to read, read, and draw.

(2) The process of mechanical drawing surveying and mapping with the introduction of three-coordinate measurement technology

In order to enable students to learn new techniques, master new methods, and develop new ideas from the surveying and mapping process, traditional measurement tools (vernier calipers, calipers, etc.) are still used to measure the data of regular features in the surveying and mapping teaching process, while non-regular features are measured with the help of Advanced measuring tools (coordinate measuring machine) realize the acquisition of the coordinates of the surface points of the parts.

The part has simple regular features such as cylindrical holes, planes, etc., as well as irregular features that cannot be measured or easily measured by traditional measurement methods. During the measurement process, the teacher guides the students to first use the vernier caliper to complete the measurement of the regular features of the part shape [7]. Secondly, use the coordinate measuring machine to complete the measurement of the space surface, obtain a series of space point positions and export the IGES format file, and then import the IGES format file of the space point coordinates into the UG software, and use the surface fitting function to complete the construction of surface features. Finally, the three-dimensional model of the finished part is constructed according to the size data of the regular features on the surface of the part obtained by the traditional measurement method and the surface features measured by the coordinate measuring machine [11].

After the introduction of three-coordinate measurement technology, the surveying and mapping of mechanical parts and components uses new measurement ideas to replace some of the traditional measurement methods, improve measurement accuracy and efficiency, and solve the measurement problem of irregular surface features of complex models. In addition, students are guided to find The combination of the cartography surveying and mapping theory learned and the application of modern industrial technology will help stimulate students' interest in learning and cultivate students' sense of innovation.

(3) Integration of ideological and political elements in the curriculum

According to the course nature, characteristics and training objectives of the "Mechanical Drawing Surveying and CAD Training" course, and according to the educational purpose of Lide Shuren, the teaching implementation adopts the hierarchical teaching concept to divide the teaching content into several teaching tasks. In a single task Relevant curriculum ideological and political elements are incorporated into the curriculum, and a generative internal fit relationship is established in a seamless and organic way. In addition, in the process of group task implementation, students cultivate their communication and expression and teamwork awareness through reasonable division of labor and effective organization, so as to achieve the effect of moisturizing things silently.

3.2 The Reform of Teachers' Teaching Methods

According to the teaching content of each item, the knowledge points of the theoretical part and the skill points of the practice part are decomposed, and the learning of the knowledge points and skill points is refined according to the three parts before class, during class, and after class [8]. ① Before class, first group the class members. Make relevant PDF files, PPT files and micro-videos that require students to prepare before class and upload the knowledge points and skills points to the network platform, publish the pre-class learning task requirements, and set up online assessment questions for the pre-class learning content, with multiple question types For multiple-choice questions and true or false questions, it assesses students' learning before class. ② In the class, first comment on the students' online preview before class, and guide the students to complete the offline learning of the knowledge points and skill points in the project through the project implementation process PPT produced before the class. Aiming at the internal structure of the abstract and difficult parts in the course, starting from the teaching carrier itself, it can be displayed intuitively and vividly through the use of on-site materials or three-dimensional animation simulation to cultivate students' spatial imagination ability through observation, operation and imagination. ③ After class, according to the different learning situation of students, after class expansion exercises are arranged in different levels. In order to give students more opportunities to combine theory and practice to carry out innovation and entrepreneurship competitions or practices, the "National College Students Advanced Graphics Technology and Product Information Modeling Innovation Competition" expression and configuration-related competition content are incorporated into the teaching process, and the competition is combined to give Students have more opportunities to combine theory and practice to stimulate students' interest in learning and innovation ability, improve their professional skills, train students' professional skills throughout the teaching process, and improve students' innovation and entrepreneurship capabilities. In addition, attract students to participate in teachers' scientific research projects, provide technical services for enterprises, and let students learn more about the importance of technical drawings in the current manufacturing process of enterprises.

3.3 The Reform of Students' Learning Methods

Students learn according to the three parts of learning content before class, during class, and after class. ① Before class, according to the learning tasks released by the teacher, complete the pre-class preview: including project task analysis, task group decomposition, online and offline learning of learning content, online search of learning materials, upload and submission, pre-class test questions Finish etc. ② In the class, according to the teacher's guidance in the class to complete the study of related project tasks, a group of students complete the surveying and mapping tasks together. When the project tasks encounter difficulties in offline learning, members of the group are preferred to collaborate, discuss and analyze, and then seek help from the teacher for guidance. In the process, they cooperate with each other, learn from each other, summarize and acquire knowledge, and give full play to the students' independent analysis and resolution capabilities while enhancing their initiative. ③ After class, students will carry out selective expansion and improvement of homework according to their own learning situation.

3.4   Reform of Assessment Methods

The ultimate goal of practical teaching assessment is to test students' mastery of the knowledge and skills they have learned, and it is also an assessment of teachers' teaching effects [9]. Establishing a complete process assessment system is necessary to improve the effectiveness of students' training. Assessment not only plays a supervisory role for students, but also has a positive guiding role. This course research mainly evaluates students from the following aspects: (1) Assessment of professional knowledge. Including surveying and mapping ability, drawing quality, etc., the assessment of national standards such as the layout of the drawing paper, the selection of line type and line width, the layout of the drawing, etc.; the assessment of the selection of assembly drawings and part drawing expression schemes, etc.; through stage inspections, real-time understanding of students The actual mastery of the task learning knowledge points. (2) Assessment of classroom performance. In the process of guiding surveying and mapping, teachers should have a good grasp of students' learning attitudes and surveying and mapping progress. Through classroom performance, they should understand the students' ability to find and use information, the ability to flexibly apply the knowledge they have learned, and the team's collaboration and harvest. The comprehensive evaluation of the aspect gives the results fairly. (3) The assessment and monitoring of the outstanding process. Mainly include: ① Diversification of assessment content: The assessment content includes knowledge points, skill points, learning attitudes, subject performance, etc., as well as safety awareness, quality awareness, responsibility awareness, and team awareness in the ideological and political content. ② Diversification of evaluation themes: student self-evaluation, mutual evaluation, teacher comments, etc., highlighting the dominant status of students. ③ Diversification of evaluation methods: online pre-class test questions, offline skill assessment, achievement display, etc.

The specific implementation assessment is as follows: the course assessment is divided into two parts: process assessment and summative assessment [10]. The procedural assessment adopts the form of classroom performance, usual homework, stage inspection, etc., and the summative assessment adopts the form of large-scale homework. Course assessment results = process assessment results × 50% + summative assessment results × 50%. The detailed plan is shown in Table 1.

Table 1. Course assessment composition and assessment standards

Assessment form		Points	Assessment criteria
Process assessment results account for 50%	Classroom performance	10	According to the student's usual attendance rate, class performance and interaction, and depending on their attitude and activeness towards the course learning, 10% will be included in the total course grade
	Usual homework	15	According to the completion and quality of students' homework, and their understanding and mastery of course knowledge, 15% will be included in the total course grade
	Stage inspection	25	According to the students' drawing completion status, completion quality, and completion progress, 25% will be counted into the total course grade
Final assessment results account for 50%	Big homework	50	It is divided into two parts: sketch drawing and computer drawing. According to the completeness of the drawing, the dimensioning, the correctness and standardization of the view expression, etc., 50% is counted into the total score of the course

The stage check accounts for 25% of the final grade. The total individual score is derived from the intermediate performance and results completed by the students in each project or even each task. This score is derived from group self-evaluation, team member mutual evaluation, and teacher evaluation. Specific inspection the standards are shown in Table 2.

Table 2. Scoring criteria for stage inspection

Class: ***		Name: ***		Student ID: ***		
Phase inspection items		Content and requirements	Points	Self-evaluation	Mutual evaluation	Teacher comment
Use of tools and instruments		During the drawing process, the tools and instruments can be placed correctly and used in a standard manner without damage.	15			
Draw assembly diagram		Can clearly and correctly express the assembly relationship, the layout is reasonable and beautiful, the numbering is standardized, and there is no omission.	15			
Draft drawing	View representation	The view is complete and clear, the size measurement is basically complete, and the ratio is basically correct.	30			
	Dimensioning	The dimensions are correct, clear and reasonable.	15			
	Surface cleanliness	The drawing surface is neat and beautiful, without alteration.	5			
	Words	The technical requirements are filled in without errors, the fonts are neat and beautiful, and the digital writing meets the standard requirements.	10			
	title	The title bar and text writing comply with national standards, and the content should be filled in accurately.	5			
	Number of drawings	Complete all drawing tasks assigned by the teacher without omissions.	5			
Scoring by the rater:						
calculation process			100	×20%	×20%	×60%
Total score						

4   Conclusions

Through many times of teaching reform and practice of mechanical drawing surveying and mapping, good teaching results have been received. Students have greatly improved their hands-on ability, problem-analyzing and problem-solving ability, drawing (free-hand drawing, manual drawing, computer drawing) skills, engineering quality, unity and cooperation ability, innovation ability and other aspects. Zero breakthroughs have also been achieved in innovation competitions and corporate technical services.

The machinery manufacturing industry has become a pillar industry of China's national economy. I hope this article will serve as an inspiration to promote the improvement of machinery training courses.

Acknowledgements. Fund Project: This article is the research result of the first semester classroom teaching reform pilot course "Mechanical Product Surveying and CAD Training" at Shandong Huayu University of Technology (Project Number: 2021KG-06).

This article is the research result of the 2021 school-level ideological and political demonstration course "Mechanical Drawing Surveying and CAD Training" of Shandong Huayu University of Technology.

References

1. Woo, J.-C., Lam, A.D.K.-T.: The study on engineering-type parametric 3D CAD used in training art & design major students. J. Internet Technol. **21**(7), 2011–2017 (2020). https://doi.org/10.3966/160792642020122107015
2. Hamade, R.F., Artail, H.A., Sikstrom, S.: Correlating trainee attributes to performance in 3D CAD training. J. Eur. Ind. Train. J. HRD Spec. (2) (2007)
3. Witvrouwen, I., Pattyn, N., Gevaert, A.B., et al.: Predictors of response to exercise training in patients with coronary artery disease - a subanalysis of the SAINTEX-CAD study. Eur. J. Prevent. Cardiol. **26**(11), 1158–1163 (2019). https://doi.org/10.1177/2047487319828478
4. Cho, A.: Training inmates in autodesk CAD programs gains traction. Eng. News-Rec. (17) (2017)
5. Sali, L., et al.: Computer-based self-training for CT colonography with and without CAD. Eur. Radiol. **28**(11), 4783–4791 (2018). https://doi.org/10.1007/s00330-018-5480-5
6. Gerald, D.: CAD props prove useful in training, verifying, illustrating. Fabricator (8) (2020)
7. Pezeshk, A., Petrick, N., Chen, W., et al.: Seamless lesion insertion for data augmentation in CAD training. IEEE Trans. Med. Imaging **36**(4), 1005–1015 (2017). https://doi.org/10.1109/TMI.2016.2640180
8. Yan, S., Zhang, C., Xun, W.: Analysis of advantages of mechanical process design based on CAD technology. Southern Agric. Mach. 119 (2020)
9. Zhan, N.Y., Jia, W.H.: Development and realization of computer training system for mechanical drawing based on AutoCAD 2007. Mech. Electr. Eng. Technol. (2012)
10. Abdulrasool, S.M., Mishra, R., Khalaf, H., et al.: The effect of using computer technology tools to enhance the teaching-learning process in CAD-CAM-CNC module in mechanical engineering subject area. Glob. Learn **2010**(1) (2010)
11. Mccosker, D.P.: Guidelines for the use of electronic survey field notes and CAD computation files as legal records (2006)

English Learning Patterns for Mobile Users Based on Multiple Data Mining Algorithms

Wei Liu[✉]

School of English Language and Literature, Xi'an Fanyi University,
Xi'an 710105, Shaanxi, China
zl_8583@163.com

Abstract. The analysis of online learning models is of great significance for the configuration and development of a distance learning platform with a better learning experience, and the development of more effective and accurate online learning evaluation and guidance. The purpose of this article is to study the extraction and recognition of English mobile user learning patterns based on multiple data mining algorithms. First, the research foundation is explained, and on this basis, the importance of using data mining technology to analyze the learning model is put forward. Subsequently, the research status and related theories and technologies of online learning behavior analysis and data mining technology at home and abroad are introduced. Taking the mobile users of the "Smart English" platform as the main research object, the relevant data mining algorithms are used to extract and analyze the characteristics of their learning methods. Experimental results show that there are significant differences in behavior in the study of the number of active days and forum posts of mobile users with different education levels (P < 0.05).

Keywords: Data mining · Mobile users · English learning · Mining and recognition

1 Introduction

Learning data collection is an effective means of diagnosing teaching effects. Through the collection and analysis of the data generated by classroom teaching, more hidden data and relevance can be explored. Influence, whether there is a causal relationship [1, 2]. The collection of learning data is helpful to explore students' learning attitudes and even learning motivations. From the various learning behaviors of students, they reflect their personal attitudes towards classroom learning. Combining the learning situation of students reflected by learning data can better characterize their individuality. The creation of a personalized learning environment provides a data basis, guides teachers to conduct personalized education, and assists students' personalized development [3, 4]. At the same time, it can provide a scientific reference for adjusting teaching decision-making and precise intervention [5].

The intelligent learning system provides students with customized learning resources based on their teaching needs and personal preferences. Xhen Chen introduced a learning style model to represent the characteristics of online students. An

© The Author(s), under exclusive license to Springer Nature Switzerland AG 2022
V. Sugumaran et al. (Eds.): ICMMIA 2022, LNDECT 138, pp. 521–528, 2022.
https://doi.org/10.1007/978-3-031-05484-6_65

improved recommendation method based on adaptive recommendation based on online learning style (AROLS) is also proposed, which realizes the adaptation of learning resources by extracting student behavior data. First, AROLS creates student groups based on their online learning style. Second, it applies collaborative filtering (CF) and mining association rules to extract the preferences and behavior patterns of each cluster. Finally, it creates a set of personalized variable-size recommendations. Some experiments used real-world data sets. The results show that our online learning style model is conducive to student data mining, and AROLS is significantly better than the traditional CF method [6]. It usually takes a long time to run data mining algorithms, but none of the existing data mining software provides important progress indicators. Gang L studied the problem of providing progress indicators for the construction of machine learning models and the execution of data mining algorithms. Discuss the inherent goals and challenges of this issue. They then described the initial framework for implementing such progress indicators and two possible high-level uses to facilitate future research on the subject [7]. Therefore, extracting and analyzing the learning behavior characteristics of online students, and applying the results to the distance learning platform, has a good effect on improving the quality of distance education [8].

This paper studies and analyzes the association rules and methods of sequential pattern extraction in data mining. After analyzing the performance and limitations of Apriori and AprioriAll algorithms, they are optimized as a method of analyzing learning pattern data. In the research on the mobile users of the "Smart English" platform, we use mining and learning methods to analyze the differences in the learning methods of different types of mobile users, as well as the differences in the learning methods of different types of mobile users. So as to effectively improve the scientific nature of education and teaching prediction and decision-making.

2 Research on Mining and Recognition of English Learning Patterns of Mobile Users Based on Multiple Data Mining Algorithms

2.1 Data Mining Technology

Before starting data mining, the processing of business data roughly determines some available training algorithms [9].

Data preparation mainly provides the basis for subsequent data processing through three stages of data integration, selection and promotion. Therefore, the data preparation process is very important. Through these services, denoising, deduplication, conversion type, etc. [10].

(1) Data integration: Through the conversion of data generated by different data sources, it is integrated into a specific format [11, 12].

(2) Data selection: After processing the retrieved data into a database for data tracking, not all the collected data are effective for data mining. Therefore, from a practical perspective, it is necessary to select as many relevant data as possible.

(3) Data compilation: Data orientation includes copying, duplicating and modifying data types, mainly through data conversion, integration, cleaning and calibration, to make these data more effective in the data mining system, and the data has been processed in a certain way.

Data mining analyzes the needs of users, determines the source of data mining, selects appropriate and effective data mining methods according to relevant requirements, and tries to obtain many results, trends, templates, etc.

The expression and interpretation of the results is that the data mining algorithms, related technologies and data quality that have been developed will have an impact on the data mining results. If the wrong method or wrong data is used, the results of data mining will deviate from human expectations. Therefore, it is necessary to complete the evaluation of the final results, remove unnecessary or unnecessary results, and finally present them to users in various ways to provide relevant explanations for the results.

2.2 Discovery of Learner Learning Patterns

Learn how to learn to refer to the extraction and maintenance of network training rules, including the extraction and analysis of technical procedures, the extraction and analysis of courses, and the establishment of training process libraries. Take knowledge mining as an example, explore the relationship between knowledge fields, and finally recommend it to different students. The steps are:

(1) First, understand the students' academic position on the technical aspects of the online training platform at this time.

(2) Calculate the learning time of each student's learning space and summarize the learning areas on the page.

(3) Use the built-in model extraction algorithm of the object for extraction and analysis. In addition to canceling the standard learning space to adapt to the training methods of students, you can also publish real-time tutorials to students in multiple locations on the training website and recommend pages they are interested in, including conference posts and training sources. The program will create many training routines, and can put some effective rules into the teacher's teaching database with high confidence. The data contained in the database will be continuously updated during the training process. When new team rules are in place, they can be stored in the student data. Once individual students can access the distance learning platform, they can map and insight from the relevant training resources in the process database.

3 Research and Investigation of Mobile User English Learning Pattern Mining and Recognition Based on Multiple Data Mining Algorithms

3.1 Mobile Platform

This experiment selects the "Smart English" platform, which is a product of Education Technology Co., Ltd. Its main feature is that it can be taught anytime and anywhere. Any mobile device or PC can achieve the purpose, which is convenient for students' immediate learning needs. In the process, through human-computer interaction and memory engine management optimization, the system automatically selects students' new words (sentences), new words (sentences), familiar words (sentences), and checks for defects at multiple levels such as memory, dictation, dictation, and example sentences. Mending leaks, mobilize the eyes, ears, hands, and mouth to quickly remember. The platform also has a student terminal, a teacher terminal and a parent terminal.

3.2 Learning Pattern Mining and Recognition

Association rules can get the brief connections between data, can mine the deep-level relationships between data, and get the connections and rules between them through analysis. Under normal circumstances, support is used to measure the importance of group rules. The support degree of group rules $X => Y$ can represent the probability of X objects and an object set Y appearing simultaneously in all applications. As shown in formula 1:

$$\sup(X \geq Y) = \sup(XY) = P(XY)$$
$$= \left(\frac{\text{Number of transactions}}{\text{total number of transactions containing X and Y}}\right) \times 100\% \quad (1)$$

The greater the scale of support, the greater the frequency, and the importance of group rules. Reliability is often used to describe the probability condition of the result of a group rule. Refer to conf($X => Y$), and its expression is shown in formula 2:

$$\text{conf}(X \geq Y) = \sup(XY)/\sup(X) = P(Y|X) \quad (2)$$

Confidence refers to the probability that an event in the object set Y will occur under the condition that the object set X occurs. It is a general purpose for measuring the accuracy of group rules.

4 Research and Analysis of Mobile User English Learning Pattern Mining and Recognition Based on Multiple Data Mining Algorithms

4.1 Differences Between Different Types of Mobile User Learning Modes

Through the curriculum research of the intelligent English platform, different types of mobile users will be able to analyze the differences in learning methods. Initially, one-way analysis of variance (ANOVA) was chosen to analyze different types of mobile user learning methods. The significant difference learning method is the least significant difference (LSD) method that controls the difference between categories. One-way analysis of variance of the changes in different types of learning modes for mobile users shows that there are significant differences between the four learning modes: active days, forum posts, learning chapters, and module completion rates. In order to study the specific differences between different types of mobile users in these learning modes, we conducted a Least Significant Difference (LSD) test under different learning modes, as shown in Fig. 1.

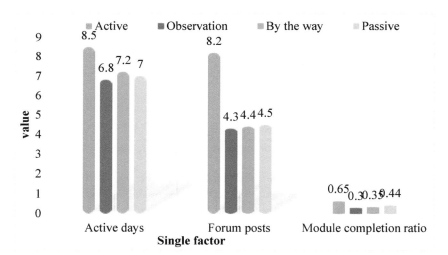

Fig. 1. LSD test of different types of mobile user learning patterns

In terms of active daily behavior, there are significant differences between active mobile users and observation, occasional visits and passive mobile users ($P < 0.05$), and there are also significant differences between observation mobile users and passive mobile users ($P < 0.05$), there is no significant difference between other types of students ($P > 0.05$), in terms of the number of forum posting behaviors, there are significant differences between active mobile users and observational, accidental and passive mobile users ($P < 0.05$), other types There is no significant difference between mobile users ($P > 0.05$). In the way of learning tutorials, there is only a significant difference in observation between active mobile users and mobile users (P < 0.05). There is no significant difference between the other two types of mobile users. There is a significant difference (P > 0.05) between the observed mobile users and other types of mobile users (P < 0.05), and there is no significant difference (P > 0, 05). In terms of unit completion rate learning behavior, only active learning observation and occasional visits to two mobile device users have significant differences (P < 0.05). There are two types of mobile users for observation visits and occasional visits. There is a significant difference with passive mobile users (P > 0.05). Judging from the average learning behaviors of various mobile users, it is obvious that the average of the three behaviors expected to become active mobile users is higher in the number of active days than other types of mobile users, the number of posts in the forum and the number of learning chapters. And the average value of passive mobile users' learning behavior per unit completion rate is higher than that of other types of mobile users.

4.2   The Differences Between the Learning Modes of Mobile Users with Different Educational Levels

Based on the research of intelligent English platform courses, the differences in learning behaviors of mobile users with different educational backgrounds are analyzed, and one-way variance (ANOVA) is used to analyze the learning behaviors of mobile users with different educational backgrounds. Educational level in univariate analysis. The analysis results showed that the two learning behaviors only had significant differences in the number of active days and forum posts (P < 0.05), but there was no significant difference in other learning methods (P > 0.05). Perform statistical analysis on the average value of the significant difference between the number of active days and the number of forum posts. The average learning behaviors of mobile users with different educational backgrounds in terms of active days and forum posts are shown in Table 1.

Table 1. Mean values of different learning behaviors of learners at different educational levels

Learner's educational level	Average (number of forum posts)	Mean (number of active days)
Below middle school	6.88	7.52
Middle School student	6.55	6.28
Bachelor	5.46	5.77
Master	4.86	4.55
PhD	7.91	4.01

The average number of posts posted by the mobile doctoral user forum is the highest. As shown in Fig. 2, the average number of posts by mobile doctoral users is the lowest in the active day mode, indicating that doctoral students rarely have time to participate. The average number of forum posts in high school and below is also relatively high, and the average learning style in the number of active days is also the

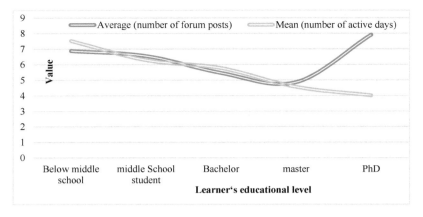

Fig. 2. Mean values of different learning behaviors of learners with different educational levels

highest, which shows that elementary and middle school students have relatively more time to participate.

5 Conclusions

The current social technology is updated very quickly, and artificial intelligence is slowly integrating education. This requires teachers in the new era to constantly update and explore new teaching methods to improve teaching effects. The hybrid teaching mode based on the intelligent platform is a brand-new teaching mode, which is a combination of technology and modern teaching. This research aims to collect various learning mode data of mobile users under the "smart English" platform, and use statistical analysis, text mining and other methods to reveal the correlation between data and data, and to mine the effective information hidden in the data. Guide teaching improvement, and provide reference significance for student learning data research.

References

1. Tseng, C.J., Lu, C.J., Chang, C.C., et al.: Integration of data mining classification techniques and ensemble learning to identify risk factors and diagnose ovarian cancer recurrence. Artif. Intell. Med. **78**(may), 47–54 (2017)
2. Tahmasebi, P., Javadpour, F., Sahimi, M.: Data mining and machine learning for identifying sweet spots in shale reservoirs. Expert Syst. Appl. **88**(11), 435–447 (2017)
3. Naik, B., Nayak, J., Behera, H.S.: A TLBO based gradient descent learning-functional link higher order ANN: an efficient model for learning from non-linear data. J. King Saud Univ. – Comput. Inf. Sci. **30**(1), 120–139 (2018)
4. Guezzaz, A., Asimi, Y., Azrour, M., et al.: Mathematical validation of proposed machine learning classifier for heterogeneous traffic and anomaly detection. Big Data Min. Anal. **4**(1), 18–24 (2021)

5. Liu, T., Shi, K., Li, W.: Deep learning methods improve linear B-cell epitope prediction. BioData Min. **13**(1), 1–13 (2020)

6. Chen, H., Yin, C., Li, R., et al.: Enhanced learning resource recommendation based on online learning style model. Tsinghua Sci. Technol. **25**(3), 348–356 (2020)

7. Gang, L.: Toward a progress indicator for machine learning model building and data mining algorithm execution: a position paper. ACM SIGKDD Explor. Newsl. **19**(2), 13–24 (2017)

8. Zhang, X., Zhu, X., Bao, W., et al.: Distributed learning on mobile devices: a new approach to data mining in the internet of things. IEEE Internet Things J. **11**(99), 1 (2020)

9. Coulibaly, L., Kamsu-Foguem, B., Tangara, F.: Rule-based machine learning for knowledge discovering in weather data. Future Gener. Comput. Syst. **108**(11), 861–878 (2020)

10. Dorgo, G., Abonyi, J.: Learning and predicting operation strategies by sequence mining and deep learning. Comput. Chem. Eng. **128**(SEP.2), 174–187 (2019)

11. Kinnebrew, J.S., Killingsworth, S.S., Clark, D.B., et al.: Contextual markup and mining in digital games for science learning: connecting player behaviors to learning goals. IEEE Trans. Learn. Technol. **10**(1), 93–103 (2017)

12. Ma, G., Ahmed, N.K., Willke, T.L., Yu, P.S.: Deep graph similarity learning: a survey. Data Min. Knowl. Disc. **35**(3), 688–725 (2021). https://doi.org/10.1007/s10618-020-00733-5

Exploration and Online Sharing Network Teaching Platform Based on Information Technology

Weiwei Zhang[✉]

College of Mechanical Engineering, Shandong Huayu Institute of Technology,
Dezhou 253000, Shandong, China
zhang8412860@163.com

Abstract. Nowadays, where undergraduate education is highly valued, reforming the teaching model, building a student-centered online and offline hybrid teaching platform, exploring the use of new and more effective teaching methods and methods, has become our higher education work The first task of the person. Therefore, the "Mechanical Drawing" course strives to be an online open course with more complete content, more diversified resources, rationalized functions, and more convenient learning, which can share the teaching resources of high-quality courses through modern information technology methods and train students Comprehensive ability.

Keywords: "Mechanical Drawing" · Blended teaching · Online open courses

The development of modern communication and network technology has profoundly changed the way of life and learning of human beings, and has also promoted the reform and innovation of education and teaching. Mobile phones and computers have not only become necessities of life, but also enable "everyone can learn and learn everywhere" It's easier to achieve. Therefore, the "Mechanical Drawing" course teaching can rely on the online teaching platform to build the "Mechanical Drawing" course into an online open course, and strive to create a new model of organic integration of inside and outside classes, full integration of online and offline, and deep integration of teacher lectures and student learning [1, 2], to promote students' independent learning to cultivate students' comprehensive abilities.

1 Curriculum Reform Research Ideas

Starting from the graduation requirements of mechanical design and manufacturing and automation majors, combined with the students' academic conditions, determine the training goals of this course, determine the key and difficult points of this course according to the key and difficult problems to be solved, and reorganize and order the teaching with the student as the center. Content, according to the characteristics of teaching content, design student-centered online and offline mixed classroom teaching, select effective teaching methods and means according to student characteristics and teaching content, and establish a process evaluation system to reflect student-centeredness, Continuous improvement through feedback from students, peers and supervisors.

2   Implementation Steps of Curriculum Reform

2.1   Determine the Course Objectives

According to the graduation requirements of mechanical design and manufacturing and its automation major:

Knowledge goal: master the basic knowledge of engineering graphics.
Capability goal: have the ability to identify and express the relevant technical elements of mechanical engineering; have the ability to develop and design mechanical products.
Quality goals: love the motherland, abide by laws and regulations, establish a correct outlook on the world, life, and values; noble professional ethics and good ideological and political qualities; a correct outlook on learning and employment, and a spirit of innovation. Be able to consciously abide by professional ethics, norms and fulfill responsibilities in engineering practice [2, 3].

Determine the training objectives of this course in detail:

Knowledge goal: understand the national standard regulations, understand the principle of orthographic projection, master the methods and steps of reading and drawing parts and assembly drawings, and reserve the theoretical knowledge of graphics for subsequent courses.
Ability goal: able to correctly use drawing tools and instruments, independently analyze and solve the problem of drawing representation of actual parts, and have the practical ability to solve complex problems.
Quality goals: Blended teaching cultivates students' independent learning ability, honest learning attitude, positive sense of innovation, and teamwork awareness [4].

2.2   Reorganizing Teaching Content

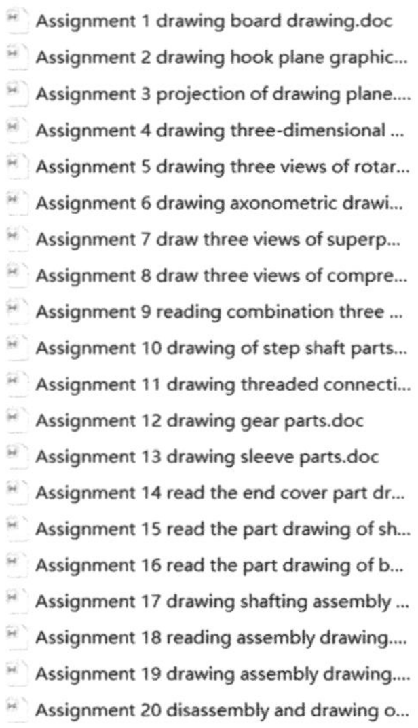

Assignment 1 drawing board drawing.doc
Assignment 2 drawing hook plane graphic...
Assignment 3 projection of drawing plane....
Assignment 4 drawing three-dimensional ...
Assignment 5 drawing three views of rotar...
Assignment 6 drawing axonometric drawi...
Assignment 7 draw three views of superp...
Assignment 8 draw three views of compre...
Assignment 9 reading combination three ...
Assignment 10 drawing of step shaft parts...
Assignment 11 drawing threaded connecti...
Assignment 12 drawing gear parts.doc
Assignment 13 drawing sleeve parts.doc
Assignment 14 read the end cover part dr...
Assignment 15 read the part drawing of sh...
Assignment 16 read the part drawing of b...
Assignment 17 drawing shafting assembly ...
Assignment 18 reading assembly drawing....
Assignment 19 drawing assembly drawing....
Assignment 20 disassembly and drawing o...

Fig. 1. Teaching content organization

The teaching content is reorganized and sequenced based on the students' academic conditions and the difficult problems to be solved. According to the actual work of the enterprise and the cognitive laws of the students, all teaching content is integrated into seven modules for online knowledge teaching; seven practical teaching modules are set up in the order from flat to three-dimensional, first simple and then complex [5]. It is further refined into 20 specific tasks, as shown in Fig. 1. So that theoretical learning has a carrier, and skill training has an entity.

2.3   Curriculum Resource Construction

Establish online courses on mechanical drawing and enrich course resources, including electronic teaching plans, courseware, micro-class videos, exercise library, test paper library, etc. In order to solve the problem of difficult training of students' spatial imagination, three-dimensional model library and moving library are added to help students gradually build spatial thinking ability; to solve the problem of difficult training of students' ability to draw and read patterns, a pattern is established by selecting typical parts and components [6, 7]. Library. As shown in Figs. 2, 3, and 4.

Fig. 2. Micro lesson video   **Fig. 3.** Model animation library   **Fig. 4.** Pattern library

2.4   Course Ideological and Political Integration

Throughout the teaching process, moisturizing things silently incorporate ideological and political elements, such as:

Begin with the development of graphics, carry out patriotic education to students, and enhance national pride;

Starting from the standardization and rigorousness of drawing, cultivate the spirit of craftsmanship;

Take configuration design as the main line and carry out innovative thinking training;

Using actual engineering cases as materials, cultivate the spirit of craftsmanship, and cultivate students' working attitudes of earnestness, truth-seeking, and excellence.

2.5   Implementation of the Teaching Process

The implementation of the specific teaching process is divided into the following four steps with the student as the center:

The first step (before class): advance knowledge and learn online before class. In order to ensure the cultivation of practical application ability in the classroom, the knowledge teaching is moved to the pre-class, and the students learn independently online, and enter the classroom with the knowledge base and questions [8].

The second step (in class): task-driven, classroom practice application. Classroom teaching focuses on the cultivation of practical application ability, takes practical engineering cases as the carrier, and uses task-driven mode to inspire and guide students to analyze problems, discover problems, and solve problems. The themes of the class hours are What (knowing the reason), Why (knowing the reason), and the teacher-led How (teaching them to fish).

(1)   Classroom internalization.

Offline classrooms are mainly based on research teaching methods and classroom discussions, and other teaching methods are used flexibly. Through the evaluation and understanding of the online learning effect of students before class, teachers carefully screen the content of classroom teaching, choose its key points and difficulties, especially the typical learning difficulties encountered by students in the early online learning, adopt problem-oriented and comprehensive quality For the purpose, to guide students to apply the knowledge points of online learning to drawing, image reading and innovative design practice, instead of memorizing a few concepts and doing a few questions, and learning for the test [9, 10].

Secondly, design the teaching links within the classroom, and cultivate the students' ability to analyze and solve problems as much as possible within the limited classroom teaching time. Teachers flexibly adopt three forms of teaching according to the content of the classroom. The first is classroom teaching. After the teacher logically sorts out the knowledge points, please have mastered the students to introduce their own learning experience on the difficult problems that the students have reported before, and the teacher will comment and inspire. The second type is group discussion. The teacher is divided into groups according to the learning ability of all students. The students in the group conduct discussion and mutual aid learning. The teacher observes the students' discussion and problem-solving situation, and intervenes to solve the difficult problems of the group. The third type is classroom training. Part of the exercises is selected from the exercise set, and students are allowed to do their homework in class, which will be completed by teacher guidance or discussion among students.

Adopting mixed teaching has greatly improved the initiative of students in learning, strengthened the conscious management of students' learning process, created a good learning atmosphere, and significantly improved the teaching effect.

(2) Give full play to the role of teachers in interaction.

In the process of interacting with students, five steps are generally used:

Teacher introduction-students think-teacher guidance-student discussion-teacher talk.

To give full play to the guiding role of teachers, teachers must first be good at asking questions and let students think positively; secondly, they must be good at assigning key points to the students for discussion. On the basis of full discussion by the students, the teacher must summarize and summarize the important and difficult points [11, 12].

The third step (after class): Consolidate the ability and complete the homework after class. According to the different levels of students' mastery, drawing exercises are arranged at different levels to strengthen and consolidate the knowledge they have learned. So that students at every level have improved through hard work.

The fourth step (summarizing migration): summarizing knowledge, transferring and extending. Through the staged assignment of major assignments, the development of subject competitions, etc., the knowledge is summarized, digested and absorbed, so as to achieve the ability to draw inferences and transfer applications.

2.6 Process Assessment System

In order to reflect the student-centered thinking, cultivate students' independent learning ability and honest learning attitude, the process of assessment has been increased, from the original 30% to 50%.

Course evaluation results = usual evaluation results * 50% + final exam results * 50%, the specific process evaluation indicators for usual results are shown in Table 1:

Table 1. Process evaluation content and evaluation index

The composition of the usual performance assessment	The average score is calculated as a percentage, with a total of 100 points. It consists of three items: online learning, ordinary homework, and large-scale homework. Each assessment item accounts for 30%, 40%, and 30% of the usual total score respectively								
Normal performance assessment and scoring standards	Online learning	The total score of online learning is 100 points, which is composed of video learning and online testing, which respectively account for 10% and 20% of the usual scores	Video learning	Video completion rate	All the course videos are fully completed, and the scores of a single video are evenly distributed, with a full score of 100 points				
				Chapter study times	Chapter learning times up to 100 times as a perfect score				
			Online test	Online discussion	Publish a topic accumulates 5 points, reply to a topic accumulates 5 points, (multiple responses under the same topic do not repeat the score), get a like accumulate 5 points, the maximum is 100 points				
					Online test scores are taken as the average score of multiple test scores, and counted into the usual scores on a hundred-point system				
	Usual homework	The total score is 100 points. Normal homework will be graded according to the quality of completion and whether it is turned in on time. Each homework will be scored according to the percentile system. The total score of the homework will be the average score of each homework and will be counted into the normal grade based on the percentile system								
	Big job	The total score is 100 points, and the major assignments are scored based on the four observation points of the work progress, the mastery of basic concepts, the correctness of problem solving, and the drawing specifications. Each major assignment is scored according to the 100-point system, and the total score of major assignments will be averaged, which will be included in the usual grades on the basis of the 100-point system	Observation point	80−100 points	70−80 points	60−70 points	40−59 points	0−39 points
			Work progress (Weight 0.2)	Completed ahead of schedule	Complete on time	Delayed completion	Make up	Not all done
			Mastery of basic concepts (Weight 0.2)	>85% have a clear concept	>65% concept clear	> 45% concept clear	>25% concept clear	>10% concept clear Clear
			Correctness of problem solving (weight 0.3)	>85% correct	>65% correct	>45% correct	>25% correct	>10% correct
			Drawing specifications (Weight 0.3)	The drawing conforms to the national standard, and the drawing surface is neat and tidy	The drawing can meet the national standard, and the drawing surface is not tidy	The drawing basically conforms to the national standard, and the drawing surface is not tidy	The drawing line type is not divided, and the drawing surface is untidy	Does not meet the national standard, the drawing is too bad

3   Features and Highlights of Curriculum Reform

3.1   Advance Knowledge, Focus on Ability, and Integrate Professional Quality into the Classroom

Move knowledge learning forward and use online resources to allow students to use fragmented time to learn. Classroom teaching focuses on the cultivation of knowledge application ability, carefully select parts or models as carriers, cultivate students' drawing practice ability, and through continuous practical training and evaluation feedback Let students obtain meticulous, rigorous, serious and responsible professional qualities [13, 14].

3.2   Student Center, Teacher Guidance, Mixed Teaching to Improve Learning Effect

The whole teaching process is "student-centered" divided into three stages: before class, during class and after class. To give full play to the guiding role of teachers and the dominant position of students, teachers must first be good at asking questions and make students think actively; secondly, they must be good at assigning key points to the students for discussion. On the basis of full discussion by students, teachers should focus on important and difficult points. Summarize and summarize, so as to mobilize students' enthusiasm and improve students' learning effect [15, 16].

3.3   Hierarchical Teaching and Personalized Guidance to Eliminate Students' Fear of Difficulty

In view of the reduction of class hours and the continuous content of teaching, students attending classes such as "eating raw meal"; the heavy burden of the course, the students cope with the situation, etc., the use of hierarchical assignments, superstar learning platform video demonstration guidance and QQ group full-course Q&A to eliminate students The emotion of fear of difficulties can improve the learning effect of students [17, 18]. The micro-video guidance method is shown in Fig. 5:

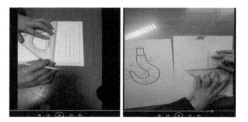

Fig. 5. Schematic diagram of the micro-course teaching

Acknowledgements. This article is the research result of the first semester classroom teaching reform pilot course "Mechanical Drawing" of Shandong Huayu Institute of Technology 2021–

2022 (project number: 2021KG-05). This article is the research result of the 2021 school-level ideological and political demonstration course "Mechanical Drawing" of Shandong Huayu Institute of Technology.

Fund Project. This paper is the research achievement of mechanical drawing, a first-class undergraduate construction course of Shandong Huayu Institute of technology in 2020.

References

1. Chand, A.A., Lal, P.P.: Remote learning and online teaching in Fiji during COVID-19: the challenges and opportunities. Int. J. Surg. **01**, 1 (2021)
2. Miguel, A.: General considerations for online teaching practices in bioinformatics in the time of COVID-19. Biochem. Mol Biol. Educ. Bimonthly Publ. Int. Union Biochem. Mol. Biol. **5**, 1 (2021)
3. Cheriguene, A., Kabache, T., Kerrache, C.A., Calafate, C.T., Cano, J.C.: NOTA: a novel online teaching and assessment scheme using blockchain for emergency cases. Educ. Inf. Technol. **27**, 1–18 (2021). https://doi.org/10.1007/s10639-021-10629-6
4. Abid, T., Zahid, G., Shahid, N., Bukhari, M.: Online teaching experience during the COVID-19 in Pakistan: pedagogy–technology balance and student engagement. Fudan J. Humanit. Soc. Sci. **14**(3), 367–391 (2021). https://doi.org/10.1007/s40647-021-00325-7
5. Sisson, A.D.: Distant glowing screens: lessons learned from adapting to online instruction. J. Hospitality Tourism Res. **5**, 57 (2021)
6. Beason-Abmayr, B., Caprette, D., Gopalan, C.: Flipped teaching eased the transition from face-to-face teaching to online instruction during the COVID-19 pandemic. Adv. Physiol. Educ. **45**(2), 384–389 (2021)
7. Megan, R., Rebecca, R.: Perceptions of preparedness for online teaching due to the COVID-19 pandemic as a graduate of an education program at a university in the Midwest. J. Digital Learn. Teacher Educ. **2**, 24 (2021)
8. Downer, T., Gray, M., Capper, T.: Online learning and teaching approaches used in midwifery programs: a scoping review. Nurse Educ. Today **103**, 104980 (2021)
9. Seetal, I., Gunness, S., Teeroovengadum, V.: Educational disruptions during the COVID-19 crisis in Small Island Developing States: preparedness and efficacy of academics for online teaching. Int. Rev. educ. Int. Z. Erziehungswissenschaft Rev. Int. pedagogie **67**(1), 185–217 (2021). https://doi.org/10.1007/s11159-021-09902-0
10. Monasch, E.M., Wadell, P.M., Baumann, S., Hopkins, M., Hou, M.Y.: An interdisciplinary flipped classroom module on postpartum depression using telemedicine and online teaching. Acad. Psychiatry J. Am Assoc. Directors Psychiatr. Residency Training Assoc. Acad. Psychiatry **46**, 172–174 (2021). https://doi.org/10.1007/s40596-021-01475-2
11. Joana, P., Ariadna, L., Francesc, S., Marc, A., Daniel, A.: A methodology to study the university's online teaching activity from virtual platform indicators: the effect of the Covid-19 Pandemic at Universitat Politècnica de Catalunya. Sustainability **13**(9), 14 (2021)
12. Sirojiddinova, S., et al.: The using Up-Date requirements in classroom during online teaching. Asian J. Multidimension. Res. **10**(5), 54 (2021)
13. Roach, A.V., Attardi, S.M.: Twelve tips for applying Moore's theory of transactional distance to optimize online teaching. Med. Teacher **01**, 147 (2021)
14. Michael, H.: Creating accessible online instruction using universal design principles: a LITA guide. J. Hosp. Librariansh. **21**(2), 2 (2021)

15. Roseni, E., Koroshi, A.S.: Udents' challenges and their professional development during online teaching process in Albania. Engl. Lang. Teaching Linguist. Stud. **3**(2), 2 (2021)
16. Hofer, S., Nistor, N., Scheibenzuber, C.: Online teaching and learning in higher education: lessons learned in crisis situations. Comput. Hum. Behav. **121**, 106789 (2021)
17. McComas, M.J., Barragato, A.K., Kinney, J.: No skills left behind: online teaching and learning in preclinical dental hygiene. J. Dent. Educ. **85**(S3), 1936–1937 (2021)
18. Li, A., Wang, H., Vijayalakshmi, S.: An artificial intelligence recognition model for English online teaching. J. Intell. Fuzzy Syst. **40**(2), 78 (2021)

Design of Interactive Music Intelligent Database System Based on Artificial Intelligence Algorithm

Chengxi Cai[✉]

Lanzhou Resources and Environment Voc-Tech University Academy
of Marxism, Lanzhou 730021, Gansu, China
ccxl8993111936@163.com

Abstract. With the rapid development of Internet technology, artificial intelligence plays a more and more important role in society. Many functions have been developed based on intelligent algorithms. This paper takes interactive music database as an example. Firstly, the definition and characteristics of interactive music are introduced. Then the artificial intelligence algorithm is analyzed and applied. Secondly, the existing system is analyzed in detail, and its evaluation and design implementation scheme are made. According to the selected scheme, a simple and practical model with strong scalability, easy operation and easy maintenance is constructed. MySQL is used as the development tool to complete the construction of interactive music database. Finally, the performance of the system is tested. The test results show that the operation time and system delay time of the intelligent database system are about 2 s. The CPU occupation rate of the system is about 2%, and the memory occupation is about 3500K. The system uses the same function and has unique certainty for users with different permissions, and can normally start and complete corresponding operations under different playback modes.

Keywords: Artificial intelligence algorithm · Interactive music · Music intelligence · Intelligent data

1 Introduction

With the rapid development of the Internet and information industry, people have higher and higher requirements for the quality of life. In daily life, we can obtain a variety of data through a variety of ways [1, 2]. It also puts forward higher-level and wider requirements for these problems, such as massive, complex, huge and easy to lose. Solving such problems based on machine intelligence algorithm (AI) has become a research hotspot [3, 4].

Many scholars have done relevant research on artificial intelligence technology. Abroad, in the 1950s, American artificial intelligence technology began to be studied. By the 1960s, there had been a large number of relevant research results. At present, the United States has a mature, advanced and reliable algorithm library. It includes describing and analyzing human intelligent behavior based on different models such as knowledge and model, rule-based and unstructured methods, and using these data to

© The Author(s), under exclusive license to Springer Nature Switzerland AG 2022
V. Sugumaran et al. (Eds.): ICMMIA 2022, LNDECT 138, pp. 538–545, 2022.
https://doi.org/10.1007/978-3-031-05484-6_68

establish corresponding solutions (such as neural network) to promote social progress and national development plan, which has attracted extensive attention all over the world [7, 8]. The development of interactive music system in China started late. So far, there have been many excellent research and development based on user behavior model (SCM). At present, Lenovo, Alibaba, Tencent and other companies are widely used [9, 10]. The above research has laid the foundation for this paper.

Based on the current mainstream algorithms, this paper analyzes the solutions to this problem. Firstly, by summarizing and combing the existing technologies, problems in principle and solutions, a design idea of interactive database system based on hash function is proposed in theory, and MySQL is used as the background management platform.

2   Discussion on Interactive Music Intelligent Database System Based on Artificial Intelligence Algorithm

2.1   Interactive Music

1) Definition

The concept of interactive music comes from the 1980s. American scholars put forward the term "interactive experience". It is an interactive behavior that takes users as the center and relies on data analysis technology, multimedia processing and other related skills to create and perform. With the advent of the Internet era, people's demand for information resources is also increasing. Various types of music media websites came into being, and are sought after by the public, becoming one of the mainstream trends. Interactive music refers to the intelligent application based on machine. In the music works database, machine-based programming realizes the interaction with users through data mining technology and artificial intelligence algorithm. It includes two parts: user and content, including multimedia data such as text, audio and video. In the process of interaction, the content needs to be processed. First, load and store these information into the database system, and then store them and calculate the corresponding parameter values to generate the required output file, or directly generate other requests, and get the results for further use or modification [11, 12].

2) Features

Based on the basic characteristics of users, interactive music integrates the data stored in the traditional library to form the characteristics of interactive function and activity authority management. It mainly includes the following aspects: (1) immediacy. Interactive music has strong real-time interaction ability. Users can listen to and watch songs at any time and anywhere through mobile phone software or computers. This enables digital music works to receive timely feedback. At the same time, it also allows the audience to adjust the works according to their own needs. (2) Emotional expressiveness. In the current era, people have shown great dependence on network information and are willing to deal with these emotional problems through computers. Interactive music can make users have a high degree of experience under the existing computer application environment, so that users can express their emotions through music interaction with others, so as to alleviate contemporary people's psychological anxiety. (3) Fast propagation speed. The direct communication between music and lyrics is realized on the network. Because the audio information can be transmitted through the computer, it can quickly convert the audio and video content into pictures and other forms for users to share. At the same time, it also makes the playback mode change from traditional mode to digital mode, and gives special meaning to each field, so as to meet the needs of different users.

2.2   Artificial Intelligence Technology

1) Definition
Artificial intelligence has a wide definition. It is a complete algorithm system, and through its combination with human intelligence, it forms a new model. In the late 1960s, artificial neural networks appeared. The definition of artificial intelligence refers to the use of computer technology to simulate the human brain and under-stand the knowledge in the human brain neural information processing system by identifying, learning and expressing control rules. Under this system, people can judge complex problems independently. In essence, intellectualization is the research process of using a limited number of machines to replace the existing unstructured data sets and carry out calculation, analysis and reasoning. It refers to the emergence and utilization of computer technology to meet the needs of the development of human society. For this model, we can combine machine learning, memory and computing technologies, which is the so-called AI (arural network). Artificial intelligence is an interdisciplinary science, which involves the fields of natural science, social science and agricultural society. The most important research in human machine intelligence is human-computer interaction theory. At present, what we understand is that artificial intelligence and computer programming are an important stage in the development of artificial intelligence. It is not only simply using algorithms to solve practical problems or deal with real situations, but also completing work tasks through sensing devices. The controller is responsible for the instructions required for the execution of various functions in the whole system, while the driver provides services for the realization of machine control, processes data information and sends corresponding commands to the microcomputer or terminal equipment.
2) Advantages
Intelligent algorithms have advantages in processing data. With the continuous development of artificial intelligence technology, people can use the existing computer foundation to identify, store and retrieve a large amount of information content. At the same time, they can also effectively improve the accuracy and computational efficiency of identification. Therefore, we need to obtain more rel-evant rule data from a large number of unstructured or semi-structured transaction sets and analyze them to draw better conclusions and serve decision-making. In the current society, artificial intelligence technology has been widely used in various industries and has achieved great success. The machine is highly intelligent and the

algorithm efficiency is high. Because the traditional data processing method is manual operation, the workload is large, the speed is slow and time-consuming, and the use of AI algorithm can effectively solve this problem. Computers can replace people to complete a lot of repetitive work and machines with insufficient judgment ability to replace people to do some information that we can't touch in our daily life. Artificial intelligence can improve the interaction between algorithms and users. Through the simulation of the algorithm, we can understand what common points exist among users at different levels, and what services they need to make the machine work better, or make more effective use of existing resources to meet people's needs, so as to improve the running speed of the whole network system and reduce costs.

3) Algorithm

Genetic algorithm is a parallel optimization method of global search and local problems. It can be applied to solve other technologies such as stability and rapidity in complex systems. It can effectively use the law of population evolution in nature to adapt to environmental conditions quickly and efficiently, and obtain the optimal solution by adjusting these process parameters. In theory, the optimization process of genetic algorithm is random generation, crossover and mutation operators. It starts from an initial solution and then continues to improve. In this step, we need to select the constraint function and give a series of fitness values to form an optimal combination to determine the interaction force and interaction relationship between the new individual and local variables. For a given population PR = {1, 2, 3,.., a}, the fitness value of individual a $\in$ P is f (a), and its selection is roughly as follows:

$$P(a_j) = \frac{f(a_j)}{\sum_{i=1}^{n} f(a_i)} \tag{1}$$

The formula determines the probability distribution of individuals in the offspring population. A mating pool for reproduction is generated through the selection operation, in which the expected number of individuals in the parent population is:

$$P(a_j) = n^* p(a_j), j = 1, 2...n \tag{2}$$

When the difference of individual fitness in the group is very large, the ratio of the probability that the best individual and the worst individual are selected, that is, the selection pressure will also increase exponentially. The survival opportunities of the best individuals in the next generation will increase significantly, while the survival opportunities of the worst individuals will be deprived. At present, the best individual in the population will quickly fill the whole population, resulting in the rapid reduction of population diversity and the premature loss of evolutionary ability of genetic algorithm. This is also an easy problem in the selection of fitness ratio.

3   Experiment

3.1   Interactive Music Intelligent Database System Structure

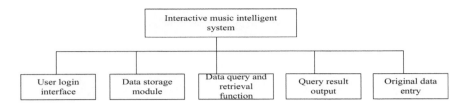

Fig. 1. Music intelligent database system structure

The system is mainly realized by artificial intelligence algorithm, and its core algorithm is based on genetic algorithm, which is designed on this basis (as shown in Fig. 1). (1) User login interface: when registering, enter the account password and click the login button to enter the registration page Before logging in, you need to verify the account information and fill in the relevant information again to complete the login operation. (2) Data storage module, data query and retrieval function: according to different permissions, you can select the corresponding search method to realize the access to all records and history in the database. (3) Original data entry: preliminary analysis of the original data to be stored in MySQL database. (4) Query result output: after entering the target data, perform the above operations to obtain the query result.

3.2   System Requirements Analysis

The main function of this design is to realize an interactive music data storage and sharing based on artificial intelligence algorithm. The system is based on artificial intelligence algorithm. Through the analysis of the data, we can get the different needs of users for lyrics, music and emotion. Therefore, it is necessary to analyze and study the song types, lyrics and historical playback records uploaded by users. At present, the mainstream audio websites have their own independent and complete, and have a high-quality network system, so this paper first needs to consider whether the types of songs will vary from person to person. Secondly, the function module must be able to realize. Thirdly, the platform also needs to have a perfect security mechanism, safeguard measures, good use experience, simple operation, easy learning, easy maintenance and upgrading. Finally, the authority allocation is added when the user logs in and registers. When there is an error or the account and password are illegally stolen, it will be prompted and processed in time to ensure the stable and reliable operation of the system and data integrity.

3.3   System Performance Test Steps

Before completing the design of interactive music intelligent database system based on artificial intelligence algorithm, we need to test whether its data table meets the user's requirements for logical function. If it does not meet the user's modification opinions, it needs to be recalculated. The first thing to do is to determine the output format and the relationship between the fields in the data table. Then, all the contents are arranged in order into a table form, stored in the server, saved and connected with other modules. Finally, the corresponding results are automatically generated according to the information recorded in the form for everyone to use, view and learn. In the test process, the parts that do not meet the needs of users are removed, and then the unnecessary parts are redefined and designed, new data are added and the normal operation of the system is verified. Finally, the conclusion will be drawn through the experimental results.

4 Discussion

4.1 System Performance Test

Table 1 is the performance test data of interactive music intelligent database system based on artificial intelligence.

Table 1. Performance test data

Number of tests	Operate time (s)	Delay time (s)	Utilization rate of CPU (%)	Accounting for memory (k)
1	2	2	2	3541
2	3	4	3	3124
3	2	3	1	3686
4	2	1	2	3245
5	3	2	3	3845

In the test process, firstly, the user's identity information and login password are encrypted, and then the data flow diagram and field length are subtracted respectively. As can be seen from Fig. 2, the operation time and system delay time of the intelligent database system are about 2 s. The CPU occupation rate of the system is about 2%, and the memory occupation is about 3500K. Through the above analysis, it can be seen that users with different permissions can use the same function and have unique certainty (such as music category), and can normally start and complete corresponding operations under different playback modes.

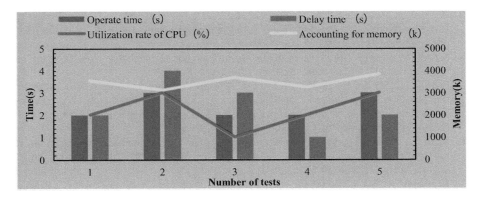

Fig. 2. Performance test contrast

5   Conclusion

With the development of artificial intelligence technology, the design of interactive music intelligent database came into being. This paper studies it. Firstly, it introduces the current research results and relevant theoretical basis at home and abroad. Secondly, the algorithm, principle and the factors that need to be considered when the user model realizes the function on different platforms are analyzed and compared in detail. Finally, aiming at the defects of the current mainstream algorithms, this paper puts forward a feasible scheme and completes the software and hardware design of the whole system, including MySQL database server cluster memory and its interface technology.

References

1. Jia, L., Li, L.: Research on core strength training of aerobics based on artificial intelligence and sensor network. EURASIP J. Wirel. Commun. Netw. **2020**(1), 1–16 (2020). https://doi.org/10.1186/s13638-020-01785-3
2. Otani, S., Yamanishi, R., Iwahori, Y.: Generation of web image database based on hybrid noise removal method of visual and semantic features. Trans. Jpn. Soc. Artif. Intell. **32**(1), WII-N1-10 (2017). https://doi.org/10.1527/tjsai.WII-N
3. de Andrade, P.H.M., Villanueva, J., Braz, H.: An outliers processing module based on artificial intelligence for substations metering system. IEEE Trans. Power Syst. **35**(5), 3400–3409 (2020)
4. Arriba-Mangas, A.D., Fukuda, R., Aoyama, H.: Development of model identification methodology based on form recognition for computer-aided process planning. J. Adv. Mech. Des. Syst. Manuf. **11**(5), JAMDSM0054 (2017)
5. Vijay Kumar, T.V., Arun, B.: Materialized view selection using HBMO. Int. J. Syst. Assur. Eng. Manag. **8**(1), 379–392 (2015). https://doi.org/10.1007/s13198-015-0356-4
6. Asemi, A., Ko, A., Nowkarizi, M.: Intelligent libraries: a review on expert systems, artificial intelligence, and robot. Libr. Hi Tech **26**(June), 1–23 (2020)

7. Zhao, W., Luo, J., Li, S., Qi, J., Meng, H., Li, Y.: Design of dynamic calf weighing system based on moving-IIR filter algorithm. J. Electr. Eng. Technol. **16**(2), 1059–1069 (2020). https://doi.org/10.1007/s42835-020-00604-5

8. Schedl, M., Yang, Y.H., Herrera-Boyer, P.: Introduction to intelligent music systems and applications. ACM trans. Intel. Syst. **8**(2), 17.1-17.8 (2017)

9. Pietro, G.D., Gallo, L., Howlett, R.J., et al.: Big data security on cloud servers using data fragmentation technique and NoSQL database. In: Smart Innovation, Systems and Technologies Intelligent Interactive Multimedia Systems and Services, vol. 98, pp. 5–13 (2019). https://doi.org/10.1007/978-3-319-92231-7(Chapter1)

10. Mark, M., Oliviero, S., Wolfgang, W.: Intelligent technologies for interactive entertainment. Lect. Notes Inst. Comput. Sci. Soc. Inform. Telecommun. Eng. **9**(3), 47–48 (2017). https://doi.org/10.1007/978-3-319-73062-2

11. Bembenik, R., Skonieczny, P.G., et al.: Classification of music genres by means of listening tests and decision algorithms. In: Studies in Big Data Intelligent Methods and Big Data in Industrial Applications, vol. 40, pp. 291–305 (2019). https://doi.org/10.1007/978-3-319-77604-0

12. Hondorp, H., Nijholt, A., Reidsma, D.: Intelligent Technologies for Interactive Entertainment. Lect. Notes Inst. Comput. Sci. Soc. Inform. Telecommun. Eng. **9**(3), 47–48 (2017). https://doi.org/10.1007/978-3-319-73062-2

Comprehensive Evaluation System of Sports Tourism Resources Development Based on Data Mining Algorithms

Chang Chen[✉], Guoquan Wang, Yongchao Chu, Jianmin Ding, Tiantian Dong, and Yijun Cai

Department of Physical Education, Guangzhou XinHua University, Guangzhou, Guangdong, China
chenchang_2016@126.com

Abstract. Today, with the rapid development of information, people have put forward higher-level requirements for things. Sports tourism has become a new era trend. With the rapid development of sports tourism, our people's demand for recreation and participation is increasing. This article intends to use data mining algorithms (DMA) to study the comprehensive evaluation system of sports tourism resources (STR) development, with the purpose of rectifying the province's STR. This article mainly designs and researches the comprehensive evaluation system for the development of STR by means of investigation and qualitative and quantitative methods. Through investigation and calculation, it is found that the service system score is only 4.29, indicating that the sports tourism resource service system in this province needs to be improved.

Keywords: Data mining algorithm · Sports tourism · Resource development · Comprehensive evaluation system

1 Introduction

With the development of the Internet, sports tourism, as a new type of leisure, is rapidly emerging in China. Evaluation and analysis of scenic spots based on DMA is of certain significance. Using this method can effectively obtain tourist satisfaction and potential preferences from a large number of information sources.

There are many theoretical results of the research on the comprehensive evaluation system of STR development based on DMA. For example, some people say that Our country has abundant natural, humanistic, sports and tourism resources, but the development of resources is still in its infancy [1, 2]. Some people also use TOWS analysis to construct a scoring index system for the development of STR [3, 4]. In addition, some people said that STR are the material basis for the production and development of sports tourism, and the evaluation of STR has a fundamental role [5, 6]. So this article uses data mining (DM) to study the sports tourism resource evaluation system.

This article first studies the development of DM in the development of STR. Secondly, it describes the STR. Then the comprehensive evaluation system of tourism

© The Author(s), under exclusive license to Springer Nature Switzerland AG 2022
V. Sugumaran et al. (Eds.): ICMMIA 2022, LNDECT 138, pp. 546–554, 2022.
https://doi.org/10.1007/978-3-031-05484-6_69

resources is elaborated in detail, including the purpose and principles of the evaluation, the content system of the comprehensive evaluation, the qualitative evaluation method of tourism resources and the determination of the evaluation factor index. Then analyze the ASEB grid of sports tourism resource development. Finally, the results are obtained through investigation and evaluation.

2 Comprehensive Evaluation System of Sports Tourism Resources Development Based on Data Mining Algorithms

2.1 The Development of DM in the Development of STR

With the continuous development of social econoOur and information technology, people pay more and more attention to the participation and fun of sports activities. In this case, how to use DM technology to develop tourism resources has become a current research hotspot. In the process of developing STR, DM mainly uses the database to collect and sort information, and then obtain user needs through analysis. The application of DM in information management is mainly through sorting out large, complex and massive original texts, and using it to transform useful content into a knowledge base. Many related technologies are involved in the development of STR, such as database systems. In addition, DM also includes model algorithms, mainly through the establishment of a dynamic, multi-level and multi-dimensional number structure system for comprehensive research and evaluation. In the process of analyzing the needs of tourists, the main purpose is to collect and sort out tourist information, and then convert it into a model that can reflect the relationship between the problem and the data. Finally, build a corresponding system based on these characteristics for evaluation [7, 8].

DM technology can predict and analyze potential problems in the development of tourism resources, so as to provide decision-making support for the development of sports tourism. In research and practical activities, people have discovered that a lot of new information, new things, unknown events and other unstructured data exist when we come into contact with these things in our lives. But through in-depth exploration of these records or phenomena, we can get more knowledge about the field. Similarly, in sports tourism, the DM technology industry brings convenience to sports tourism. During the development of STR, some regular problems can be found through analysis and prediction of sports tourism routes and routes. The application of DM technology in the development of STR is mainly to improve the ability of tourists to obtain and evaluate the product information of the tourist attractions and scenic spots. Through the use of modern information technology such as computers and networks to achieve effective integration and analysis of these sports related resources. At the same time, it can also help managers provide more convenient, accurate and timely decision-making suggestions. In addition, with the help of the database management system, a complete and complete system data structure model can be established for the majority of users and optimized processing, thereby improving tourists' inquiries Solving ability and work efficiency when encountering problems in the process [9, 10].

2.2   Sports Tourism Resources

STR include nature, society, humanities, material resources, intangible resources, etc., and integrate their fine products. Due to the differences in geography, natural environment and customs in different regions, its resources also have different regional differences and characteristics. Most of the STR are formed naturally, are gifts from nature to mankind, and reflect the harmonious development of man and nature. Most sports tourism projects are carried out in a specific environment and at a specific time of the year [11, 12].

Although sports tourism has become a common social phenomenon in people's way of traveling, the concept of sports tourism is still advocated by sports scientists with their own views and attitudes, and there are many, many explanations. Some sociological analysis shows that sports tourism is not only a combination of tourism and sports, but also an ordinary branch of tourism econoOur, but a qualitative change after the combination of sports and tourism. From a marketing perspective, sports tourism is an industry that provides professional tourism services that integrate fitness, entertainment, leisure sightseeing, and adventure challenges. From the perspective of tourism, tourism is the driving force of sports tourism, so sports tourism must meet the conditions of tourism. From the perspective of physical education, it is a form of outdoor sports, leisure and outdoor sports. With the continuous evolution of travel modes and people's travel needs, the number of people participating in sports tourism is increasing, and the scope of activities, activity fields and types of activities related to sports tourism are also constantly evolving.

2.3   Comprehensive Evaluation System of Tourism Resources

The evaluation of tourism resources is an extremely complex task. To make an objective and fair evaluation of it, certain principles must be followed in the actual operation.

(1) Content system of comprehensive evaluation

The basic value of tourism resources reflects the quality of the resources themselves. The use of tourism resources should meet certain development conditions, so that the geographical location, infrastructure, accessibility, diffusion and combination degree of the scenic spot, regional economic development, environmental quality, and the support and participation of local governments and community residents are affected by the value of tourism resources. The analysis of its development value can better reflect the total value of the assessed tourism resources and make a more objective assessment. The development of tourism resources must conform to market conditions and be based on market development. In other words, market demand also determines the value of tourism resources. Therefore, the market scope of tourism resources and the quality of tourism resources in the market are also important basis for comprehensive evaluation of tourism resources. In addition to its own public value, various tourism resources also have their own unique value. Characteristics represent the vitality of different types of tourism resources. When evaluating resources, the value of resource characteristics should be carefully evaluated, rather than a general evaluation independent of the characteristics of the resource itself.

(2) Determine the index of evaluation factor level

The viewing value of STR is reflected in the attraction to sports tourists. Ornamental sports activities mainly include ethnic folk sports observations, sports competitions, sports cultural festivals, and ecological STR.

According to the different functions of physical activity experience, it can be divided into three evaluation factors: fitness, leisure and entertainment, and adventure and stimulation. The educational value of sports tourism is related to the educational function of STR. Sports activities can have a certain spiritual impact on sports tourists, and more often they have a positive impact on tourism. The higher the educational value of STR, the stronger the attraction to sports tourists. The natural environmental conditions of STR are related to the geographical environment, natural resources and climatic factors on which the STR are based, including three evaluation factors of natural environment quality, wealth resource integrity and climate suitability. According to the content and influencing factors of sports activities, the historical and cultural conditions for the existence of STR mainly include three evaluation factors: ethnic folk culture, competitive sports culture and sports tourism atmosphere. Site conditions are an important basis for the development of STR, which are mainly related to the geographical location and traffic conditions of the area where the STR are located. The development of STR must be guided by market demand. Only by accurately positioning the target market can we develop sports tourism products that meet the needs of sports tourists.

2.4 ASEB Raster Analysis of the Development of Sports Tourism Resources

ASEB is a new type of analysis method that integrates travel experience into the analysis system and is consumer-oriented.

The comprehensive development of STR takes market-oriented resources as the core, the development of STR that are more attractive to tourists, and the process of organizing and integrating tourism resources. The comprehensive development of STR should be combined with ASEB grid analysis and follow the basic principles of comprehensive development of tourism resources to maximize the social, economic and environmental benefits of the development and utilization of STR.

Integrate the spatial layout and in-depth excavation of STR. According to the development theory of tourism space structure, combined with factors such as the distribution of STR, development and utilization conditions, tourism infrastructure and accessibility, the design of sports tourism development space is designed.

Integrate sports tourism products to create a well-known brand. High-quality routes are the key to the integration and development of tourism resources, and wise route design promotes the formation of regional tourism networks. Efforts to create attractive tourist routes can better and more directly enhance the image of sports tourism.

Integrate the image of tourist destinations and enhance regional competitiveness. It is necessary to break the geographical boundaries, be resource-oriented, serve tourists as the goal, and focus on regional city centers to build a tourism geographic information system and network platform to provide information for sports tourism. On the basis of tourist satisfaction, improve the service awareness and service level of persons in charge, operators and service personnel of sports tourist attractions, and establish an effective tourist talent training system. Integrate tourism education resources, rely on rich academic resources, take advantage of sports-related degrees, tourism colleges and other educational resources, and accelerate the cultivation of high-quality sports tourism talents.

3   Quantitative Evaluation of Sports Tourism Resources

3.1   Quantitative Evaluation Steps

The popularity of sports tourism viewing value refers to its popularity in the region. Strangeness refers to the peculiarity of the appearance of STR. Pleasure is the degree of instructing tourists to be pleasing to the eye. The fitness of participation value refers to the value of enhancing tourists' physical fitness and promoting health. Recreation refers to the value of regulating the body and mind and being relaxed and happy. Adventure excitement is the value of seeking thrills. The safety of sports activities refers to the safety status of sports activities. The evaluation levels of this type of indicators are divided into five levels: extremely high, high, normal, low, and extremely low, and points are assigned in turn. The sports humanistic spirit of sports tourism education value refers to the positive spirit experienced by tourists in the process of participating in sports tourism activities. Sports cultural knowledge refers to the sports-related knowledge contained in sports activities. Sports skills refer to the state of skills displayed in sports activities. Natural environmental quality refers to the environmental quality of the area where STR are located. Resource abundance and completeness refer to the abundance and completeness of STR. Climate suitability refers to the degree of climate suitability for sports tourism activities. Ethnic folk culture is the richness of regional ethnic folk sports culture. Sports competition culture is the strength of the Olympic culture of competitive games. The sports tourism atmosphere is the overall atmosphere for the development of sports tourism in the region. The tourist source market of sports tourism is the number and scope of the tourist source that a region can attract. Regional geographic location refers to the superiority of the geographic location of the region. Investment capacity conditions refer to the level of regional economic development and investment and financing capabilities. The physical activity infrastructure condition is the completeness of the regional sports infrastructure construction. The sports tourism service support system refers to the status of talents specializing in sports tourism services. The guarantee of policies and regulations refers to the conditions of policies and regulations that promote the development of sports tourism in a region. The difficulty of construction refers to the degree of difficulty of the project where the resource is located. This kind of index is divided into five grades of excellent, good, medium, poor, and poor for scoring. The capacity of tourism environment refers to the capacity of tourism activities under certain conditions. It is divided into five grades: maximum, large, large, general, and small.

3.2   Questionnaire Survey Method

The target of the questionnaire: tourists and sports tourism practitioners in the province.

Questionnaire validity test: This article evaluates the rationality of the questionnaire design, measurement indicators, content, and structure. The higher the degree of rationality, the higher the validity of the questionnaire. Reliability test of the questionnaire: The repeated test method is adopted to obtain the agreement rate of the statistical data of the two questionnaire answers.

A total of 1,000 questionnaires were issued this time, and 800 copies were effectively recovered. The on-site distribution, filling, and return of the questionnaire effectively improved the accuracy of the questionnaire.

3.3   Determine the Weight of the Evaluation Index

First construct the judgment matrix. The judgment matrix is the core of the AHP method, and it is an important part of transforming the opinions of subjective judgments into objective values. The judgment matrix should satisfy:

$$\begin{aligned} & \text{xmn} > 0 \\ & xmn = \tfrac{1}{xmn} \end{aligned} \tag{1}$$

The value of the judgment matrix should be determined by many factors. Calculate the product of each row element of the matrix, normalize the vector, and calculate the weight vector:

$$w = (w_1, w_2, \ldots, w_a)^s \tag{2}$$

Among them, w represents the weight, and a is the number of items. Then, the sports tourism resource evaluation model can be expressed as:

$$S = \sum\nolimits_m^a W_m Q_m \tag{3}$$

X is the comprehensive evaluation value of STR, and W_m is the weight of the m-th index. The full score of the comprehensive evaluation value is 10 points, using the weighted average method, and then multiplying by the corresponding weight value of each indicator to obtain the comprehensive evaluation value.

4   Analysis of Evaluation Results

4.1   The Value of Sports Tourism Resources

According to the calculations in this article, the value of the STR in this province is mainly judged from its viewing value, tourism value, educational value and humanistic value. The specific situation is shown in Table 1:

Table 1. Sports tourism resources of their own value

	Weight	Score value	Comprehensive evaluation
Ornamental value	0.0472	5.74	0.2657
Participation value	0.0865	6.79	0.3458
Educational value	0.0314	6.27	0.2019
Human value	0.0211	5.76	0.1025

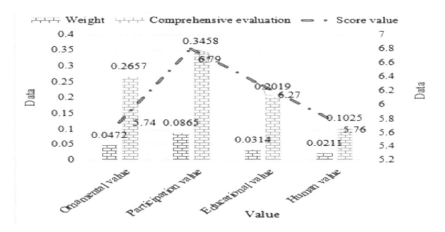

Fig. 1. Sports tourism resources of their own value

As shown in Fig. 1, we can see that the scores of these four items are all between 5 and 7, which shows that the STR in this province are still lacking in characteristics on the whole and need to be further explored and developed. At the same time, in contrast, sports tourism is highly participatory and its educational function is fully demonstrated.

4.2   Conditions for the Development of Sports Tourism Resources

Aiming at the development conditions of STR in this province, this article conducts research from four aspects: product demand, source market, regional location and service system.

As shown in Fig. 2, we can see that the highest score is the regional location, and the service system has the lowest score. In addition, the comprehensive evaluation of product demand is the highest, and the weight value is also the largest. The lowest is the service system. This shows that the province's physical activity infrastructure conditions and service support system have not reached the standard level. Therefore, the province's sports tourism needs to strengthen the construction of the service system.

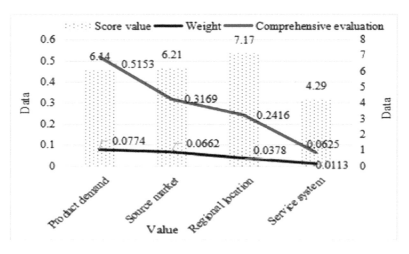

Fig. 2. Conditions for sports tourism development resources

5   Conclusions

Our country's sports industry is developing rapidly, but due to the lack of scientific planning and management, a series of problems have been caused. Through the evaluation of the development of STR, it can be understood that there are many types of STR, but there are some problems. The province's STR lack a suitable service system, and the characteristics of sports tourism are not obvious and the attractiveness is not high. Through the data analysis of the sports tourism resource development evaluation system, the province's sports tourism needs to strengthen the construction of the service system.

Acknowledgements. This work was supported by Research on spatial Layout optimization and sustainable Development of Sports tourism in National Whole-region Tourism Demonstration Area (GDSS2020N176).

A project approved by General project of Guangdong Provincial Sports Bureau in 2020.

References

1. Hossain, M.A., Ferdousi, R., Hossain, S.A., Alhamid, M.F., El Saddik, A.: A novel framework for recommending data mining algorithm in dynamic IoT environment. IEEE Access **8**, 157333–157345 (2020)
2. Awan, S., Dadan, R.: Graduate rate analysis of student using data mining and algorithm apriori. Int. J. Soft Comput. **12**(5–6), 287–293 (2017)
3. Mohadab, M.E., Bouikhalene, B., Safi, S.: Automatic CV processing for scientific research using data mining algorithm. J. King Saud Univ. Comput. Inform. Sci. **32**(5), 561–567 (2020)

4. Al-Muqrashi, A.,Sosoaimanickam, A., et al.: A comparative study of the efficient data mining algorithm for forecasting least prices in Oman fish markets. Int. J. Appl. Eng. Res. 13 (11 Pt.1), 8751–8758 (2018)

5. Abboud, Y., Brun, A., Boyer, A.: C3Ro: an efficient mining algorithm of extended-closed contiguous robust sequential patterns in noisy data. Expert Syst. Appl. **131**, 172–189 (2019)

6. Roy, J., Chatterjee, K., Bandyopadhyay, A., Kar, S.: Evaluation and selection of medical tourism sites: a rough analytic hierarchy process based multi-attributive border approximation area comparison approach. Expert Syst. **35**(1), e12232.1-e12232.19 (2018)

7. Hajizadeh, F., Poshidehro, M., Yousefi, E.: Scenario-based capability evaluation of ecotourism development – an integrated approach based on WLC, and FUZZY – OWA methods. Asia Pac. J. Tourism Res. **25**(6), 627–640 (2020)

8. Stangl, B., Pesonen, J.: Consumer evaluation of hotel service robots. In: Information and Communication Technologies in Tourism 2018,308−320 (2018). https://doi.org/10.1007/978-3-319-72923-7(Chapter24)

9. Hughes, K.A., Convey, P., Turner, J.: Developing resilience to climate change impacts in antarctica: an evaluation of antarctic treaty system protected area policy. Environ. Sci. Policy **124**(7), 12–22 (2021)

10. Roy, J., Chatterjee, K., et al.: Evaluation and selection of medical tourism sites: a rough analytic hierarchy process based multi-attributive border approximation area comparison approach. Expert Syst. Int. J. Knowl. Eng. **35**(1), e12232.1 (2018)

11. Gil-Alana, L.A., Henrique, D., Wanke, P.: Structural breaks in Brazilian tourism revenues: unveiling the impact of exchange rates and sports mega-events. Tourism Manag. **74**, 207–211 (2019)

12. Valente, L.: The FC Porto museum project and the challenges of a new reality. Worldwide Hospitality Tourism Themes **12**(6), 767–770 (2020)

Machine Translation of British and American Literature Based on Parallel Corpus

Yansen Xu[1(✉)] and Hongjiao Li[2]

[1] College of Foreign Languages, Bohai University, Jinzhou, Liaoning, China
bhdxxys@126.com
[2] Jinzhou Senior High School, Jinzhou, Liaoning, China

Abstract. Machine translation is a practical technology that uses electronic computers to translate between different languages. With the development of artificial intelligence technology and the advancement of brain science research, machine translation will replace human translators for most of the translation work. The corpus not only contains a rich corpus, but also has powerful retrieval and statistical functions of the computer, which can easily and efficiently discover the potential patterns of the content covered by the corpus and improve the quality and efficiency of machine translation. With the rapid development of computer software and hardware technology, as well as the continuous improvement of corpus construction, machine translation of British and American literature based on parallel corpus has become a new research direction, which improves the quality of British and American literature translation and promotes the dissemination of British and American literature through the knowledge mining of bilingual parallel corpus.

Keywords: Parallel corpus · Machine translation · Translation process · Inverted index

1 Introduction

With the accelerated integration of the world economy, the rapid expansion of online information and the increasing frequency of international exchanges, how to overcome language barriers has become a common challenge for the international community. Translation is not only about language conversion, but also has a more important role in cultural communication, serving as a bridge between people of different nationalities, contexts and living backgrounds. As the way of manual translation is far from being able to meet the needs of a large amount of information that needs to be translated, the use of machine translation technology to assist people to obtain information quickly has become an inevitable trend. Machine translation emerged in the 1940s, and there was a boom in the research of machine translation technology in the US and other developed countries. Thanks to the rapid development of machine translation technology, especially the emergence of neural network machine translation in recent years, the accuracy of machine translation has been greatly improved, the quality of translation output has achieved a qualitative leap, and the application scenarios have become increasingly diversified [1, 2].

© The Author(s), under exclusive license to Springer Nature Switzerland AG 2022
V. Sugumaran et al. (Eds.): ICMMIA 2022, LNDECT 138, pp. 555–561, 2022.
https://doi.org/10.1007/978-3-031-05484-6_70

Machine translation technology is the process of transforming text and speech from one language into another using computer equipment and language processing functions, and the software that completes this process is called a machine translation system. Machine translation is the branch of natural language processing that switches natural languages between different languages and is inextricably linked to linguistics and machine learning. Machine translation technology involves the disciplines of computing, cognitive science and linguistics and has been recognized by the scientific community as one of the most difficult topics in the field of artificial intelligence. Machine translation systems have massive dictionaries and translation models that can cover multiple domains and industries and can translate in different scenarios. Human translation can only be refined to a specific field or industry. Machine translation is bringing all kinds of convenience to people's lives [3, 4]. People can use machine translation to solve the language difficulties they encounter in food, clothing, housing and transportation. From travel abroad and translation of scientific and technical literature, to international trade and cross-language cultural exchange, the need for multilingual information connectivity makes machine translation invaluable.

Literary translation facilitates cultural exchange between countries, especially as foreign literature is prevalent in the domestic market and the translation of foreign literature is particularly important. Although current translations of foreign literature ensure the integrity of the story, they are more likely to use the linguistic logic of Chinese and to perceive the work in the context of Chinese culture, failing to truly reflect the ideological connotations that the author wants to convey. British and American literature is an important branch of foreign literature, which is not only numerous but also of excellent quality, and has had a profound influence on the literary creation of China. With the rapid development of computer software and hardware technology, as well as the continuous improvement of corpus construction, machine translation of British and American literature based on parallel corpus has become a new research direction. Through knowledge mining of bilingual parallel corpus, the quality of British and American literature translation has been significantly improved, and the dissemination of foreign literature in China has been promoted.

2   The Basic Model of Machine Translation of British and American Literature Based on Parallel Corpus

Parallel corpus can extract both the original text and the corresponding translation at the same time, using query translation based on parallel corpus as a linguistic boundary approach across source and target languages, and using English bilingual dictionaries as the main source of knowledge to achieve query translation processing [5]. The basic model of machine translation of British and American literature based on parallel corpus is shown in Fig. 1.

Not only does the corpus contain a rich corpus, but it also has powerful computer search and statistical functions that make it easy and efficient to discover potential patterns in the content covered by the corpus. With the development of Internet technology, the use of bilingual parallel corpus is becoming more and more widespread and has become an important resource for many machine translators. Only when the

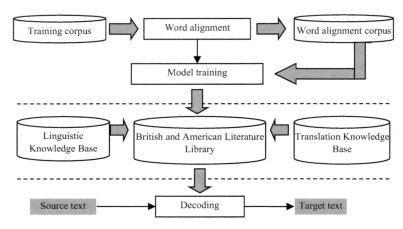

Fig. 1. The basic model of machine translation of British and American literature based on parallel corpus

training corpus reaches certain requirements will its manifest better translation effects. The training phase focuses on solving the word alignment problem, as well as the design and training of various related models, so as to provide suitable translation knowledge for the subsequent decoding phase. Corpus training entails pre-training the language model, initializing the translation model and iteratively optimizing the initial translation model. To address the problem that it is not possible to achieve high translation accuracy using a limited number of training samples, migration learning can be used to share lexical-level and sentence-level features from various source languages, into a single target language. This approach is premised on the assumption that source languages include both multi- and low-resource languages, and the main goal is to be able to share learning models across multiple languages. In addition to combining migration learning methods with semi-supervised learning methods to achieve machine translation between languages with sparse parallel corpus, pairwise learning methods are also very effective [6–8]. Pairwise learning is a new machine learning paradigm that can reduce the dependence on large-scale annotated data. Transfer learning uses several related tasks to assist in the learning of the main task; whereas several tasks in pairwise learning have no priority and are superior learning algorithms.

3   Construction Process of Parallel Corpus

With the development of corpus linguistics, the role of parallel corpus is becoming increasingly important in areas such as natural language processing, information retrieval and machine translation. he large amount of bilingual interlingual information in the parallel corpus provides authentic original texts and translations, accurate translations of unlisted words, many new words and meanings for the dictionary, accurate interpretations of commonly used and familiar words, and the ability to match the longest existing translation examples. The construction of a parallel corpus is divided into a 'requirements layer, foundation layer, data layer and application layer', which correspond to 'corpus planning, corpus acquisition, corpus organization and corpus application', as shown in Fig. 2.

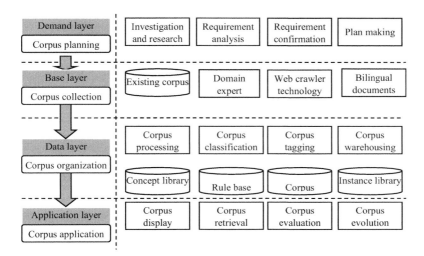

Fig. 2. Construction process of parallel corpus

3.1   Requirements Layer, Corresponding to Corpus Planning

The main task of the requirements layer is to carry out requirements analysis and solve the problem of user requirements for the parallel corpus. The requirements analysis is based on research, which can be done using the field survey method. The contents of the survey and analysis will be written into a requirements analysis report, which will be reviewed and confirmed by the relevant personnel as a basic document for the construction of the corpus and used as the basis for the corpus construction planning.

3.2   Foundation Layer, Corresponding to Corpus Acquisition

The main task of the foundation layer is to carry out corpus acquisition and to solve the problem of sourcing the corpus for the parallel corpus. The corpus is firstly based on the existing corpus, and then an appropriate collection method is used based on the advice of domain experts. Web crawlers are currently the most widely used corpus acquisition technique, allowing a wide collection of the vast resources of the Internet. Bilingual documents are also an important source of corpus.

3.3   Data Layer, Corresponding to Corpus Organisation

The main task of the data layer is to perform corpus organization and solve the problem of corpus entry for parallel corpus. Semantic types are established, the hierarchical relations between each semantic type are determined, attributes are set for each semantic type, and term instances are constructed. The setting of semantic types and their relations constitutes the concept base, the setting of attributes constitutes the rule base, and the concept base and the rule base together constitute the knowledge ontology.

3.4   Application Layer, Corresponding to Corpus Applications

The main task of the application layer is to apply the corpus and solve the problem of using the corpus of the parallel corpus. The core issue of corpus application is corpus retrieval, which is also a kind of information retrieval and has a very important status and role in the construction of the corpus. It is related to whether the corpus can meet the user's usage needs, whether the value of the corpus can be realized, and also marks the construction level of the corpus.

4   Case-Based Machine Translation Process for British and American Literature

With the increase in computer processing speed and storage capacity, as well as the emergence of a large number of electronic bilingual corpus, the field of machine translation has given rise to instance-based machine translation methods, which imitate similar instances based on the similarity principle of natural language processing and overcome the weaknesses of traditional rule-based machine translation methods. The example-based approach is able to make full use of parallel corpus resources, imitating the basic patterns in the human process of translating a foreign language, by transforming and substituting the source and target languages in the corpus and finally obtaining the translation result. The advantage is that a large number of rules are avoided and the speed of translation is increased. An example-based machine translation process for British and American literature is shown in Fig. 3.

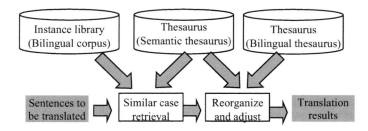

Fig. 3. Case-based machine translation process for British and American literature

Instance-based machine translation was first proposed by Japanese machine translation experts. The basic idea is to find out the translation instance that is most similar to the part to be translated in the bilingual instance database that has been collected, take the previous translation instance as the main knowledge source, and then complete the translation work by a series of deformation operations such as replacing, deleting or adding to the translation text of the instance. However, limited by the size of the corpus, it is difficult for instance-based machine translation to achieve a high matching rate. Instance-based machine translation also suffers from recognition errors caused by intra-lingual lexical items, intra-lingual structural relations and extra-lingual factors, requiring manual post-translation re-organization and adjustment functions to effectively capture common machine translation errors and further optimize machine translated texts. Instance-based machine translation requires a large amount of corpus retrieval work, and the speed of corpus retrieval is dependent on indexing, requiring the creation of inverted indexes, a simple index structure is shown in Fig. 4.

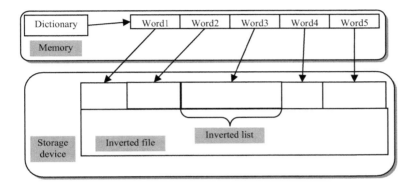

Fig. 4. Simple structure of inverted index

For the simple structure of the inverted index shown in Fig. 4, a specific storage form of the "word-document matrix" is implemented. The inverted index consists of a "word dictionary" and an "inverted file", which is the physical file in which the inverted index is stored. The inverted index is already a very complete indexing system, which not only improves the retrieval performance of the system, but also improves the scalability of the system [9, 10].

5   Conclusion

Machine translation frees translators from some of the mechanical and tedious tasks of translation to engage in more creative work. Machine translation has evolved from rule-based machine translation techniques, to instance-based machine translation techniques, to statistical-based machine translation techniques, and finally to neural network-based machine translation techniques. Whether it is text translation or voice translation, machine translation is currently difficult to reach the level of human translation. However, with the development of artificial intelligence technology and the advancement of brain science research, machine translation will replace human translators in most of the translation work. British and American literature translation plays an important role as a bridge for mutual communication between countries. In the process of translation, the differences between Chinese and Western cultures lead to various problems that translators often encounter when understanding and translating. With the rapid development of the Internet as well as big data, Internet translation based on big data has become a breakthrough in the practicalization of machine translation technology, providing more path options for the translation of British and American literature. The translation resources are mined from the entire Internet, on the basis of which statistical translation models and deep learning models are trained using a cloud computing platform to build a bilingual translation corpus, ultimately allowing machine translation to reach a truly practical level, improving the translation of British and American literature and promoting the dissemination of British and American literature in China.

References

1. Bogush, A.M., Korolova, T.M., Popova, O.V.: Teaching machine translation to the students majoring in the humanities. Inf. Technol. Learn. Tools 71(3), 122 (2019)
2. Sorami, H., Matt, P., Kevin, D.: Membership inference attacks on sequence-to-sequence models: is my data in your machine translation system? Trans. Assoc. Comput. Linguist. 8 (1), 49–63 (2020)
3. Gunarto, H.: Apps-based machine translation on smart media devices - a review. IJCCS (Indonesian J. Comput. Cybern. Syst.) 13(1), 95–104 (2019)
4. Ahmadnia, B., Serrano, J.: Employing pivot language technique through statistical and neural machine translation frameworks: the case of under-resourced Persian-Spanish language pair. Int. J. Nat. Lang. Comput. 6(5), 37–47 (2017)
5. Mantas, J., et al.: Towards a bilingual Alzheimer's disease terminology acquisition using a parallel corpus. Stud. Health Technol. Inf. 180(1), 179–183 (2012)
6. Curry, N.: Academic discourse as a social, global, and multilingual discourse. In: Academic Writing and Reader Engagement: Contrasting Questions in English, French, and Spanish Corpora, pp. 1–272. Routledge, London (2021)
7. Yun, P.: Corpus linguistics approaches to trainee translators' framing practice in news translation. Transl. Interpreting 12(1), 90–114 (2020)
8. Guo, M.: Analysis of Non-English major students' English writing vocabulary based on corpus. Int. J. Intell. Inf. Manag. Sci. 9(3), 39–46 (2020)
9. Tonellotto, N., Macdonald, C.: Using an inverted index synopsis for query latency and performance prediction. ACM Trans. Inf. Syst. 38(3), 1–33 (2020)
10. Andrzej, C., Paweł, L.: Efficient inverted index compression algorithm characterized by faster decompression compared with the golomb-rice algorithm. Entropy 23(3), 296 (2021)

Using Computer Multimedia Technology to Construct Physical Education Teaching System

Zhou Yang[1(✉)] and Cheryl[2]

[1] Physical Education Department, Dalian Polytechnic University,
Dalian 116034, Liaoning, China
yz18900997788@163.com
[2] Physical Education Department, University of Toronto,
Toronto M5S 2E8, Canada

Abstract. With the progress of Internet and computer technology, the scope of use of modern information technology has been continuously expanded, and a large amount of data has been obtained as an adjunct of development. Mankind has entered this era of information explosion. Now all industries in the society need the support of data, data has penetrated into all industries, people's development and utilization of big data (BD), play a more and more important role, it has become an important factor for the development and promotion of the industry. In the era of BD, the education industry is facing unprecedented opportunities and challenges. Applying BD technology to teaching activities and giving full play to its role can enrich teaching methods and learning methods, promote the reform of traditional model of instruction and improve the quality of education and teaching. This paper takes BD as the background for analysis to build a new sports model of instruction, and use the formula for calculation, the efficiency and quality of new sports teaching pattern on the efficiency and quality of data with 13% and 31% respectively than traditional sports model of instruction, prove that BD context (BDC) using CMT to build a sports model of instruction is of great significance.

Keywords: Big data · Computer · Multimedia technology · Physical education · Model of instruction

1 Introduction

The Outline of Curriculum Reform for Basic Education (Trial Implementation) clearly supports the universal application of information technology in the teaching process. Under the background of scientific and technological development, the traditional education and teaching methods are actively reformed, the advantages of information technology are brought into play, and the teaching quality of teachers and the learning efficiency of students are improved [1]. BDC, it is of great speak volumes to integrate the information technology into the traditional PE naturally and appropriately, and to construct a new PE model by using the CMT, so as to complement each other's advantages.

© The Author(s), under exclusive license to Springer Nature Switzerland AG 2022
V. Sugumaran et al. (Eds.): ICMMIA 2022, LNDECT 138, pp. 562–569, 2022.
https://doi.org/10.1007/978-3-031-05484-6_71

CMT is progressing under the influence of many factors and network education has become the direction, trend and future of modern education. Computer multimedia has a variety of functions, which can fully mobilize students' learning enthusiasm in visual and auditory aspects [2]. Therefore, using multimedia to optimize the teaching process and build a new model of instruction plays an important role in improving the quality of teaching activities. Since entering the 21st century, as the widespread application of CMT, the traditional physical education (PE) teaching method is challenged more and more severely. This exposes the inconvenience of traditional PE [3]. The particularity of PE lies in the fact that PE class should not only impart physical knowledge, but also impart correct techniques to students meanwhile. In the course of learning, students can listen to the teacher's action essentials to explain, and can more intuitively see the teacher's demonstration actions and establish visual performance.

It takes a long time for CMT to combine with other subjects, but its combination with sports is mostly in sports competition and measurement, and its combination with sports teaching is not close enough. Chinese scholars wu feng, Fang Bin College PE Curriculum and Integration dimension calculation and practical thinking "and Yang Pinliang Liao Min" CMT and PE curriculum integration research on "education field" build into the CMT in the sports teaching thinking and research, put forward to build a positive effect with the teachers, students and teaching process of PE model of instruction [4]. But these studies are old, flawed, and not put in the context of BD.

In this paper BDC using the CMT to build a new sports model of instruction to explore, based on the direction histogram, construct local coordinate system, calculate the new sports model of instruction and the difference between traditional PE model of instruction, with the help of a clear data suggest that new sports teaching in the important value of modern education.

2   Big Data and Computer Multimedia Technology

2.1   Big Data

In the Internet era, BD has been applied in the field of education as a new way to promote education reform. BD technology is a powerful tool to promote the rapid development of various fields. All fields can feed back information, count information, analyze information, share information and transmit information in real time through BD technology. BD technology greatly facilitates people's life and study. BD technology has the advantages of individualization, sharing, high efficiency and large capacity, and its application to education can break through the bottleneck of current education. For example, through BD technology, the basic information of students and teachers can be dynamically managed and analyzed [5]. Teaching, the teacher can use BD platform, according to the students' learning situation for sorting, statistics, and real time control of the students' learning situation, can recruit object on humanization education mode, the common progress of the different levels of students, promote the students' learning ability in science class, objective, personalization, stratification education strategy, create a good foundation for the education teaching.

2.2   CMT

Great advances in computer technology has brought great impetus to the continuous progress of human society. Under the help of this system, modern multimedia technology is combined with communication technology and network platform [6]. In addition, intelligent simulation systems and other technologies continue to upgrade the form of computer multimedia conversion. CMT is mainly video compression, network TV, streaming media, covering a variety of technologies, through continuous improvement and development, the technology is the production of social operation, can effectively help, education, military, medical and other fields, respectively can play a different role. The application of computer multimedia can help users to simplify the processing objectives and improve the efficiency of computer application. For example, 5G communication technology, cloud storage technology and other key technologies do not interfere with each other, but they share common characteristics such as interactivity and complexity [7].

With the rapid development and application of computer, the application field of CMT is expanding. Multimedia technology can process text, sound, pictures and other media information. It is now used in all kinds of household appliances. It has the characteristics of integration, interaction and practicability. In the field of education, multimedia technology closely combines the three kinds of information processing technology, namely acoustics technology, computer and communication technology, which promotes the better development of information processing technology. Teachers use multimedia technology to teach, and with the support of vast amounts of data, teaching is more convenient and students' understanding ability is improved.

3   Construction of Computer Multimedia College Physical Education Teaching System

3.1   Strategy of Using CMT to Build a New PE Model of Instruction in BDC

In the context of comprehensive promotion of quality-oriented education concept, education reform is also deepening. BDC, CMT is used to build a new PE model of instruction, and innovative talent cultivation model is used to enable students to achieve all-round development and improvement in PE class [8]. The key point of mode reform is to take full advantage of computer multimedia calculation: first, to break through the traditional model of instruction. Change the traditional teaching activities are limited to the classroom, can not really realize people learn sports, people learn useful sports phenomenon. The teacher will make the important and difficult actions in the teaching into micro lessons, so that students can learn according to their own needs at all times and anywhere to promote students to master the learning content, alter the way you study, break the fixed model of instruction, and truly realize dynamic teaching.

Secondly, teachers should realize dynamic teaching. PE teachers in the use of multimedia computer teaching, teachers can use information technology with the dynamic advantage to static sports teaching content into a dynamic teaching content, so the whole PE learning won't get boring because of this, the PE class will become more vitality and vigor, and thus truly effective optimization of teaching.

Finally, teachers should implement personalized teaching. Teachers should make use of the background of BD to analyze and process students' learning conditions and data, so as to understand students' learning level of PE, as well as the PE foundation and ability of each person, as well as the characteristics and actual needs of PE learning [9]. After mastering these, we can better implement personalized teaching. Under the condition of meeting the overall needs of the students' PE learning, the individual needs should be counselling, or the students should find resources for independent learning on the network platform.

3.2 System Construction

The characteristics of the structure are as follows: the environment used by the client is a standardized common Web browser, and all applications are stored on the Web server, which can be directly downloaded when needed. Easier to manage and maintain, because the client does not require dedicated software, you only need to update the software on the server when upgrading network applications.

The working principle of the system: teachers and students through the browser to access the multimedia network teaching platform, students use personal computer devices through the browser and connected to the server, related sports teaching content learning, sports resources information query, timely communication between teachers and students, personal data upload and other operations. Administrators of multimedia network teaching platform and PE teachers in colleges and universities can update and maintain the contents stored in the server through browsers, and upload the latest PE teaching resources to the server. Meanwhile, they can answer questions online with students and give sports guidance to students. The interactive results are downloaded to the browser in the form of Web pages, and the user can observe the request results.

3.3 Data Calculation Formula

The direction histogram signature descriptor is calculated by establishing a local coordinate system to code the normal vector of the feature points. Compared with the ordinary three-dimensional coordinates, the local coordinate system can better reflect the local features of the model surface. The SHOT descriptor firstly defines the local coordinate system of the 3D model, and enhances the recognition ability of the descriptor by introducing the geometric information of the feature points. The local histogram at the given point is then computed to form a descriptor with mixed features.

In view of the important influence of the establishment of local coordinate system on the calculation of 3D model descriptor, the establishment of an appropriate local coordinate system is the primary task of the calculation of 3D model descriptor. The specific calculation is shown as follows:

$$M = \frac{1}{k}\sum_{i=0}^{k}(p_i - \hat{p}) \tag{1}$$

Wherein, the total least squares estimation of the normal vector is obtained by eigenvalue decomposition of the covariance matrix M. It can be determined that the sign of Z axis of the local coordinate system with the origin of the characteristic point P is the same as the mean of the nearest neighbor normal vector k, and Z axis is the normal vector of the characteristic point P. This method can describe the global characteristics of the model, but can not describe the local characteristics of the model. When there are multiple models or the models are deformed, the descriptor must have strong recognition and local symbol disambiguation must be carried out. Thirdly, based on the combination of feature points for improvement, the formula can be:

$$R = \frac{1}{\sum_{i:d \leq R}(R - d_i)}(R - d_i) \tag{2}$$

To solve the problem of descriptor discrimination in the symmetric part of the model, the points farther away from the feature point P are assigned a smaller weight, while the points closer to the feature point P are assigned a larger weight, which improves the accuracy of quality and efficiency of the calculation of PE model of instruction.

4   Using Computer Multimedia Technology to Construct New Physical Education Model of Instruction Data Analysis in Big Data Context

4.1   Advantages of Using CMT to Construct a New PE Model of Instruction in BDC

From the perspective of the influence of BD information technology on PE teaching, the flexible application of computer multimedia can improve students' learning enthusiasm. In order to truly change the classroom atmosphere and completely change the traditional model of instruction of "teachers speak, students practice", we must use advanced teaching means. Audio-visual multimedia through the powerful function and sound, light, color, shape of the lamp affect students' psychology, for classroom teaching provides images, video, video, sound and other intuitive realistic dynamic change process of technical action. Attracting students' hearing and vision, inducing students' thinking, satisfying their strong thirst for knowledge and strong psychology of curiosity, stimulating their interest in learning, and achieving the purpose of students' learning motivation. In addition, under the application of CMT, sports knowledge can guide students to study easily and quietly, and enhance students' understanding ability. Understanding is an important stage for students to master knowledge. Use the computer to visualize, concretize, silently change, faster and slower. Help students to understand complex abstract knowledge. Before the demonstration, ask the students to look at it with questions: which parts of the action are made up of, and what attention points each part has. Used to the content is not easy to master, the teacher spent a lot of time into the model, using multimedia technology to the specific means of image, let the students make clear the teaching content of teachers, with the aid of clear sweet voice, bright colours, let the students feel fresh, encouraging them to actively participate in the classroom, in the best state of learning, wholeheartedly, positive thinking, promote the internalization of knowledge, learning sports knowledge and action, to improve their sports level.

Base on the investigation of participation of students in traditional and new PE model of instructions in a school, we can see the different attitudes of students towards the traditional and new PE model of instructions. The specific situation is shown in Table 1:

Table 1. Participation of students in traditional and new PE model of instructions

Project	Class 1	Class 2	Class 3	Class 4	Class 5
Traditional PE model of instruction	67.2%	59.3%	66.7%	64.3%	68.5%
The new PE model of instruction	87.4%	85.3%	87.9%	87.5%	88.7%

BDC using the CMT to build a new sports model of instruction embodies the development trend of information technology, highlighting the pursuit of modern education, class of traditional sports model of instruction with different class participation of new sports teaching pattern, to see that BDC using the CMT to build a new sports teaching pattern to the positive role to the classroom teaching effect has been improved, can effectively activate the student thought spark, promote better educated numerous for brief, to help students perceptual knowledge and develop students' creative thinking ability.

4.2 Analysis of Calculation Results

According to the direction histogram signature descriptor and SHOT descriptor theory, calculate the efficiency and quality of the traditional PE model of instruction and the BD background using CMT to build a new PE model of instruction, and compare the relevant data.

According to experiment, select students in a school as data samples to calculate the efficiency of the PE model of instruction by using CMT under the background of traditional PE model of instruction and BD. Figure 1 can be drawn as follows:

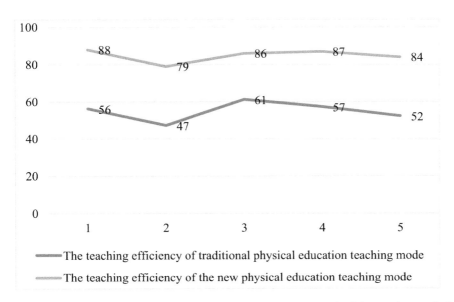

Fig. 1. The efficiency comparison chart of traditional PE model of instruction and the construction of new PE model of instruction by using CMT BDC

In the meantime, the corresponding formula is used to calculate the quality of the traditional PE model of instruction and the new PE model of instruction constructed by using CMT in BDC. Figure 2 is as follows:

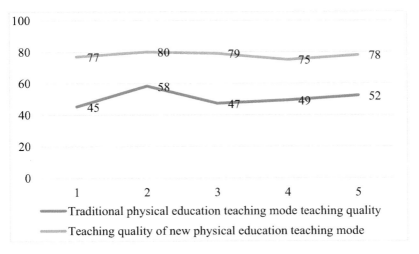

Fig. 2. The quality comparison chart of the traditional PE model of instruction and the new PE model of instruction constructed by using CMT BDC

5   Conclusions

As an important part of quality-oriented education, PE should be developed in an all-round and balanced way. The wide application of CMT playing an become increasingly prosperous important role in the compilation of computer multimedia courses and the interaction with PE [10]. BDC, the use of CMT, build a new sports education pattern, is the manifestation of the necessity of modern education technology development, the optimization of PE model of instruction, meet the students' PE learning motive and interest, deepen the students for a variety of sports movement and the concept of the understanding of the theoretical knowledge, improve the quality of PE and improve students' performance.

References

1. Baral, R., Murphy, D.C., Mahmood, A., Vassiliou, V.S.: The effectiveness of a nationwide interactive ECG teaching workshop for UK medical students. J. Electrocardiol. **58**, 74–79 (2020)
2. Liu, H.Y., Wang, I.T., Chen, N.H., Chao, C.Y.: Effect of creativity training on teaching for creativity for nursing faculty in Taiwan: a quasi-experimental study. Nurse Educ. Today **85**, 104231 (2020)
3. Kim, R.H., Mellinger, J.D.: Educational strategies to foster bedside teaching. Surgery **167** (3), 532–534 (2020)
4. Grabski, D.F., et al.: Compliance with the Accreditation Council for Graduate Medical Education duty hours in a general surgery residency program: challenges and solutions in a teaching hospital. Surgery **167**(2), 302–307 (2020)
5. Huddleston, C.B., Fiore, A.C.: Commentary: "WhenI was in training"—the phrase we hate to hear, but love to say. J. Thorac.Cardiovasc. Surg. **159**(4), e279–e280 (2020)
6. Massoto, T.B., et al.: Mesenchymal stem cells and treadmill training enhance function and promote tissue preservation after spinal cord injury. Brain Res. **1726**, 146494 (2020)
7. Alyami, M., et al.: Standardizing training for pressurized intraperitoneal aerosol chemotherapy. Eur. J. Surg. Oncol. **46**(12), 2270–2275 (2020)
8. Garratt, E.: Feasibility of a self-developed online training tool to deliver specialist training - a review. Physiotherapy **107**, e16 (2020)
9. Bu, L., et al.: Effects of physical training on brain functional connectivity of methamphetamine dependencies as assessed using functional near-infrared spectroscopy. Neurosci. Lett. **715**, 134605 (2020)
10. Van Hecke, A., Duprez, V., Pype, P., Beeckman, D., Verhaeghe, S.: Criteria for describing and evaluating training interventions in healthcare professions – CRe-DEPTH. Nurse Educ. Today **84**, 104254(2020)

Data Mining of Swimming Competition Technical Action Based on Machine Learning Algorithm

Yuhang Chen[1(✉)], Lijun Zhu[1], and Deepmala Karki[2]

[1] Physical Education College, Bohai University,
Jinzhou 121000, Liaoning, China
956952077@qq.com
[2] LBEF Campus (in Academic Collaboration with APU Malaysia),
Kathmandu, Nepal

Abstract. With the rapid development of computer technology, the information revolution is rapidly changing our lives. Swimming technical data is also growing rapidly. For these large data sets, only a small part of it is what we care about and need. In order to control the game in time and implement effective management, advanced computer technology must be used to analyze technical movements. This article aims to study the data mining of swimming competition technical movements based on machine learning algorithms. Taking breast-stroke as an example, based on the analysis of machine learning principles, data mining tasks and breaststroke technical movements, five athletes from the province are selected as the research objects. Naive Bayesian method classifies breaststroke technical movements and analyzes the related data of athletes' breaststroke technical movements. The analysis results show that the stroke time ratio of A and D is relatively large, indicating that their characteristic is dominated by the action of the stroke technique; the stroke time of B, E, and C is relatively small compared to the overall breaststroke movement, indicating that it is not a breaststroke athlete whose rowing technique is the leading role.

Keywords: Machine learning algorithm · Swimming competition · Swimming technique action · Data mining

1 Introduction

In the information age, due to the rapid development of computer technology and the increasing popularity of database technology applications, people are facing a rapidly increasing amount of data [1, 2], which has become one of the focuses of many IT workers. One of the main functions of data mining is to quickly and efficiently extract the answers needed by questions from a large amount of actual data [3, 4].

In recent years, data mining has achieved tremendous development and success. In sports, data mining technology is widely used in the technical and tactical analysis of many sports events, and the use of tools is becoming more and more mature [5, 6]. A large amount of mining data is used for technical and tactical analysis, providing new methods and ideas for sports technical and tactical analysis. Research has also developed more and more systematic vocabulary [7, 8]. In general, current data mining research mainly focuses on three aspects, data mining algorithm research, data mining theory research, and data mining application research.

Based on consulting a large number of relevant references, this article combines the principles of machine learning, data mining tasks and breaststroke technical movements, selects five athletes from the province as the research objects, uses the Naive Bayes method to classify the breaststroke technical movements, and data related to athletes' breaststroke techniques are analyzed.

2 Data Mining of Swimming Competition Technical Action Based on Machine Learning Algorithm

2.1 Principles of Machine Learning

(1) Naive Bayes algorithm

Naive Bayes calculations can generally be divided into such three processes. The first one is to get the next training sample first, then get a certain number of training sample sets, arrange the attributes and export them to the elements of the training set, and finally get the next training sample. The second step is to train the classifier. Through big data analysis of a training sample, the probability and generation probability of each type are determined, and the classifier is trained based on this. The third step is to classify and replace all training samples. The type with the highest probability is obtained from all samples classified in the training classifier, and it is used to calculate the probability of different types of generation, which is the training classification result [9, 10]. However, because the independence assumption of Naive Bayesian calculation does not conform to the actual situation in the general real world, its performance is lower than many more advanced machine learning algorithms, but it requires simpler applications and fewer Computational complexity, etc. And this is because there is very little training data in its collection.

(2) Support vector machine

Support vector machines have become a widely used classification method in machine learning algorithms. It is said that support vector machines have very powerful statistical principles, because the overall best solution for the process of creating a support vector machine model comes from the double Laga period. Moreover, support vector machines are used in many fields such as handwritten digit recognition and text sorting [11, 12].

(3) K-nearest neighbor method

The nearest neighbor method is a simple and widely used algorithm with improved machine learning performance and is an extension of the nearest neighbor method. The nearest neighbor method calculates a specific metric distance from the attribute vector of the sample points after sorting and the attribute vector of all the sample points in the training set. The sample point category with the shortest sample point placement distance in the training set is the classification result of the sample point placement.

(4) Decision tree algorithm

Decision tree algorithm is a widely used machine learning algorithm, not only can be applied to classification, but also can be applied to regression. Many other advanced algorithm systems have also used this algorithm to develop. The decision tree is usually an inverted tree structure composed of root nodes, internal nodes, leaf nodes and edges. Category association refers to the relationship between a category and a target variable of interest. As long as the association relationship is established, the category of the test data set can be predicted.

2.2   The Task of Data Mining

(1) Classification

Classification is the most widely known data mining technique. Sorting is the process of first determining the features to be split, and then using a specific algorithm to classify the sample training data set to create a ranking mining model. Finally, before exporting the model, check the operation of the model to determine the validity of the model. Rules can be used to write and predict rules.

(2) Cluster analysis

Cluster analysis comes from classification. Similar to classification, it categorizes data into different categories based on similarity. However, unlike classification, clustering is not pre-defined, but based on the characteristics of actual data. They are defined by the similarity between them. That is, if the data attributes are in the same category, they are similar, and vice versa. Cluster analysis can generate macros and discover possible relationships between data distribution patterns and characteristic data. In the social sciences and natural sciences, there are also many grouping problems.

(3) Association analysis

Association mining analysis refers to having these specific rules among the attribute values of multiple variables. Association rules usually have a more direct application to retail business, but they can also be used for other purposes, such as predicting communication or network failures. The main purpose of association analysis is to find the hidden data in the real data and the relational network between the data. The research will evaluate the correlation between the analyzed characteristics and the association rules through two maximum support and reliability thresholds, which will make the rules obtained by mining more in line with actual needs.

(4) Forecast

Forecasting can also be seen as a special way of classifying the mining industry in a strict sense. Use the past data for mining and analysis, discover the law of change from the variability of early data development, build a model according to a specific algorithm, and replace and test the actual data in the model. If the test result is higher than the specified value, the model is valid and operable, and the model can predict the future evolution of the data to a certain extent. Forecasting applications include machine learning, pattern recognition, flood forecasting, and language recognition.

(5) Timing mode

The time series model is the same as the forecast, but also uses previous data for mining and analysis. The difference is that changes in data are usually associated with changes over time. If their time intervals are equal, these data are usually available.

First, check the line structure of the time series graph to determine the behavior of the time series pattern, and then measure the similarity between them in distance. Then, you can use the time series pattern graph to predict future data values. Time series mode applications mainly include financing and equity analysis.

(6) Deviation analysis

In data mining, deviation analysis is a mining method for mining obvious changes between the historical record of the data and the current state or template. For example, project management difference analysis usually refers to the difference between the actual completed work and the scheduled work, which can be analyzed from the changes to the database. If these changes are found over time, the deviation can be successfully completed.

2.3 Breaststroke Technical Movements

(1) Arm movement

  1) Outbound

When the arm is extended and the arm is extended, the hands start to move forward. The palm of the hand is first outwards, or draw outwards with a slightly downward curve. Then, slowly turned both hands downwards and started to catch water. In the final stage of the downward stroke to the outside, the elbow flexion angle is generally about 20°–30°, so that the arms and palms can be turned back into position to form a water-grabbing posture. Before forming a water catching gesture, the distance between the two hands to underdraw is generally between 50–70 cm. When you start to draw outwards, first move your palms downwards, and then when you draw to the left shoulder, slowly turn your palms outwards, and then lower your palms to a water-holding posture. At the same time, the little finger side of the hand should also be swiped out in advance to reduce the water retaining surface. When starting the down stroke, the palm of the hand should be turned to the outer side, so that after the water catching action, the palm of the hand should be turned outward to face the water surface. When the arm outstretching movement turns into an outside painting action, the stroke speed is slightly slower, and the outside painting speed should gradually increase during the outside painting process to before the water catching action. Once the catching movement is completed, the outer stroke is terminated and the inner stroke is started.

  2) Inline

After the water catch is completed, our hands first go out and back, and then go in and down to make a large arc inward stroke. At this time, you should maintain a high elbow posture, but the movement of the elbow joint does not exceed the movement of the shoulder joint. The knee joint should be used as the axis, and the fingertips and forearms should be rotated inward and downward to do the stroke. When the hands are drawn together under the chest, the stroke is completed. In the process of inlining, bend your elbows gradually. When the stroke is completed, bend the elbow roughly 80°–90°. When grabbing the water, turn your palms back out. When swiping inward, the palm of your hand gradually turns inward. When the stroke is completed, gradually turn the palm of your hand to the back. Compared with the butterfly stroke, maintaining an external rotation posture before the hands are stroked under the elbow, the driving force is greater. After the hands stroke under the chest, continue to turn inward until the stroke is completed. During the entire stroke, keep your palms and upper arms vertical, and gradually increase the stroke speed. Before the stroke is completed, you should reach the top of the maximum stroke speed.

(2) Leg movement

1) Eversion

The effect of kicking in breaststroke depends largely on the technique of valgus soles. When the foot-retracting action is ready to proceed, the two feet start to do the valgus action: that is, the two feet are rotated forward, and the heels are divided and separated as much as possible, while the toes are facing left and right, and the knee joints are also rotated slightly inward, and finally the most favorable water-confrontation state is created between the sole of the foot and the side of the calf. The kicking action needs to be completed when the foot's valgus movement has not been completed, but because the final valgus action or the underwater posture is usually performed shortly after the kicking action is completed, it cannot wait until the palm flipping or arranging action is completed. Finish the kick after the stance toward the water.

2) Pedal clamp

The effect of breaststroke leg action is completely determined by the accuracy of the kicking technique. After the feet have turned out, keep pushing backwards to pinch the water until the feet are all close together. The action of kicking and pinching water is mainly realized with the help of hip extension and hip extension. The kicking action should pay special attention to the function of extending the congenital hip joints, so that the legs can be raised and maintained at a better kicking section than the calf, otherwise downward kicking will occur.

3   Experiment

3.1   Research Objects

This article selects five athletes from the province as the research objects, and uses cameras, underwater cameras and a tank swimming technique test system to take pictures of the underwater breaststroke techniques of these five athletes. On-site observation and monitoring of athletes and competitions.

3.2   Data Mining Method of Swimming Technique Movement

Naive Bayes Classification (NB for short) is a simple and very effective classification method. The NB method is based on Bayesian assumptions. That is, the technical movements of swimming are independent of each other when determining the types of technical movements. During training, the previous probabilities of characteristic technical actions belonging to each category are calculated. According to the characteristics, when a new technological action appears, the successor probability of the energy belonging to each category is calculated according to the first probability of the scientific action, and finally the category with the highest probability of inheriting the energy as the classification result is obtained. Suppose di is an arbitrary action system, which belongs to a certain category in action category 1. According to the NB classification:

$$p(d_i) = \sum\nolimits_{j=1}^{n} p(c_j)p(d_i/c_j) \tag{1}$$

$$p(d_i/c_j) = \frac{p(c_j)p(d_i/c_j)}{p(d_i)} \tag{2}$$

To classify technical actions is to calculate the probability of all sample classes under the given di according to the above formula. The class with the largest probability value is the class where di is located.

4   Discussion

4.1   The Time Phase Analysis of 5 Athletes in the Rowing Phase

Table 1. An analysis of the action phase of breaststroke athletes' paddle skills

	Outside time	Inline time	Reach time	Rowing time	An action cycle
A	0.31	0.15	0.13	0.61	1.15
B	0.47	0.15	0.43	1.07	1.71
C	0.35	0.19	0.15	0.71	1.55
D	0.53	0.19	0.45	1.19	1.55
E	0.53	0.17	0.29	1.01	1.61

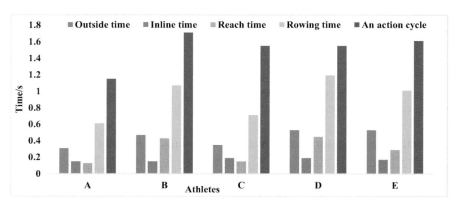

Fig. 1. An analysis of the action phase of breaststroke athletes' paddle skills

It can be seen from Table 1 and Fig. 1 that in the time phase of the stroke technique, the time ratio of the strokes A and D is relatively large, indicating that its characteristic is dominated by the role of the stroke technique; B, E and C strokes time is relatively small compared to the overall breaststroke action, indicating that he is not a breast-stroke athlete whose rowing technique is the leading role.

4.2   5 Athletes' Leg Technical Movement Phase Analysis

Table 2. Analysis of 5 athletes' leg technical movement time phase

	Closing time	Kick time	Water kick time	Kick time	An action cycle
A	0.31	0.15	0.13	0.61	1.15
B	0.47	0.15	0.43	1.07	1.71
C	0.35	0.19	0.15	0.71	1.55
D	0.53	0.1	0.45	1.19	1.55
E	0.53	0.17	0.29	1.01	1.61

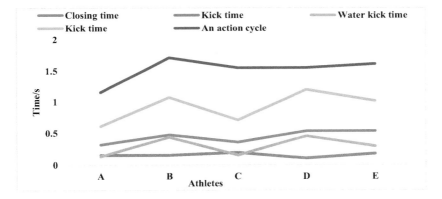

Fig. 2. Analysis of 5 athletes' leg technical movement time phase

Table 2 and Fig. 2 show that in the technical phases of the leg movements, D and E have a large proportion of the overall breaststroke movement due to the slow retraction of the legs and the inconsistent movement of the kick and the water kick stage, which is not conducive to the overall breaststroke rhythm. The compactness of B; B due to the stagnation of the action during the kick and water phase, resulting in a larger proportion of the leg action time in the overall breaststroke action; C is outstanding in the side kick and side clip technique, and the proportion of A's leg action time is small, which is conducive to making the overall rhythm of breaststroke is more consistent and compact.

5   Conclusions

In recent years, social life information has greatly improved the ability to use infor-mation technology to collect data. We are facing the challenge of data explosion. Data mining technology can further process and analyze real-world data, so that people can find corresponding laws and rules, and make scientific predictions and decisions. The study of swimming techniques is one of the most important topics in the swimming field in sports in recent years. Exploit and analyze a large amount of game data to find out some rules in the game to facilitate targeted training by the coaching staff.

References

1. Díaz, Z., Segovia, M.J., Fernández, J., del Pozo, E.: Machine learning and statistical techniques. An application to the prediction of insolvency in Spanish non-life insurance companies. Int. J. Digit. Account. Res. 5(9), 1–45 (2005)
2. Pavlova, A., et al.: Machine learning reveals the critical interactions for SARS-CoV-2 spike protein binding to ACE2. J. Phys. Chem. Lett. 12(23), 5494–5502 (2021)
3. Marozzo, F., Talia, D., Trunfio, P.: A workflow management system for scalable datamining on clouds. IEEE Trans. Serv. Comput. 11(3), 480–492 (2018)
4. Lee, J., Ohba, N., Asahi, R.: Discovery of zirconium dioxides for the design of better oxygen-ion conductors using efficient algorithms beyond data mining. RSC Adv. 8(45), 25534–25545 (2018)
5. Wilk-Kolodziejczyk, D., Regulski, K., Gumienny, G., Kacprzyk, B., Kluska-Nawarecka, S., Jaskowiec, K.: Data mining tools in identifying the components of the microstructure of compacted graphite iron based on the content of alloying elements. Int. J. Adv. Manuf. Syst. 95(4), 3127–3139 (2018)
6. Islam, M.S., Hasan, M.M., Wang, X., Germack, H.D., Noor-E-Alam, M.: A systematic review on healthcare analytics: application and theoretical perspective of data mining. Healthcare 6(2), 54 (2018)
7. Ma, E.Y., Kim, J.W., Lee, Y., Cho, S.W., Kim, H., Kim, J.K.: Combined unsupervised-supervised machine learning for phenotyping complex diseases with its application to obstructive sleep apnea. Sci. Rep. 11(1), 4457 (2021)
8. Al Hussien, S.S., Mohamed, M.S., Hafez, E.H.: Coverless image steganography based on optical mark recognition and machine learning. IEEE Access 9, 16522–16531 (2021)
9. Al-Saud, M., Eltamaly, A.M., Mohamed, M.A., Kavousi-Fard, A.: An intelligent data-driven model to secure intravehicle communications based on machine learning. IEEE Trans. Industr. Electron. 67(6), 5112–5119 (2020)
10. Khan, A.A., Jamil, A., Hussain, D., Taj, M., Jabeen, G., Malik, M.K.: Machine-learning algorithms for mapping debris-covered glaciers: the Hunza Basin case study. IEEE Access 8, 12725–12734 (2020)
11. Djenouri, Y., Belhadi, A., Belkebir, R.: Bees swarm optimization guided by data mining techniques for document information retrieval. Expert Syst. Appl. 94, 126–136 (2018)
12. Shousha, H.I., Awad, A.H., Omran, D.A., et al.: Data mining machine learning algorithms using IL28B genotype and biochemical markers best predicted advanced liver fibrosis in chronic hepatitis C. Jpn. J. Infect. Dis. 71(1), 51–57 (2018)

Music Note Segmentation Recognition Algorithm Based on Nonlinear Feature Detection

Yue Shi[(✉)]

College of Music, Capital Normal University, Beijing 100048, China
yuegao2021@126.com

Abstract. In the current note recognition methods, different optimal kernel functions are used for different note features, and the weight of each kernel function in the classification is obtained through learning, but the baseband structural features of note signals can not be obtained, and the deviation of baseband extraction results is large, which has the problem of large error of recognition results, and recognition takes a long time. In this paper, a method of segmented recognition of music notes based on nonlinear feature detection is proposed. The vector space matrix is used to analyze and process the music notes, so as to reflect the correlation of music notes and obtain the music retrieval results; The segmentation line is determined by setting the constraint conditions. The kernel function method in nonlinear feature detection is introduced into music note recognition. According to the definition of kernel function and kernel matrix, FFT is used to convert the low-pass component in audio frequency domain, so as to realize fundamental frequency recognition. The baseband extraction result obtained by this method is more accurate, the recognition accuracy is high, and it is practical.

Keywords: Nonlinear feature detection · Note recognition · Hidden Markov model · FFT

1 Introduction

With the continuous development of computer network, various media data are also emerging, in which text, image, audio, video and other information are growing rapidly, which has brought great convenience to people's life. In terms of music information, massive music data and various types of music forms can effectively alleviate people's psychological state and improve people's thinking consciousness [1]. When people search for a specific music or track in the network, they can query according to the limited notes to improve the application value of music recognition. In the field of music signal recognition, melody, rhythm and harmony are the basic contents of music and the focus of research. They are widely used in music classification, music analysis and so on. However, a large amount of music data is also easy to cause redundancy of network music information. Therefore, the baseband structure characteristics of note signals can be extracted by means of music note retrieval to effectively store and identify multi-type music information. Music information database is established by using the characteristics of notes, melody and beat to recognize notes [2].

V. Sugumaran et al. (Eds.): ICMMIA 2022, LNDECT 138, pp. 578–585, 2022.
https://doi.org/10.1007/978-3-031-05484-6_73

Since the beginning of this century, relevant scholars at home and abroad have done a lot of research on music recognition. The current research direction mainly focuses on neural network and recognition algorithm. Younis [3] and others use bat algorithm to initialize music features and use various programs to remove music lines, so as to obtain more accurate recognition results. Allabakash [4] and others use the optimized neural network to build a note classification framework to identify the note pitch or fundamental frequency of the audio signal, effectively complete the preprocessing of many applications such as note separation and score, and finally achieve the feature recognition and classification of music notes. DENG [5] and others proposed an automatic recognition algorithm of block simplified spectrum image based on the combination of PCNN (Pulse Coupled Neural Network) and DNN (Depth Neural Network). HUANG [6] and others designed an end-to-end note recognition model based on deep learning. The model uses deep convolution neural network to directly output the time value and pitch of notes with the whole score image as the input. In terms of data preprocessing, the music score image and corresponding label data required for model training are obtained by parsing the musicxml file. The label data is a vector composed of note pitch, note time value and note coordinates. Therefore, the model learns the label vector through training and transforms the note recognition task into detection and classification task. Then data enhancement methods such as noise and random clipping are added to increase the diversity of data and make the trained model more robust; In the model design, based on darknet53 basic network and feature fusion technology, an end-to-end target detection model is designed to recognize notes. The deep neural network darknet53 is used to extract the music score image feature map, so that the notes on the feature map have a large enough receptive field, and then the upper feature map of the neural network is spliced with the feature map to complete the feature fusion, so that the notes have more obvious feature texture, so that the model can detect small objects such as notes. The model adopts multi task learning, and simultaneously learns the classification task of pitch and time value and the regression task of note coordinates, which improves the generalization ability of the model. Sun [7] proposed a digital audio tone recognition method via time-frequency domain information extraction, collected audio signals, extracted time-frequency domain information, obtained the overall standard audio spectrum sequence by Goertzel algorithm, and normalized the energy of the obtained sequence. The filter is used to match the unrecognized digital audio spectrum sequence with the standard tone template to remove the interference of external factors. The corresponding musical notes are obtained from the best matching results to realize the efficient recognition of digital audio music.

Although the above method realizes the recognition of music notes, it can not obtain accurate baseband structural features of note signals, and the deviation of baseband extraction results is large, which has the problem of large recognition error, and will consume a lot of recognition time. In order to solve the above problems, a note segmentation recognition method based on nonlinear feature detection is proposed in this paper.

2 Method

2.1 Music Note Retrieval

In order to improve the efficiency of music note recognition, music notes are retrieved first. Music notes can be retrieved through the similarity of vectors and matrices, but this retrieval can only obtain similar results on the surface. The quality of retrieval depends on the distance definition of vectors and matrices. In order to make the search results reflect more potential correlation, this paper uses vector space matrix to analyze and process music notes to reflect the correlation of music notes [8, 9].

Let the musical note expression matrix be S, which can perform singular value decomposition:

$$S = UAV^* \tag{1}$$

Multiply both sides of formula (1) by q_i and G_s, where G_s is a diagonal matrix, the first s diagonal elements are 1, and the remaining elements are 0:

$$D_s = UAG_s \tag{2}$$

Suppose the note to be queried is set to L, and the similarity between the query note and the music matrix is defined as:

$$\text{sim}(S, L) = \frac{1}{q_s - (1 - D_s)\varepsilon_i} \tag{3}$$

Among them, q_s represents the music matrix in the query space; ε_i represents the note query vector.

The formula (3) can be further organized as:

$$\text{sim}(S, L) = \frac{2q_s(c_i - c_j)}{max(\varepsilon_i) - min(\varepsilon_i)} - 1 \tag{4}$$

Among them, ε_i can not only reflect the autocorrelation of the matrix, but also obtain the cross-correlation between S and L. Therefore, the complete relevance of the music can be obtained by formula (4). Using this relevance to sort the music can produce satisfactory retrieval results for users.

2.2 Music Note Segmentation

Through obtaining the position of the dividing line by dynamically setting the threshold and determining the dividing line by setting the constraint conditions, the adaptability and accuracy of note recognition was improved [10].

The amplitude function definition in the traditional segmentation algorithm is expressed as follow:

$$f(n) = 1/\sqrt{n} \sum\nolimits_{k=1}^{N} f(k) \qquad (5)$$

Among them, $f(n)$ denotes the waveform amplitude function; $f(n)$ denotes the amplitude of the h-th sampling point; N denotes the window length; n denotes a certain frame of the input signal.

Then we have the amplitude difference function of $f(n)$:

$$E(s_i) = \frac{f(n+1) - f(n)}{E(f_x)} \qquad (6)$$

Applying $E(s_i)$ is more obvious than applying $f(n)$ alone to the dividing line of a single note.

The restriction conditions are set to find Optimal solution [11, 12].

By setting the minimum number of frames and the maximum number of frames, remove those obviously inappropriate divisions:

$$W = \begin{cases} 1 E_{min} \leq E < E_{max} \\ 0 \text{ otherwise} \end{cases} \qquad (7)$$

Among them, W represents a set of Booleans whether the partition is valid; E_{min} represents the smallest frame of the signal; E_{max} represents the largest frame of the signal; E represents the number of frames occupied by a certain note in the partition.

(2) The notes are divided evenly. When a single note is greater than or less than 1.5 times the number of frames occupied by the adjacent note, the division is judged invalid:

$$W = \begin{cases} 1 0.5 \leq E' < 1.5 \\ 0 \text{ otherwise} \end{cases} \qquad (8)$$

Among them, E' represents the number of frames occupied by adjacent notes.

According to the above judgment conditions, the segmentation of music notes is realized.

2.3   Music Note Recognition

Currently, the more commonly used nonlinear feature detection methods include the kernel method and the manifold method, and the kernel method has become its theoretical research hotspot due to its efficient and flexible calculation. Traditional feature detection is aimed at linear problems, and the detection effect for non-linear problems is not good, and music notes have non-linear characteristics. Therefore, this paper introduces the kernel function method into music note recognition [13]. The kernel function method is a data re-encoding technology that uses kernel mapping to process non-linear data. Compared with other non-linear feature detection methods, the advantage of the kernel method is that only the "kernel function" is used instead of the original data. Linear mapping has the advantage of simple calculation process.

The kernel function is defined as: $V(a, b)$, where a represents a point in a low-dimensional space, and b represents a point in a high-dimensional space. Define the kernel matrix according to the definition of the kernel function: if the kernel function satisfies the Mercer condition, the matrix is the kernel matrix of the kernel space. It can be seen from the definition that the kernel function defines a non-linear feature space. This method can effectively simplify the non-linear relationship between space mappings. Only the inner product transformation of vectors is used without the need for the coordinates of sample points in the data space.

According to the definition of kernel function and kernel matrix, FFT is used to perform frequency domain conversion on low-pass components in audio to realize fundamental frequency identification. When choosing DWT decomposition scale, pay attention to the fact that the highest output frequency of FFT is actually only half of the highest input frequency. If the highest frequency in the input time-domain signal of the FFT can be controlled to be between 2 and 4 times the frequency of the single tone to be identified, and then FFT is performed, the output will only contain the fundamental component of the single tone. That is, the highest frequency F of the DWT low-pass component is required to meet:

$$2F_i < F < 4F_i \tag{9}$$

DWT adopts binary low-pass decomposition, and its low-pass decomposition scale can be calculated by formula (10):

$$\partial = 1/n \sum_{i=1}^{n} f(k) + \sum_{j=1}^{m} (x_i - x_j)^2 \tag{10}$$

Among them, ∂ represents the decomposition scale; $f(k)$ represents the signal sampling frequency; x_i represents the pitch frequency of the single tone to be recognized contained in the signal; m represents the largest integer not exceeding x_i. According to formula (10), the upper limit and scale of the wavelet low-pass decomposition frequency of all single musical sounds can be obtained, that is, the recognition of music audio can be realized.

3   Simulation Results and Analysis

To validate the feasibility of the music note segmentation recognition algorithm based on nonlinear feature detection, the comparison method and comparison index are determined, and the experimental results are obtained.

3.1   Experiment Preparation

In the experiment, the average AP accuracy of the notes to be recognized was 0.87, the timing accuracy was 0.96, the pitch accuracy was 0.98, the highest frequency was

5421 Hz, and the sampling frequency was 45.8 kHz. The experiment was performed with Matlab. The note recognition method based on the bat algorithm and that based on deep learning are used as the contrast method, and the error of the recognition result and the recognition efficiency are used as the experimental indicators to verify the effect of the method in this paper.

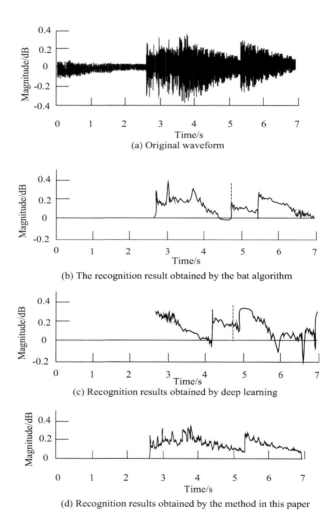

(a) Original waveform

(b) The recognition result obtained by the bat algorithm

(c) Recognition results obtained by deep learning

(d) Recognition results obtained by the method in this paper

Fig. 1. Error comparison of recognition results

3.2 Results

(1) Error of recognition result

The recognition effects of different methods are compared with the error of the recognition result as an indicator, as shown in Fig. 1.

As shown in Fig. 1, the bat algorithm has a high recognition accuracy for notes, the peak amplitude is large, and the recognition effect is better. But for low rhythms and mixed sounds, the peaks of low-frequency signals are not obvious, and the recognition results of the deep learning method are not as good as the bat algorithm. The method in this paper can make up for the shortcomings of the traditional method. It can be seen that the peak of the low-frequency signal is very obvious, indicating that the method in this paper can effectively identify low rhythms and mixed sounds.

(2) Note recognition efficiency

The recognition effects of different methods are compared with note recognition efficiency as indicator, as shown in Fig. 2.

From the data in Fig. 2, it can be seen that the time used for music note recognition in multiple iterations of the method in this paper is less than 2.0 s, which is much lower than the time used by bat algorithm and deep learning method for music note recognition in multiple iterations. It can be seen that the recognition efficiency of the method in this paper is higher.

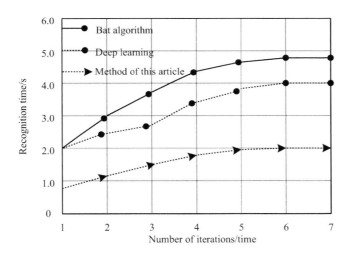

Fig. 2. Comparison of recognition efficiency

4   Conclusion

This paper presents a music note segmentation recognition algorithm based on non-linear feature detection. After experimental verification, it can be concluded that the recognition accuracy of the method in this paper is higher, and it has higher efficiency, which is fully verified The application value of this method.

References

1. Kholjurayevich, M.N., Madaminovich, F.K., Yuldashev, J.G.: The role of music culture in inculcating the idea of national independence in the minds of the younger generation. ACADEMICIA: Int. Multidiscip. Res. J. **11**(5), 71–74 (2021)
2. Jirsa, T., Korsgaard, M.B.: The music video in transformation: notes on a hybrid audiovisual configuration. Music Sound Mov. Image **13**(2), 111–122 (2019)
3. Younis, A.N., Ramo, F.M.: A new parallel bat algorithm for musical note recognition. Int. J. Electr. Comput. Eng. Syst. **11**(1), 558–566 (2021)
4. Allabakash, I.T., Rajendra, D.K.: An effective optimization-based neural network for musical note recognition. J. Intell. Syst. **28**(1), 173–183 (2019)
5. Castellanos, F.J., Gallego, A., Calvo-Zaragoza, J.: Automatic scale estimation for music score images. Expert Syst. Appl. **158**(15), 113590 (2020)
6. Mahanta, S., Khilji, A., Pakray, D.: Deep neural network for musical instrument recognition using MFCCs. Computacion ySistemas **25**(2), 351–360 (2021)
7. Malik, S.: Automatic mood recognition system of Indian popular music and its applications. A Peer Rev. Int. J. **3**(3), 443–446 (2015)
8. Calvo-Zaragoz, J., Toselli, A.H., Vidal, E.: Handwritten music recognition for mensural notation with convolutional recurrent neural networks. Pattern Recogn. Lett. **128**(1), 115–121 (2019)
9. Marzik, G., Sato, S., Girola, M.E.: Compressive sensing for perceptually correct reconstruction of music and speech signals. Appl. Acoust. **183**(2), 108328 (2021)
10. Tsai, W., Tu, Y.M., Ma, C.H.: An FFT-based fast melody comparison method for query-by-singing/humming systems. Pattern Recogn. Lett. **33**(16), 2285–2291 (2012)
11. Kolokolov, A.S.: Preprocessing and segmentation of the speech signal in the frequency domain for speech recognition. Autom. Remote Control **64**(6), 985–994 (2003)
12. Sangeetha, R., Nalini, N.J.: Recognition of musical instrument using deep learning techniques. Int. J. Inf. Retrieval Res. **11**(4), 41–60 (2021)
13. Hizlisoy, S., Yildirim, S., Tufekci, Z.: Music emotion recognition using convolutional long short term memory deep neural networks. Int. J. Eng. Sci. Technol. **24**(3), 760–767 (2020)

The Practice of MySql in Realizing the Background Data Management of the Hybrid Teaching Management System

Wenwu Miao[✉]

School of Marxism, Kunming University, Kunming 650214, Yunnan, China
wenwum@163.com

Abstract. With the rapid development of the Internet and information technology, traditional private education classrooms have gradually been unable to meet the different individual learning needs of today's students due to problems such as unified teaching methods and poor teacher-student interaction. The hybrid learning management system makes full use of the Internet's high-speed information transmission capabilities and multiple display methods to conduct online teaching of traditional classroom content through Internet multimedia, which can fully meet the needs of learners for individualization, diversification and rapid learning. This article aims to study the practice of MySql in implementing the background data management of the hybrid teaching management system. Based on the analysis of the characteristics of MySql, database design principles, and personalized learning recommendation algorithms, a hybrid teaching management system and its database are designed, and tested the performance of the system. The test results show that the performance of the system in this paper is good, and the actual response time of the system is within the expected response time, which meets and meets the needs of this paper.

Keywords: MySql · Blended teaching · Management system · Background data management

1 Introduction

The rapid development of information technology has enabled science, network and communication technology to enter every link and field of society. Countries all over the world are setting off a wave of informatization construction, and the influence of information technology on social development is increasing [1, 2]. Informatization has also had a great impact on education, and it has had a great impact on educational concepts, teaching methods, educational content, learning methods, and even educational models [3, 4].

Compared with domestic Internet technology and multimedia technology, foreign Internet technology and multimedia technology are more mature, and the education industry is also more mature and perfect. The completion and development speed of hybrid learning mode is faster than that in China. One scholar believes that hybrid learning is a simple integration of traditional education mode and online education. Another scholar said that blended learning mainly has four basic requirements. First, we need to combine different teaching methods and technologies to improve educational achievements; Secondly, we need to achieve the goal of network learning. Third, it is necessary to combine network teaching with teachers' actual teaching. Fourth, teachers' actual teaching tasks and network teaching objectives should be consistent [5, 6]. Some researchers believe that blended learning is a combination of traditional and distance learning methods. Use hybrid learning methods to overcome the limitations of traditional teaching methods on teaching content and maximize the breadth of educational knowledge and online educational content [7, 8]. In order to meet the current personalized learning needs of students, some scholars have proposed a system based on hybrid learning management education. The system first analyzes and matches students' personality, and then uses Bruce learning model to provide learning content related to different students. Experiments show that this hybrid learning can improve students' learning effect [9, 10]. Some scholars analyzed that the sustainable development, construction, management and maintenance of education resource platform is the primary solution for classroom education in the future. The author proposes a hybrid learning that combines the network education management system and the education resource platform. The education management platform optimizes the existing teaching methods and improves the efficiency of school resource management [11, 12]. The previous theoretical results provide a theoretical basis for the research of this paper.

Based on consulting a large number of references related to MySql and the hybrid teaching management system, this article combines the characteristics of MySql, database design principles, and personalized learning recommendation algorithms to design the modules in the hybrid teaching management system. The system contains four modules, namely login and registration module, system management module, teaching resource management module and user personal center module. In addition, this article also designed the MySql database of a hybrid teaching management system to manage the back-end data of the system.

2   The Practice of MySql in Realizing the Background Data Management of the Hybrid Teaching Management System

2.1   Features of MySql

(1) Good ease of use. MySQL is a simple and easy-to-use high-performance database system. Compared with the complicated configuration and management of mainstream database systems such as Oracle and DB2, the configuration and management of MySQL is much simpler, and it is relatively simple and effective. (2) Program language support. SQL (Structured Query Language) is the language of choice for all modern databases. MySQL and other commercial databases have some things in common, because MySQL can support it well, and MySQL also supports the use of ODBC. (3) Support to the network. MySQL can also be used in various Internet environments, whether it is LAN, WAN or Internet, users can use the Internet to access the MySQL database, and at the same time can also use the Internet to share data with anyone on other sites. Of course, MySQL also has its own access control mechanism, which can

control the specific data that someone queries in designated places. (4) Portability. MySQL can run on different popular operating systems, such as different Unix, FreeBSD, Linux, Windows, OS/2, so you can easily transfer the database from one operating system to another without changing the application. (5) Openness. This is also the advantage of open source databases. If you are not familiar with some specific principles or algorithms, you can directly consult the source code and analyze it. If you find that some content does not meet the requirements of this application, you can modify it yourself. At the same time, open source license users can use it for free, as long as it conforms to the GNU General Public License.

2.2   Principles of Database Design

(1) Fully understand the use of the target database. Too many data fields in a data table is not simple, so you must consider the I/O unit of the database and choose the type of logical data element. If it is an integer type, do not change it to a floating point type. Set indicators by measuring strengths and weaknesses, but too many indicators will weaken server performance, so do not easily modify indicator entries while the system is running. In addition, relational databases also need to meet the data integrity constraints of related models, stored procedures, effective access to stored procedures, and data default values. (2) The integrity of the database. The so-called consistency is the consistency and validity of the data in the database. Common terms include entity integrity, regional integrity, and report integrity. Each column in the entity integrity definition table can usually be uniquely identified by the master key. In MySQL, you can define the unique pointer UNIQUE constraint, PRIMARYKEY constraint and ID attribute to achieve entity integrity.

2.3   Personalized Learning Recommendation

Since the content obtained by the user is closely related to the keywords, the ratio of the keyword weight to the total weight is set to μ, and the keyword weight is adjusted. If the number of input keywords is k, the weight of the input keywords is as follows:

$$W_{in} = \frac{\mu \sum_{i=1}^{m} W_i}{k} + W_{key} \tag{1}$$

The value of W_{key} is its original characteristic value.

Calculate the correlation Sim between resource document vector D_i and query keyword vector Q_j, using the angle cosine formula to calculate:

$$Sim(D_i, Q_j) = \cos \theta = \frac{\sum_{k=1}^{n} W_{ik} \bullet W_{jk}}{\sqrt{(\sum_{k=1}^{n} W_{ik}^2)(\sum_{k=1}^{n} W_{jk}^2)}} \tag{2}$$

3 Experiment

3.1 System Function Module Design

Login and Registration Module: The main user roles of the hybrid education management system are the system administrator, user teacher, student and tourist user. Therefore, when implementing the login and registration modules, carefully consider the functions that each role must complete. Enter the login and registration interface. The interface should be configured with the guest user login system function, which facilitates the visitor user, through classroom video learning, classroom information display and other educational function modules. Next, the login or registration page must implement the registration input box, password input box, registration button, and login button, while other functions allow system users to directly enter the registered user name or password in the corresponding login page or registration interface, thereby Jump directly to the corresponding landing page or registration interface. In order to ensure the security of the system user password, the login box password uses * to display the password; in addition to the registration function, the registration interface also has the registration information function, such as a user account completed by a combination of numbers, letters, and symbols, and the length is 6 to 12 user password and repeated password input, password verification and other functions. Secondly, in order to ensure that users will not forget their passwords, the system also provides password recovery functions on the login and registration pages, allowing users to reset user passwords by entering corresponding personal identification information.

System Management Module: The system management module is mainly divided into two modules: application management and system resource management. User management functions are mainly to modify user accounts, user passwords, user e-mail addresses, mobile phone numbers used, etc., to facilitate users to update and protect personal information in a timely manner. System course resource management mainly refers to teachers obtaining or uploading relevant course resources through course resource management to complete the update and maintenance of course resources. Students can download, learn and use corresponding course resources through this module. It is stored in the form of a file on the back-end server, and a course resource table is created through an Oracle database to complete course resource completion. If the system administrator needs to manage many course resources, just use the course resource table to perform corresponding operations.

Teaching Resource Management Module: After logging in, the administrator can retain educational resource information and control new resources. After logging in, teachers and students can upload, download, collect, retrieve educational resources, play audio and video resources, update resource information, delete personal resources, and other management of the uploaded resources. Video tutorials are played through FIvPlayer. The play mode can be switched by changing the value of the stage. If the user stops playing in the middle of the play, the system can record the current interruption time, and directly locate the last interruption point when it is played next time.

User Personal Center Module: User-centered personal unit is one of the most commonly used modules in software system development. After the user is registered and logged in, there will be a personal user unit for centralized user information management. The user personal center mainly includes user basic information management, user learning records, user download records, user favorite records, etc. The main interface uses Linear Layout for linear Layout, by reading the contents of the database to display basic information in the following format: tutorial items, downloads, favorite items, etc., using GridView elements for menu display, menu bar setting styles and monitoring events. Clicking on the corresponding menu item will jump to the corresponding interface through the start Activity() function, read the corresponding content of each program content from the database through SQL statements, and use a custom List View to complete.

3.2   Database Design

Database Logic E-R Diagram Design: The E-R diagram is an image representation of a relational conceptual model, which is used to describe the relationship between objects through the type of entity, some attributes of the entity, and the relationship between the entity and the entity. The main shapes of E-R diagrams are rectangles, ellipses, diamonds, and connecting lines. Rectangular figures represent real objects, elliptical figures represent certain characteristics of real objects, and diamond shapes represent the relationship between real objects and entities. The main relationship is "submission" and "management", "use" and so on. In the student's functional module, the entities used include student personal information, system announcement information, course announcement information, teacher information, course list, content resources, test questions, homework, grades, and questions submitted by users. The relationship between each entity and the student is represented by the E-R diagram of the student user, as shown in Fig. 1.

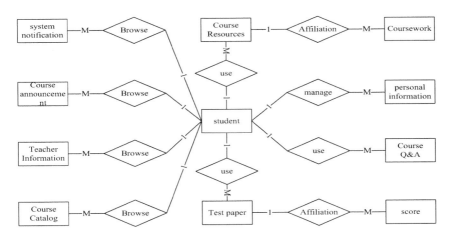

Fig. 1. Student user E-R diagram

Database Logical Structure Design: The ultimate goal of database logical structure design is to transform the E-R diagram from the conceptual design mode into the logical architecture of the software system, that is, the relational system design mode. Before converting the ER diagram into a relational model, we must standardize the data to prevent a large number of overlapping data elements in the same data structure and reduce data redundancy. Object names and formats must also be standardized. Correspondence between entities such as 1:1, 1:n, m:n, the same code relationship transformation, through the relationship conversion rules between entities, the corresponding entity structure and attributes are changed.

4   Discussion

The performance test of the system mainly tests the system response speed, operating speed, network service speed, etc. It mainly tests the system startup time, the response time of each functional module, the network access speed and the system CPU/memory usage, and the hybrid teaching management system performance test. The main test cases and test results are shown in Table 1:

Table 1. System response time

Number of concurrent users	Expected response time/s	Actual response time/s
100	1	0.75
300	1.8	1.12
500	2.5	1.87
1000	3.6	2.86

It can be seen from Table 1 and Fig. 2 that when the number of concurrent users in the system is 100, the expected response time is 1 s, and the actual response time is 0.75 s; when the number of concurrent users in the system is 300, the expected response time is 1.8 s, and the actual response time is 1.12 s; when the number of concurrent users of the system is 500, the expected response time is 2.5 s, and the actual response time is 1.87 s; when the number of concurrent users of the system is 1000, the expected response time is 3.6 s, and the actual response time is 2.86 s. It

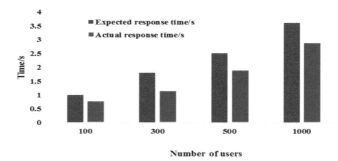

Fig. 2. System response time

shows that the performance of the system in this paper is good, and it meets and satisfies the needs of this paper.

5   Conclusions

By using MySQL to manage the basic data management in a hybrid education management system, it can be found that MySQL is still quite excellent in performance. It has super scalability and flexibility, as well as powerful data protection capabilities. MySQL has good connectivity, speed and stability, making it easy to access network databases.

References

1. Kong, Q., Qin, C.: LasDB: a collective database for laboratory animal strain resources. Anim. Models Exp. Med. **1**(4), 24–29 (2018)
2. Specht, G., Roetschke, H.P., Mansurkhodzhaev, A., et al.: Large database for the analysis and prediction of spliced and non-spliced peptide generation by proteasomes. Sci. Data **7**(1), 146 (2020)
3. West, A.W., Prettyman, S.: Practical PHP 7, MySQL 8, and MariaDB Website Databases (A Simplified Approach to Developing Database-Driven Websites) ‖ Take a Brief Look at Oracle MySQL 8. Chapter 12, pp. 479–505 (2018). https://doi.org/10.1007/978-1-4842-3843-1
4. Miranskyy, A.V., Al-zanbouri, Z., Godwin, D., Bener, A.B.: Database engines: evolution of greenness. J. Softw.: Evol. Process **30**(4), e1915.1–e1915.25 (2018)
5. Li, S., Gong, W., Wang, L., et al.: A hybrid adaptive teaching–learning-based optimization and differential evolution for parameter identification of photovoltaic models. Energy Convers. Manage. **225**(4), 113474 (2020)
6. Ashraf, M.M., Malik, T.N.: A hybrid teaching–learning-based optimizer with novel radix-5 mapping procedure for minimum cost power generation planning considering renewable energy sources and reducing emission. Electr. Eng. **102**(4), 2567–2582 (2020). https://doi.org/10.1007/s00202-020-01044-0
7. Li, K., Xie, X., Xue, W., Chen, X.: Hybrid teaching—learning artificial neural network for city-level electrical load prediction. Sci. China Inf. Sci. **63**(5), 1–3 (2019). https://doi.org/10.1007/s11432-018-9594-9
8. Li, K., Xie, X., Xue, W., Chen, X.: Hybrid teaching–learning artificial neural network for city-level electrical load prediction. Sci. China Inf. Sci. **63**(5), 212–214 (2020)
9. Cho, J.H., Hong, W.P.: Optimal design study of stand-alone hybrid energy system using TLBO-CS algorithm. J. Korean Inst. Illum. Electr. Installation Eng. **32**(5), 18–29 (2018)
10. Hoad, K., Kunc, M.: Teaching system dynamics and discrete event simulation together: a case study. J. Oper. Res. Soc. **69**(4), 517–527 (2018)
11. Ochia, R.: A hybrid teaching method for undergraduate biomechanics lab. Biomed. Eng. Educ. **1**(1), 187–193 (2021). https://doi.org/10.1007/s43683-020-00033-w
12. Khamari, D., Sahu, R.K., Gorripotu, T.S., Panda, S.: Automatic generation control of power system in deregulated environment using hybrid TLBO and pattern search technique. Ain Shams Eng. J. **11**(3), 553–573 (2020)

College Students' Emotion Analysis and Recognition System Based on SVM Model

Shuting Liu[(✉)]

College of Engineering Technology, Xi'an Fanyi University, Xi'an 710105,
Shaanxi, China
xafy0l@l26.com

Abstract. This paper takes social network text as the research object, and introduces SVM model into emotion analysis and recognition system to effectively improve the accuracy of emotion classification of network text, master the development law of network public opinion, effectively study and judge college students' social network information interaction behavior, and intervene in high-risk network information interaction behavior in real time, Guide college students to make correct use of network information resources and better grasp the ideological and behavioral state of college students and the law of network information interaction. Establish a fast and efficient crisis early warning and feedback system to make the psychological crisis prevention work achieve early detection, early prevention and early treatment, cultivate students' psychological capital, strengthen psychological quality education, and effectively improve students' mental health level.

Keywords: Social networks · Emotion analysis · Recognition system · SVM model

1 Introduction

With the popularity of social networks, weibo and wechat have gradually become the largest social platforms in China. College students can easily acquire, share, communicate and interact with network resources, and a large number of online texts containing subjective opinions have been generated [1]. In this study, the emotion classification model constructed by SVM algorithm was used to conduct emotion analysis on the weibo posts published by students of Xi'an Fanyi University from August 2021 to December 2021, focusing on the emotional responses of users in different time periods and under different event backgrounds, in order to provide timely and effective feedback for university management departments [2].

2 Research Design

2.1 Social Networks

Social network, also known as social network, is a website and technology that can support people to write, share, comment, discuss and communicate. Social networking is based on Internet technology, especially web2.0 technology based interactive community. Social network gives everyone the ability to create and disseminate content. It is a network for social interaction. It is a way of social communication through ubiquitous communication tools. It can give users great participation space, which can not only meet the needs of Internet users to store their personal basic data, but also meet the psychological feeling needs of users to "be found" and "be worshipped", as well as the needs of users to "establish relations" and "exert influence" [3]. College students are keen to publish their opinions and opinions on the social network platform. The text data on the social network can truly reflect the emotional state of college students to a certain extent. There are many kinds of social networks. This study takes the mainstream network "microblog" in social networks as an example. Social network emotion is shown in Fig. 1.

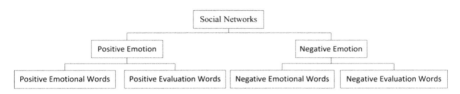

Fig. 1. The original image

2.2   Text Emotion Analysis

There are several common views on the division of emotion at home and abroad: some studies divide emotion into positive emotion and negative emotion; Some studies divide emotion into three categories: positive, negative and neutral; In order to express the intensity of emotion, some studies divide emotion into five levels, 0 is the largest negative emotion and 4 is the largest positive emotion; According to people's emotional expression, some studies divide emotions into "happiness, sadness, praise, criticism, confidence and accident". This study adopts two-dimensional classification: positive emotion and negative emotion. Positive emotion refers to positive emotion, and negative emotion refers to negative emotion. The most commonly used methods of emotion analysis are machine learning method and emotion dictionary method. Compared with affective dictionary method, machine learning method does not need too much manual intervention and cost investment, has less deviation, and occupies a certain advantage in update speed. This study adopts the supervised algorithm of machine learning to build an emotion classification model to analyze the emotional tendency of microblog text.

2.3   Research Ideas

The experimental idea of this study is mainly divided into two parts. The first part is the construction of sentiment classification model. The open corpus is divided into training set and test set, text preprocessing and text quantization are performed on the test set, and then the algorithm function is used to train the sentiment classification model [4].

By comparing the model precision, recall and F value of the current mainstream machine learning algorithms, the model with the best classification effect is selected. The second part is the collection and analysis of big data, crawling the information data of relevant users in microblog through crawler technology, carrying out text preprocessing and text Vectorization in turn, and then using emotion classification model to analyze the information data. Some network emotional words are shown in Table 1.

Table 1. Part of the network emotion words

Network emotion words	Example
Positive	Good, Give A Like, Great, Can, Like
Negative	Fault, Tragedy, Uncomfortable, Boring, Want To Cry

3 Construction of Emotion Classification Model

3.1 Experimental Data Set

This experiment uses the public corpus on the code hosting platform "GitHub" for model training. The corpus is a data set of positive and negative emotion tagging for some comments. Its construction time is relatively new. During the data tagging period, multi person verification is adopted to ensure the data quality. The corpus is divided into two parts: training set and test set [5]. The training set is used to evaluate the training of emotion classification model; The test set is used to verify whether the constructed model can accurately obtain the correct classification label.

3.2 Pretreatment

Preprocess the text data to enable the computer to recognize the corpus in the form of natural language. Common preprocessing includes the following contents: information extraction and removing the "noise" contained in the text. The extracted useful data is used for Chinese word segmentation with the help of existing word segmentation tools. With the help of the existing stop words dictionary, the stop words in the text are filtered out to reduce the impact of stop words on the analysis results. This study mainly uses word segmentation and de stop word processing of preprocessing technology. Use the word segmentation tool "nlpir" developed by Beijing University of technology to segment the collected microblog text data; Using the stop word thesaurus of Harbin Institute of technology to remove the stop words in the interference experiment. The flow chart of emotion analysis is shown in Fig. 2.

Fig. 2. Emotion analysis flow chart

3.3   Text Vectorization

Text vectorization is to transform the text into a series of semantic vectors that can represent the text through the algorithm model. In the research, text vectorization is usually carried out with words as the basic unit. The size of data capacity can directly affect the accuracy of algorithm model, calculation efficiency, calculation time and so on. Text vectorization not only plays an important role in improving model efficiency, reducing computing time and spatial dimension, but also plays an important role in improving algorithm performance.

This study uses the neural network algorithm "word2vec" to generate word vector. By learning from a given corpus, the algorithm can generate word vector spaces with different dimensions. The algorithm analyzes and processes words based on context, so it can achieve the purpose of emotional tendency classification. These word vectors can be placed in various NLP computing tasks. "Word2vec" adopts two models: CBOW and Skip-gram. The components of both have input layer, projection layer and output layer. The principle of CBOW model is to predict the current words according to the words before and after the current words, while Skip-gram model predicts the words before and after the current words in the word vector space.

On this basis, the Skip-gram method in "word2vec" is used to vectorize the text data in the training set, so that each word has a specific value to represent its features, and the feature space is constructed [6]. Skip-gram method mainly uses the current words in the word vector space to predict the words before and after them, so as to lay a foundation for model training. SVM commonly used kernel functions are as follows: Linear kernel function.

$$K(x, y) = x \cdot y \tag{1}$$

The kernel function.

$$K(x, y) = [(x \cdot y) + 1]^d \tag{2}$$

kernel function

$$K(x, y) = \tanh(a(x \cdot y) + b) \tag{3}$$

This study tracked 174 accounts located in Xi'an Fanyi University within half a year, and manually screened and obtained the IDs of 127 students. Use the sina weibo crawler Program published on "GitHub" to continuously update config JSON, and then use Spyder to execute weibo spider Py program to crawl the microblog text of students with known microblog ID from August 1, 2021 to December 17, 2021.

The crawling microblog data has achieved good results. A total of 12478 pieces of data from 127 students were crawled and stored in 127 CSV files. It is mainly composed of microblog ID, text, release time, location, number of likes, etc. The text and release time are the main experimental data of this paper.

4 Data Analysis

In this study, 12478 text data of 127 students from August 2021 to December 2021 were preprocessed and vectorized by word segmentation and stopping words, and the emotion classification model based on SVM algorithm was selected to classify their emotional polarity. In this study, the emotion classification model constructed by SVM algorithm was used to determine the emotional orientation of the experimental data set. Output the text with the judgment result of "−1", namely the negative emotion text; Output the text with the judgment result of "1", namely the positive emotion text [7]. Emotional classification is shown in Table 2.

Table 2. Emotional classification

Experiment design	Evaluation index			
	Accuracy	Precision	Recall	F1-Score
SVM	0.8129	08286	0.8158	0.8254

4.1 Positive and Negative Emotion Text Quantity Statistics

In this study, the number of texts with positive and negative emotions was counted by month, and the number of positive texts in each month was significantly more than negative texts, indicating that the texts posted by students in the university on weibo social network were relatively positive in emotional tendency.

4.2 Proportion Distribution of Negative Emotional Text

This study analyzes the proportion of negative emotion in each month. It can be found that the proportion of group negative emotion is the smallest in October 2021 and the largest in November 2021. The word frequency of college students on social networks is shown in Table 3.

Table 3. College students' word frequency on social networks

Keyword	Word frequency	Keyword	Word frequency
Association	298	COVID-19	594
Retest	184	Join the party	357
Excellent class	325	Military training	502
Make-up exam	65	Final exam	574

4.3 Negative Text Content Analysis

This study analyzes and discusses the negative text content according to the time period. During August 2021, the frequency of "COVID-19" in the text content of negative emotion and positive emotion is relatively high. It is speculated that the research group pays more attention to "COVID-19" during this period. During September 2021, the frequency of "Military Training" in the text content of negative emotion and positive emotion is relatively high [8]. It is speculated that the research group pays more attention to "Military Training" during this period. In November 2021, in the negative emotion text content, the three topics of "schoolwork", "junior college to undergraduate", "postgraduate entrance examination" appeared more frequently, indicating that these three topics may be related to the negative emotion of the research group at this stage. In contrast, the frequency of "schoolwork", "junior college to undergraduate", "postgraduate entrance examination" in the positive text at the same time is low. It is speculated that during this period, "schoolwork", "junior college to undergraduate", "postgraduate entrance examination" have a direct impact on the negative emotion of the research group. In the negative text content in December 2021, the two topics of "Final Exam" and "COVID-19" appear more frequently, and the topic basically coincides with the time period in which students' negative emotions appear. It is speculated that the negative emotions of the research group in this time period are related to "Final Exam" and "COVID-19" events to a certain extent. The emotional tendency of the research group on social networks is generally positive, while the proportion of negative emotional tendency is small. There are two main periods of negative emotion concentration, namely, August 2021 and November and December 2021. In these two periods, the events that students are more concerned about are "COVID-19", "Military Training" and "Final Exam". Among them, "COVID-19" is an emergency, and "COVID-19" and "Final Exam" are common topics for this group. This shows that this group will pay attention to social events in addition to their daily life [9].

5   Influence of Social Network on College Students

5.1   Positive Impact

Social network communicates and interacts with others through virtual network, which is an important supplement to the social way in the real world, expands the scope of interpersonal communication and promotes the connection between college students and social communication. Social network leads college students to expand from an acquaintance society to a relative stranger society. For outgoing and pleasant college students, it is easy to establish good interpersonal relationships with others on the network, and then produce subjective well-being. Social support communicates the intimate relationship between people, which makes them get help when they need it [10].

The positive energy of social network communication improves college students' sense of morality and honor. Good people and good deeds that are courageous and helpful are spread through social networks every day. College students receive more information to promote social integrity and justice morality, which will naturally improve their moral sense. The positive energy on social networks infects contemporary college students all the time and cultivates pure moral sentiment. Social network is also an important position to create a sense of honor for college students [11].

5.2   Negative Impact

Some college students are addicted to social games. Different from traditional games where characters are set by computers, opponents and friends in social games are all real people. The happiness brought by the sense of reality makes many college students become more and more addicted to social games. If this goes on for a long time, the inner loneliness of college students is more intense, and their spirit is gradually depressed, which seriously affects their daily study and life.

Social network information is complicated, and college students browse social network frequently every day. The negative information on social network causes a lot of negative emotions. College students are disturbed by wrong information with ulterior motives, and have self-doubt, dilute patriotic and national consciousness, and even have a pessimistic mood towards the society, which causes extremely serious harm [12]. Social network violence makes people angry and hatred, which has a very negative impact on college students' emotions.

6   Guide Social Network to Increase Positive Emotion of College Students

6.1   Educators Should Widely Carry Out Social Networking Education

Special education on the cognition and use of social networks is often carried out to enable college students to understand the characteristics of social networks, treat virtual phenomena correctly and learn how to solve problems. Encourage the positive feelings brought by social network to college students, pay more attention to the negative emotional problems, through regular social network knowledge popularization education, make college students realize the disadvantages of social network, prevent addiction to social network and game addiction; By presenting facts and reasoning, combining theory with practice to persuade and guide college students to enhance their right and wrong and political discrimination, so that they can firm up correct beliefs and beliefs, and not be confused by fallacies and rumors; Often through psychological counseling mechanism, teach college students healthy psychological adjustment mechanism, can control and deal with their emotions, avoid the outbreak of extreme emotions. Educators should observe and understand the state of students, be good at communicating with students, and actively encourage the expression of positive energy [13].

6.2   College Students Should Adjust Their Emotions Reasonably and Reduce Negative Emotional Conflicts

College students should learn to self adjusting when using social networks, appropriate use of social network social, communicate with netizens resolve stress study and life, etc., release the negative emotions, at the same time attention should be paid to avoid adverse social, away from the network of violence and abuse, improve network awareness, to prevent the emotional hurt, learn to emotional processing method, lower have negative emotions. Pay attention to improve self-control, enhance the pursuit of self-value, set up a grand goal in life, unremitting struggle in the real world, put an end to Internet addiction, overcome emptiness and confusion. In the face of the emotional depression caused by the negative information filled with social networks, emotional catharsis, transfer and sublimation can be adopted to relieve, overcome emotional impulse with rational thinking, express emotions reasonably, control emotions and reduce negative emotional conflicts.

7 Conclusions

This study constructs an emotion classification model based on SVM algorithm. The model is used to classify the emotion of the collected text data, and the time period and related events of negative emotion are obtained [14]. For the topic of negative emotion, relevant school management departments can take some targeted measures to alleviate students' negative emotion. For example, on the one hand, for the daily life of this group, such as military training and final exam, psychological training can be appropriately organized to cultivate students' positive attitude towards college life; Organize lectures on professional learning to enhance the learning interest and professional confidence of this group; Carry out positive group day activities to help the group establish ideals and beliefs; Actively carry out physical exercise activities to help students enhance their physique. On the other hand, in the face of social emergencies such as "COVID-19", relevant management departments should pay close attention to the psychological status of students, actively carry out mental health education and training, and improve students' psychological stress resistance; For emergencies, strengthen the popularization of corresponding prevention and control knowledge.

Acknowledgements. This work was supported by the Scientific Research Project of Xi'an Fanyi University (Project Number: 20A06). Horizontal Scientific Research Projects (Project Number: 21XYH156).

References

1. Zhang, D., Xu, H., Su, Z., Xu, Y.: Chinese comments sentiment classification based on word2vec and SVMperf. Expert Syst. Appl. **42**(4), 1857–1863 (2016)
2. Alim, M.M.F.: A sentiment analysis study for twitter using the various model of convolutional neural network. J. Phys. Conf. Ser. **1981**(4), 042136 (2021)
3. Mikolov, T., Chen, K., Corrado, G., Dean, J.: Efficient estimation of word representations in vector space. Comput. Sci. **1**, 47–61 (2013)
4. Li, H., Zhang, L.: Deep learning based text sentiment analysis during epidemic. Int. Core J. Eng. **7**(7), 467–472 (2021)
5. Toye, G., Sheppard, S., Chen, H.L.: Data sharing and reuse within the academic pathways study. Adv. Eng. Educ. **5**(2), 1–19 (2016)
6. Berka, P.: Sentiment analysis using rule-based and case-based reasoning. J. Intell. Inf. Syst. **55**(1), 51–66 (2020)

7. Conijn, R., Snijders, C., Kleingeld, A., Matzat, U.: Predicting student performance from LMS data: a comparison of 17 blended courses using Moodle LMS. IEEE Trans. Learn. Technol. **10**(1), 17–29 (2017)
8. Sharma, A., Ghose, U.: Sentimental analysis of twitter data with respect to general elections in India. Procedia Comput. Sci. **173**, 325–334 (2020)
9. Kazanidis, I., Pellas, N., Fotaris, P., Tsinakos, A.: Facebook and Moodle integration into instructional media design courses: a comparative analysis of students' learning experiences using the Community of Inquiry (CoI) model. Int. J. Hum.-Comput. Interact. **34**(10), 932–942 (2018)
10. Patil, S., Gupta, A.: A survey on understanding students learning experiences by mining social media data. Imp. J. Interdiscip. Res. **2**(8), 161–164 (2016)
11. Costello, K.L., Veinot, T.C.: A spectrum of approaches to health information interaction: from avoidance to verification. J. Assoc. Inf. Sci. Technol. **71**(8), 871–886 (2020)
12. Yada, S., Ikeda, K., Hoashi, K., Kageura, K.: A bootstrap method for automatic rule acquisition on sentiment cause extraction. In: 2017 IEEE International Conference on Data Mining Workshops (2017)
13. Shofiya, C., Abidi, S.: Sentiment analysis on COVID-19: related social distancing in Canada using twitter data. Int. J. Environ. Res. Public Health **18**(11), 5993 (2021)
14. Bayrakdar, S., Yucedag, I., Simsek, M., Dogru, I.A.: Semantic analysis on social networks: a survey. Int. J. Commun. Syst. **33**(11), e4424 (2020)

Data Analysis University Teaching Management Based on Association Rule Algorithm

Wenjun Yu[1,2(✉)]

[1] Yunnan Technology and Business University, Kunming, China
shirly201402@163.com
[2] Jose Rizal University, Metro Manila, Philippines

Abstract. With the development of social informatization, it has become an inevitable trend for university project management to use informatization tools to organize teaching decisions. Administrators can more efficiently arrange educational affairs for schools by analyzing teaching data. However, university data includes not only teacher management data but also student learning data. Faced with a huge amount of data and information, manual analysis methods will consume a lot of time and cost. And with the help of association rule algorithm, data centralized processing can be realized. This article mainly uses student learning performance information as a data set, and puts forward an analysis and application of data mining for teaching management (TM) in colleges and universities (CAU) around the basis of Apriori algorithm and algorithm ideas. In this application process, the aim is to use the Apriori algorithm to obtain some association rules (AR) in the student achievement database that can help us in teaching in CAU, so as to better arrange courses and class hours for students.

Keywords: Association rules · Apriori algorithm · Teaching management · Data analysis

1 Introduction

In the wake of my country's higher education, the number of students in CAU is increasing, and teaching resources are increasingly scarce. Under this teaching situation, it is particularly important to arrange scientific and reasonable teaching organization and management mechanisms for students and teachers. The teaching management system after the reform of the teaching model in CAU also needs to be optimized, and effective analysis and investigation of the teaching and informatization process of the teachers and students of the whole school can further integrate teaching resources and improve the quality of teaching.

There are many related literatures on the research of university teaching management data analysis based on association rule algorithm, and many scholars have achieved good results in this research direction. For example, a scholar believes that correlation analysis as a management tool can effectively improve the quality of university teaching and improve its management level. Actively promote relevance analysis in colleges and universities, and realize it by helping CAU to carry out a series of relevance analysis activities and issuing a large number of analysis statements and guidance opinions [1]. A research team studied high-efficiency management activities to find out the management methods suitable for school operations and the quantitative rules of management measures. The study found that the correlation analysis of university teaching management has obvious benefits in optimizing resource allocation, improving management models, and enhancing competitiveness [2]. Although the use of association rule algorithms to analyze university teaching management data has achieved good results, the analysis of teaching data in universities in my country still uses traditional analysis methods. In order to achieve efficient teaching management, we should boldly try association rule analysis methods.

This article expounds the concept of AR and several types of AR data mining (DM) algorithms, analyzes the DM process of college students' performance, and then uses the AR algorithm to analyze the degree of support and confidence of college academic performance to support the smooth development of teaching work.

2 Association Rule Algorithm and Its Data Mining Process

2.1 The Concept of AR

AR mining is to find the relative relevance and connection degree of each item in a large amount of information. The form of association rules is simple, people can easily understand its conceptual theory, and through data analysis, it is possible to discover important connections between data. Although association rules were originally derived from commercial shopping analysis, as time progressed, many researchers continued to develop and expand them. Association rule algorithm models have been fully applied in many fields, and rapid data processing has been achieved. At present, the application range of AR technology is very wide, such as the application in the communication field and the application in the business field [3, 4]. The association rule data mining algorithm can be classified from the following perspectives.

(1) Mining of basic AR

The basic rule of the association algorithm is the Apriori algorithm, and the most used layered processing technology based on the creation of data elements. Some scholars have discovered a DHP algorithm that can save data processing time. The algorithm decentralizes display and data set compression technology to reduce the number of candidate data sets and the amount of compression [5]. Some scholars also proposed the Partition algorithm based on the idea of hierarchical processing and concentrated analysis. This algorithm is a partitioning algorithm, which is more convenient when processing data sets [6]. The above several algorithms use hierarchical analysis as the data processing idea. The disadvantage of hierarchical analysis is that the data needs to be scaled multiple times into multiple candidate data sets. Due to its low processing efficiency, many scholars develop new data processing methods to make up for the levels. Analysis of deficiencies. For example, the FP growth algorithm does not need to scale the data frequently, and only scales the data once to achieve rapid processing and free up the storage space of the database. This is a time-saving and labor-saving processing method [7].

(2) Mining complex type association rules

Complex association rules are evolved on the basis of basic association rules. The concept of multi-step association governs the collective naming of genetic association rules and multi-step association rules. At the bottom of the rules, it is difficult to find multiple scattered data spaces with strong correlations. The use of higher-level implicit concepts is usually beneficial to find more representative rules, and at the same time, it can meet the needs of different users with different conceptual knowledge. For this reason, the multi-level concept of association rules is proposed. Generally speaking, Boolean association rules are very difficult to deal with continuous variables with attribute values, so a method to detect certain information in quantitative features is proposed [8].

(3) Parallel mining algorithm

The current association rules are mostly serial mining algorithms, and parallel mining algorithms are rarely used. The use of multiple processes for parallel data distribution has attracted people's attention, so the development of parallel mining algorithms has gradually started [9].

2.2 Application of AR Algorithm to Analyze the Data Mining Process of Student Performance

When applying the AR algorithm to analyze the student's performance, we must first determine the mining object, and the mining object in this article is the student's academic performance. After the object is determined, the relevant data is collected. This step is the key to data mining. Although the mining results are unpredictable, the mining goals should be predictable. The prerequisite of data mining is to have data. If there is no data, there is no need to carry out subsequent steps, and the data must be rich and complete. Collecting data requires a lot of work and time-consuming. It is necessary to collect students' basic information and performance data in various ways during the teaching practice. The second is to pre-process the data. Data preprocessing is to reduce data noise and filter out unstable data, and retain data useful for the analysis process, which is an important part of data retrieval. The data information from the data collection stage is incomplete. For example, the data set itself has missing information or data inconsistencies. The data with these problems must be eliminated. Therefore, the process of preprocessing the data is a necessary condition for DM. At the same time, data preprocessing can simply filter the data in advance to obtain a more complete original information database, which is conducive to the smooth development of the next level of DM. The last is the DM process. Choose the appropriate association rule DM algorithm or use DM tools to DA. The key to this step is to select the appropriate algorithm and evaluate the interest of the associated model. Figure 1 shows the DM process [10, 11].

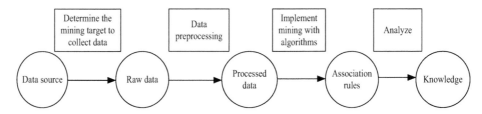

Fig. 1. Schematic diagram of the mining process

2.3 Association Rule Algorithm

Let $I = \{i_1, i_2, \cdots, i_m\}$ be the element set of m, and the data job $D = \{t_1, t_2, \cdots, t_n\}$ consists of a series of individual attributes, and each object t_i (i = 1, 2,..., n) corresponds to a subset of I. X and Y are non-empty subsets of I, and the percentage of transactions that contain X or Y in D is called the support of X or Y on D [12]. Association rules are denoted as $X \Rightarrow Y$, and $X \cap Y = \varnothing$. The conditions for the establishment of the association rule $X \Rightarrow Y$ under the traditional support-confidence framework are:

(1) It has a degree of supports. That is, at least s% of the data in data job D contains X, which is generally defined by the user, and it is an indicator that measures the statistical significance of all data defined by the association standard. It can be measured by the ratio of the number of all transactions in data x and D, denoted as support (X). Let D be the total number of transactions in the transaction database, namely:

$$support(X) = \| \{t \in D \mid x \subseteq t\} \| / \| D \| \tag{1}$$

(2) It has a confidence level of c. That is, at least c% of the data that contains X in D also includes Y. The confidence level c here is also generally specified by the user, and cooperation rules are used to measure the degree of trust. The measurement ratio of data X and Y includes the existing quantity X in database D, which is recorded as confidence (X = Y), namely:

$$\text{confidence } (X \Rightarrow Y) = support(X \cup Y)/support(X) \tag{2}$$

Among them, $X, Y \subseteq I, X \cap Y = \varnothing$.

3 Analysis and Research of Association Rule Algorithm in University Teaching Management Data

3.1 Research Significance

The analysis of TM data in CAU is the core work of teaching, and students' performance is one of the standards used to evaluate the quality of teaching, and it is also a measure of the effect of students' learning. The use of data mining technology to analyze the characteristics of student performance and find out hidden performance information provides an important basis for school principals to supervise teaching work, encourage teaching management, and further deepen teaching depth.

3.2   Research Methods

This paper takes the data related to the performance of economics students in a university as the research object, first uses the Apriori algorithm of association rules to mine the student performance data, then converts the continuous data into discrete data, and finally obtains the support and confidence of the data.

4   Application of Algorithm Based on Association Rules in Data Processing of University Teaching Management

4.1   Data Selection and Conversion

For this experiment, the credits of the subject, the student's student number, and the name have no effect on the data analysis. This experiment mainly focuses on the analysis of economics courses, so only the course scores in this area are kept in the data source, which is the courses are usually calculated on a hundred-point system.

Table 1. Select six subject performance data

	Advanced mathematics	University English	Microeconomics	Macroeconomics	Game theory	Behavioral economics
A	87	92	89	76	82	91
B	90	94	85	81	78	88
C	93	90	90	87	85	89
D	84	95	92	82	88	86
E	85	88	83	79	90	90

The experiment uses the Apriori algorithm in the association rule algorithm to conduct DM on the results of five students majoring in economics. The results of the six selected subjects are shown in Table 1. In order to transform the data into a suitable mining model, it is necessary to convert the scores of subsequent students into a Boolean data (0,1) pattern. Due to the different grading standards of different courses, the conversion method is adopted for the score data this time. If the score is greater than the average score (AS) of the subject, the score will be set to 1; if the score is less than the AS of the subject, the score will be set is 0. The AS for advanced mathematics is 85, the AS for college English is 90, the AS for microeconomics is 86, the AS for macroeconomics is 80, and the AS for game theory is 82. Behavioral economics the AS is 84 points. Therefore, student A's scores in macroeconomics, student B's microeconomics and game theory scores, student D's advanced mathematics scores, student E's college English, microeconomics, and macroeconomics scores should be converted to 0, all other scores converted to 1. The results after data conversion are shown in Table 2.

Table 2. Grade conversion

	Advanced mathematics	University English	Microeconomics	Macroeconomics	Game theory	Behavioral economics
A	1	1	1	0	1	1
B	1	1	0	1	0	1
C	1	1	1	1	1	1
D	0	1	1	1	1	1
E	1	0	0	0	1	1

4.2   Analysis of Data Mining Results

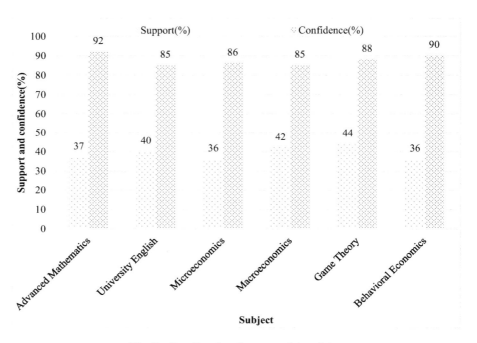

Fig. 2. Results of performance data mining

The subject grade type is set to discrete, and the AR Apriori model is imported, so that the minimum support is set to 35%, and the minimum confidence is set to 85%. The data mining results are shown in Fig. 2. The support and confidence of the five students' advanced mathematics scores were 37% and 92%, respectively, the support and confidence of the college English scores were 40% and 85%, and the support and confidence of the microeconomics scores were 36% and 86%, the support and confidence of macroeconomics performance are 42% and 85%, respectively, the support and confidence of game theory performance are 44% and 88%, the support and confidence of behavioral economics performance These are 36% and 90% respectively.

5   Conclusions

Traditional TM methods are no longer applicable to the college TM and teaching work, and they do not meet the needs of social development. The processing of student information and grade data in CAU is still in the initial data statistical analysis stage, and the hidden information of the data cannot be mined. Therefore, based on the AR algorithm, this paper filters the academic performance information of a certain college student, converts the continuous performance score into a discrete $(0,1)$ form, and uses the AR Apriori model to analyze the confidence level of the student's performance. Management staff provide reference basis for teaching decision-making.

References

1. Chiclana, F., Kumar, R., Mittal, M., Khari, M., Chatterjee, J.M., Baik, S.W.: ARM–AMO: an efficient association rule mining algorithm based on animal migration optimization. Knowl.-Based Syst. **154**, 68–80 (2018)
2. Kabir, M., Xu, S., Kang, B.H., et al.: A new multiple seeds based genetic algorithm for discovering a set of interesting Boolean association rules. Expert Syst. Appl. **74**, 55–69 (2017)
3. Can, U., Alatas, B.: Automatic mining of quantitative association rules with gravitational search algorithm. Int. J. Software Eng. Knowl. Eng. **27**(3), 343–372 (2017)
4. Liu, Y.: Research on association rules mining algorithm based on large data. Inf. Technol. **40**(1), 192–194 (2017)
5. Cai, Q.: Cause analysis of traffic accidents on urban roads based on an improved association rule mining algorithm. IEEE Access **8**, 75607–75615 (2020)
6. Lu, P.H., Keng, J.L., Tsai, F.M., et al.: An Apriori algorithm-based association rule analysis to identify acupoint combinations for treating diabetic gastroparesis. Evid.-Based Complement. Altern. Med. **2021**(17), 1–9 (2021)
7. Subbulakshmi, B., Deisy, C.: An improved incremental algorithm for mining weighted class-association rules. Int. J. Bus. Intell. Data Min. **13**(1–3), 291–308 (2018)
8. Qin, L., Zhang, Y., Kang, H., et al.: Mining association rules between stroke risk factors based on the Apriori algorithm. Technol. Health Care **25**(S1), 197–205 (2017)
9. Nguyen, L.T.T., Vo, B., Nguyen, L.T.T., Fournier-Viger, P., Selamat, A.: ETARM: an efficient top-k association rule mining algorithm. Appl. Intell. **48**(5), 1148–1160 (2017). https://doi.org/10.1007/s10489-017-1047-4
10. Talebi, B., Dehkordi, M.N.: Sensitive association rules hiding using electromagnetic field optimization algorithm. Expert Syst. Appl. **114**, 155–172 (2018)
11. Chunjie, L., Wang, Y.: Survey and analysis on the students' satisfaction with undergraduate teaching management in local agricultural colleges and universities: a case study of Tianjin Agricultural University. Asian J. Agric. Res. **12**(08), 73–75+86 (2020)
12. Li, X., Wang, Y., Li, D.: Medical data stream distribution pattern association rule mining algorithm based on density estimation. IEEE Access **7**, 141319–141329 (2019)

Optimization and Application of Particle Swarm Algorithm in Software Engineering

Jing Chen[✉]

Yunnan Land and Resources Vocational College, Kunming, Yunnan, China
chenjing84@woomail.cn

Abstract. With the extensive development and application of software technology tools and software technology collaboration environments in business, the current focus of software technology research has shifted from the scale and scale of software technology to how to analyze and improve it. The purpose of this work is to study the optimization and application of swarm particle optimization in software technology. A new fitness construction method is proposed and implemented to automatically create multiple test cases based on the MPRPSO algorithm. Creating a larger test set improves the efficiency of creating test cases to a certain extent. The experimental results show that the improved MPRPSO algorithm has about 10 iterations, which is better than the comparison algorithm. The advanced MPRPSO is not only more efficient in creating software technical tests, but also has better algorithm performance and is more suitable for various system-based software. The test method determines the case algorithm.

Keywords: Particle swarm algorithm · Software engineering · Optimization and application · MPRPSO algorithm

1 Introduction

For software companies, the software process is the most complex and important business process of the entire company. The most important thing to improve the business process of each software company is to improve the software [1, 2]. In recent years, people have realized that in order to develop high-performance, high-quality, and low-cost software, one must focus on improving the software process and using software technology and quality control methods on a global scale [3]. As for the situation in China, the reason for the decline of China's software industry is not the decline of technology, but the decline of software management [4, 5]. Promote the development of my country's software industry, enhance the innovation capability and global competitiveness of the information industry, promote the transformation and upgrading of traditional enterprises, and further accelerate technological development with language, speed and health, and software optimization. Technology plays an important role in managing all software businesses [6, 7].

Software and more (SE) Software developers use search-based optimization techniques to solve SE problems with multiple conflicting sites. These technologies usually use CPU-intensive evolutionary algorithms to detect changes in the competitor's solution population. Another approach proposed by Chen J is to start with a larger team

© The Author(s), under exclusive license to Springer Nature Switzerland AG 2022
V. Sugumaran et al. (Eds.): ICMMIA 2022, LNDECT 138, pp. 609–615, 2022.
https://doi.org/10.1007/978-3-031-05484-6_77

and then outline the best solution. They call this method "SWAY", short for "Sample Method". In the research of many software technology models, Sway is very easy to implement. We found that this sampling method is competitive with the corresponding image evolution algorithm, and the computational cost is very low. Considering the flexibility and efficiency of Sway, they recommend this method as the basic method of software-based software models, especially for models with very slow execution speeds [8]. Henrique JP must prove that the EMSO software is as good as the class-room course and has not been understood (additional subjects) and the tenth level of process engineering. Different case studies have been proposed, including different programming concepts, and different methods have been used to solve them in EMSO [9]. Research on the optimization and application of swarm particle algorithm in software technology has practical application value.

This article first understands the development of software technology, and optimizes and analyzes the software technology based on the research of software technology at home and abroad. The existing swarm particle algorithm is optimized, and a multi-role swarm optimization process is proposed. The algorithm has received intensive training. Compare and analyze the progress of swarm particle algorithm, such as the advantages and disadvantages of particle swarm algorithm, and further evaluate the effect of particles. According to the different effects of particles, based on the modification of multi-particle effects, the particles are divided into three groups of optimal particle swarm algorithm, and the MPRPSO algorithm is proposed in the experiment. The MPRPSO algorithm has a certain degree of reliability.

2 Research on Optimization and Application of Particle Swarm Algorithm in Software Engineering

2.1 Software Engineering

In recent years, simulation software technology has also been frequently studied. Most widely used for advanced and older scripts. The old design listed first here refers to the original technology when the code was last written [10]. Advanced prototyping refers to a series of repetitive processes that develop scripts into products. The most important and successful development in this area is the introduction of many primitive languages that can be used to define a possible system model without the need to define a specific system algorithm. Use conversion modeling technology and recycling code technology. So far, software recycling technology has been one of the most important research topics in software technology research projects [11]. Although some practical results have been achieved in some specific fields, software reuse technology is still scientific. The relationship between rare circumstances and legal and economic issues has not yet been fully resolved. Therefore, software reuse technology has not really achieved success [12].

2.2 Particle Swarm Algorithm

The PSO algorithm is an excellent algorithm that compares the trajectories of birds looking for food in nature to reach the best possible solution. In the process of searching for food, not all birds know how to find food and how far away from it. Close to food sources and areas, and through collaboration, constantly adjust and adjust the position and direction of the aircraft until the food is found. This cognitive behavior provides a natural basis for people to learn the PSO algorithm. PSO-based algorithms are most suitable for social research to enable particles to complete their local training. The algorithm compares particles with no size or population size with a group of birds, similar to the biological behavior of a group of birds seeking food, exchanging information and constantly adjusting their conditions to find always better solutions.

Each atom in the particle swarm is called a particle in the algorithm. For each particle in it, it is a possible solution in the solution space. In the best case, particles are based on their own optimization experience and information shared between groups.

The swarm particle algorithm is simple, easy to implement, few parameters, and strong search capabilities. From its design point of view, it has successfully achieved many optimizations and implementations, and can be applied to many optimization problems. What is important is the combination of the optimal position and condition of each object in the algorithm. The current situation and situation can be optimized and aggregated, and the current situation and the best situation can be determined. The best position in human history always leads him in a direction that also encourages his progress.

The algorithm is as follows: first cluster the particles, then add the starting value of each endpoint and the length value of each particle, and then calculate the new value of the velocity and position of each particle according to the optimization. Finally, we judge whether the algorithm meets the final requirements.

The swarm particle algorithm itself can only solve the problems in optimization applications, so it cannot solve many performance problems. Therefore, in order to solve the problem of multi-task optimization, on the basis of particle optimization swarm, combined with the characteristics of multi-task optimization, an advanced particle swarm algorithm, that is, a multi-functional optimization algorithm, is created. The swarm particle mass algorithm has been successfully implemented in many multi-modal sites.

3   Investigation and Research on Optimization and Application of Particle Swarm Algorithm in Software Engineering

3.1   The Structure of the Test Environment in Software Engineering

The construction of the test area is an important function of the design case method and the basic link of the automatic generation of use cases. Its main functions include: system code static analysis, system application, input and output, test drive generation and parameter performance, etc. The program application is used to monitor the execution of the test program. It is important to put some relevant information that does not affect the program execution and execution process into the test program, and is only used to obtain the execution plan. Appropriate inspectors obtain the area information of the test method corresponding to the test system through these application details, and use this as the basis for evaluating the quality of the system.

3.2   Application Test

In the Windows operating system, the interval calculation part of the calculator pro-gram is composed of multiple branch paths, and the program logic is more compli-cated. The actual experiment of automatically generating multi-path test cases is to use this part of the program as test cases and discover all test cases.

Use the iteration and execution time of all test cases covering the entire target industry as an experiment. Since year1 = year2 and month1 = month are important tests, it is difficult to generate test cases, and it is usually necessary to unilaterally add test cases. The test data generation and application functions in this document are based on total path coverage rather than specific test paths. For the sake of clarity of the experimental results, the corresponding branch paths are ignored.

3.3   Particle Swarm Optimization Algorithm of Multi-particle Role Synergy (MPRPSO)

This paper proposes a novel particle swarm hybrid algorithm optimization based on the synchronization of multiple particle effects in a swarm. Advanced algorithms separate the particles in the group and isolate different effects, combine different particle effects with their respective detection modes, and collaborate in the development of solution areas. This obvious division of labor can avoid the risk of swarm particle optimization falling into the optimal area to a certain extent.

Aiming at the feature that the particle swarm algorithm is easy to integrate into the optimal region, the concept of PP based on turbulence detection is produced. At present, the confusion model is usually mainly used to distinguish between science and self. The mathematical model of the Logistic world map is shown in formula 1:

$$y_{n+1} = \mu y_n(1 - y_n); n = 0, 1, 2, \ldots, 0 < y_0 < 1 \tag{1}$$

The number of repeated k of the finest particle k is used as the trigger point for PP information injection, and gbt is the optimal position for the tth release. This process is performed as shown in Eq. 2:

$$\begin{cases} k = k+1, gb_t = gb_{t-1} \\ k = 0, gb_t \neq gb_{t-1} \end{cases} \tag{2}$$

When the number of repetitions of the global optimal evolution k reaches the predetermined end, we can assume that the population has lost its potential. The increased variability of EP helps the algorithm eliminate local optima. From the analysis of the complexity of the algorithm, the time complexity of particle formation is O (TmaxND), and the complexity of the algorithm mainly depends on the size of the calculation problem and the proportion of each particle.

4 Investigation and Analysis of Particle Swarm Optimization and Application in Software Engineering

4.1 Improved Particle Swarm Algorithm's Total Path Coverage in Software Engineering

The experimental parameters are set as follows: Tmax is 100, the optimal fitness threshold of the population β is 0.5, and the optimal attenuation coefficient α is 0.5. The rest of the parameter settings are the same as before, and each test is run 10 times. Table 1 shows the comparison of the improved MPRPSO algorithm with the RAPSO algorithm and the HC-MARPSO algorithm on the total coverage path.

Table 1. Comparison of coverage rate of each algorithm (/%)

Population size	Improve MPRPSO		RAPSO		HC-MARPSO	
	Best	Mean	Best	Mean	Best	Mean
10	100	99.0	100	94.6	100	91.8
20	100	99.2	100	91.4	100	95.4
30	100	98.7	100	94.6	100	93.3
40	100	98.5	100	93.0	100	92.5
50	100	99.6	100	93.2	100	91.3

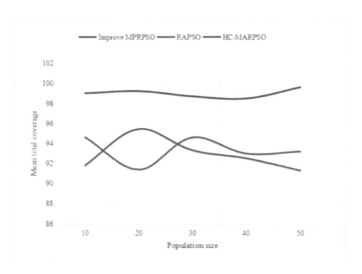

Fig. 1. Comparison of coverage rate of each algorithm (/%)

The total path coverage of each algorithm in software engineering is shown in Fig. 1. It can be found that the best output setting test case, in which all the experimental algorithms can reach the total local area, but the average local data system is different. When the granularity is low, the advanced MPRPSO algorithm occasionally cannot be integrated into the optimal solution, but when the granularity is large (≥ 100), the generated test package can cover the entire process. Although the comparison algorithm can achieve a complete and unstable path, the average area size is between 90% and 100%. The results show that the advanced MPRPSO algorithm has certain advantages in the production quality set test. The average number of iterations of the algorithm can be seen, the average number of iterations of the comparison algorithm is about 20, while the improved MPRPSO algorithm is about 10, which is better than the comparison algorithm and shows better use case generation performance.

4.2 The Execution Time of the Improved Particle Swarm Algorithm in Software Engineering

The average execution time of each algorithm is shown in Fig. 2.

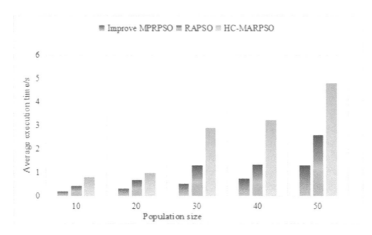

Fig. 2. Algorithm execution time

It can be seen that the improved MPRPSO has some advantages over the average running time. As the population increases, the average execution time of advanced MPRPSO and comparison algorithms will increase. Under all experimental population sizes, the execution time of advanced MPRPSO is shorter than other comparison algorithms, and as the population increases, the execution time advantage becomes more and more obvious, which shows that advanced MPRPSO is not only performed at the beginning of the test box. Excellent accuracy and excellent algorithm performance, well as a set of multiple test cases.

5   Conclusions

In today's enterprise software development environment, software engineering tools and software engineering collaboration environments have been extensively developed and implemented. The software industry has developed into a pillar industry of my country's information industry. The government, banks and other large institutions vigorously promote IT business, and the complexity and volume of application software development continue to increase. The importance of introducing software engineering supervision into software engineering projects is self-evident. This paper improves the particle swarm algorithm and proposes a multi-particle collaborative particle swarm optimization algorithm. Research the optimization and realization of algorithms in software engineering. The improved MPRPSO has some advantages over average running time and overall path coverage.

References

1. Delice, Y., Kızılkaya Aydoğan, E., Özcan, U., İlkay, M.S.: A modified particle swarm optimization algorithm to mixed-model two-sided assembly line balancing. J. Intell. Manuf. **28**(1), 23–36 (2017)
2. Langazane, S.N., Saha, A.K.: Effects of particle swarm optimization and genetic algorithm control parameters on overcurrent relay selectivity and speed. IEEE Access **10**, 4550–4567 (2022)
3. Toritani, S., Shauri, R.L.A., Nonami, K., Fujiwara, D.: Numerical solution using nonlinear least-squares method for inverse kinematics calculation of redundant manipulators. J. Robot. Mechatron. **24**(2), 363–371 (2012)
4. Balicki, J.: Many-objective quantum-inspired particle swarm optimization algorithm for placement of virtual machines in smart computing cloud. Entropy **24**(1), 58 (2022)
5. Zhu, Q., Lin, Q., Chen, W., et al.: An external archive-guided multiobjective particle swarm optimization algorithm. IEEE Trans. Cybern. **47**(9), 2794–2808 (2017)
6. Farhang, Y., Afroozeh, A., Jahanbin, K.: Improved particle swarm optimization algorithm in k-means. Autom. Electr. Power Syst. **538–541**(7), 2658–2661 (2017)
7. Shami, T.M., El-Saleh, A.A., Alswaitti, M., et al.: Particle swarm optimization: a comprehensive survey. IEEE Access **10**, 10031–10061 (2022)
8. Chen, J., Nair, V., Krishna, R., et al.: "Sampling" as a baseline optimizer for search-based software engineering. IEEE Trans. Software Eng. **99**, 1 (2018)
9. Henrique, J.P., Sousa, R.D., Secchi, A.R., et al.: Optimization of chemical engineering problems with EMSO software. Comput. Appl. Eng. Educ. **26**(1), 141–161 (2018)
10. Yuan, F., et al.: Optimization design of oil-immersed iron core reactor based on the particle swarm algorithm and thermal network model. Math. Prob. Eng. **4**, 1–14 (2021)
11. Prajapati, A., Chhabra, J.K.: A particle swarm optimization-based heuristic for software module clustering problem. Arab. J. Sci. Eng. **43**(12), 7083–7094 (2017). https://doi.org/10.1007/s13369-017-2989-x
12. Yuan, F., et al.: Optimization design of oil-immersed iron core reactor based on the particle swarm algorithm and thermal network model. Math. Prob. Eng. **4**, 1–14 (2021)

The Application of Genetic Algorithm in University Mobile Office System Design

Yefei Lei[✉]

Tianjin Vocational Institute, Tianjin, China
leiyefei88@163.com

Abstract. With the development of information technology, the information management system has penetrated into the management work of all walks of life, and has played an unparalleled advantage in management practice, which has effectively improved the efficiency of management work. However, in the field of university office, China's university office management is still in the traditional management mode. Not only has the time period been relatively long, but the efficiency of management is also relatively low. In the face of the increasing demand for business volume, the traditional management mode has been seriously inconsistent with development. In this regard, this paper studies the application of genetic algorithms in the design of college mobile office systems.

Keywords: Genetic algorithm · University · Mobile · Office system · System design

1 Introduction

The application of a new technology may profoundly change the way people live and work. As a new application technology, mobile office needs to be gradually improved under the continuous exploration, development and hard work of basic telecom operators, industry application staff and technical personnel. In recent years, mobile office has been gradually put into application in some unit management fields, but the overall architecture design of the platform and the development of business applications still need to be standardized [1]. How users can quickly access the mobile communication system, build a basic platform, and complete business application development has become the primary problem to be solved in mobile office. Therefore, a standardized and reasonable infrastructure design plan also has practical application significance. The infrastructure introduced in this article fully considers the system's advanced nature, openness, data security, and system consistency, compatibility and scalability, and can meet current application needs [1].

2 Genetic Algorithm

Using genetic algorithm to solve the maximum value of multimodal function is one of my coursework [1]. The whole content is as follows:

© The Author(s), under exclusive license to Springer Nature Switzerland AG 2022
V. Sugumaran et al. (Eds.): ICMMIA 2022, LNDECT 138, pp. 616–624, 2022.
https://doi.org/10.1007/978-3-031-05484-6_78

2.1 Problem Description

Program the genetic algorithm and solve the maximum value of the multimodal function. The expression of the multimodal function is as follows:

$$\max f(x, y) = 21.5 + x\sin(4\pi x) + y\sin(20\pi y) \tag{1}$$

$$\text{s.t.} -3.0 \leq x \leq 12.1 \tag{2}$$

$$4.1 \leq y \leq 5.8 \tag{3}$$

The image of the function made with MATLAB is as follows (see Fig. 1):

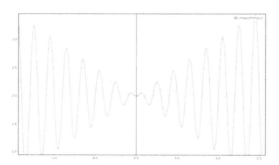

Fig. 1. Function image made by MATLAB

2.2 Overview of Genetic Algorithm

Genetic Algorithm (GA), also known as evolutionary algorithm. Genetic algorithm is a heuristic search algorithm that is inspired by Darwin's theory of evolution and draws on the process of biological evolution. Its main feature is to directly operate on structural objects. Therefore, unlike other algorithms for solving optimal solutions, genetic algorithms do not have derivation and limitation of function continuity [2]. Probabilistic optimization methods are used, and no definite rules are needed. It can automatically obtain and guide the optimized search space, and adjust the search direction adaptively.

2.3 The Flow of the Algorithm

Through the above explanation, how to simulate natural evolution to find the optimal solution of the multimodal function in the problem has been relatively clear [2]. Here I will list the main steps of genetic algorithm and analyze them one by one:

Step 1: randomly generate a population as the first generation solution of the problem (usually, the first generation solution may be quite different from the optimal solution, which is tolerable, as long as the first generation solution is randomly generated to ensure the diversity of individual genes);

Step 2: Find a suitable coding scheme to encode individuals in the population, you can choose common coding schemes such as floating-point number coding or binary coding (it needs to be pointed out that different coding schemes directly affect the realization of subsequent genetic operators detail);

Step 3: Taking the function value of the multimodal function as the fitness of the individual, calculate the fitness of each individual in the population (the calculated fitness will provide a basis for subsequent individual selection);

Step 4: Choose the parent and the mother to participate in reproduction according to the level of fitness. The principle of selection is that individuals with higher fitness are more likely to be selected (in order to continuously eliminate individuals with low fitness);

Step 5: Perform genetic operations on the selected father and mother, that is, copy the genes of the father and mother, and use crossover, mutation and other operators to produce offspring (on the basis of retaining excellent genes to a greater extent, Mutation increases the diversity of genes, thereby increasing the probability of finding the optimal solution);

Step 6: Determine whether to continue to execute the algorithm according to certain criteria, or find out all the children with the highest fitness to return as the solution and end the program (the criteria for judgment can be the set threshold of the solution, the specified number of iterations, etc.).

2.4   Calculate Fitness

The fitness function is also called the evaluation function, which is usually used to distinguish the quality of individuals in a group. Highly adaptable, that is, excellent individuals have a greater chance of participating in reproduction and inheriting their own genes. Generally, the fitness function is determined according to the objective function, and sometimes the value of the objective function is directly used as the fitness [3]. Here, considering the multimodal function to be solved, the peak distribution is dense and the peak diameter is very narrow, which is not conducive to the convergence of the genetic algorithm. Therefore, this article does not directly use the multimodal function value as the fitness, but uses the logarithmic function to increase the peak function is smoothed, and the smoothed function value is used as the objective function. The specific method is to calculate the logarithm of the multimodal function twice [3]. Therefore, the relationship between the multimodal function and fitness can be expressed as follows:

$$f(x, y) = 21.5 + x\sin(4\pi x) + y\sin(20\pi y) \tag{4}$$

$$\text{fitness} = \log(\log(f(x, y))) \tag{5}$$

The fitness function image made with MATLAB is shown in Fig. 2:

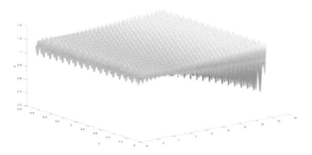

Fig. 2. MATLAB makes a function image of fitness

2.5   Selection Operator

The selection algorithm in this article uses the very commonly used "roulette algorithm", and the principle of the betting algorithm is very simple and clear [4]. When creating a game, we sum the fitness of all individuals in the population, and the result obtained may be called the total fitness. Then, divide the fitness of each individual by the total fitness, and then accumulate the obtained quotients one by one [4]. After the accumulation is completed, the total is 1. The following Fig. 3 can show the principle of betting more vividly:

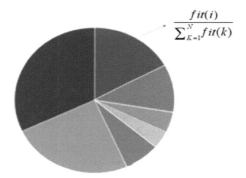

$$\frac{fit(i)}{\sum_{K=1}^{N} fit(k)}$$

Fig. 3. Principles of selection operator betting

Going back to the selection operator, the selection operator needs the market as the basis. When it runs, it will generate a random number from 0 to 1, and then find the interval in which the number is located in the market. The individual corresponding to this interval is the bet [4]. The selected individual. Therefore, individuals with higher fitness are more likely to be selected, which is reasonable.

2.6   Crossover Operator

The crossover operation of genetic algorithm is essentially to exchange part of the parent and mother genes in a certain way. Common crossover operators include single-point crossover, two-point crossover, multi-point crossover, uniform crossover, and arithmetic crossover [5]. In this paper, the two-point crossover method is selected. The realization process is neither complicated, but also has good randomness. This method can be illustrated by the following Fig. 4:

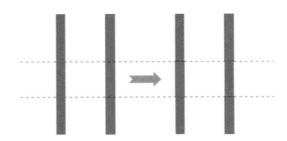

Fig. 4. Operators randomly generated by two intersections

2.7   Mutation Operator

In nature, genetic mutation can increase the diversity of individuals, which for genetic algorithms increases the randomness of individuals and can increase the probability of finding the optimal solution. The operation of the mutation operator used in this article is to randomly select a certain bit of the gene for inversion.

2.8   Reproduction Function and Evolution Function

The main operators of the genetic algorithm have been analyzed above, the next thing to do is to integrate these operators according to the flow of the genetic algorithm to realize the algorithm function [5]. In this article, two key functions are involved, namely reproduction function and evolution function. The reproductive function includes gene duplication, crossover and mutation. At the meantime, this paper also adopts the offspring competition strategy, that is, the two offspring individuals produced by the parent only retain the highest fitness.

3   In the Design of the University Mobile Office System Based on Genetic Algorithm Logic

The logical architecture of the system consists of a three-tier system structure, namely the business access layer, the information processing layer, and the user interface layer, as shown in Fig. 5.

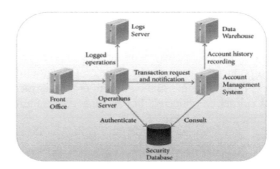

Fig. 5. Logical layer of mobile office system

3.1   Business Access Layer

Mainly responsible for connection and communication with basic telecom operators' systems. Including the protocol communication with China Mobile and China Unicom's short message gateway, as well as the connection of the following new services, such as fixed network short message, MMS and other system platforms. The main consideration at the access layer is the issue of scalability. Since the currently used CMPP and SGIP protocols are all proprietary protocols that are simplified and formulated on the basis of the international standard protocol SMPP, the core support system must be implemented to face different services [6]. The interface has good compatibility and requires that the internal protocol of the core system should be a relatively advanced and mature protocol standard. Therefore, it is recommended that the system adopt the simplified SMPP protocol for internal data communication to ensure better compatibility at the service access layer of the system, and complete the conversion of different external connection protocols to the internal protocol of the system on this layer. In the initial stage at this level, you can first consider connecting with China Mobile's short message gateway through the CMPP protocol, and communicating with China Unicom's short message gateway through the SGIP protocol [6]. At the meantime, the protocol extension interface entity module for the business connection is reserved in this layer to facilitate the smooth and seamless upgrade of the protocol module at this layer in the future. The business access layer provides the connection and protocol conversion of the system to different short message gateways, so that users can make full use of various existing service systems [7].

(1) CMPP protocol processing: provide the connection between the system and the mobile short message gateway, and perform the conversion between the CMPP protocol and the SSTP protocol.
(2) SGIP protocol processing: provide the connection between the system and China Unicom's short message gateway, and carry out the conversion between SGIP protocol and SSTP protocol [8].
(3) SMGP protocol processing: provide the connection of the system to the Netcom (telecommunications) short message gateway, and carry out the conversion between the SMGP protocol and the SSTP protocol.

(4) Protocol extension entity: At present, the realization of short message is mainly realized through the short message center of China Mobile and China Unicom. With the rapid development of mobile data, there will be more data operators and new services [7]. They may different protocol processing mechanisms are adopted, so providing system extension entities will provide good service and support for system upgrades.

3.2   Information Processing Layer

Mainly realize the core data processing and centralized management of the system. This layer has nothing to do with the specific connection business content, and the processing related to the proprietary connection business is completed by the corresponding interface layer. This layer is centered on store and forward work, which can support interactive, push, and customized information services, and At the meantime complete the authentication, security and encryption of the user side and the filtering of data information [8]. This layer also includes billing and management for core store and forward data. The billing module can perform billing and corresponding billing statistics for operators and user terminals based on different billing strategies and billing types (monthly or billing, etc.). The management module mainly realizes connection user management, internal configuration maintenance management, data statistical report management, and maintenance log management. This layer should mainly focus on the stability and efficiency of data [9]. Therefore, it is recommended that the bottom layer of the system adopt mature C++ voice for development work, because in the development process of large-scale telecommunications systems, C++ is the most efficient and stable underlying support language, and it also has a good Portability. The information processing layer is the core part of the system. It provides different processing mechanisms for the different needs of users [9].

(1) Message storage and forwarding: All information entities are stored and forwarded, so that the system can manage information, and At the meantime, it can provide mobile data service operators and users with charging basis and query records. Information can be deleted from the system after a certain period of time [10].

(2) Business processing: different designs are used to satisfy different user requests. User requests are generally divided into three categories, corresponding to three processes: interactive business processing, customized business processing, and push business processing.

(3) Management control: user management, configuration management, statistical management and log management.

(4) Billing processing: A two-level billing mechanism is adopted to complete the billing of telecommunications and the billing of users.

(5) Information content filtering: to block illegal information that may have a negative impact on society or disrupt public order.

(6) User authentication and security encryption: In order to prevent unauthorized users from accessing and malicious interference; the system should adopt a safe transmission mechanism and authentication method to ensure the stable operation of the system and the security of information [10].

3.3   User Interface Layer

Mainly face different access users for classification processing, distribution and flow control. The access users faced are mainly enterprise users, industry users, and ISP/ICP users [11]. In addition, interface extension entities are reserved to facilitate access for users with special protocol requirements. The needs of users are constantly increasing and changing, and users are also developing towards diversity, as shown in Fig. 6. The refinement of social division of labor makes any set of systems unable to meet the needs of all users. Therefore, the system is designed to take into account the general needs of users and develop client software that meets the needs of most user groups. At the meantime, it is aimed at different user groups [11]. The special needs of different users provide development kits for secondary development to achieve individualized needs.

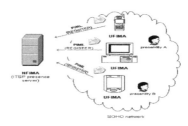

Fig. 6. Hierarchical structure diagram of system logical architecture

(1) Industry users receive and distribute; industry users generally have their own industry's special requirements and cannot develop systems for them, so try to choose the method of providing API functions frequently used in the industry to provide users with an open interface.
(2) Enterprise user access and distribution: The C/S structure is adopted to provide enterprise users with client software to achieve access.
(3) ICP/ISP access and distribution: Provide a way to embed plug-ins in the webpage to make ICP/ISP access to people to ensure that it can serve more individual users.

4   Summary

Based on this research background, this paper analyzes the relevant theories of workflow in detail based on the investigation and analysis of the office management process of colleges and universities, and practically applies the workflow theory to the office process of the system. And in-depth analysis and design of the university office and mobile office system, and use the Web framework to realize the function of the system, through the design of this article, the efficiency of university office management has been effectively improved. Through the analysis and introduction of the application of the legacy algorithm, it is applied to the design of the university mobile office system, which greatly improves the efficiency of the university office.

References

1. Khadir, K., Guermouche, N., Guittoum, A., Monteil, T.: A genetic algorithm-based approach for fluctuating QoS aware selection of IoT services. IEEE Access **10**, 17946–17965 (2022)
2. Emami, S.A., Roudbari, A.: Multimodel ELM-based identification of an aircraft dynamics in the entire flight envelope. IEEE Trans. Aerosp. Electron. Syst. **55**(5), 2181–2194 (2019)
3. Wang, Q., Zhao, H.L., Dai, Z.Y.: Genetic algorithm optimization of linear induction accelerator transport magnetic field. High Power Laser Part. Beam **9**(4), 313–317 (2019)
4. Cha, Y.J., Agrawal, A.K., Kim, Y., Raich, A.M.: Multi-objective genetic algorithms for cost-effective distributions of actuators and sensors in large structures. Expert Syst. Appl. **39**(9), 7822–7833 (2012)
5. Raghavan, A., Maan, P., Shenoy, K.A.: Optimization of Day-Ahead energy storage system scheduling in microgrid using genetic algorithm and particle swarm optimization. IEEE Access **8**, 173068–173078 (2020)
6. Zhou, K.R.: Design and implementation of a unified portal in university information system integration. Wirel. Internet Technol. **16**, 62–65 (2019)
7. Nagowah, S.D., Ben Sta, H., Gobin-Rahimbux, B.: A systematic literature review on semantic models for IoT-enabled smart campus. Appl. Ontol. **16**(1), 27–53 (2021)
8. Beyer, J., Yang, Y., Pfister, H.: Visualization design sprints for online and on-campus courses. IEEE Comput. Graph. Appl. **41**(6), 37–47 (2021)
9. Yao, G., Qi, H.Y.: Design and implementation of official document statistics automation in university office automation system. J. Ningbo Vocat. Tech. Coll. **4**(5), 26–30 (2019)
10. Ceccarini, C., Mirri, S., Salomoni, P., Prandi, C.: On exploiting data visualization and IoT for increasing sustainability and safety in a smart campus. Mob. Networks Appl. **26**(5), 2066–2075 (2021). https://doi.org/10.1007/s11036-021-01742-4
11. Basheer, S., Alluhaidan, A.S., Mathew, R.M.: A secured smart automation system for computer labs in engineering colleges using the internet of things. Comput. Appl. Eng. Educ. **29**(2), 339–349 (2021)

Design and Implementation of an Intelligent Education Management Platform Based on Face Recognition

Yuanyuan Li[✉]

Sanya Aviation and Tourism College, Sanya, Hainan, China
yuanyuan86@ishuobo.cn

Abstract. The development of artificial intelligence has greatly changed people's work and lifestyle. Intelligence has appeared in various fields, which has brought great convenience to people. Education is considered to be one of the most important topics for a family. With the family's emphasis on education, the number of students is increasing year by year, and there is a serious imbalance between the number of teachers and the number of students. Teachers cannot guarantee the quality of education while taking into account the classroom status of all students, assisted by artificial intelligence. Education helps to realize the fairness of education. In recent years, face recognition technology has achieved remarkable results in security, attendance and other fields, providing ideas for intelligent education. This paper studies the intelligent education management platform based on face recognition design and implementation.

Keywords: Face · Face recognition · Intelligent education · Management platform

1 Introduction

With the gradual popularization of higher education in China, the number of students in colleges and universities is increasing. The traditional education management model often has problems such as waste of resources, easy errors, and easy omissions due to duplication of work and low efficiency. Education is an important place for students to live after school. On the one hand, the density of middle school students is high, which is easy to produce education group effect; on the other hand, students are in adolescence and belong to an unstable group [1]. Traditional educational management concepts and methods cannot quickly adapt to the needs of digital smart campus development, and many problems cannot be obtained in a timely and effective manner [1]. Therefore, it is necessary to design and develop an intelligent education management platform based on face recognition technology.

2 The Key Technology of Intelligent Education Management Platform

2.1 Principles of Face Recognition

Principles of Face Recognition Technology

The intelligent supervision system is based on the face recognition technology in the field of artificial intelligence, and the system technical architecture is divided into three levels from bottom to top: the underlying API, the interface API, and the application, as shown in Fig. 1.

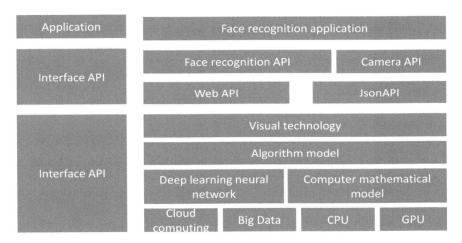

Fig. 1. Face recognition technology framework

The underlying API is the underlying implementation of face recognition technology, involving software and hardware, especially various algorithm model technologies [2]. The performance of CPU and GPU determines the speed of face recognition. When the performance of CPU and GPU is stronger, the speed of face recognition is faster. Cloud computing is a kind of distributed computing that uses large-scale computer clusters to complete data processing in a short time and provide powerful computing services [1]. Through the establishment of a mathematical model, the deep learning algorithm is continuously optimized, combined with visual technology, so that the face recognition technology reaches the commercial standard.

Interface API realizes the encapsulation of low-level APIs, including some commonly used camera call APIs, network APIs and JsonAPIs. It provides high-level calls to face recognition APIs and is the core functional interface of the system [2].

2.2 Face Detection Algorithm

Literature [2] proposed a new object detection method named "YOLO" (it is an object recognition and positioning algorithm based on deep neural network, its biggest feature is fast running speed and can be used in real-time system). The method simplifies the classification problem of object detection and can locate the input image category through only a neural network [3]. The face recognition algorithm used in this study is improved based on the above-mentioned documents, and its design idea is: First, extract HOG (Histogram of Gradient Directions) from the collected face images, and at the same time, use the improved "YO-LO" convolutional neural network algorithm to extract face image features; secondly, fuse the extracted face features with HOG and classification detection; again, determine the facial features.

2.2.1   HOG Extraction of Facial Image Features

The main function of the HOG algorithm is to quickly extract the contour information of the image, and effectively reduce the degree of influence of external conditions such as color and light in the process of object detection, especially human detection. First, convert the face image into a JPG file with a pixel of 448×448. And then calculate the gradient value of each pixel in the horizontal and vertical directions of the collected image through the convolution kernel [4], the specific formula is as follows:

$$g_x(x, y) = h(x + 1, y) - h(x - 1, y) \tag{1}$$

$$g_x(x, y) = h(x, y + 1) - h(x, y) \tag{2}$$

Among them, $g_x(x, y)$ represents the direction gradient of the horizontal coordinate point of the image; $g_x(x, y)$ represents the direction gradient of the vertical coordinate point of the image; is the pixel value of the current coordinate point on the image. Through this formula, the current the direction and gradient of the pixel.

Second, according to the direction and gradient mode of each pixel calculated in the first step, the entire picture is converted to HOG. The specific method is as follows: first, the obtained pixels are divided into 196 units according to the size of 32×32, and then count the direction and gradient mode of the pixels contained in the unit [4]. Among them, the abscissa of HOG is 9 intervals divided into 196 units according to the gradient direction, and the ordinate is the mode of HOG of all pixels in each interval.

Third, after obtaining the HOG of all units, use the L2 norm normalization method to traverse the entire face image with a sliding window with a pixel size of 64×64 and a step size, and then use the normalization method to process each unit.

2.2.2   Feature Map Fusion

Use the "YOLO" convolutional neural network to fuse the feature maps. The collected face avatars are processed through a series of convolutional neural networks based on the "YOLO" theory. The collected size is 448×448 pixels and the color mode is RGB JPG face images are processed for 6 times of convolution and 4 times of pooling [5]. Each time of convolution and pooling, operations such as convolution, down sampling, biasing and activation function processing will be performed on the input image to extract feature values of face images.

2.2.3 Classification and Detection

Through feature extraction and feature map fusion, the facial feature map of the neural network can be obtained, but to realize the recognition and detection of the face image collected by the hardware, the classifier needs to be set, trained, and detected [5].

First, define the loss function. In the actual application process, the non-face area contained in the image is often much larger than the face area, so it is necessary to define the loss function to show the error between the calculated value and the labeled value [6]. This error is generally including positioning error and classification error, different weights can be assigned to the two errors in the calculation.

Second, test and train the entire detection network. According to the feedback results during the test and training process, adjust and detect the main parameters of each layer in the network, especially the fully connected layer, in time, and adjust the parameters to make the calculation result very high [6].

2.3 OpenCV Technology Platform

The technology platform used for image acquisition, input and storage involved in this research is OpenCV (Open-Source Computer Vision Library). This platform is an open source, cross-platform computer vision library composed of many C++ libraries. It is used in computer image processing and vision [7]. It provides many algorithms, and provides C#, Java, Python, MATLAB, Ruby and other language interfaces, which has good portability.

3 Analysis and Design of Intelligent Education Management Platform

3.1 System Analysis

The intelligent education management platform takes the actual needs of efficient digital education as the starting point, takes students, teachers, administrators, and system maintainers as the research objects, and uses digital means to systematically analyze various business needs and realize the informatization of education management around the daily affairs of education [7]. And intelligence, and finally combine the school's basic network and database resources to gradually establish a "smart campus" system platform.

3.2 Face Recognition

Human face is an important part of human body structure and forms unique biological characteristics. Compared with other biological characteristics, such as fingerprints, palm prints, voice, iris, gait, etc., human face does not require additional action coordination [8]. Face recognition technology is a typical application in the field of computer vision. Especially in recent years, as the technology has matured, it has been widely used in various application fields such as access control, time attendance, hotels, airports, and even self-service checkout counters in supermarkets.

Generalized face recognition includes image acquisition, face detection, face alignment, feature representation and other processes [8]. The typical process of face recognition mainly includes face detection, feature location, preprocessing, feature extraction, feature comparison, decision-making judgment, and other stages.

The core of face recognition application is feature extraction. In the past few decades, researchers all over the world have proposed many very effective feature extraction and face recognition algorithms, which have greatly promoted the development of face recognition technology [8]. The most classic algorithms are LBP (Local Binary Patterns), HOG (Histogram of Oriented Gradient), PCA (Principal Component Analysis), LDA (Linear Discriminant Analysis)), SVM (Support Vector Machines, support vector machine) and so on. Some of these algorithms are used alone for face recognition, and some algorithms such as PCA, LDA, and SVM are combined for feature extraction and recognition [4], and some newer methods such as combining curvelet transform and contourlet transform are used to extract features.

3.3   Face Detection

Face detection is the first step. The mobile terminal App obtains the video stream through the camera (front). First, the face can be detected through the video stream before subsequent actions can be performed. Live body detection and recognition verification without face detection is meaningless. And waste time and resources.

Face detection is an application of target detection, mainly based on facial features such as eyes, nose, mouth, chin, and eyebrows [9]. OpenCV mainly uses Haar-Cascade-based face detection, which can work on the CPU in real time, but detection errors will often occur, such as detecting non-human faces for human faces (see Fig. 2).

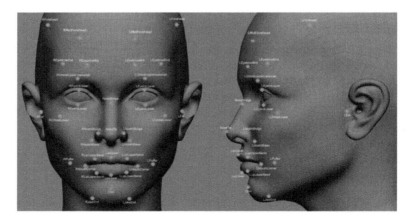

Fig. 2. 68 feature points of face in dlib

Most modern smartphones can use the face detection mode supported by their own camera hardware to detect faces. Taking Android as an example, the new camera API has been introduced in Android 5.0 version. Compared with the old camera API, it works with HAL3. There are more advanced API architecture and camera experience. The face detection mode is built into the camera API, as shown in Fig. 3.

Fig. 3. Face detection mode of camera API

The face detection part of this system uses the combination of camera API and OpenCV. When the mobile phone has a simple or fully supported mode, the camera API is used for the fastest and most efficient detection, otherwise it switches to the OpenCV detection mode [9].

3.4 Live Detection

Nowadays, when security has become a global issue, people are paying more attention to the security of the system. Face recognition is as vulnerable to attacks as traditional information systems, and live detection is used to prevent face attacks. It can be imagined [10]. For financial applications such as Alipay, it would be a terrible thing if someone who is interested in embezzling other people's information to log in and pay with a fake face.

3.5 Identification and Verification

After the face detection and the living body detection are all passed, the identification verification is finally carried out.

The face detection and living body detection of this system are all carried out through the mobile terminal App but limited by the computing resources of the mobile phone, the real identification verification must be placed in the cloud (or server side), so it is better to use dlib-based open-source people. The face recognition library face recognition builds a face recognition environment, face recognition encapsulates dlib calls, provides simple and easy-to-use commands and interfaces, it has a recognition rate of up to 99.38% tested on the LFW face database, which can be said to be powerful and easy an open-source face recognition library to get started [10]. The process of identification verification is shown in Fig. 4.

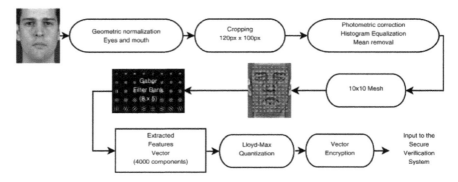

Fig. 4. Face recognition verification process

4 Conclusion

Through the research, design and development of the intelligent education management platform based on face recognition technology, it not only improves the work efficiency of education management, but also reduces the work intensity of education management personnel. To a certain extent, it solves the various aspects of education management in colleges and universities. It is difficult and can effectively integrate and analyze the collected data and information in a timely manner, which is of great significance to the resource scheduling and safety management of colleges and universities.

References

1. Deepa, V., Sujatha, R., Mohan, J.: Unsung voices of technology in school education-findings using the constructivist grounded theory approach. Smart Learn. Environ. **9**(1), 1–25 (2021). https://doi.org/10.1186/s40561-021-00182-7
2. Bangqi, L.: The development form and practice path of smart education. Mod. Educ. Technol. **10**(5), 110–113 (2019)
3. Bulut, D., Samur, Y., Cömert, Z.: The effect of educational game design process on students' creativity. Smart Learn. Environ. **9**(1), 8 (2022). https://doi.org/10.1186/s40561-022-00188-9
4. Riekki, J., Mämmelä, A.: Research and education towards smart and sustainable world. IEEE Access **9**, 53156–53177 (2021)
5. Wong, B.: Research landscape of smart education: a bibliometric analysis. Interact. Technol. Smart Educ. **19**(1), 3–19 (2022)
6. Mahmud, M.M., Freeman, B., Bakar, M.S.A.: Technology in education: efficacies and outcomes of different delivery methods. Interact. Technol Smart Educ. **19**(1), 20–38 (2022)
7. Lan, N.: The subdivision of smart education is accelerating. Commun. World **49**, 921–925 (2019)
8. Akhrif, O., Benfaress, C., Jai, M.E.L., et al.: Completeness based classification algorithm: a novel approach for educational semantic data completeness assessment. Interact. Technol. Smart Educ. **19**(1), 87–111 (2022)

9. Huiru, Z., Yanhua, L., Wang, S.: Smart education and improvement of teaching methods in colleges and universities. J N. China Electric Power Univ. Soc. Sci. Ed. **9**(10), 67–70 (2019)
10. Gurrala, G., Challa, K.K., Rajesh, K.B.: Development of a generalized scaled-down realistic substation laboratory model for smart grid research and education. IEEE Access **10**, 5424–5439 (2022)

English Teaching Ability Evaluation Algorithm Based on Bayesian Network Classification Model

Guanghua Liang[✉]

Henan Vocational College of Industry and Information Technology,
Zhengzhou 454000, Henan, China
liangguanghua666@sina.com

Abstract. The education industry is one of the ways to cultivate talents in our country. In order to cultivate better leaders, it is necessary to test the ability of English teachers. Due to the in-depth development of network technology, it is no longer a problem to use advanced technology to evaluate the ability of English teachers. To this end, this paper uses the relevant knowledge of Bayesian network classification to construct an English teaching ability assessment model. Its purpose is to facilitate the testing of English teachers' abilities by constructing models and to recruit more excellent teachers. This article mainly uses experimental and comparative methods to evaluate the teaching ability of English teachers. The results show that the system is based on the Bayesian network classification model, and the score of English teachers is above 70 points. Although it is lower than the traditional teaching evaluation score, it is more authentic and convincing.

Keywords: Bayesian network · Classification model · English teaching · Ability assessment

1 Introduction

Improving teaching efficiency and promoting the realization of teaching effects is an important problem currently faced by schools. How to train English teachers to become talents with professional skills is a way to improve the teaching effect. Based on data mining algorithms and pattern recognition theory, the Bayesian classification model can be used to establish a complete set of teaching evaluation system.

There are many people who study Bayesian network classification models and teaching ability assessment methods. For example, Liang Cong said that Bayesian networks are a hotspot in data mining research and an effective tool to determine the uncertainty dependence between things [1]. Zhao Yiming established an information security evaluation data topology structure based on hierarchical protection and Bayesian networks, and qualitatively identified hierarchical protection evaluation data through expert experience [2]. Wang Qing uses a combination of quantitative and qualitative analysis to test the effectiveness of formative assessment for higher education English teaching [3]. Therefore, this article combines Bayesian network and English teaching evaluation system in order to achieve the goal of system optimization and selection of excellent teachers.

© The Author(s), under exclusive license to Springer Nature Switzerland AG 2022
V. Sugumaran et al. (Eds.): ICMMIA 2022, LNDECT 138, pp. 633–640, 2022.
https://doi.org/10.1007/978-3-031-05484-6_80

This article first studies the basic theories and related algorithms of Bayesian networks. The second study is the English teacher classification system based on the Naive Bayes algorithm. After that, it analyzes the current problems faced by the methods of evaluating the ability of college English teachers. Then the overall realization of the system is designed and explained. Finally, the relevant results are obtained through experiments.

2   English Teaching Ability Assessment Algorithm Based on Bayesian Network Classification Model

2.1   Bayesian Network

The goal of Bayesian networks is to use probabilistic techniques to solve uncertain thinking problems. The directed edges represent the causal relationship between variables. If all nodes in the Bayesian network receive a parent node, they are independent of all non-descendant nodes [4, 5]. The general probability formula of the whole network is (1)

$$p(a_1, a_2, \cdots, a_m) = \prod_{i=1}^{m} p(a_1 \tau(a_i)) \tag{1}$$

If a node does not have a parent node, the conditional probability will degenerate into an edge probability $p(a_i)$.

In fact, Bayesian classification does not simply assign data objects to a certain category, but calculates the probability of data objects belonging to different categories and compares the category labels with the highest probability. The category label with the highest probability is true. According to the needs of users, it may be required to be classified into certain categories. When using the Bayesian method to classify the sample set, all attributes should participate in the classification to avoid ignoring some attributes with larger weights. In Bayesian classification, the sample data type can be discrete, continuous or mixed [6, 7].

Given the proof variable F = f, calculate the combination of the maximum value of a set of query variables B under this condition:

$$B^* = agrmaxp(B = b \mid F = f) \tag{2}$$

It can be seen that the computational complexity is an exponential function of the number of query variables.

The naive Bayes classification algorithm is mainly used to evaluate the conditional probability of a sample belonging to a certain class [8, 9].

2.2   English Teacher Classification System Based on Naive Bayes Algorithm

Constructing an English teacher classification system based on the naive Bayes algorithm can realize the scientific classification of the differences in English teacher difference test results, and realize the establishment of classifiers. In the later stage, English teachers are automatically classified, which overcomes the problems in English. Subjective factors have a greater impact on teachers' ratings, and individual ratings should be avoided. Avoid grouping blindly without knowing English teachers [10, 11].

The classification system is based on classification theory and naive Bayes difference theory. The classification system is based on the Naive Bayes algorithm, and the steps are as follows:

Describe the differences between English teachers by examining the characteristics of the differences between English teachers. The former English teachers are ranked based on past data and expert experience. The naive Bayes classification algorithm is used to construct an English teacher classifier, and the English teacher to be classified by the classifier calculates the conditional probability of belonging to each category. Select the maximum value that the English teacher belongs to each category, that is, the category of the English teacher to be classified [12].

The hardware environment created by the system program is composed of machines and network resources provided by the laboratory. The software environment is Eclipse, mysql, Tomcat, and the development language is JAVA. The system has scientific and reasonable research problems and goals, rigorous and achievable research content, and technical solutions.

2.3 The Current Problems Faced by College English Teachers' Ability and Quality Evaluation Methods

(1) At present, the main problems in the assessment of English teachers for teaching positions in colleges and universities in my country are as follows:
Evaluation methods tend to quantify individual indicators excessively or unilaterally. The evaluation work did not play a leading role. Although the principle of effective performance appraisal should be respected, performance appraisal is often a mere formality due to ineffective support measures and poor execution. The introduction and selection of talents still use traditional methods. There are many contradictions between teaching and learning, and the quality of teaching is not high. In terms of talent development and protection, there is a lack of effective methods to evaluate.
(2) The evaluation calculation process is:
First, comprehensively evaluate the factors of the third-level indicators, and then conduct a comprehensive evaluation of the second-level indicators. Finally, a comprehensive assessment of the skills and qualities of English teachers.
The meaning of each factor in the evaluation factor set is different. In the comprehensive evaluation, determining the importance of each element in the element concentration is the decisive link, and whether the value is correct directly affects the result of the comprehensive evaluation.
(3) Evaluation index system
At present, the assessment of English teachers for teaching positions in colleges and universities is mainly the assessment of teaching quality. The assessment system cannot fully reflect the development needs of higher education and the common aspirations of

teachers and students. The assessment methods are mainly qualitative. Based on the competency model studied by the predecessors, this paper designs a questionnaire survey, from the perspective of students and excellent teachers, clarifies the competencies and quality elements of ideal teachers in the minds of students, which represent extraordinary achievements. Fixed English teacher: Based on the principle of pair analysis, according to the important substitute teacher's skills and quality items, as well as the ability and quality evaluation system, calculate the attention ranking model and the clue interception model. Teaching positions created by college English teachers.

The evaluation of English teachers' teaching ability is an important factor in teaching evaluation, and it is also a necessary guarantee for cultivating outstanding talents. This article is guided by developmental and multiple intelligence evaluation theories, paying attention to the differences in the skills of English teachers in various aspects, and has formulated a variety of evaluation standards. At the same time, after the evaluation, the development of English teachers should be encouraged. Quickly feedback the evaluation results to English teachers and put forward suggestions for improvement to help English teachers improve and guide the development of English teachers.

Teaching ability refers to the general term used to describe the ability of English teachers to carry out teaching activities. It refers to the potential ways for English teachers to achieve class goals and perform well in class. The teaching ability of English teachers is the basic skills of the quality characteristics that English teachers should have, which can directly and effectively reflect whether individual English teachers can successfully cope with the "education" task. With growth of pedagogy and the reform of various educational theories, psychology is particularly important in the development of these theories, especially the growth of cognitive psychology. These theories can make the research on the teaching ability of English teachers more effective and in-depth. Scholars have their own views on the composition of English teaching skills. However, teaching skills can basically be divided into two categories, namely classroom teaching skills and off-campus research and design skills.

2.4   Overall System Realization

The English teacher evaluation system implemented in this document uses ASP programming to provide a comprehensive set of teacher teaching evaluation solutions that can follow the evaluation process and authoritative framework, specific and comprehensive student evaluation, teacher mutual evaluation, supervisor and teacher mutual evaluation. The system administrator assesses teachers in four ways, calculates the final grades of teachers based on weights, generates assessment reports, and comprehensively analyzes the assessment results of English teachers through horizontal comparison and historical data analysis. The whole process is shown in Fig. 1:

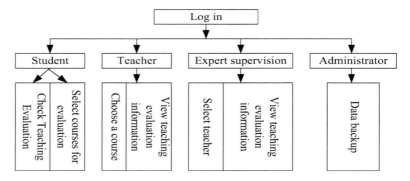

Fig. 1. The overall process of the teaching evaluation system

The system can operate normally after being opened and tested, and supports the entire management of the teaching evaluation process. After the project is packaged, it can be deployed and run on multiple operating system platforms. Once the network application server is started, the user's client software can quickly and easily access the university course evaluation system. In the education evaluation information system, system administrators mainly carry out maintenance work such as addition, deletion, modification, and checking of education evaluation information. The main users of the system are students, teachers, and qualified supervisors.

3   System Implementation

3.1   System Development Environment

This system selects MyEclipse platform as the development environment. MyEclipse is an excellent JAVA development tool that integrates all components of the J2EE framework. It has very powerful functions, supports many applications, and supports various plug-ins. -ins open source. PowerDesigner is a set of tools developed by Sybase, whose main purpose is to realize the overall management of the database modeling process. Using PowerDesigner, various databases of the system can be modeled, including data flow, conceptual model, physical model, etc., which can shorten the development cycle and improve the development efficiency of the system. MYSQL is currently one of the systems that support network databases widely. This system chooses MYSQL as the data carrier, and uses MYSQL to provide various optimization mechanisms to achieve rapid data access. Navicat is a simple MYSQL database management and development tool. Navicat supports a graphical user interface, allowing developers to design and implement database applications quickly and easily.

3.2   Test Plan

When testing software systems, test cases cannot cover all possible situations. According to the characteristics of the use of this system, in the actual testing process, developers are mainly responsible for unit testing and integration testing, organizing

English teachers, students, supervisors, system administrators and other users to conduct tests. For actual use, including alpha testing in the development environment and beta testing performed by multiple users in one or more users' actual use environments.

3.3 Test Function

The main elements of the test system include the following parts, and testers evaluate each part. System test items include: requesting English teacher information, requesting student information, requesting supervisor information, and requesting system administrator information. In addition, this article also invites English teachers to conduct experimental research to compare their teaching skills with traditional teaching evaluation methods and methods based on Bayesian network classification models.

4 Result Analysis

4.1 Evaluation Results of English Classroom Teachers' Teaching Ability

Five teachers used two evaluation methods to evaluate the teaching ability of English classroom teachers. After processing the data, the results are shown in Table 1. The teaching ability of teachers in the traditional evaluation method is relatively high, and the result of the test using Bayesian classifier is lower.

Table 1. Evaluation results of teachers' teaching ability in English class

	Traditional method	Bayesian method
1	79.1	73.5
2	78.6	79
3	75.6	71.2
4	77.5	76.1
5	75.4	76.3

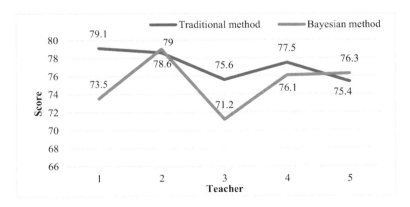

Fig. 2. Evaluation results of teachers' teaching ability in English class

It can be seen from Fig. 2 that the results of the two evaluation methods are very close, the maximum difference of the evaluation results is 5.6 points, and the smallest difference is 0.4 points. It is possible and accurate for English teaching based on Bayesian classification to use an index system to evaluate teachers' teaching skills.

5 Conclusion

With the popularization of the Internet, computer technology continues to develop. Bayes has developed rapidly, widely used and mature in the field of statistics and prediction. Based on the Bayesian network classification model, this paper studies and analyzes the behavior of students in the course of English teaching in the classroom, and obtains relevant information through data mining and semantic tree methods. Comprehensive scoring is based on the exchanges and interactions between students and teachers. Through the research of this article, it is found that it is feasible and accurate to use the teacher's teaching ability evaluation index system based on Bayesian classification method in English classroom.

References

1. Leguey, I., Bielza, C., Larranaga, P.: Circular Bayesian classifiers using wrapped cauchy distributions. Data Knowl. Eng. **122**, 101–115 (2019)
2. Zia, T., Ghafoor, M., Tariq, S.A., et al.: Robust fingerprint classification with Bayesian convolutional networks. IET Image Proc. **13**(8), 1280–1288 (2019)
3. Bressan, G.M., Azevedo, B., Santos, H., et al.: Bayesian approach to infer types of faults on electrical machines from acoustic signal. Appl. Math. Inf. Sci. **15**(3), 353–364 (2021)
4. Gupta, A., Arora, P., Brenner, D., et al.: Risk prediction using bayesian networks: an immunotherapy case study in patients with metastatic renal cell carcinoma. JCO Clin. Cancer Inform. **5**(5), 326–337 (2021)
5. Erarslan, A.: Strengths and weaknesses of primary school english language teaching programs in Turkey: issues regarding program components. Eurasian J. Appl. Linguist. **4**(2), 325–347 (2018)
6. Gershon, S.K., Ruipérez-Valiente, J.A., Alexandron, G.: Defining and measuring completion and assessment biases with respect to english language and development status: not all moocs are equal. Int. J. Educ. Technol. Higher Educ. **18**(1), 1–21 (2021). https://doi.org/10.1186/s41239-021-00275-w
7. Bajwa, M.N., et al.: Confident classification using a hybrid between deterministic and probabilistic convolutional neural networks. IEEE Access **8**, 115476–115485 (2020). https://doi.org/10.1109/ACCESS.2020.3004409
8. Yang, C., Mott, J.H.: HFACS analysis of U.S. general aviation accidents using Bayesian network. Proc. Hum. Factors Ergonomics Soc. Annu. Meet. **64**(1), 1655–1659 (2020)
9. Kawabe, R., Ito, H., Yamashita, H., et al.: Hierarchical structure learning in a Bayesian network for the analysis of purchasing behavior. Total Qual. Sci. **4**(3), 99–108 (2019)
10. Surkamp, C., Viebrock, B.: Teaching English as a foreign language institutionalised foreign language learning—teaching English at different levels, 17–37 (2018). https://doi.org/10.1007/978-3-476-04480-8(Chapter2)

11. Lestari, M., Arono, A.: Evaluation of English teaching materials used at sd it ummi in kota bengkulu based on ktsp (kurikulum tingkat satuan Pendidikan). JOALL J. Appl. Linguist. Lit. **3**(2), 88–102 (2019)
12. Pitua, B., Wlalak, G.: Academic teachers self-evaluation of English language competences and teaching methodology. New Educ. Rev. **59**(5), 35–46 (2021)

Analysis on Influencing Factors of Happiness of Rural Empty Nest Elderly Based on Health Internet of Things Data Mining

Daofang Yu[1], Ping Zhang[2], and Sulan Long[3(✉)]

[1] College of Traditional Chinese Medicine, Nanchang Medical University,
Nanchang 330004, Jiangxi, China
yudaofang@foxmail.com
[2] Department of Critical Medicine, Jiangxi Hospital of Traditional Chinese
Medicine, Nanchang 330006, Jiangxi, China
zpzv@sina.com
[3] Faculty of Nursing, Nanchang Medical University,
Nanchang 330004, Jiangxi, China
lsl19820929@163.com

Abstract. In recent years, with the attention of our government and academia to the quality of life of residents, the problem of well-being has also attracted the attention of many domestic researchers. Happiness is an important indicator of social and individual quality of life, which has been studied in the West for more than half a century. The overall process of health IOT data processing is to receive the massive heterogeneous data from the intelligent collection terminal of health IOT, store and filter the data in a distributed manner, and then conduct distributed data mining. This study investigated the influencing factors of happiness of rural empty nesters.

Keywords: Well-being · Empty nesters · Data mining · Internet of things

1 Introduction

Research on the influencing factors of happiness of the elderly since the end of last century, there have been a large number of studies on the influencing factors of happiness of the elderly. In terms of demographic variables, Diener found that married elderly men were happier than young and high-income elderly men. In terms of social support, most studies show that social support can play a positive role in the prediction of well-being. In terms of personality factors, Tian Lin believes that personality is the main influencing factor of well-being, in which introversion and introversion have a positive impact on well-being, while neuroticism has a negative effect [1].

Under the rapid development trend of Internet of things and cloud computing, with the increasing attention of society to health problems, application research for intelligent medical treatment and health Internet of things has been formed. The social medical and health service network based on Internet of things is mainly for community and rural grass-roots medical units, using personal physiological information

© The Author(s), under exclusive license to Springer Nature Switzerland AG 2022
V. Sugumaran et al. (Eds.): ICMMIA 2022, LNDECT 138, pp. 641–648, 2022.
https://doi.org/10.1007/978-3-031-05484-6_81

intelligent perception and collection equipment, Through the Internet of things application middleware platform and access gateway, realize the docking between the target population and the background medical information service and medical service resources, and truly realize the comprehensive information exchange and service network between things, things and people, and people and people. The middleware data processing platform is a development platform and basic framework designed for the application characteristics of the Internet of things. Its main function is to isolate the underlying sensor network from the upper application, shield the differences of communication protocols and data formats between different sensing and identification devices at the bottom, and provide a unified data processing, network monitoring, application development and service scheduling interface for the upper application development, so as to simplify the deployment of sensor networks and related application development [2]. The following Fig. 1 shows the application development related to the Internet of things network.

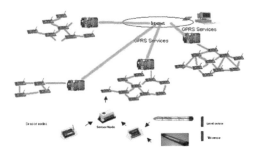

Fig. 1. Internet of things network related application development

At home and abroad, most of the factors affecting the well-being of the elderly are studied objectively by quantitative analysis, but most of the research results are not unified [3]. This paper will start from a small sample, from a micro perspective, through in-depth interviews and descriptive statistics, pull in the distance from the empty nest elderly, so as to study the influencing factors of the happiness of the empty nest elderly in rural areas.

2   Relevant Technical Analysis

2.1   Data Mining Algorithm

Data mining is a process of automatically discovering useful information in large data repository. It is an essential part of knowledge discovery in big data. Knowledge discovery is a process of converting unprocessed data into useful information [4]. Facing the problems of scalability, multidimensional, heterogeneity and complexity, data ownership and dispersion caused by data analysis, it is necessary to establish corresponding models according to statistics, artificial intelligence, optimization, information theory and other knowledge to mine, analyze and process the data.

$$I(s_1, s_2, ..., s_a) = -\sum_{i=1}^{m} p_i \log_2(p_i) \tag{1}$$

$$E(X) = \sum_{J=1}^{Y} \frac{S_{1J} + ... + S_{MJ}}{S} J(S_{1J}, ..., S_{MJ}) \tag{2}$$

$$I(S_{1J}, ..., S_{MJ}) = -\sum_{i=1}^{m} p_{ij} \log_2(p_{ij}) \tag{3}$$

Data mining can be divided into two categories: prediction task, which predicts the value of a specific attribute according to the value of other attributes; Describe the task, which is a pattern for exporting potential connections in summary data.

The healthy Internet of things has the characteristics of interconnection, collaboration, prevention, popularization, security and innovation [5]. Connectivity means that no matter where the patient is, the management doctor can view the patient's health files and medical records through the Internet of things, and consult with other experts to provide the best treatment and nursing services for the patient. Collaboration refers to recording, integrating and sharing medical information and resources through a proprietary medical network, realizing information exchange and collaborative work between different medical departments, and providing integrated services for patients. As shown in Fig. 2 below, healthy Internet of things realizes healthy Chinese LAN.

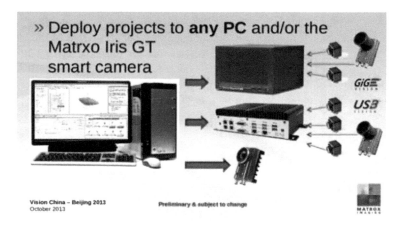

Fig. 2. Healthy Internet of things to achieve healthy Chinese LAN

Preventive health means that the Internet of things can timely find the signs of major diseases according to the mining and prediction of patients' historical physical signs, so as to make a rapid and effective response [6]. Universality refers to the ability to break through the conceptual limitations of large hospitals and ordinary hospitals and provide universal high-quality medical services. Security refers to ensuring the security of personal medical information without authorization. Innovation means that the health Internet of things can innovate the traditional medical model and stimulate more innovative development in the field of health. We will accelerate research to realize the healthy Internet of things with these characteristics and realize the Chinese dream of healthy Chinese people.

2.2   Current Situation of Empty Nest Elderly in Rural Areas

(1)   There are few sources of income and poor self-sufficiency. The economic income of the 50 empty nesters mainly comes from the land rent income, while the government subsidies and the supply of children and relatives are not much, barely meeting the basic life. They lack the ability to make money except for agricultural work. 80% of the elderly are financially dependent on their children and relatives when they are ill, which increases the unhappiness of the elderly.

(2)   It is difficult to engage in labor services and take care of themselves. 30% of the elderly in good health will engage in farm work, household activities and some labor with little exercise; while most of the elderly in poor health mainly take care of household activities, and even some elderly in worse health need the care of others, but they have no children around them, so they feel that life is difficult [7].

(3)   It is difficult to realize self-worth. With the growth of age, empty nesters gradually retire from jobs such as work and farm work, and their children are not around, so they have nothing to do all day. 66% of empty nesters think their life is boring, so they can only count their days. "The increase of emptiness, loneliness and gratitude is not conducive to the physical and mental health of empty nesters.

(4)   The educational level is generally low. Only one of the 50 elderly has attended junior high school, the rest are at or below the level of primary school, and even more than two-thirds of the elderly have never attended school. In the face of the changing world, empty nesters are at a loss [8].

(5)   Lack of social care. There are few activities in the community, and the government will not visit frequently. The number of children returning home is very limited. If there is no wife, the elderly are very lonely. 72% of empty nesters think they have no sense of existence.

3   Analysis of Influencing Factors of Happiness of Rural Empty Nesters Based on Healthy Internet of Things Data Mining

The file data storage of the healthy Internet of things data processing platform adopts cloud computing technology and is stored by the distributed file system HDFS based on Hadoop cluster. Because the data collected at the massive heterogeneous intelligent collection terminal may have redundancy and abnormal data, it is necessary to clean the distributed data of the data. The cleaned data can be designed and distributed based on

Hadoop data processing platform This paper mainly studies the distributed data mining algorithm to obtain the information needed for the happy application of the rural empty nest elderly [9].

In addition to the physical health status, the mental health status of the elderly also deserves attention. Although more than half of the elderly have a strong sense of satisfaction and security in life and less negative and lonely feelings, about 10% of the elderly are often in a lonely and negative state. The life satisfaction of these elderly is only 7.15, which is significantly lower than that of the elderly group Average level. In order to establish a data cleaning framework in the healthy Internet of things data processing platform, it is coded based on Hadoop framework. By analyzing and filtering the data collected by the intelligent collection terminal, the original cleaning process is transformed into MapReduce model for parallel data cleaning.

3.1 Health Status

Nearly 58% of empty nesters have serious diseases, and less than 10% of empty nesters are completely healthy. In addition, 46% of empty nesters whose medical expenses account for more than 30% of their living expenses every year. Medical expenses have become a great economic burden for empty nesters who originally had little income. In order to save money, they do not treat them in time, resulting in serious minor diseases [10]. Empty nesters lack the care of their children, They are more sensitive to their own physical conditions. At the same time, they lack the popularization of medical knowledge and are prone to wishful thinking. Due to the decline of self-care ability, the elderly are more dependent on others in life, especially the widowed empty nesters in rural areas.

3.2 Marital Status

As the saying goes, "young couples always come together", the presence or absence of spouses is also an important factor affecting the happiness of empty nesters. The happiness of empty nesters with spouses is significantly higher than that of empty nesters without spouses. Spouses play an indispensable role in the life care and spiritual comfort of empty nesters, and the love between husband and wife is conducive to promoting physical and mental health [11].

3.3 Child Support

Generally speaking, the most important life support for the elderly is their children, and the support from their children plays an important role. The concern from their children can fundamentally improve the problem of the elderly's lack of security and prone to depression. Because most of their children go out to work, it is difficult to go home all year round, but 98% of the empty nesters still hope that their children can often come back to visit themselves, be taken care of by their children and return home Sense of belonging and security.

4   Design of Happiness Data Mining Algorithm for Rural Empty Nest Elderly

The data mining of the health Internet of things is mainly aimed at the long-term health files of the rural empty nest elderly. These health file information is similar to our files in reality, but here through the body area network technology, the information records the physiological information characteristics of people at the corresponding stage, including basic body health information data, such as height, weight, living habits, disease history, family history and medication History, treatment, disease evolution process, long-term law of various health indicators, etc. combined with the change law of various health indexes, compared with the established disease model in the database, timely and scientifically give early warning of possible diseases, and put forward prevention and health care suggestions. Provide help for the clinical treatment of existing diseases, and patients can clearly see their health It will provide the possibility for early diagnosis and early treatment, and avoid repeated examination, repeated medication and wrong judgment of the disease [12]. As shown in Fig. 3 below, the development code design of the system is shown.

```
ListNode * findNode(LinkedList *ll, int index);

int insertNode(LinkedList *ll, int index, int value);

int removeNode(LinkedList *ll, int index);

void push(Stack *s, int item);

int pop(Stack *s);

int isEmptyStack(Stack *s);

void enqueue(Queue *q, int item);

int dequeue(Queue *q);

int isEmptyQueue(Queue *s);
```

Fig. 3. System development code design

Because the key of the massive data mining parallel algorithm based on Hadoop data processing platform is to face a large amount of data, the core research direction is the distributed research of data mining algorithms. Hadoop processing platform mainly uses MapReduce programming model to analyze and process data. It is necessary to transform the existing data mining algorithms [13]. According to the specific characteristics of each algorithm, the analysis is The distributed massive data mining algorithm based on Hadoop platform mainly includes distributed association rule algorithm, distributed classification algorithm and distributed clustering algorithm. These algorithms are used to effectively find association rules, classify and cluster the data, Find out dependency models, find anomalies and trends, etc., so as to provide data and model support for the analysis of rural empty nesters [14]. Data mining algorithm inevitably needs iterative processing. Hadoop data processing algorithm adopts different iterative methods, and the output of each iteration is the input of the next iteration. By managing intermediate data and the location of each round of tasks, we can To obtain higher efficiency.

5   Conclusion

Based on the analysis of the influencing factors of the well-being of the rural empty nest elderly based on the data mining of the healthy Internet of things, the rural empty nest elderly have an urgent demand for the necessary health education knowledge, free physical examination and door-to-door service. We should take a variety of countermeasures against the factors affecting SWB and rely on the joint efforts of the elderly themselves, family and society to provide some necessary health services for the elderly, such as free physical examination Blood pressure measurement, free physical examination and free health knowledge, so as to achieve early detection, early diagnosis and early treatment of diseases, so as to gradually improve and improve the quality of life and SWB of the empty nest elderly in the process of healthy aging and active aging.

Acknowledgements. Planning project of humanities and social sciences research program in colleges and universities of Jiangxi province in 2020, project number: GL20143.

References

1. Tao, H.: Analysis of fluctuation factors of healthy exercise based on machine data mining and internet of things. Environ. Technol. Innov. **3**, 101647 (2021)
2. Guan, W., Lu, H.-J., Chen, J.-J., Wu, J.: Research on the distributed data mining cloud framework oriented internet of things, **8** (2022)
3. Mkrttchian, V., Gamidullaeva, L., Finogeev, A., et al.: Big data and internet of things (IoT) technologies Influence on higher education: current state and future prospects. Int. J. Web-Based Learn. Teach. Technol. (IJWLTT) **16**(5), 135–157 (2021)
4. Dong, C., Xiuquan, Q., Gelernter, J., Xiaofeng, L., Luoming, M.: Mining data correlation from multi-faceted sensor data in internet of things, **1** (2022)
5. Sukumaran, S C.: DNA-based authentication to access internet of things-based healthcare data (2021)
6. Rashid, M., Singh, H., Goyal, V., et al.: Big data based hybrid machine learning model for improving performance of medical internet of things data in healthcare systems-sciencedirect. Healthc. Paradigms Int. Things Ecosyst., 47–62 (2021)
7. Wang, R., Dai, M., Ou, Y., et al.: Residents' happiness of life in rural tourism development. J. Destin. Mark. Manag. **20**(1), 100612 (2021)
8. Xu, T., Han, H.: Two-week prevalence of disease among the rural elderly–6 provinces, China, 2018−2019, **2**(38), 4 (2020)

9. Pan, Y., Liu, P.X., Hong, F.F., et al.: The impact of mindfulness therapy on life happiness of the elderly in nursing homes. Psychosom. Med. Res. **2**(2), 39–45 (2020)
10. Wang, N.: An analysis of the demand for education for the elderly in rural areas: a case study of the suburbs of hangzhou. Asian Agric. Res. **12**, 69–74 (2020)
11. Sørensen, J.F.L.: The rural happiness paradox in developed countries. Soc. Sci. Res. **98**, 102581 (2021)
12. Yang, T., Zhang, C.M., Wang, R.R., et al.: The prevalence of depression among the empty-nest elderly in China:a meta-analysis of cross-sectional studies, **3**(1), 11 (2021)
13. Ivanova, K.: My children, your children, our children, and my well-being: life satisfaction of "Empty Nest" biological parents and stepparents. J. Happiness Stud. **21**(2), 613–633 (2020). https://doi.org/10.1007/s10902-019-00097-8s
14. Rahman, J.R., Sanshi, S., Ahamed, N.N.: Health monitoring and predicting system using internet of things & machine learning. In: 2021 7th International Conference on Advanced Computing and Communication Systems (ICACCS) (2021)

Video Analysis and Diagnosis System
of National Traditional Sports Confrontation

Wei Li[✉]

Yunnan University of Business Management, Yunnan 650033, China
wangtianxinq8@163.com

Abstract. At present, the analysis of national traditional sports confrontation training is mainly based on the subjective experience judgment of coaches, which is lack of objectivity and timeliness. This study builds an arm + DSP + CPLD technology framework platform, uses video image intelligent analysis and processing technology to record and diagnose athletes' training and competition in real time, and obtains the processing results that can be used as a reference for coaches and athletes, so as to improve training efficiency and competition performance.

Keywords: National traditional sports · Confrontation projects · Embedded system · Video image analysis

1 Introduction

National traditional competitive sports play a very important role in Chinese national traditional sports. For example, Sanda martial arts, Chinese wrestling, individual martial arts, etc. For a long time, the training level of Chinese traditional competitive sports is relatively low. The coach said, "I only believe in my experience". Players can acquire skills through the coach's speaking, personal training and repeated practice. This backward training method limits the improvement of training level and efficiency.

The introduction of digital image analysis technology into sports training is a new research content to improve the scientificity and efficiency of training. Because machine vision has better memory ability and accuracy than human eyes, it can quickly capture any subtle movements of athletes' training, and analyze, summarize, compare and save these subtle technical movements through video image analysis algorithm [1]. Training data will not lie, it provides a more intuitive way to judge the quality of athletes' actions.

Combined with embedded chip hardware technology, video image advanced analysis and processing technology and database technology, the research group has developed a set of "video analysis and intelligent diagnosis system" for national traditional sports confrontation projects "Whether the system can provide a new training mode for coaches: by recording high-resolution training video, after analyzing the two-dimensional and three-dimensional information of athletes' technical actions, obtain the real data of athletes' technical actions, and then scientifically decompose, compare and synthesize subtle actions through video image analysis algorithm to scientifically evaluate the training level and purpose It can also compare the action images of different athletes, so as to help athletes find the gap with excellent athletes, so as to master the technical essentials of action as soon as possible, reduce unnecessary repetition, achieve the purpose of intuitive teaching and rapid feedback; in addition, it can greatly reduce the possibility of injury to athletes, which is of great significance to improve the training level and competitive level of athletes It has positive significance. Figure 1 below shows the process of improving training level and competitive level.

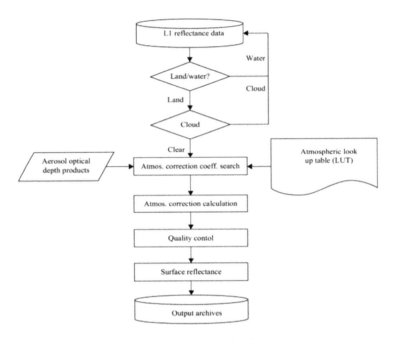

Fig. 1. Training level and competitive level improvement process

2   Related Work

2.1   National Traditional Sports

Zhou Weiliang (2003) defined the sports that emerged and developed before modern China, influenced the Chinese nation and spread to the present. "Hu Xiaoming (2003) believes that sports with Chinese traditional characteristics are inherited from the life history of the Chinese people". Zhang xuanhui (2006) Diao Zhendong (2009) believes that it generally refers to sports activities handed down from generation to generation and has no impact on other nationalities in a certain area. "Tu chuanfei (2009) Chen Qing (2010) defined it as a kind of sports popular in various countries or regions and rich in ethnic and national cultural characteristics [2].

In short, the regionality and nationality of national traditional sports are the key to determine the definition of national traditional sports. The above scholars mainly pay attention to its sports and relatively lack cultural characteristics. On this basis, the author believes that "National Traditional Sports" It is recognized by one or several ethnic groups and carried out, applied and spread in fixed areas. It is a sports and cultural activity with obvious national cultural characteristics, which is still used and spread today. As shown in Fig. 2 below, sports and cultural activities with national cultural characteristics are shown.

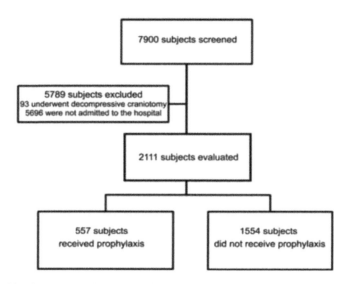

Fig. 2. Sports cultural activities with national cultural characteristics

2.2   Characteristics of National Traditional Sports

On the study of the characteristics of national traditional sports, scholars have summarized different conclusions from different angles and different methods, but their understanding of the characteristics of national traditional sports, such as nationality, tradition, fitness and entertainment, is generally the same [3].

As we all know, the sports created by both Han and ethnic minorities are affected by their religious culture and customs, and show a significant national style in the ideological connotation and expression form of sports. Scholars have the same opinions on the summary of the characteristic of "nationality". Some people think that "regionality" is the cause of the characteristic of "tradition" Characteristics are exactly a factor causing the "traditionality" of national traditional sports. It is considered that the geographical environment of a certain region is the condition for a nation to live and reproduce for a long time [4]. It is in this environment that the fixed mode of production formed by people is handed down and followed from generation to generation, showing obvious traditionality. However, most mathematicians discuss the two separately, and believe that "traditionality" and "Regionality" is two independent characteristics.

About "competitiveness" Characteristics, a more representative view holds that human beings have their own group consciousness and self-consciousness. When we separate our group from other groups and self from others, competition consciousness will be born. Competition consciousness is characterized by competitiveness in sports activities. Some studies believe that the germination of sports competition has appeared in China's primitive society, and competition activities have always been to repay the blessings of gods The important content of religious celebrations. This obviously ignores the factors affecting culture and summarizes it purely from the biological point of view. Among thousands of national traditional sports activities, they are more dependent on production, life, military training, entertainment and ethical activities. There is no sports form with independent form and independent value, which is incompatible with Olympic competitive events, lack of competitive characteristics based on individual competition.

Based on the general guiding ideology of national traditional sports, some scholars put forward that national traditional sports have the characteristics of "scientificity" based on the overall optimization theory of life "double cultivation of life and simultaneous cultivation of mind and body". This is derived from China's early philosophical thought - yin-yang theory, with health preservation Qigong as a typical representative. Others summarized that national traditional sports have the characteristics of "backwardness" And "conservatism" It is considered that Chinese traditional sports culture is mainly formed and developed in the long-term feudal society, and its backwardness is determined by the economic foundation and political system of the feudal society [5]. The feudalism of Chinese traditional sports culture is mainly reflected in its long-term publicized and standardized hierarchy, power consciousness and servility psychology. It should be said that these views are of great significance to development It is of certain significance to examine the National Traditional Sports from multiple angles and further carry out theoretical research.

3   Overall Scheme Design and Implementation of the System

"Video analysis and intelligent diagnosis system" "It is an arm + DSP + CPLD technology framework platform built with athletes in competition or training as the research object. It uses video image processing technology to automatically shoot, collect, analyze and manage the continuous video image frame sequence, and finally obtains the processing results that can be used as a reference for coaches and athletes, so as to improve the training efficiency and ratio The purpose of competition results [6].

3.1   System Function Module Design

The national traditional sports video analysis and intelligent diagnosis system consists of four modules: competition video front-end acquisition and analysis module, competition video management module, technical and tactical data analysis and expert diagnosis module, data information management module (athlete information, competition information, trainer information, etc.).

(1) Game video front-end acquisition and analysis module

The function of the competition video front-end acquisition and analysis module is to comprehensively and completely record the whole process of the competition, collect the competition video data into the system, and provide an accurate data source for the follow-up video analysis [7]. Taking sanda competition as an example, the video data recorder takes a confrontation unit of athletes in Sanda competition as the data acquisition unit, and a Sanda Competition divides it into dozens In each confrontation unit, observe the technical and tactical position, posture and gain and loss points of athletes on both sides, record and save them; then the video analysis module encodes and processes different technical index data, and each information in each confrontation unit is transformed into a determined code and stored in the database; during statistical analysis, search the code in the database, i.e. It can calculate the relevant indexes of Sanda, and then convert them into meaningful statistical analysis data output.

(2) Game video management module

The competition video management module mainly includes the functions of playing, editing and synthesizing the competition video. Guide the competition video to be analyzed into the video editing video frame for multiple previews, and then record the effective technical and tactical information of the athletes in the video into the database according to the competition rounds [8]. If you need to move the collected video data, it can be automatically saved for the competition items through the video synthesis system Objective technical and tactical analysis database. The specific functions of this module are as follows:

1) Video playback: in addition to the general functions of the player, in order to facilitate observation and analysis, the system also adds 1/4, 1/2, 1 (normal speed), 2-speed playback settings, single frame fallback of video playback and other functions.

2) Video clip: each video clip is completed according to the beginning and end of the confrontation unit. The corresponding technical and tactical index information is recorded by manual button.

3) Video synthesis: automatically classify and synthesize the same attributes in the technical and tactical intelligent analysis system. According to different requirements for technical and tactical characteristic information, the synthetic technical and tactical analysis video can be recorded manually [9].

4) By reviewing the game video, collect and analyze the technical and tactical data at the same time, including recording the points of gain and loss, rounds, tactical analysis and summary, etc.

(3) Technical and tactical data analysis and expert diagnosis module

The technical and tactical data analysis and expert diagnosis module is one of the core modules of the system, which can obtain various statistical data, charts and diagnosis results to meet the needs of coaches and athletes. Taking Chinese wrestling or Sanda events as an example, when the competition information is recorded in the system (real-time recorded on the spot through the competition video front-end acquisition module or detailed recorded after reviewing the competition video), the system can automatically analyze the application information of attack, defense, movement, hitting effect, gain and loss point area and other technologies, and can also auto-matically make statistics according to the personalized needs of coaches and athletes' technical and tactical analysis [10]. The system not only records the athletes' main gain point technology, loss point and other technical information, but also records the use times and gain points of each fight round The technical indexes such as point acquisition rate and success rate can be accurately counted. The final conclusion is drawn based on the expert diagnosis conclusion automatically output by the system, plus the coaches' own experience and knowledge (Fig. 3).

3.2 Realization of System Technical Scheme

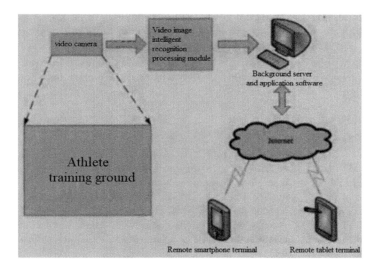

Fig. 3. Frame diagram of video analysis and intelligent diagnosis system

"Video analysis and intelligent diagnosis system" "The framework of the system is shown. It is mainly composed of four parts: CCD high-definition camera, video image intelligent recognition and processing module, background server and corresponding software, and thousands of remote clients. The workflow of the system is as follows: first, set up CCD high-definition camera at the training site to ensure that the shooting range can cover all the activity range of athletes [11]. The video map is obtained by camera shooting After imaging, the image data is transmitted to the video image intelligent recognition processing module through the video cable, which runs the special video image processing algorithm for real-time analysis and processing, and then the processed image data is sent to the background server through the wired network or WiFi wireless network [12]. The final image is displayed, saved and managed through the software, which is convenient for athletes and coaches to view and analyze at any time, Improve technical action. If the background server can be connected to the intemnet network, other users can also see the training video images online in real time through remote clients such as smart phones and tablets. However, due to the network speed, the remote client can only browse some videos after compression processing.

4 Conclusion

In order to improve athletes' training performance more effectively and help coaches make full use of scientific and technological achievements to improve training means, this study introduces the embedded video image intelligent recognition and processing hardware system and its image recognition algorithm, and gives the design schematic diagram of the hardware platform and the implementation of the image processing algorithm From the experimental results, this technology platform is feasible and effective. With the in-depth development of the system, more practical athlete action recognition and accurate positioning, multi lens video technology action overlap and solidification will be developed in terms of function. It is believed that the subsequent improvement and application of this system will provide greater help for coach scientific training.

References

1. Nan, Y.M., Kong, L.L.: Advances in diagnosis and treatment of liver cirrhosis with integrated traditional Chinese and Western medicine (2018)
2. Xia, M., Yang, X., Jing, F., Teng, Z., Lv, Y., Lixia, Y.: Application of chromosome microarray analysis in prenatal diagnosis. BMC Pregnancy Childbirth 20(1), 1–11 (2020)
3. Keishi, S., Hajime, O., Yusuke, M., et al.: The role of video-assisted thoracoscopic surgery in the diagnosis of interstitial lung disease. Sarcoidosis Vasculitis Diffuse Lung Dis. Official J. WASOG (2021)
4. Mi, K.M., Tae-Yong, P., Ah, L.J., et al.: A study of tongue and pulse diagnosis in traditional korean medicine for stroke patients based on quantification Theory Type II. Evidence-Based Complement. Altern. Med. eCAM (2020)
5. Delorme, N, DM, X., Croué, A., et al.: Unusual presentation of infantile myofibromatosis with an ulcered plaque. Ann. Dermatol. Venereol. (2019)
6. So, K.: Developing a reliable and versatile system for the diagnosis of metastatic vertebral fracture: challenges and future directions. J. Nat. Compr. Cancer Netw. JNCCN (2021)
7. Usher-Smith, J.A., Thompson, M.J., Sharp, S.J., et al.: Factors associated with the presence of diabetic ketoacidosis at diagnosis of diabetes in children and young adults: a systematic review. BMJ (Clinical Res. Ed.) (2019)
8. Kedir, E., Worku, J., Takele, T.: Developing a prototype knowledge-based system for diagnosis and treatment of diabetes using data mining techniques. Ethiop. J. health Sci. 30(1) (2020)
9. Karthik, A., Jiang, B., Li, Y., et al.: A Web-based System to Assist With Etiology Differential Diagnosis in Children With Arterial Ischemic Stroke. Topics Magn. Reson. Imag. 30(5), 253–257 (2021)
10. Yang, F., Tang, Z.R., Chen, J., et al.: Pneumoconiosis computer aided diagnosis system based on X-rays and deep learning. BMC medical imaging 21(1), 1–7 (2022). https://doi.org/10.1186/s12880-021-00723-z

11. Fei, X., Deruo, L., Yongqing, G., et al.: Survival rate and prognostic factors of surgically resected clinically synchronous multiple primary non-small cell lung cancer and further differentiation from intrapulmonary metastasis. J. Thorac. Dis. (2020)
12. Israel, F.P., Seims, A.D., Lisa, V.H., et al.: Modified uniportal video-assisted thoracic surgery versus three-port approach for lung nodule biopsy in pediatric cancer patients. J. Laparoendosc. Adv. Surg. Tech.. Part A (2019)

The Integration and Optimization of British and American Literary Information Resources Based on Web 3.0

Mingyi Ling[✉]

Sanya Aviation and Tourism College, Sanya 572000, Hainan, China
bowenfriend@163.com

Abstract. The rapid development of information technology is an important context for the development of contemporary literature and culture. Based on Web 3.0, this paper analyzes the necessity and feasibility of the optimization and integration of British and American literature curriculum resources for English Majors by using mash up, open ID, open API, semantic web technology, intelligent search, cloud computing and other technologies. Relying on Web 3.0 technology, this paper constructs the curriculum resource platform of British and American literature, which can make the construction of curriculum resources of British and American literature systematic, scientific and personalized, so as to change the lagging and messy status of teaching resources of British and American literature.

Keywords: Web 3.0 · British and American literature · Information resources · Integration · Optimization

1 Introduction

In his masterpiece the third wave, Alvin Toffler once said that human society is experiencing the reconstruction of civilization brought by the third wave after the impact of the two waves of agricultural revolution and industrial revolution. "The third wave of building a new information field" has made information technology the core force of civilization reconstruction, and information processing has thus become an unavoidable basic problem in many fields outside the information industry [1].

The combination of information technology and foreign language education has a long history. With the rapid development of information technology, foreign language education has gone through the stage of computer-aided education and network-aided education, and is moving towards the integration of standard chemistry and information technology. Some scientists believe that "information technology and foreign language education" or "ecological integration of information technology and foreign language curriculum" has laid a theoretical foundation for foreign language education under the new technology. Discipline [2].

British and American literature is an important part of foreign language education. At present, with the information explosion and the maturity of network technology, the information sources of British and American literature are increasing rapidly.

V. Sugumaran et al. (Eds.): ICMMIA 2022, LNDECT 138, pp. 657–664, 2022.
https://doi.org/10.1007/978-3-031-05484-6_83

Classification, integration and optimization have become problems to be solved [3]. With the development of Internet technology, Web 1.0 and Web 2.0 are gradually mature and perfect, and are rapidly giving way to Web 3.0. We are committed to intelligence, accuracy and personalization Web 3.0, a network service, provides a strong technical support and a new network platform for the integration and optimization of British and American literary information resources, which has been deeply criticized.

2   Overview of Web 3.0

2.1   Concept of Web 3.0

With the development of the Internet, more and more people surf the Internet. Therefore, as long as people who have come into contact with computers and the Internet know, the web is what we call the network every day. Of course, its role is to provide services for many network users. Users can also be called netizens. The purpose of surfing the Internet is to find the information they need or purely for entertainment, no matter what users surf the Internet for We can classify it into the spiritual and material levels [4]. According to the content of network services and the spiritual and material feelings brought to users, we can completely classify the network, and the service quality provided by different levels of networks is different. Therefore, there are Web 1.0, Web 2.0 and Web 3.0. Compared with Web 1.0, Web 2.0 has different functions It is to meet more spiritual needs of users, so that users can have certain interactive rights on the Internet, such as news reading and manufacturing. Users can have their own home page space, showing a certain personalization. In this, users can communicate and communicate with their netizens or readers. Although Web 2.0 has brought vitality and development to the network, Web 2.0 is not an end With the existence of extreme, especially the higher and higher requirements of users for the network, the existing functions of Web 2.0 show some limitations [5].

According to the limitations of Web 2.0, the emergence of Web 3.0 is an inevitable result. At present, Web 3.0 is still very abstract, and its concept has not been customized in academic circles, which makes many researchers' research on Web 3.0 only exist in marginal areas. In the minds of scholars, the idea of Web 3.0 is to realize a real state of democracy and equality. Network users have higher information control, In addition to the interaction in the Web 2.0 era, we will also gain higher control over information. We know that for a more intelligent Internet, search may not be important, because users completely turn their thinking to the computer at this time, and what we need is to carry out spiritual communication convenient for legal supervision [6]. No matter how powerful Web 3.0 is, its focus is to make Internet users not worry But they should have a sense of belonging and be their own masters, which requires network users to use real identity information and supporting game rules, that is, network legal system and network authentication.

2.2   Features of Web 3.0

By comparing Web 3.0 with Web 2.0 and Web 1.0, we can at least conclude that Web 3.0 has the following characteristics:

(1) The information source is individualized, changing from authoritative information source to individual information source. It has good personalized settings, supports the user's personal skin template, and the size and position of the window can be adjusted freely.

(2) The communication channels are diversified. Before the Internet, the communication channels of information were mainly newspapers and television broadcasting. After the emergence of the Internet, it has become the fifth media. In the Web 3.0 era, mobile phones will change from communication tools to the main channel of information communication, which is likely to become the sixth media.

(3) Customized information has become the mainstream, the individualization of information and the diversification of communication channels will inevitably lead to information flooding [7]. Human energy is limited. Therefore, how to facilitate users to customize their favorite information has become one of the key problems to be solved in the Web 3.0 era.

(4) The credibility of information and how to supervise it will become an urgent problem to be solved. Web 1.0 is the era of mainstream media, and the credibility of information is still questionable. In the Web 3.0 era, the proliferation and individualization of information will inevitably lead to the difficulty of distinguishing true and false information, and the authoritative media will be challenged as never before.

(5) The individualization of information sources and the diversification of communication channels determine that the future information will change from large pieces of information to incisive information. The micro blog represented by Sina Weibo has the characteristics of Web 3.0, and its sentences are short and concise, which makes those who have never used the blog to make a long speech or dive forever in the forum have the impulse and ability to speak [8].

(6) The massive amount of information causes the excessive proliferation of network information. The 21st century will be the century of information explosion, which we have heard a long time ago. With the development of the network, Web 3.0 will usher in the real era of information explosion.

3   Processing of British and American Literature Information Resources Based on Web 3.0

3.1   British and American Literature Information Resources

The article holds that books, teaching materials, courseware, electronic materials, multimedia CD-ROM and other network resources constitute the information resources of foreign language teaching in a narrow sense, while the collection of information activity elements such as information, information creator and information technology accumulated in the process of foreign language learning belongs to foreign language teaching in a broad sense Therefore, this paper uses the concept of narrow foreign language teaching information resources to explore the integration and optimization of Web 3.0 on narrow British and American literary information resources.

As we all know, literature is deeply and multi dimensionally related to philosophy, history, culture, sociology and other disciplines, and literature itself covers writers, works, literary theory, literary criticism and other aspects. These internal and external information of British and American literature together constitute important information resources that British and American Literature Teaching relies on [9]. These information resources can be divided into two levels Times:

(1) Core resource level: the core resources of British and American Literature Teaching refer to the information resources containing the ontological content of literature, such as writers, works, literary theory, literary criticism, literary thought, etc. the core resources of literature teaching are the central content of literature teaching at all levels and the clue and basis for the use of literary information resources.

(2) Level of auxiliary resources: the auxiliary resources of British and American Literature Teaching refer to the information resources related to the composition of literature and scattered in philosophy, sociology, history, culture and other disciplines. They are an important reference for literary learning and play a role in supplementing, deepening and expanding the understanding of literary works, literary theories and literary phenomena.

3.2 Necessity

In 2012, at the "national academic seminar on college foreign language curriculum and teaching reform", the participating scholars reached a consensus that it is the only way for College foreign language teaching reform to make full use of information technology and realize the informatization of College English teaching.

Although British and American literature teaching is different from college English teaching, as a curriculum at different levels of foreign language discipline, there are indisputable similarities between them. Moreover, under the situation of information technology penetrating into literary and cultural life, the existence form and relationship of literary information have undergone fundamental changes. Under the traditional economic and cultural form, literary resources are numerous, However, with books as the main storage form, its collection and sorting mainly rely on manpower. However, in the historical context of knowledge economy and information explosion, there are more and more literary resources stored in the form of electronic information, and there is multi-dimensional interconnection between them [4]. In the face of such a huge information system, without the strong support of information technology, manpower can not realize information The development of web technology, especially the emergence of Web 3.0 technology, provides key technical support for the integration and optimization of British and American literary information resources; without this support, multi-form and diverse information resources will not be able to form a system and effectively transform into available discipline resources.

4   Integration and Optimization Platform of British and American Literature Information Resources Based on Web 3.0

4.1   Platform design

Web 3.0 really plays a role in the teaching of British and American literature. It must rely on the network environment and related technologies it provides to build an efficient platform for British and American literature curriculum resources, that is, the integration and optimization platform of British and American literature information resources. The platform is represented by a diagram, as shown in Fig. 1.

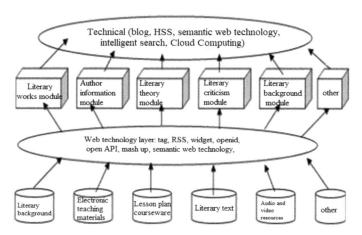

Fig. 1. Integration and optimization platform of British and American literature curriculum resources based on Web 3.0

5   Integration of Source Information

Relying on the powerful information processing function of Web 3.0, the above platform is first committed to the integration of source information. The so-called source information refers to the information resources in various forms that belong to the category of British and American literature or have a certain correlation with British and American literature. They have a wide variety and various storage forms, which need to be sorted and optimized by web technology. Firstly, these source information are classified and aggregated based on their own attributes through tag, semantic web, Mashup and RSS, and re edited into interrelated network system by using widget, SNS, open API and other technologies, so as to complete the preliminary integration of source information. Generally speaking, based on the specific content of literary research, the integrated results can be preliminarily divided into literary works module, author information module, literary theory module, literary criticism module, background knowledge module, etc. on this basis, they can be further divided into smaller modules according to actual needs, such as "background knowledge module of medieval literature" "Renaissance literature background knowledge module", "medieval religious background knowledge module", etc. the information module formed by this integration will aggregate literature information with homogeneous characteristics, and pay attention to the relationship between literature and other related disciplines, so as to form an information system centered on literature and involving other disciplines, so that information users can not only make overall consideration, but also get what they need [10]. As shown in Fig. 2.

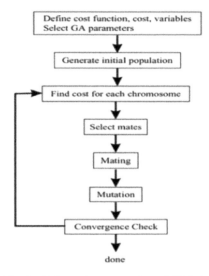

Fig. 2. Information processing of Web 3.0

6   Optimization of British and American Literature Information Resources

The improvement of British and American literature information resources is the process of supplementing, revising and perfecting the relevant information of British and American literature. It is divided into two aspects: the improvement of individual information and the improvement of information system. We know that British and American literature information resources are a dynamic information system, which changes constantly with the changes of the times, discipline development and practical needs. In view of this, we are isolated The information resources of Web 3.0 need to be constantly updated, revised and improved, and the information system itself also has problems in structure adjustment, content improvement and so on. The semantic web, tag. RSS, widget, mash up. Semantic web technology, intelligent search and other technologies of Web 3.0 can not only continuously improve the accuracy of information, but also gradually improve the construction and integration of information system Information resource optimization is an important basis.

It should be pointed out that the integration and optimization of British and American literature curriculum resources are one and two sides of the processing of British and American literature curriculum resources, which are closely related and inseparable. The integration must be the integration for the purpose of optimization, otherwise the integration will not have practical benefits; the optimization must also be the optimization based on integration, otherwise the optimization will become a tree without roots and cannot have a deep-seated and long-term impact. In short, the curriculum resource integration and Optimization Platform Based on Web 3.0 aims to effectively integrate and improve the high-quality information of British and American literature, so as to realize the scientificity and systematicness of information resource processing. As shown in Fig. 3.

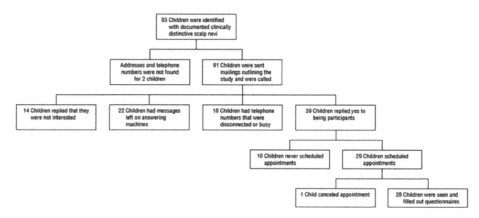

Fig. 3. Based on Web 3.0's curriculum resource integration and optimization platform

7 Conclusion

The rapid development of information technology in the second decade of the twentieth century has created important conditions for the development of literature and culture. Information retrieval, integration and optimization are not industry-specific phenomena, but have a certain universality. The integration and use of resources depends on strong support for the development of information technology. According to Web 1.0 and Web 2.0, Web 3.0 realizes intelligent, correct and personalized network services through powerful technical means, It provides new technical support for the integration and optimization of British and American literary information sources. On this basis, the British and American literature information processing platform is English-American literature.

References

1. Grene, C.W.: The prodigal son in english and american literature: five hundred years of literary homecomings, Alison M. Jack, Oxford University Press, 2019 (ISBN 978-0-19-881729-1), viii + 182 pp. hb 55. Rev. Relig. Theol. **27**, 238–239 (2020)
2. Chung, U.: Asian american literary forms. In: Belasco, S., Gaul, T. S., Johnson, L., Soto, M. (eds.) A Companion to American Literature, pp. 398–413. Wiley, Hoboken (2020). https://doi.org/10.1002/9781119056157.ch87
3. Martín, U.S., Aarons, V. (ed.) The new Jewish American literary studies. Mod. Judaism, (1), 1 (2022)
4. Qiu, X.: Intelligent classification of logistics multi-distribution resources based on information fusion. Int. J. Inform. Technol. Manage. **20**(3), 250 (2021)
5. Mhalciuc, C.C., Grosu, M.: The contribution of the information provided by the management accounting in the decision-making process. Eur. J. Acc. Financ. Bus. **16**(26), 117–131 (2021)
6. Atzori, M., Koutrika, G., Pes, B., Tanca, L.: Special issue on "Data Exploration in the Web 3.0 Age." Future Gener. Comput. Syst. **112**, 1177–1179 (2020)
7. Gao, J.: A brief analysis of the literature network resource mining and data development. In: Zheng, X., Choo, K.-K.R., Dehghantanha, A., Parizi, R., Hammoudeh, M. (eds.) CSIA 2019. AISC, vol. 928, pp. 319–324. Springer, Cham (2020). https://doi.org/10.1007/978-3-030-15235-2_48
8. Lagios, N., Méon, P.G.: Experts, information, reviews, and coordination: evidence on how literary prizes affect sales. Working Papers CEB (2021)
9. Bougioukas, K.I., Bouras, E.C., Avgerinos, K.I., et al.: How to keep up to date with medical information using web based resources: a systematised review and narrative synthesis. Health Inf. Libr. J. **37**(4), 254–292 (2020)
10. Gannon, T.C.: Native american literary forms. Companion Am. Lit. **3**, 382–397 (2020)

Research on the Design of a Traditional Music Push System for Big Data

Ying Liu[✉]

XianYang Normal University, Xianyang 712000, China
MAOgen2007_a@163.com

Abstract. Since the development of the concept of big data, they have been worrying. Society has had extensive discussions about its perspectives, applications and benefits Great information technology has penetrated all areas of people's lives and will certainly affect the preservation, innovation and development of traditional Chinese music culture in the future. In the "Big Data Age", Feinberg believes that big data has changed people's actions, thinking and values. Big data has changed the way people solve problems; in fact, the data is only real access, and the core data is cadres. Different topics have different ideas and thoughts This article discusses and analyzes the importance, benefits and application of big data and hopes More attention is devoted to the research of traditional music and big data, To promote the more effective dissemination and development of traditional music culture in the world.

Keywords: Traditional music · Big data · Digitalization · Cultural ecosystem

1 The Meaning of Big Data

The definition of big data has many different descriptions. The core of this value space is the value of the data itself. Therefore, the big data technology system is developed with the core purpose of improving the value of the data. In short, cloud computing provides users with various computing resources and services through the Internet. Blockchain itself is a technology, but it is also a model. Supported by technology, modu cloud focuses on computing and storage databases. When enterprises go to the cloud, they need to choose a safe path to the cloud [1]. In this context, major cloud manufacturers and third-party cloud security companies have launched corresponding cloud technology architectures in recent years. The cloud system also gives its customers the ability to continue to provide all-round services [2]. Both schools have their own characteristics, but from the perspective of discovering data value, the value of data and analysis needs to further integrate business scenarios to finally realize data value.

Introduce industry knowledge and build a unique industry experience model to maximize data value. "For example, how to solve the problems of mass customization, supply chain and globalization are the practical challenges that these enterprises need to face". In order to seize this series of opportunities and meet challenges, digital transformation has become an "essential choice" for more and more companies [3]. The example of moduyun is based on the data center and integrated with DSP in Omni

V. Sugumaran et al. (Eds.): ICMMIA 2022, LNDECT 138, pp. 665–672, 2022.
https://doi.org/10.1007/978-3-031-05484-6_84

channel, member management, CRM and marketing management to form the core application of the consumer asset operation analysis platform. Technology precipitation and development to meet market challenges, and use technology accumulation and development to meet market challenges. Relying on technology accumulation and strong cloud computing R & D capability, we have created rich industry solutions, opened a win-win cloud ecosystem, promoted the construction of industrial Internet, and helped realize digital upgrading in various fields. As shown in Fig. 1.

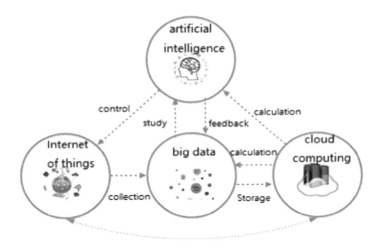

Fig. 1. Big data mode

At the same time, the large music data is still scattered, so it is necessary to classify the large music data of different terminals. There are two types of big data. One of them is the static data structure Static data structure means the transition from paper document to digital document. The other is interpersonal. It is called dynamic data structure [4]. In other words, after the digitization of paper documents, after the establishment of a static data structure in the age of big data, people in the digital age pay attention to the interaction between culture and other human activities. Static and dynamic data structures generate interactive activities, an entire process model of social and human cosmetic activities, and a complete data source for analyzing new lifestyles [5]. Data help deepen scientific research, transfer more relevant experiences, and computer can research direction, research direction, etc. predict. D. Of course, data reporting will continue. If more data is available, different analysis results of different music elements can be obtained. Therefore, data alone is not sufficient. In terms of traditional music culture, it is important to find an understanding of what is hidden in big music data, and the ultimate purpose is to analyze, use and extract data.

2   Advantages of Music Big Data

In the Internet era, music big data not only has the ability to predict the future, but also has the advantage of in-depth speculation and digging for the use of corresponding music resources. For music researchers, the advantages of music big data are as follows.

2.1   Provide a Multi-dimensional Research Strategy

According to the real-time dynamic of music behavior, the researchers dig and analyze the massive data of traditional music behavior generated by the research objects on the Internet, reveal the content of the law of the development of traditional music culture, and put forward the conclusions and Countermeasures of traditional music research, which has greatly changed the methods and means of music research [6]. For example, through the wechat public platform of traditional music, we can understand the needs of traditional music lovers for traditional music and other situations, and master the different answers and discussions of traditional music researchers for various problems through information interaction. The real-time dynamics of music behavior is shown in Fig. 2 below.

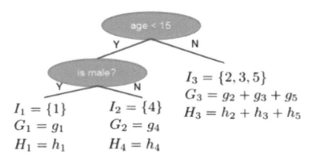

Fig. 2.   Real time dynamics of music behavior

2.2   Creating the Value of Cultural Clustering

Cloud computing provides storage, management and analysis for massive and diversified music big data from different sources, and feeds the results back to the application of music resources, thus creating great cultural and social value. Among them, researchers can find music data points which are naturally combined together, and use music database to classify and associate them. In addition, using big data as a rest tool, we can query and compare data sets vertically and horizontally according to personal needs [7]. For example, by comparing and analyzing the music of the same era, the same region and the same mode, we can achieve the cultural clustering of music type, music cluster, music system and music circle.

2.3   Insight into the Context Between Data

This paper explores the relationship between data in music database, reconstructs researchers' research thinking, and finds the association, affiliation and causality between music data. For example, relationship mining in traditional music education research. The relationship between students' achievement and teaching mode, teaching method and teacher-student interaction; the relationship between traditional music inheritance and inheritor, ecological environment, history and culture, field distribution, etc. [8]. If the average query delay is an important index to measure the performance of multidimensional query engine, then the expansion rate is an important index to measure the pruning results of multidimensional cube. The expansion rate is the ratio of the size of the materialized cube to the size of the original data, which well reflects the complexity of the pre calculation. Similarly, the expansion rate also needs to be estimated, and the estimation method is as follows:

$$Exp(\Omega) = \frac{\sum queryLatency(req)}{n}, req \in \Omega \tag{1}$$

$$Exp(M) = \frac{\sum scale(m) \times column(m)}{scale(M) \times column(M)}, m < M \tag{2}$$

3   Application of Music Big Data

We are here to enjoy the convenience of big data. In fact, relying on big data technology can provide innovative means and methods for all kinds of database applications and needs, making full use of digital resources collection, collation, induction, summary, analysis, etc. This is what big data brings us. Therefore, the existing music and cultural resources data and digitization is also a part of big data.

First, Music Association. Its purpose is to find the related tracks and find the correlation, sequence and belonging relationship among the songs. The analysis of music association mainly involves the following dimensions: music style, related musicians, creation time, music language, music communication, music distribution, music evolution, etc. [9]. Therefore, through the analysis of music big data, we can find out the relationship and ethics of different dimensions of music. For example: first, to explore the historical context of the evolution of the same folk song by gathering the same music from different regions; second, to find the music with higher degree of research and then to find a larger research group; third, to find the music of the same author and study the author's creation process, so as to provide conditions for the research of music oral history; fourth, to find out the relevant music sounds, and to use machine learning to further study the similarity of music; fourth, to find out the music sound of the same origin and to use machine learning to further study the similarity of music; For example, "listening to music and recognizing music" in music playing software, that is, "when you hear the sound of music, you can find the music", and so on. As shown in Fig. 3.

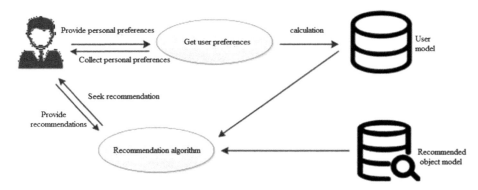

Fig. 3. Big data music recommendation

4 Music Field Track and Simulation

As a special cultural phenomenon, music has its own special historical facts. Its particularity lies in the fact that music embodies its social significance in many kinds of things. Musicologists record, process, analyze and interpret the traditional music resources in the field investigation. Combined with the analysis of music big data, researchers have a deeper understanding of the real-time dynamics of the relevant interactive information and research of traditional music, such as the investigation object, case study, regional division, music form, music skills, music behavior, music concept, ecological environment, development trend, etc. [10]. Through the phenomenon, we can also analyze: the collection of field investigation objects, the collection of interviewed people, the collection of regional distribution of people, etc. In this way, researchers can use big data to analyze which people have been interviewed, which people have been interviewed more, and which researchers have interviewed these people [11]. Of course, they can also analyze the data according to the reverse thinking: the research method of a certain research group, the relationship between the research direction and the music types of a certain area [12]. In addition, we can track the geographical distribution of music clubs with the help of GLS geographic information system, and researchers can make full use of the relevant knowledge, such as geography, knowledge, etc., to present a lot of music events of traditional music research into the information of time and space crisscross.

The track of music also presents different curves with different geographical distribution, and its curve characteristics are different with different frequency and time. The specific simulation is shown in Fig. 4.

Recommendation algorithm itself is actually just an efficient technical tool, which can quickly realize the accurate mining of user needs. In short, algorithmic recommendation technology will analyze people's behavior, habits and preferences by capturing users' daily use data, and then accurately recommend and distribute information. The core purpose of the company's application of algorithmic recommendation technology is to keep users' attention and turn it into revenue while providing more accurate push [13].

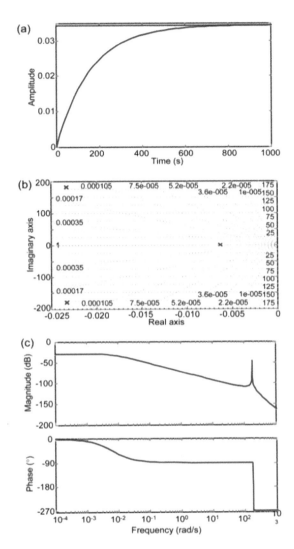

Fig. 4. Music track and simulation with frequency and time

Personalized music recommendation technology based on big data, education and practical application value in the field of social network have gradually become an effective method to alleviate the problem of information overload. Personalized recommendation in the context of big data collects a large number of information that users are interested in, and uses big data analysis technology to mine users' preferences. As shown in Fig. 5 below, the preference mining model of big data analysis technology mining users is shown.

Algorithm 2: Approximate Algorithm for Split Finding

---

for $k = 1$ **to** m **do**
  Propose $S_k = \{s_{k1}, s_{k2}, \cdots s_{kl}\}$ by percentiles on feature k.
  Proposal can be done per tree (global), or per split(local).
end
for $k = 1$ **to** m **do**
  $G_{kv} \leftarrow = \sum_{j \in \{j | s_{k,v} \geq x_{jk} > s_{k,v-1}\}} g_j$
  $H_{kv} \leftarrow = \sum_{j \in \{j | s_{k,v} \geq x_{jk} > s_{k,v-1}\}} h_j$

Fig. 5. Big data analysis technology mining user preference mining model

5 Concluding Remarks

How to make use of big data is not impossible, only you can't think of it. Of course, there are advantages and disadvantages for any new thing. For example, the current relevant laws are not perfect, user privacy, security has become a problem. Therefore, the application of traditional music big data goes beyond computer science, cultural ecology, music anthropology and other disciplines, music education, library science, cartography, statistics, sociology, psychology, information technology, etc. D. It solves the practical application problems of education and scientific research. Finally, we must formulate effective tourism policies with the aim of developing traditional music.

References

1. He, T., Lin, S., Chen, X.: Research on the training mode of innovative and entrepreneurial thinking of design talents a case study of digital media arts (2020)
2. Li, S., Huang, T., Xia, Y.: Research on application value of traditional cultural elements in visual design. World Sci. Res. J. **6**(3), 176–179 (2020)
3. Hu, B.: On the chinese traditional clothing innovation preliminary discussion. Int. J. Soc. Sci. Educ. Res. **2**(11), 117–119 (2020)
4. Hu, C., Chen, C., Zhou, T., et al.: The design and research of a new pharmaceuticals-vending machine based on online medical service. J. Ambient. Intell. Humaniz. Comput. **3**, 1–10 (2020). https://doi.org/10.1007/s12652-020-02482-1
5. Underberg-Goode, N.: Participatory research and design in the portal to peru. Ann. Anthropol. Pract. **44**(1), 119–125 (2020)
6. Haug, M., Camps, P., Umland, T., et al.: Assessing differences in flow state induced by an adaptive music learning software, 1−4 (2020)
7. Zijin, W.: Research on automatic classification method of ethnic music emotion based on machine learning. J. Math. **2022**, 1–11 (2022)
8. Lili, Y., Zhao, D., Xue, Z., Gao, Y.: Research on the use of digital finance and the adoption of green control techniques by family farms in China. Technol. Soc. **62**, 101323 (2020)
9. Tang, Z., Lu, A., Yang, Y.: Design research in the practice of memory place-making. Open House Int. **45**(1/2), 55–68 (2020)

10. Wang, Z., Liu, S.: Preliminary research on the evolution of traditional Chinese kitchen utensils design (2020)
11. Gu, X., Gu, X.Z., Hao, W.Y., et al.: Research into factors influencing continued use intention of music educational apps based on e-learning (2020)
12. Yang, T., Nazir, S.: A comprehensive overview of AI-enabled music classification and its influence in games (2022). https://doi.org/10.1007/s00500-022-06734-4
13. Phatnani, K.S., Patil, H.A.: Music footprint recognition via sentiment, identity, and setting identification (2022). https://doi.org/10.1007/s11042-021-11430-w

Integration and Utilization of Art Education Resource Platform Under Big Data

Jingxian Liu[✉]

Shandong Management University, Jinan 250300, China
liujingxian201711@163.com

Abstract. In order to integrate and utilize the art education resource platform on the basis of big data, this paper will carry out relevant research, mainly discuss the basic concepts and application methods of big data, and then put forward the platform design scheme and resource integration and utilization methods. Through this study, with the help of big data, the art education resource platform can be fully integrated and utilized. Various resources can provide help for art education and promote the improvement of education quality and efficiency.

Keywords: Big data · Art education · Art education resource platform

1 Introduction

With the combination of information technology, network technology and education, people begin to get used to using data resources for education, which is no exception in the field of art education. But under this development trend, people gradually found that the education work of the resource demand is more and more big, how to integrate the resources use became a big problem, the difficulty is that if you want to meet the demand of large resources, you have to get huge data, and then integrate these data resources, and human ability is limited, it is hard to a even sleepless for integrating all the data, Therefore, the integration and utilization of data resources are in a deadlock. How to break the situation is a concern of the art education field, and it is necessary to carry out relevant research.

2 Basic Concepts and Application Methods of Big Data

2.1 Basic Concepts

Big data is a huge data integration, internal data scale is huge, and update speed, at the same time each there are intricate relations between the data, if people are able to sort out these data, can have a better understanding to the real problem, in order to make reasonable decisions, both for the application of large data value. But it is important to note that objectively big data is pure data volume, does not have the initiative role, therefore must through the role of the large data related tools for mining, integration, and artificial in big the huge scale of large, complex data relationships of data, data, the

V. Sugumaran et al. (Eds.): ICMMIA 2022, LNDECT 138, pp. 673–679, 2022.
https://doi.org/10.1007/978-3-031-05484-6_85

characteristics of quick update, in the face of these work, this has hindered the use of big data, As a result, a series of technical tools have been developed to help people mine and integrate big data.

2.2   Application Methods

At present, the technical tools mainly used in big data mining and integration are intelligent technology + neural network scheme: First of all, smart technology is a kind of reality can be feedback, which can identify the data and then according to the result of recognition to the real situation to judge, finally make a decision a technology, this technology is similar to human logical comparison on the operation process [1, 2], the most prominent characteristic is changed according to the situation, different from traditional automation technology will only according to the mode of operation of the default process, Therefore, it has higher application value. At the same time, the data recognition speed of intelligent technology is very fast, and it can identify millions or even a large amount of data in a very short time, indicating that this technology can be used as a big data application tool. Secondly, the neural network mainly acts as the operation logic of the intelligent technology in the technical scheme, that is, the intelligent technology can recognize the real data, all rely on the neural network. Neural network is mainly rely on to drive the operation of machine learning algorithm, the algorithm, which can identify characteristics of the data to the data definition, and then the data will become a base node, then from each basic nodes, will be given basic nodes and other defined data, calculate the correlation degree between each other, similarity, etc., to build a network structure, This structure is the neural network model (see Fig. 1 for details), which can operate in a continuous cycle, accept every piece of data for training, and finally output the results to promote the operation of intelligent technology, and also let the artificial understand the actual situation through the data, so as to make accurate decisions [3–5].

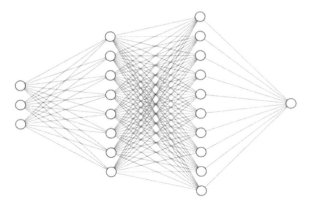

Fig. 1. Neural network model

Combined with Fig. 1, the basic layer, hidden layer, output layer and data result are respectively from left to right. The basic layer is composed of several basic nodes, and each node is a basic data. Then, the connection between each basic node and other data nodes jointly constitutes the hidden layer, which is generally not displayed to the public. Eventually got through the calculation of algorithm each node with other data, generate the output layer, output layer can choose from several results as one of the most optimal output as a result, often directly according to the results to make a decision, but if the result is not satisfactory, you can find other not output in the output layer, this does not affect the operation of neural network [6].

3 Art Education Resource Platform Design and Resource Integration and Utilization Methods

3.1 Platform Design

At present, a lot of art education to construct the corresponding resource platform, platform is mainly used to give art education work to provide all kinds of resources to help, but with the development of the era and the depth of application platform, the platform in the data shows the characteristics of big data, under the characteristics gradually formed at the same time, the original platform also appeared some problems in the applicability, The database such as the original platform is not enough to store huge data, which can easily lead to the loss of data resources, so the art education resource platform needs to be redesigned. From this point of view, the following will put forward a design scheme of art education resource platform for reference.

Overall Framework

Figure 2 is the overall framework of the art education resource platform in this paper.

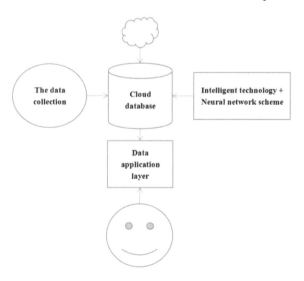

Fig. 2. The overall framework of art education resource platform

Process analysis: According to the framework in Fig. 2, all kinds of data are firstly obtained through the data acquisition layer in the system application, and then transmitted to the cloud database. The cloud database is mainly supported by the cloud resource pool and connected with the intelligent technology and neural network solution. With the help of the intelligent technology and neural network solution, all the data in the cloud database can be classified and integrated, and the integrated data will be displayed in the data application layer. Finally, users can enter the data application layer through terminal devices, see the integrated data, schedule the data according to their own needs, or use some application functions in the data application layer to make use of the data [7, 8].

System Implementation Method

Combined with Fig. 2, the implementation method of this system is as follows.

Firstly, data acquisition layer: two data acquisition methods, manual input and network transmission, are mainly selected to construct the data acquisition layer, that is, a temporary database is established in the network environment first, and the database capacity must meet the magnitude of single input data. Secondly, the database will be open internally, so that manual can input the data searched by themselves. At the same time, automatic transmission tools are developed in the network environment. As long as new data appears in the network, it will be automatically transmitted to the temporary database. Table 1 shows the configuration of the temporary database at the data acquisition layer [9, 10].

Table 1. Configuration of temporary database at data acquisition layer

The name of the	Configuration is
Category	SQL 2000
Capacity	1T

Second, cloud database: temporary database is only for storing single input data, not permanent, otherwise it will lead to insufficient capacity, and the data must be stored permanently in nature, so a new large volume database is needed to solve this problem, and this paper chooses cloud database for this purpose. The capacity of cloud database is unlimited, so no matter how the scale of data in the platform increases and changes, cloud database can be properly saved through cloud database. Therefore, this paper will connect the temporary database with cloud database, thus solving the problem. About cloud database realization method, this paper mainly through the cloud resource pool won a huge cloud resources, and then divide the part of the resources to be cloud database initial capacity, and to develop the automatic development program, the purpose is to initial capacity does not meet the demand of data stored in the cloud database automatically expand under the condition of resource, prompting increased storage capacity, So that you can continue to store data. In addition, considering that the original cloud database has certain security risks, this paper also carries out a closed design for it, leaving only the login interface in the data application layer. Anyone who wants to access the cloud database must pass login authentication. Table 2 shows the configuration of cloud data.

Table 2. Cloud data configuration

The name of the	Configuration is
Initializing capacity	5 T
The maximum capacity	Infinite

Third, intelligent technology + neural network scheme: First of all, according to the requirements of art education platform integrated utilization of resources to develop the intelligent technology tools, such as data classification, retrieval tools, analysis tools, including classification tool is responsible for all the data in the database of cloud classification, principle to obtain the features of the data, in accordance with the characteristics of the matching degree of classification, retrieval tools mainly for the convenience of data calls, Its function button in the data application layer, belongs to a kind of application function, analysis tool and retrieval tool are also application function, its function button is also in the data application layer naturally. Secondly according to the three smart technology tools, including data retrieval tools without neural network support, using automated matching tool can be realized directly, namely user input keywords in the search bar, and then the system will automatically match all data for the keywords, and then displayed to the user, and the other two tools you need to through the neural network as the support, The neural network selected in this paper is feedforward network (i.e., the neural network model introduced in Fig. 1), through which data classification tools and analysis tools are logically supported and functions can operate intelligently.

Fourth, data application layer: this layer is mainly based on network environment, this article on the Web page and Internet construction the foundation of data application layer, and then through the interface development, design, etc., realize the data visualization interface and toolbar options, contains all applications within toolbar options, after the user clicks can choose according to their own needs. At the same time, the page skipping design is carried out for each tool. When the user clicks the tool button, it will jump to the dedicated page of the tool. The purpose can be realized by completing the operation in this page.

3.2   Resource Integration and Utilization Methods

Based on the design results of art education resource platform, there are two ways to integrate and utilize resources in art education, namely, resource sharing and resource distribution. The details are as follows.

Resource Sharing and Utilization

Educational organizations can take the art education resource platform as a resource sharing center, establish cooperative relations with other educational organizations, and then concentrate the resources mastered by each organization, complete resource integration through the classification of the platform, so that the resources can be used by each educational organization, and realize resource sharing. Figure 3 shows the basic flow of shared resource utilization. Resource sharing and utilization can well solve the problem of insufficient resources of educational organizations, and the integration function of the platform can avoid the retrieval difficulties and isolated information problems after the huge data stream enters the platform.

Fig. 3. Basic flow of resource sharing utilization

Resource Distribution and Utilization

The distribution and utilization of resources mainly acts on the interaction between teachers and students, and there are two forms of utilization: First, teachers can distribute relevant resources to students as learning tasks according to education needs. In the process, the platform can help teachers to obtain relevant resources and undertake the distribution tasks. For example, teachers need students to read relevant art theory materials, so they can consider using resources in this way. Secondly, because students will generate relevant data when learning through the platform, the data will also be transmitted to the cloud database, so teachers can learn about the data of students' learning through the cloud database. These data represent the students' learning situation, from which students' learning needs can be understood. For example, students often refer to the materials related to piano art on the platform during self-study activities, which indicates that students like piano art. At this time, teachers can assign tasks to students through the first form, or provide learning materials to students, and so on, which is conducive to the pertinence and quality of art education. In this process, the platform mainly plays a role of data analysis, that is, the platform can analyze students' preferences according to the frequency of students' data retrieval, so that teachers can directly conduct teaching according to their preferences.

4   Conclusion

To sum up, big data can provide resources for art education, but to integrate and utilize these data resources, educational organizations should reconstruct the art education resource platform. The new platform integrates all kinds of advanced technologies and algorithms, which can replace human beings to integrate resources and promote human beings to make better use of resources.

References

1. Lu, X., Petrov, V., Moltchanov, D., et al.: 5G-U: conceptualizing integrated utilization of licensed and unlicensed spectrum for future IoT. IEEE Commun. Mag. **57**(7), 92–98 (2019)
2. Zambare, V.P., Christopher, L.P.: Integrated biorefinery approach to utilization of pulp and paper mill sludge for value-added products. J. Clean. Prod. **274**(10), 122791 (2020)
3. Turilli, M., Merzky, A., Naughton, T., et al.: Characterizing the performance of executing many-tasks on summit. In: 2019 IEEE/ACM Third Annual Workshop on Emerging Parallel and Distributed Runtime Systems and Middleware (IPDRM).ACM (2019)
4. Nima, S., Nualdaisri, P., Tolaema, I., et al.: Research knowledge utilization on integrated municipal solid-waste management and community participation for public policy in the deep south of Thailand. Int. J. Bus. Adm. Stud. **5**(1), 28–36 (2019)
5. Siting, Z., Michael, C., Bingcao, W., et al.: Utilization of high-cost interventions for targeted clinical conditions during the early stages of ACO development in a commercially insured population. Popul. Health Manag. **22**(2), 377–384 (2019)
6. Laliberté, F., et al.: PS1244 real-world Healthcare Resource Utilization (HRU) of patients diagnosed with Classical Hodgkin Lymphoma (CHL) treated with anti-pd1 checkpoint inhibitors in the United States (US). HemaSphere **3**(S1), 567 (2019)
7. Tumenova, S.A.: Management of integrated utilization of industrial waste storage facilities. IOP Conference Ser. Mater. Sci. Eng. **1079**(6), 062048 (2021)
8. Suheri, A., Cahyani, R.W., Hardjana, A.K., Sulystiawati, E.: Integrated utilization of land and vegetation in secondary forest areas. 3BIO J. Biol. Sci. Technol. Manag. **1**(1), 7 (2019)
9. Rajpal, S., Joe, W., Subramanyam, M.A., et al.: Utilization of integrated child development services in india: programmatic insights from national family health survey, 2016. Int. J. Environ. Res. Public Health **17**(9), 3197 (2020)
10. Monkman, S., Kline, J., Cail, K.: Integrated capture and utilization of cement kiln CO_2 to produce more sustainable concrete. In: 15th International Conference on the Chemistry of Cement (2019)

Enterprise Financial Risk Prediction Based on BP Neural Network Algorithm

Yukun Deng[✉] and Jian Ouyang

Hunan Software Vocational and Technical University, Xiangtan 411100, China
1106362908@qq.com

Abstract. In order to effectively improve the quality of enterprise financial risk prediction, this paper will focus on the BP neural network algorithm related research, first discusses the basic concept of BP neural network algorithm, financial risk prediction status quo, and then on the basis of BP neural network algorithm for financial risk prediction system design. Through the analysis, the system can give full play to the role of BP neural network algorithm, improve the accuracy and depth of enterprise financial risk prediction results.

Keywords: BP neural network algorithm · Enterprise · Financial risk prediction

1 Introduction

Financial management is an important link in the enterprise internal management, and financial risk prediction is the main task of financial management, it is directly related to the stability of financial operations, and indirectly to the whole enterprise operation has certain influence, if there is a problem can lead to enterprise under large economic loss, even makes the enterprise. On this basis, modern enterprises attach great importance to financial risk prediction, but limited by a variety of factors, the quality of this work has not been improved, and the difficulty is becoming more and more difficult, so enterprises began to seek a breakthrough, thus paying attention to the BP neural network algorithm. Related fields that BP neural network algorithm can help enterprises to deal with the increasingly complex financial risk prediction problem, not only can improve the accuracy of predicted results, also can more in-depth analysis of the risk situation, so enterprises should introduce the algorithm to reform financial risk prediction work, in order to help enterprises to achieve this, this article will conduct related research.

V. Sugumaran et al. (Eds.): ICMMIA 2022, LNDECT 138, pp. 680–687, 2022.
https://doi.org/10.1007/978-3-031-05484-6_86

2 Basic Concept of BP Neural Network Algorithm and Current Situation of Financial Risk Prediction

2.1 Basic Concepts of the Algorithm

BP neural network algorithm is the product of artificial intelligence technology, its main feature is to simulate the human neural operation mode to calculate and analyze real problems, the analysis process is mainly based on various artificial neural network models. The more common artificial neural network models are feedforward model and feedback model, so algorithms can also be divided into feedforward algorithm and feedback algorithm. The following two kinds of algorithms will be discussed.

Feedforward Algorithm
The core of the feedforward algorithm is the feedforward neural network model, which is mainly composed of input layer, hidden layer and output layer. The input layer is composed of several base nodes, and then each base node is connected with the hidden node in the hidden layer unidirectionally and independently, so that the base node is updated. Through this process, we can know the similarity, matching degree and correlation degree between the base node and the corresponding implicit node, from which we can get some results. Hidden layer is mainly composed of a number of hidden nodes, and the hidden layer is not a single layer, its itself and can be divided into several layers, the layer is the key of the feedforward neural network model for problem analysis layer, respectively, in the input layer and output layer is a one-way connection, the relationship between basic nodes through this layer will generate the corresponding results, but because the basis of the input layer node number >1, Therefore, the number of results obtained from this layer analysis is also >1. Finally, the output layer, the layer allows only output a result, so the foundation through the hidden layer nodes generated after several results on the results of the output layer to screen and select the most optimal results as the only output, and the results of screening often rely on machine learning algorithms to realize, such as gradient descent method (see is part of the feedforward algorithm, formula 1) [1–3]. Figure 1 shows the basic structure of the feedforward neural network model.

$$\Delta\theta = \frac{n}{m}\sum_{i=1}^{M} G_i \tag{1}$$

where, n is the number of features on each sample; m is the amount of data on each node; M is the number of nodes segmented according to dimensions; G_i is the squared loss function. This algorithm is suitable for sample segmentation or data dimension segmentation machine learning algorithm.

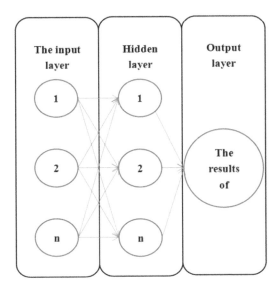

Fig. 1. Basic structure of feedforward neural network model

Feedback Algorithm

Compared with feed-forward neural network model of feedforward algorithm, feedback algorithm of neural network model is more complicated, but it also makes the feedback algorithm can deal with more complex problems, at least in the enterprise financial risk forecast and feedback algorithm application is more common (feedforward algorithm in the enterprise financial risk prediction has been widely used, but because of the increasingly complicated modern financial risk, Therefore, feedback algorithms are becoming more and more popular. Feedback neural network model is also composed of input layer, hidden layer and output layer, but each layer are composed of a base node, and then let the layers according to certain logic connection, namely let nodes and hidden nodes cohesion, generate results from hidden nodes, the results after feedback to the input layer, and the new data node connection, Therefore, the feedback neural network model can be extended indefinitely in theory, in which each data node can not only output results, but also provide feedback to other basic nodes as input signals. Figure 2 is the basic structure of the feedback neural network. It is worth noting that the feedback neural network model can be extended indefinitely in theory, so when solving practical problems, parameters of the algorithm must be set first in order to obtain results, so as to terminate the cycle iteration, otherwise the cycle will continue indefinitely. In addition, feedback neural network is a typical memory model, and each cycle will generate a sample result, which means that the system has learned relevant logic. Therefore, the application value of feedback neural network model will increase with the increase of the number of cycles [4–6].

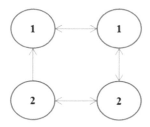

Fig. 2. Basic structure of feedback neural network

In Fig. 2, 1 and 2 in the figure are data nodes in the network respectively, and each node can be used as output layer and input layer, while the hidden layer is in the connection of each node.

2.2   Current Situation of Financial Risk Prediction

Domestic enterprise financial risk forecast work very early, early in the work performance is good, can according to various real transaction information or data to forecast the risk, but with the development of the era, all kinds of inclusion mechanism, the causes of financial risk has become more complicated, making the implementation of financial risk prediction work difficulty is more and more big, At this time, there are still many enterprises using the traditional artificial model for financial risk prediction. This case restricted by artificial ability, financial risk prediction work began to appear all sorts of quality problems, one of the main problems are reflected in the accuracy of the prediction results, such as an enterprise staff in accordance with the original way to the market price fluctuations made a judgment, that will not appear in the market price fluctuations, thus enterprise financial not at risk, But it only considers the government regulation in the process of predicting factors, ignoring the market consumers of all kinds of information, thus eventually happened to the market price fluctuations, enterprise prediction, because the staff have no time to make adjustments, so triggered the financial risk, and this is human can't change, that is too much financial risk formation, artificial even if you don't ignore a variety of factors, It is also impossible to carry out a comprehensive and in-depth analysis of all factors, so the prediction results are bound to have defects. Faced with this situation, enterprises must actively change the working mode of financial risk prediction, and BP neural network algorithm is undoubtedly an excellent option [7].

3   Design of Financial Risk Prediction System Based on BP Neural Network Algorithm

3.1   Overall System Framework

Figure 3 is the overall framework of the financial risk prediction system in this paper.

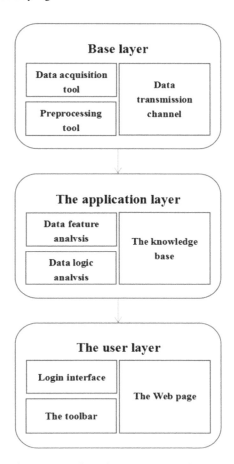

Fig. 3. The overall framework of the financial risk prediction system in this paper

3.2   System Design and Implementation

Combined with Fig. 3, this system is divided into three layers, namely basic layer, application layer and user layer. The design and implementation methods of each layer are as follows.

Base Layer

The basic layer consists of data acquisition tool, pre-processing tool and data transmission channel, and the realization method of each part is as follows: First, the data collection tools mainly USES the Java language development, the basic working principle is based on a temporary database, receive the manual entry of data on the one hand, on the other hand is to monitor the quality of the network environment by monitoring module, once the new data in the network environment, the program will automatically import it into a temporary database, Data collection is accomplished in this way. At the same time, this paper also chose the cloud database as permanent database, that is because of the limited capacity of temporary database, to store large amounts of data, so it can only be used for temporary storage, but the data itself is needs to be permanent, so to set up a permanent database, and database of cloud storage capacity can be infinite expansion, meet the demand of large financial risk forecast data storage, In the design of this paper, the temporary database is connected to the cloud database, and all the data in the temporary database will be transferred to the cloud database. Second, preprocessing tools also use Java language development, these tools are mainly used for processing low quality data, that is because the data collection tool itself does not have the ability to process the data to identify and so any data will be collected, this situation may appear duplicate data, incomplete data, nonstandard data, the data has a great influence on the subsequent analysis, It may lead to inaccurate predictions, so it is necessary to process these data before analysis. According to pre-processing requirements, the pre-processing tools developed in this paper and their functions are shown in Table 1. Third, the data transmission channel, because the system serves enterprises, and the financial management of enterprises is generally carried out in the LOCAL area network, so the data transmission channel is mainly established on the local area network. This paper chooses Wifi and Http network protocol, and realizes the communication through the signal receiving and receiving port in the physical environment. The above connection between temporary data and cloud database is based on data transmission channel [8–10].

Table 1. Preprocessing tools and functions

Tool	Role
Standard template matching tool	Delete data that does not fit the standard template (incomplete data)
Data comparison tool	Delete data with identical comparison results (sufficient data)

The Application Layer

The application layer is mainly composed of data feature analysis, data logic analysis and knowledge base, in which data feature analysis and data logic analysis are realized by algorithms. Knowledge base is a plate of cloud database in the base layer, which is specially used to store the results of data feature analysis and data logic analysis. First of all, data feature analysis is mainly realized by machine learning algorithm. Considering that there are many data types for enterprise financial risk prediction, this paper chooses gradient descent method and data feature segmentation method as two machine learning algorithms to realize data feature analysis. Data feature analysis is mainly used to obtain data features, according to which data can be classified and the system has the ability of data identification. Second data logic analysis, this paper adopted the feedback model based on BP neural network algorithm, the purpose is the comprehensive analysis on the enterprise financial data information, the results can reveal enterprise financial operation condition, in order to avoid enterprise to analyze the potential risk, at the same time clear risk outbreak, the types of risk, risk probability and risk impact, etc. In this way, enterprises can effectively prevent financial risks and ensure the stability of their financial operations.

Then the User Layer

The user layer is the level that the user uses this system, in consideration of convenience, this paper develops the user layer design work based on Web page. The system user layer is mainly composed of login interface, Web page and toolbar: Temporary page, first, this paper established a login interface for the user to enter the password to log in certification, synchronization based on cloud database to construct the account password repository, all registered password is stored in one, so when the user input password is the temporary page, the Web server will find corresponding information in the account password repository, If there is no matching information, it indicates that the account password entered by the user is wrong, and the access application will be rejected. Otherwise, the application will be approved. Second, the Web page, this paper mainly designed a variety of tools jump page, initialization page, user layer visualization; Third, toolbar, in order to facilitate users to use this system, the more rapid financial analysis, this paper designed the toolbar, main toolbar contains the characteristics of data analysis, data logic analysis of the function of the button, then the user to select the corresponding data items and click on the corresponding button is analyzed, the results will be directly from the pop-up window, so that artificial view.

4   Conclusion

To sum up, any financial risk will more or less have a negative impact on the enterprise, and if it is not timely prevented and controlled, the influence of the risk will greatly expand, and even reach the point that the enterprise cannot bear. So enterprises should introduce the BP neural network algorithm, and taking the system design of the algorithm, through the function system is perfect, and have the characteristic of BP neural network system, can carry on the whole process supervision to the enterprise financial situation, synchronization according to the data, the analysis of the financial situation, constantly play a role of financial risk prediction, And the accuracy of the prediction results will be greatly improved with the help of the algorithm, so that enterprises can better carry out financial risk prevention work.

References

1. Smith, L.N., Makam, A.N., Douglas, D., et al.: Acute myocardial infarction readmission risk prediction models: a systematic review of model performance. Circ. Cardiovasc. Qual. Outcomes **11**(1), e003885 (2018)
2. Anagnostopoulos, I., Rizeq, A.: Confining value from neural networks: a sectoral study prediction of takeover targets in the US technology sector. Manag. Financ. **45**(10/11), 1433–1457 (2019)
3. Kapdan, F., Aktas, M.G., Aktas, M.S.: Financial risk prediction based on case based reasoning methodology. In: 2019 Innovations in Intelligent Systems and Applications Conference (ASYU) (2019)

4. Bhadani, S., Verma, I., Dey, L.: Mining financial risk events from news and assessing their impact on stocks. In: Bitetta, V., Bordino, I., Ferretti, A., Gullo, F., Pascolutti, S., Ponti, G. (eds.) Mining Data for Financial Applications. LNCS (LNAI), vol. 11985, pp. 85–100. Springer, Cham (2020). https://doi.org/10.1007/978-3-030-37720-5_7

5. Sodnomdavaa T.: Corporate bankruptcy prediction model for mongolian companies. In: China International Conference on Insurance and Risk Management (CICIRM 2016) (2020)

6. Horak, J., Vrbka, J., Suler, P.: Support vector machine methods and artificial neural networks used for the development of bankruptcy prediction models and their comparison. J. Risk Financ. Manag. 13(3), 1–15 (2020)

7. Uj, A., Nmb, E., Ks, C., et al.: Financial crisis prediction model using ant colony optimization-sciencedirect. Int. J. Inf. Manag. 50, 538–556 (2020)

8. Chi, G., Uddin, M.S., Abedin, M.Z., et al.: Hybrid model for credit risk prediction: an application of neural network approaches. Int. J. Artif. Intell. Tools 28(5), 1950017 (2019)

9. Korol, T.: Dynamic bankruptcy prediction models for european enterprises. J. Risk Financ. Manag. 12(4), 1–15 (2019)

10. Moscato, V., Picariello, A., Sperlí, G.: A benchmark of machine learning approaches for credit score prediction. Expert Syst. Appl. 165(9), 113986 (2021)

Power Information Network Traffic Anomaly Detection Mechanism Based on Big Data

Mingyang Yu[✉], Zhaofeng Guo, Zhiyong Zha, Bo Jin, and Jie Xu

Information and Communication Branch, State Grid Hubei Electric Power
Co. Ltd., Wuhan 430077, China
24870875@qq.com

Abstract. In order to detect abnormal flow in power information network, this paper will carry out relevant research on the basis of big data, aiming at establishing abnormal flow detection mechanism. The research mainly introduces the application conditions of big data, and then proposes the construction method of detection mechanism. In this paper, we make full use of the role of big data to build a power information network traffic anomaly detection mechanism, which can guarantee the performance of the mechanism and accurately detect network traffic anomaly.

Keywords: Big data · Power information network flow · Traffic anomaly detection

1 Introduction

Big data is a huge data integration of body, which contains a wide variety of data, and the overall data scale is huge, reached a level of mass, and all the data from real channels, therefore represents all aspects of the related events, if able to read large data can be more comprehensive understanding of the reality, so that the prepared, or looking for development direction, This reflects the application value of big data. At the same time, the application value of big data is also reflected in the scope of application, that is, the actual work in any field will produce data, and the continuous accumulation of these data will form big data. This kind of big data can be perfectly combined with relevant fields and work, and can be used pertinently, which is applicable in any case. Because of big data has broad scope of application, so it will soon get the attention of the society from all walks of life, including the electric field, and under the unceasing development, the application of big data in the field of electric power deepening, many power related work are combined with a large data, such as electric power information network traffic detection, The job mainly because of high requirement of modern power grid communication, but communication network performance is limited, so is not enough to fully support the network communication, therefore, more or less in the process of communication can appear abnormal traffic, aiming at this kind of situation will need to test, to understand the cause of the abnormal situation, clear abnormal, in order to accurately fix. And various appearing in the large electric power communication network anomalies are often related to many factors, was very extensive, show unusual method is difficult to effectively complete testing work, and likely the defect of

© The Author(s), under exclusive license to Springer Nature Switzerland AG 2022
V. Sugumaran et al. (Eds.): ICMMIA 2022, LNDECT 138, pp. 688–694, 2022.
https://doi.org/10.1007/978-3-031-05484-6_87

test results is inaccurate, incomplete, but the big data can help people to solve these problems, make the results more accurate and complete. It is important to note that the big data while in electric power information network traffic anomaly detection has a high application value, it is necessary to around big data to establish the relevant inspection mechanism, but the big data itself does not have the initiative role, it must be under human intervention to work, at the same time human intervention method must conform to the requirements of the application conditions of big data, The reason is that big data cannot be handled by ordinary human intervention methods because of its huge data magnitude and intricately complicated data relationships. Therefore, big data can only play a role under certain conditions. On this basis, how to build abnormal flow detection mechanism of power information network based on big data has become a problem worth thinking about, and it is necessary to carry out relevant research on this problem, which will provide strong guarantee for power network communication and optimize the stability of power service.

2   Application Conditions of Big Data

Big data can only be applied under specific conditions and play its due role. Therefore, before the construction of abnormal flow detection mechanism of power information network, it is necessary to understand the application conditions of big data and complete the construction of application conditions in the actual construction work. In general, the application conditions of big data can be divided into two parts, as follows.

2.1   Intelligent Tools

Big data application's primary goal is to analyze the data, but because of big data of large scale, and the relationship between the internal data is complex, so the ordinary means of data analysis doesn't analyze the big data, in this case they desperately need to find the tools to analyze the data, thus the intelligent tool is obtained. Intelligent tool is the product of intelligent technology, the main function for data analysis, with the unusual data analysis tool is the biggest difference is that reflect on the performance and logic, namely the first performance, intelligent tool can 24-h non-stop work, can finish the work according to the preset flow rapidly at the same time, basically can in a very short period of time dealing with tens of thousands of data, In this case, other data analysis tools can not do, so intelligent tools have a great advantage in performance. The second is logic. Common data analysis tools are automatic tools, indicating that the inherent logic of these tools is automatic logic, but the inherent logic of intelligent tools is intelligent logic. This logic does not completely operate in accordance with the preset process, but can be adjusted and evolved according to the actual situation. Continuous learning in a way similar to human intelligent behavior can clarify the complex relationships in big data, and get more in-depth and comprehensive data analysis results. It can be seen that intelligent tools are a necessary condition for big data application [1–3]. Figure 1 shows the big data analysis process of intelligent tools.

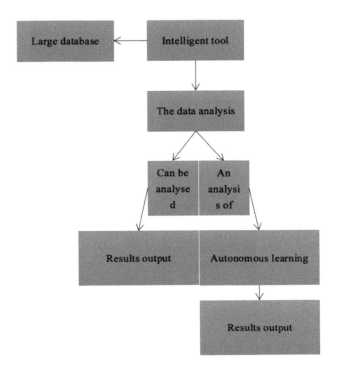

Fig. 1. Big data analysis process of intelligent tools

2.2   Data Acquisition Tool and Data Repository

Big data in addition to the order of magnitude relationship between the characteristics of the complex and huge data, also has the characteristics of the generated continuously, the reason is that the data is derived from the real event, and most real-world events are changing, so all the data, these data needs to be integrated into the large data, but the data itself does not move, Only in this way can big data correspond to real events and be used for real event analysis, indicating that data acquisition tools are also a major application condition of big data. At the same time, because of the continuous acquisition, large scale of data will be growing, at this time to the big data applications, you must first large data storage and access to scheduling, and huge, and growing data level will bring enormous data storage requirements, common database capacity is always limited, so don't meet the demand of data storage, As a result, big data cannot be applied. In view of this situation, the big data repository is proposed in relevant fields. It belongs to the cloud database in essence, has unlimited capacity, has good scalability in use, and can fully meet the needs of big data storage, which is one of the basic conditions of big data application [4–6]. Figure 2 shows how data is stored in the data repository.

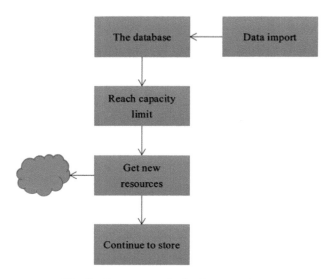

Fig. 2. Data storage in the data repository

3 Construction Method of Abnormal Flow Detection Mechanism in Power Information Network

3.1 Mechanism Framework

Figure 3 is the basic framework of abnormal traffic detection mechanism of power information network in this paper.

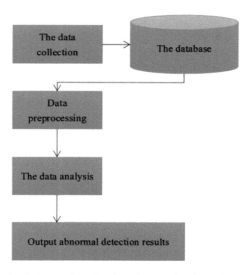

Fig. 3. Basic framework of abnormal traffic detection mechanism of power information network

In Fig. 3, data is firstly transmitted to the database with the help of data collection tools. The data collection tool selected in this paper is router, which has data transmission capability and all data in the power information network will pass through router, so it can be used as data collection tool. Secondly considering the income data are initialized data, unable to guarantee the quality of this data, in which there may be a repetition, such problems as incomplete data format, data to analysis the data directly cause disturbance to the results of the analysis, thus to the preprocessing of data in the database this paper select the preprocessing tool are programming tools, in the Java language development, It can realize data deduplication and data format completion. If there is data that cannot be repaired by pre-processing tools, it will remain in the retention state. This part of data is of small magnitude and can be processed manually. Again, the data is analyzed through three steps of screening, dimensionality reduction and standardization, and the analysis results can reveal all the reasons for the abnormal flow of power information network. Finally, the analysis results can be exported [7–9].

3.2 Construction Method

After selecting relevant data acquisition tools, databases and pre-processing tools, the power information network flow anomaly detection mechanism enters the most critical data analysis link, which is the key and difficult point of mechanism construction, and the main construction method is also reflected in this link. From this point of view, the construction method of data analysis link of abnormal flow detection mechanism of power information network will be introduced below (the construction method of other links is relatively simple, directly select tools and then set up according to Fig. 3, without further elaboration here).

Data Feature Extraction

Through large data to determine the power information network traffic for all of the exception, it must first identify all relevant data, so has the ability to identify the data to make technical system, in the construction of data analysis link need to develop data feature extraction function, the function can obtain all the relevant data characteristics of the electric power information network traffic anomaly, according to the characteristics can be analyzed. First of all, due to the large dimension of power information network flow data, in order to intuitively display, this paper adopts parallel coordinates to observe data, which can remove irrelevant dimensions and improve the purity of abnormal big data of power information network flow [10]. Through the above methods, this paper mainly extracts data features from three dimensions, as shown in Table 1.

Table 1. Three dimensions of data feature extraction

Dimension names	Explain
Business types	Differentiated business types
Distribution space of source IP addresses	Differentiating data sources
Distribution space of destination IP addresses	Distinguish data usage and transmission path

In addition, in the process of data feature extraction, in order to ensure the work efficiency, this paper limits the granularity of each feature extraction unit time, that is, the granularity of each feature extraction unit time is 5min, which is long enough for the feature extraction tool to obtain all data features for subsequent analysis. It is important to note that in the different environment data feature extraction unit time granularity, although different, but not in order to save time, improve efficiency, to reduce the time blindly size limit, because time granularity limit is too small, easily lead to subsequent detection algorithm to produce false positives, so be sure to set reasonable, usually keep within 5 min to 10 min.

Data Dimension Reduction

In view of the problem of large dimension of power information network flow data, dimensionality reduction is needed in data analysis. In terms of specific processing method, multidimensional scale analysis is chosen in this paper. Multidimensional scaling analysis is the most common dimension reduction method, is also a method of data analysis can be found in the sample space in pairs the highest similarity sample data, and then to construct the reasonable low latitudes space, at the same time guarantee sample data in a low-latitude space will not change, still can maintain high altitude characteristics, show sample data will not change because of the dimension, So the analysis results can be guaranteed.

The main process of data dimensionality reduction is as follows: First, construct t × T dissimilarity matrix for t-dimensional data that need dimensionality reduction; Second, find t vectors in m dimension, and then integrate all vectors to form analysis matrix. Third, adjust the distance between each vector in T so that it is as similar as possible to the vector distance in the analysis matrix. In addition, the distance between vectors in multidimensional scale analysis refers to Euclidean distance, which can be arbitrarily rotated and changed, but any change will not change the specific value of the distance. The main methods of data dimension reduction are shown in Formula (1).

$$\min_{\psi} \sum_{i=1}^{t} \sum_{j=1}^{t} \left(\delta_{ij}^{X} - \delta_{ij}^{X} \right)^2 \tag{1}$$

where t is the data of several dimensions, i is the standard value of the data of t dimension, j is the distance value of t vectors in the m dimension, X is the distance between each vector in the t vector matrix, which must be close to the vector distance in the analysis matrix as far as possible.

Data Standardization and Result Output

Through data dimension reduction, can start analysis of electric power information network traffic data, in the process of the first step is to standardize data processing, that is because different vector space in the basic unit of the dimension of each are not identical, such as two vectors in the input number of clusters, the number of bytes to import and export group number and the number of output bytes unit on completely different, this situation is difficult to get accurate results, In this paper, deviation standardization is adopted to solve the problem. After standardized processing, we can know all the reasons for abnormal power information network traffic. For example, the standardized data comes from router A, and the data analysis results show that the operation rate of router is 0, which indicates that the router does not work and has A high probability of failure, and the abnormal power information network traffic is related to it.

4   Conclusion

To sum up, the abnormal flow of power information network often appears in modern power network, which has a negative impact on power operation, will lead to abnormal power service, and even make the abnormal constantly worsening. In order to cope with this situation, people must establish the power information network flow anomaly detection mechanism, the construction of the mechanism should be combined with reasonable algorithm, and give full play to the role of big data, so as to quickly solve the abnormal phenomenon through the mechanism, guarantee the quality of power service.

References

1. Song, W., Beshley, M., Przystupa, K., et al.: A software deep packet inspection system for network traffic analysis and anomaly detection. Sensors **20**(6), 1637 (2020)
2. Garcia, N., Alcaniz, T., González-Vidal, A., et al.: Distributed real-time SlowDoS attacks detection over encrypted traffic using artificial intelligence. J. Netw. Comput. Appl. **173**, 102871 (2021)
3. Al-Sanjary, O.I., Roslan, M., Helmi, R., et al.: Comparison and detection analysis of network traffic datasets using k-means clustering algorithm. J. Inf. Knowl. Manag. **19**(3), 2050026 (2020)
4. Choi, H., Kim, M., Lee, G., Kim, W.: Unsupervised learning approach for network intrusion detection system using autoencoders. J. Supercomput. **75**(9), 5597–5621 (2019). https://doi.org/10.1007/s11227-019-02805-w
5. Mohamed, M.R., Nasr, A.A., Tarrad, I.F., et al.: Exploiting incremental classifiers for the training of an adaptive intrusion detection model. Int. J. Netw. Secur. **21**(2), 275–289 (2019)
6. Tamura, K., Matsuura, K.: Improvement of anomaly detection performance using packet flow regularity in industrial control networks. IEICE Trans. Fundam. Electron. Commun. Comput. Sci. **E102.A**(1), 65–73 (2019)
7. Roselin, A.G., Nanda, P., Nepal, S., et al.: Intelligent anomaly detection for large network traffic with optimized deep clustering (odc) algorithm. IEEE Access **9**, 47243–47251 (2021)
8. Haghighat, M.H., Foroushani, Z.A., Li, J.: SAWANT: smart window based anomaly detection using netflow traffic. In: 2019 IEEE 19th International Conference on Communication Technology(ICCT). IEEE (2020)
9. Ahmed, A.: Intelligent big data summarization for rare anomaly detection. IEEE Access **7**, 68669–68677 (2019)
10. Song, H.M., Kim, H.K.: Self-supervised anomaly detection for in-vehicle network using noised pseudo normal data. IEEE Trans. Veh. Technol. **70**(2), 1098–1108 (2021)

Student Mental Health Evaluation System Based on Decision Tree Algorithm

Fanglin Xie[✉], Chuhui Geng, and Qi Jiang

Harbin Institute of Technology, Weihai 264200, China
xiefanglin0307@126.com

Abstract. In order to accurately assess the psychological condition of students, students' mental health and security, this article, taking the research of the decision tree algorithm, research mainly discusses the basic principle of students' mental health assessment, the mental health assessment system is introduced in this paper the status quo, some problems are pointed out, and then introduce the decision tree algorithm, and combined with the algorithm design evaluation system. Through this study, the student mental health assessment system based on the decision tree algorithm has advantages, and the assessment results are more accurate, which can help people judge the psychological status of students and provide targeted intervention.

Keywords: Decision tree algorithm · Students · Mental health assessment system

1 Introduction

Young students grade, thought simple, life experience and social experience is not rich, so widespread psychological sensitive, the phenomenon of bear ability weak, while students in the learning process will face greater pressure, and grow as they may encounter to the larger psychological blow, or bad information, students' psychological problems are likely to be in any case, At this time, students are not psychologically healthy. At the same time, the unhealthy psychology of students is not obvious in the early stage, so it is not easy to detect, which makes it difficult for people to intervene in the first time, so that the unhealthy psychology will pass through the early stage smoothly. In the middle and later period, the unhealthy psychology of students not only deepens, but also becomes more complex, and the difficulty of intervention increases significantly. Based on this, as much as possible for the first time for students to middle and later periods of the unhealthy psychological intervention, in view of the complex psychological intervention, protect students' mental health, mental health evaluation system was put forward with the related method, aimed to quickly through the system analysis of students' psychological condition, it can improve the efficiency of psychological intervention and quality, But many current mental health assessment system is more or less, the reason lies in the deficiency in the system used by the algorithm, so the system needs to be improved, and the decision tree algorithm can be used as a new system to use, on the basis of the algorithm, can effectively solve the problem of system, improve the students' mental health, therefore, in order to complete the system improvement purpose, Relevant research is needed.

2 Basic Principles and Current Situation of Student Mental Health Assessment

2.1 Basic Principles

The basic process of student mental health assessment is shown in Fig. 1.

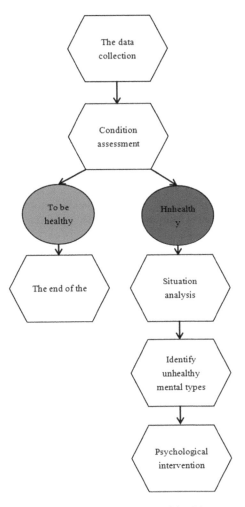

Fig. 1 Basic process of student mental health assessment

It can be seen from Fig. 1 that the basic process of student mental health assessment is highly systematic. It starts from data collection, indicating that the assessment needs to be carried out in accordance with data, while data collection needs to be carried out in accordance with the index system, so as to ensure data integrity. Relevant indicators are shown in Table 1. Can psychology analysis after data collection, analysis, the purpose is to judge the current students mental state health, if health, at the end of the assessment if unhealthy will further analysis, the analysis of the main purpose is to make sure students specific what are unhealthy psychology, the unhealthy psychological level, on the strength of the correlation between the unhealthy psychological, To clarify the direction of psychological intervention. Finally, according to the type and specific situation of unhealthy psychological, targeted intervention methods were adopted for continuous treatment, and repeated assessment after one cycle was conducted, so as to continuously understand the situation of students and more effectively guarantee their psychological health [1–3].

Table 1. Indicators of mental health status analysis

The index name	Index to explain
Environmental adaptability	Refers to the psychological adaptability of students in different environments, low ability will lead to psychological maladaptation of students, over time may appear psychological crisis, so the ability can not be lower than the standard
Mental endurance	Refers to the ability of students to bear psychological blows. The lower the ability, the lower the degree of blows they can bear. If it is different from ordinary people, it indicates that students are psychologically unhealthy
Mental self-control	Refers to the ability of students to rely on their own control of personal psychology, low ability on behalf of students psychological easily out of control, may often appear irrational ideas or behavior, such performance shows that students are not healthy
Self-confidence	Self-confidence is a necessary element of healthy psychology, if a student is not confident, then his psychology must be unhealthy
Emotional wave size	Refers to the range of students' emotional fluctuations, the range is too large or too low on behalf of students' psychological health, there may be too radical, sensitive or indifferent, numb psychology
Frequency of negative emotions	Refers to the total time that students are in a negative state in the cycle. If there are no special circumstances, the total time exceeds the standard, it indicates that students are psychologically unhealthy
Social status	Refers to the situation that students integrate into the collective, if the students can not well integrate into the collective, it means that the students may be unhealthy. The evaluation results of this index are mainly used as auxiliary suggestions for reference
Psychological sensitivity	Refers to the level of psychological vulnerability of students, the more vulnerable the more sensitive, may be angry because of some other people feel "indifferent" things, grudges, so this situation shows that students are too sensitive, in an unhealthy psychological state

2.2 The Status Quo

At present, the student mental health assessment work is in full swing in the relevant organizations, the organization has constructed a complete supporting system, such as the establishment of a complete indicator framework, the establishment of reasonable standards for each indicator, but also set up a special work team to take charge of the work. But in reality, students' mental health assessment work still exists defects, defect is mainly reflected in the evaluation system on the evaluation results of the because the job needs continuous expanding, and work to deal with large scale data analysis, face many complex problems, so can't completely rely on artificial to perform the work, Related organizations set up students' mental health assessment system, but traditional algorithm is adopted by the system itself, such as mental health rating algorithm, these algorithms are mental health assessment, is difficult to deal with complex psychological problem students, at the same time, large amount of calculation and data requirements is very high, if not completely meet the requirement of algorithm, and can lead to the result, In fact, there is no method that can completely guarantee the data quality to reach the standard. Therefore, the results of student mental health assessment are not satisfactory under the influence of the algorithm, which also indicates that the student mental health assessment needs to be reformed, and the algorithm is the key target in the reform [4–6].

3 Introduction of Decision Tree Algorithm and System Design

3.1 Algorithm Introduction

Decision tree algorithm is presented in the 1960s as a kind of algorithm, was originally called CLS algorithm, this algorithm appears to provide people with a flow analysis is in accordance with the "decision tree" problem, and get the optimal solution of train of thought, therefore received wide attention, but the initial decision tree algorithm is not perfect, it exists many defects, Such as CLS algorithm to construct the decision tree may not be able to completely cover the overall problem, people can't confirm this, it might lead to the calculation results overgeneralization, used to deal with the complex problems, so the subsequent people adhere to the basic idea of the decision tree algorithm, continuously through the research to improve the algorithm, makes the algorithm more, in the form of At present, there are more common algorithms such as ID3 algorithm and C4.5 algorithm. Although these algorithms have their own advantages and disadvantages, as long as they are correctly selected according to the actual situation, they can avoid defects and give full play to the role of decision tree algorithm to obtain the optimal solution. Therefore, even in the face of complex problems, decision tree algorithm still has high application value. Whatever the decision tree algorithm, and its essence is a kind of approximation method for calculating the discrete function value, applications only need to prioritize the data, then according to the decision tree rules in advance can use data, analyze the problem of data by decision tree, and the analysis process is mainly divided into two steps, first sample training, and each time the training will get a data node, Each node contains branches and leaves, and training sample sources constitute the main rod, so as to obtain the initial decision tree. Secondly, the decision tree pruning is carried out, that is, to replace the previously poor quality data with higher quality data or delete useless data. In this way, the decision tree algorithm is constantly updated iteratively until the optimal solution is output. The basic framework of decision tree algorithm is shown in Fig. 2 [7–9].

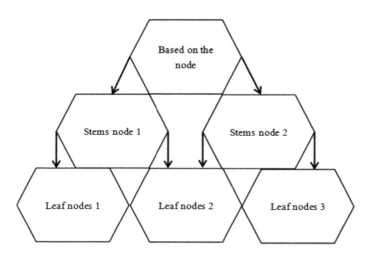

Fig. 2 Basic framework of decision tree algorithm

3.2 System Design

Combined with the decision tree algorithm, the student mental health assessment system design can be carried out. The design work can be roughly divided into X steps, as follows.

Overall Framework Design

Referring to the basic process of student mental health assessment, the overall framework of the student mental health assessment system in this paper is shown in Fig. 3.

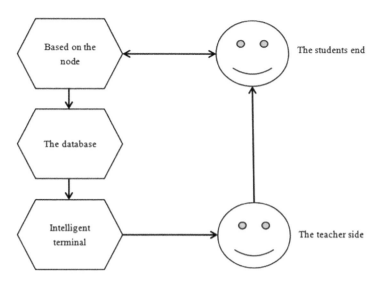

Fig. 3 Overall framework of student mental health assessment system

According to Fig. 3, this system mainly through the first data collection using data on students' psychological condition, the data will be input into the database, and then according to the cycle time, the database in the same time series data input to the intelligent terminals, intelligent terminal according to the data for training, after the completion of the decision tree, then the decision tree pruning iterate calculation, The results are then sent to the intervenor for psychological intervention [10].

Design of Data Acquisition Terminal

Data collection is the beginning of the whole student mental health assessment, and also the basic link of the system operation process, so it is very important. In order to realize data collection and ensure the integrity of collected data, this paper first designs two data collection channels: Main reference channel, the first is the questionnaire in advance to set up good mental health evaluation indicators to design the corresponding question-naire, questionnaire will be regularly sent to the students, students to fill in for the data back to the in order to ensure data integrity, all items in the questionnaire are mandatory, any blank will lead to the failure questionnaire return; The second is manual input, because there are many data items related to students' mental health, some of which cannot be collected through questionnaire survey channels, so it needs to be collected by manual observation, inquiry and other ways, and the collected data can be directly sent to the database from the data collection end. Secondly, the data acquisition method of this system refers to the description of the two data acquisition channels, so there is no further description here. In addition, considering the data quality problem, the inter-vention personnel in the application of the system should be responsible for checking the questionnaire data. When students are found to fill in randomly, they should collect the data manually again to ensure the data quality [11, 12].

Database Design

Although the amount of data involved in this system is very large, but the data mobility is very high, so the data storage requirements are not as big as imagined, just need to use the conventional large capacity database, the database selected in this paper is SQL database, its capacity is 3 T, fully meet the single data storage requirements. At the same time, SQL database has rich built-in functions, such as built-in retrieval function, which is convenient for intervention personnel to consult data, and also conducive to the subsequent intelligent terminal to identify data according to data classification.

Intelligent Terminal Design

The intelligent terminal is the core of the system, and its main function is to analyze whether students' psychological state is healthy and confirm the specific types of unhealthy psychology according to the huge data. It is also the main carrier of decision tree algorithm. The calculation method of decision tree algorithm in intelligent terminal of this system is shown in Formula (1).

$$E[\zeta] = \sum_{i=1}^{n} s(\zeta = x_i) \tag{1}$$

Type in the ζ Is the fuzzy set, E is the fuzzy entropy of the fuzzy set, x, i and n are the values of different membership degrees in the fuzzy set respectively.

Formula (1) is a kind of decision tree algorithm based on the concept of membership degree of fuzzy entropy, among the students' mental health assessment to list all healthy and unhealthy psychology, then generate the decision tree, the relationship between the construction of students' psychological data, again carries on the analysis and calculation, by decision tree to get a fuzzy approximation function, the function of membership degree if lower than expected, It shows that the students are in an unhealthy state of mind. Then, the specific unhealthy mind of the students can be confirmed by combining with the unhealthy mind of the membership degree judgment function. On this basis, the fuzzy entropy decision tree algorithm based on membership concept can be used as the intelligent logic of intelligent terminal to evaluate students' mental health. After testing in this paper, the results are accurate and have certain application value.

4   Conclusion

To sum up, students' psychology is prone to be influenced by various factors and lead to unhealthy conditions, resulting in various unhealthy psychology, which has adverse effects on students' physical and mental development, current learning and even future development. Therefore, we must pay attention to this problem. The emergence of the student mental health assessment system can provide help to the related fields, through the system can be clear whether the student mental health is unhealthy, the specific types of unhealthy psychology, among which the decision tree algorithm can improve the system assessment results, indicating that the algorithm has good application value in the student mental health assessment.

Acknowledgements. Project initiated by Ministry of Industry and Information Technology (2021): Research on the innovation of the working mechanism in school-enterprise party building and cooperative establishment (GXZY2136).

References

1. Driessens, K., Ramon, J., Blockeel, H.: Speeding up relational reinforcement learning through the use of an incremental first order decision tree learner. In: De Raedt, L., Flach, P. (eds.) Machine Learning: ECML 2001. Lecture Notes in Computer Science (Lecture Notes in Artificial Intelligence), vol. 2167, pp. 97–108. Springer, Heidelberg (2003). https://doi.org/10.1007/3-540-44795-4_9
2. Atramentov, A., Leiva, H., Honavar, V.: Multi-relational decision tree algorithm-implementation and experiments. In: Second IEEE and ACM International Symposium on Mixed and Augmented Reality vol. 8(2), pp. 296–297 (2003)
3. Luo, X., Xia, J., Liu, Y.: Extraction of dynamic operation strategy for standalone solar-based multi-energy systems: a method based on decision tree algorithm. Sustain. Cities Soc. **70**, 102917 (2021)
4. Li, W., Ma, X., Chen, Y., Dai, B., Chen, R., Tang, C.: Random fuzzy granular decision tree. Math. Probl. Eng. **2021**(1), 5578682 (2021)
5. Muniyandi, A.P., Rajeswari, R., Rajaram, R.: Network anomaly detection by cascading k-means clustering and c4.5 decision tree algorithm. Procedia Eng. **30**, 174–182 (2012)
6. Rajendran, P., Madheswaran, M.: Hybrid medical image classification using association rule mining with decision tree algorithm. Comput. Sci. **3**(10), 1173–1178 (2010)
7. Kohut, Y., Yurchak, I.: Recommendation system for purchasing goods based on the decision tree algorithm. Adv. Cyber-phys. Syst. **6**(2), 1–7 (2021)
8. Tuan, T.A., Long, H.V., Son, L.H., et al.: Performance evaluation of botnet ddos attack detection using machine learning. Evol. Intell. **13**(2), 283–294 (2020). https://doi.org/10.1007/s12065-019-00310-w
9. Priyanka, D.K.: Decision tree classifier: a detailed survey. Int. J. Inf. Decis. Sci. **12**(3), 246–269 (2020)
10. Pandey, M., Sharma, V.K.: A decision tree algorithm pertaining to the student performance analysis and prediction. Int. J. Comput. Appl. **61**(13), 1–5 (2013)
11. Saad, M.M., Jamil, N., Hamzah, R.: Evaluation of support vector machine and decision tree for emotion recognition of malay folklores. Bull. Electr. Eng. Inf. **7**(3), 479–486 (2018)
12. Gangadhar, N., Kumar, H., Narendranath, S., et al.: Fault diagnosis of single point cutting tool through vibration signal using decision tree algorithm. Procedia Mater. Sci. **5**, 1434–1441 (2014)

Vehicle Side Impact Model Optimization Based on Reliability

Qichen Zheng, Xiaowang Sun, Xianhui Wang[✉], Mengyang Wu, Tiaoqi Fu, and Di Zhou

Department of Mechanical Engineering, Nanjing University of Science and Technology, Nanjing 210094, China
wxhbeiyong@163.com

Abstract. In order to improve the reliability of vehicle side impact model, this paper will carry out relevant research. Firstly, the finite element model is established and verified, then the model optimization idea is proposed, and then the optimization is carried out according to the idea, and finally the optimized model is tested completely. The test results show that the optimized model is more reliable, and the vehicle side impact design based on this model can ensure the maximum safety.

Keywords: Reliability · Vehicle side impact · Model optimization

1 Introduction

With the continuous development of the automobile industry, cars have gradually become a common means of transportation in People's Daily life. However, driving cars must face corresponding risks, and car side collision is one of the common risks. Compared with other parts of the collision risk, car side collision is more threatening, because the side of the car is very fragile, and the impact site is closest to the occupants of the car, once the impact will be indirectly a huge impact on the occupants, resulting in injury or even death. According to statistics, the probability of serious injury caused by car side collision is as high as 90%, and the fatality rate is more than 70%, which is enough to prove the threat of car side collision. Therefore, in order to maximize the protection of the safety of the occupants and reduce the impact as much as possible, it is necessary to construct the vehicle side impact model during the design of the vehicle, and continuously optimize the model to improve the reliability of the vehicle side. Therefore, it is necessary to carry out relevant research.

2 Construction and Verification of Finite Element Model

2.1 Modeling Process

Although there are differences in side structures of different cars and side impact is a complex dynamic process, it is impossible to generalize all cars, but the finite element model construction process of side impact of any car is the same, see Fig. 1 for details.

V. Sugumaran et al. (Eds.): ICMMIA 2022, LNDECT 138, pp. 703–710, 2022.
https://doi.org/10.1007/978-3-031-05484-6_89

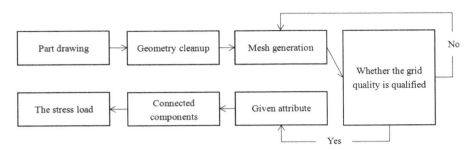

Fig. 1. Finite element model construction process of automobile side impact

According to Fig. 1, the finite element model construction process of automobile side impact has two logics: first, in the case of qualified grid quality, the material and parts are given relevant attributes, and then the parts are connected to output load conditions; Second, in the case of unqualified grid quality, return to the grid division step, continue to test after the completion of the division again, qualified into the first logic, otherwise continue to cycle, until the grid is qualified. Therefore, the finite element model of vehicle side impact obtained under this process must be qualified.

Grid division is the most critical step in the whole modeling process, and the processing must be expanded according to the modeling objectives. The grid division in this paper is as follows: The main force in the automobile side impact point B column as the goal, face-to-face contact type to meshing of B pillar, and to establish a finite element model B column, in the process of first through geometrical model cleaning to remove the smaller circle, hole, this can avoid grid deformity, fault, to ensure that the grid continuity, but also ensure complete joint geometry model and the grid, And then according to the parameters of the judge in the grid may appear larger deformation area, to control the unit length of the region around 8 mm, find out the minimum deformation area at the same time, to control the unit length of the region around 4 mm, finally choose the direction of the geometry coordinate system, uniform surface, again through conversion tools will be unified surface into the grid, After establishing the model package, mesh division can be completed and modeling can be realized at the same time. In addition, the scanning tool should be used to clean the menu after the modeling is completed to remove repeated element elements and geometric elements, and then the inspection tool should detect the abnormal reverse element of the grid. After completion, all cells and nodes in the grid should be numbered to ensure that the grid division is all effective [1–3].

2.2   Finite Element Model Establishment

According to the basic modeling process, this paper takes a domestic automobile as an example to build a finite element model for its B-pillar. The main software used in the process is Hypermesh, which is used to complete mesh division, and then CAD software is used to build a THREE-DIMENSIONAL finite element model of the automobile. Because modeling in this paper is aimed at car side B column, so using the grid of small grid, the grid except for column B, all the side of the higher safety performance can also use a smaller grid structure, such as the door, and the longitudinal beam before and after, low side crash safety, such as glass, the structure of the larger grid must be used. Focusing on the case in this paper, the B-pillar material is low carbon steel sheet with high extensibility. Therefore, the piecewide elastic-plastic material is used to define the body panels in the finite element model establishment, and the collision detection algorithm is used to calculate, as shown in Formula (1). The finite element model can be obtained after completion, as shown in Fig. 2.

$$RidDef = C + Z \bullet Vimp \tag{1}$$

In the formula, C and Z are coefficients, and their values are 6.632 and 4.705 respectively.

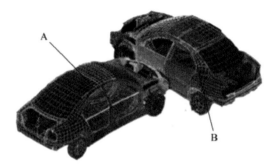

Fig. 2. Finite element model of a domestic automobile

2.3   Model Interpretation

Combined with Fig. 2, the two cars in the figure are named as car A and car B respectively. The relationship between the two is that car A crashes vertically into the side of car B, and car B is in A static state, while the speed and collision time of car A are shown in Table 1.

Table 1. Simulation parameters of the finite element model of a domestic automobile

Parameters of the item	Parameter
The velocity of car A at the time of impact	60 km/h
The time of impact when car A hit	0.105 s

In addition, the establishment of any finite element model of automobile side impact must take into account two main points: first, although the model refers to side impact, the integrity of the body frame must be guaranteed for the sake of the integrity of stress structure, so as to reflect the actual mechanical properties of automobile to the maximum extent; Second, on the basis of ensuring the integrity of the body framework of body frame is to simplify the model, the modeling of not too much to increase the number of units, security model as far as possible concise, or after the operation is complicated, the quality of the result also may not be able to get protection, so as long as the unit of quantity does not affect the simulation results, it is not necessary to use [3–6].

2.4   Model Verification

The purpose of model verification is mainly to analyze the side impact safety of the finite element model, so as to confirm whether the model conforms to the safety standard of automobile side impact, and then the accurate answer can be obtained through subsequent analysis. Aiming at the finite element model of automobile B-pillar, the thickness of the roof beam, door guard bar, door bar and other parts of the model were adjusted according to the example of the automobile, and then the side impact simulation analysis was conducted by using the model directly. The results show that there are two hidden dangers in automobile B-pillar: First, B column under the condition of the simulation parameters by the lateral deformation will happen after the collision, the deformation mode is very not ideal, but the simulation parameters in A car collision speed quickly, after contact with the B column column B first of all, the part of the deformation is the waist line, the region generally aligned with the occupants in the chest, waist, therefore can cause shock to people in the chest or the waist, Both are potentially lethal, with a high probability of lifelong damage, as confirmed by dummy tests; Second, aiming at the first hidden danger, the reinforcement plate was added, and the simulation test was conducted again. The results showed that the use of the reinforcement plate did reduce the amount of deformation and directly reduced the damage to people, but the reinforcement plate increased the complexity of vehicle assembly, and it was difficult to open and close the door after deformation, which may not be conducive to the escape of personnel [7–10]. Figure 3 shows the basic structure of b-pillar in this paper.

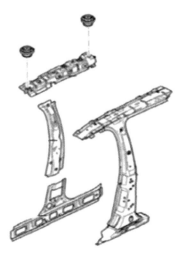

Fig. 3. Basic structure of b-pillar in this paper

3   Model Optimization Ideas and Optimization Schemes

3.1   Optimization Ideas

Combined with the hidden danger of large quantity in Fig. 3, the optimization idea of this paper is the integrated design of B-pillar strengthening plate and the optimization of b-pillar thermal forming.

3.2   Optimization Scheme

According to the optimization idea, the optimization plan is divided into two steps, the specific content of each step is as follows.

Integrated Design of B-pillar Strengthening Plate
Learned according to the hazard analysis, basic structure in Fig. 3 B column strengthened plate blessing under security has improved significantly, the necessity of reinforcing plate is used, but the unreasonable way of using can cause other hidden trouble, so the model to optimize the first step is optimizing the use of reinforcing plate B column machine, this paper chose the way of integration design. Integrated design, considering the upper reinforcement plate at both ends up and down position, section size, strength, thickness of parts and material effect on the properties of reinforcing plate actual, mainly use thermoforming technology to realize integrated design, can shift B column reinforcing plate assembly attributes, after the completion of the reinforcing plate B column can be divided into 3 3, the thickness of the paragraphs are not the same, At the same time, there is a close connection between each section, which greatly improves the strength of the parts, up to 1600 MPa, and can cope with the side impact from all angles. At the same time, due to the influence of the door structure and the frame structure of the car, the size of the strengthening plate will be designed according to the lightweight standard, and the final size of the strengthening plate is $0.5 \sim 0.7$ m.

Optimization of B-pillar Thermal Forming
Thermal forming technology is used in the integrated design of B-pillar reinforcement plate, and it is necessary to optimize the technology itself. The whole optimization process is divided into two stages, namely 2 k factor experiment and final optimization. The contents of each stage are as follows:

First, in the 2 k factor experiments, considering the cost, side impact experiment cycle mainly adopts CAE simulation software, the vehicle side impact performance evaluation in advance according to the results of optimization, the process of key analysis of the influence of the side impact properties of the critical and non-critical factors, after several times experiments for each factor, The optimal result is taken as the standard value of this factor to achieve optimization. According to this logic, the main factors of b-pillar strengthening plate in this paper are upper, middle, lower thickness and center point. Therefore, 10 experiments were conducted for each factor. Finally, the optimal result of upper, middle and lower thickness was confirmed to be 1.3 mm, and the optimal result of center point was 1, which established the optimization foundation. It is worth noting that the basic principle of 2 K factor experiment is to conduct several experiments for each factor, so the experimental workload will be doubled for each additional factor in the experiment. If the cost and period of the experiment are considered, then 2 K factor experiment is not suitable for experiments with many factors, and other methods are recommended. Through the 2 k factor experiment in this paper, the waist line of b-pillar of the vehicle has been effectively strengthened, and the resistance of the model to external impact has been increased, and the deformation mode has been significantly changed. After the impact, the shape is circular and inward. This shape has a small impact on people and will not cause too serious damage.

Secondly, according to the factor parameters of 2 k factor experiment, the process of b-pillar strengthening plate thermal forming was optimized. After optimization, the actual test was carried out. The results showed that the deformation of B-pillar after impact was consistent with the simulation results, and there was no serious damage on the dummy, and the optimized B-pillar met the lightweight design standard. It is worth noting that the car the change of acceleration is the key to the car side impact index, this paper also has carried on the special test, what is now installed on vehicle acceleration sensor, the sensor installed in the column bottom, B will not hit here, can ensure the security of sensors, to ensure the quality of the results, and then to pile foundation, according to the sensor results show that the main deformation B is not big, The main deformation position is located at the center of mass of the vehicle, where the acceleration parameter is 65 and the deformation distance is 5 cm, which is not enough to harm the human body.

4   Complete the Test

4.1   Test Scheme

Taking the same domestic automobile as an example, the above optimization method was used to optimize the b-pillar finite element model of the automobile, and the simulation test was carried out on the optimized model to confirm that the parameters and performance of the optimized model reached the standard, and then the automobile was reformed strictly in accordance with the optimized model, and the test was carried out after completion. It is confirmed that the parameters and performance of b-pillar after modification are in complete accord with the optimized model. On this basis, a dummy was installed in the driving position of the car. The distance between the dummy and b-pillar was 7 cm, the door thickness was 8 cm, and the material was elastic-plastic material with steel plate reinforcement. Then, a car weighing 1.6 T was used to carry out side impact on the dummy at a speed of 60 km/h. Keep an eye on the damage to the dummy. In addition, in order to make the test results more intuitive, the same method was used to test the transformed cars for comparison before and after.

4.2   Test Results and Discussion

Through two groups of tests, the first car hit before modification after deformation peak to 13 cm, in the vertical direction, the top is more sharp, position in the waist position, so the dummy waist has serious damage, if the transition to the human body is likely to lead to his spine fracture, belongs to the serious harm, at the same time the overall deformation pattern is very serious, the door has been unable to open. Second deformation after modified car hit peak is 5 cm, in the vertical direction, the top is split, position in the waist position, because the top of the platform, so it couldn't have direct impact to dummy the waist, so the dummies are waist injury, but the degree is not high, the transition to the human body will not lead to its spine fracture, damage to reduce, At the same time, the overall deformation mode is ideal, the door can still be opened normally. The comparison shows that the model in this paper is more reliable and has higher security guarantee after optimization.

5   Conclusion

Vehicle side impact can cause great harm to human body and may cause death. Therefore, the reliability of side impact model must be paid attention to in vehicle safety design. The initial model should be established first, and then the model should be optimized. In order to achieve the optimization goal, the staff should do a good job in strengthening plate and thermal forming technology analysis, confirm the optimization direction and parameters, and then implement to improve the reliability of the model.

Acknowledgement. The research is supported by: Open Project of Key Laboratory of Ministry of Public Security for Road Traffic Safety (018ZDSYSKFKT09).

References

1. IvysidePark, Altoona: Using simulation to estimate vehicle emissions in response to urban sprawl within Geauga county, Ohio. Ohio J. Sci. **109**(3), 52–66 (2009)
2. Witzens, J., Baehr-Jones, T., Hochberg, M.: Design of transmission line driven slot waveguide Mach-Zehnder interferometers and application to analog optical links. Opt. Express **18**(16), 16902 (2010)
3. Piccinno, F., Hischier, R., Saba, A., et al.: Multi-perspective application selection:a method to identify sustainable applications for new materials using the example of cellulose nanofiber reinforced composites. J. Cleaner Prod. **112**(JAN.20PT.1), 1199–1210 (2016)
4. Shetty, S., Nilsson, L.: Multiobjective reliability-based and robust design optimisation for crashworthiness of a vehicle side impact. Int. J. Veh. Des. **67**(4), 347 (2015)
5. Wang, Q., Zhou, W., Feng, Y.T., Ma, G., Cheng, Y., et al.: An adaptive orthogonal improved interpolating moving least-square method and a new boundary element-free method. Appl. Math. Comput. **353**, 347–370 (2019)
6. Nocera, S., Maino, F., Cavallaro, F.: A heuristic method for determining CO_2 efficiency in transportation planning. Eur. Transp. Res. Rev. **4**(2), 91–106 (2012). https://doi.org/10.1007/s12544-012-0073-x
7. Wijngaarden, B.V., Schene, A.H., Koeter, M., et al.: Caregiving in schizophrenia: development, internal consistency and reliability of the involvement evaluation questionaire-european version. EPSILON study 4. Br. J. Psychiatry **177**(39), S21–S27 (2018)

8. Tavakkoli-Moghaddam, R., Safari, J., Sassani, F.: Reliability optimization of series-parallel systems with a choice of redundancy strategies using a genetic algorithm. Reliab. Eng. Syst. Saf. **93**(4), 550–556 (2017)

9. Lukezic, A., Vojir, T., Zajc, L.C., et al.: Discriminative correlation filter with channel and spatial reliability. In: 2017 IEEE Conference on Computer Vision and Pattern Recognition (CVPR). IEEE (2017)

10. Wensing, M., Vleuten, C., Grol, R., et al.: The reliability of patients' judgementes of care in general practice:how many questions and patients are needed. Qual. Health Care Qhc **6**(2), 80 (2019)

Flipped-Blended Model Based on MOOC and Micro-lecture

Xiangyang Niu[✉] and Shan Gao

Fuyang Normal University, Fuyang 236037, Anhui, China
niuxy666@163.com

Abstract. Under the background of big data, the mechanism of massive open online course, micro-lecture and flipped class is analyzed, and the problems existing in the mathematics courses reform of computer is studied. Aiming at improving the core quality of mathematics and training the professional skills of computer, the paper excavates ideological and political elements in the probability theory course, organically integrating the traditional class, the network class and the extracurricular practice, and constructing three-dimensional flipped-blended model of computer courses based on massive open online course and micro-lecture. Taking the classical probability model as an example, the hybrid model is explored and practiced.

Keywords: Micro-lecture · Flipped class · Computer courses

1 Introduction

With the rapid development of big data, cloud computing, internet of things and other information processing technologies, higher education is facing great challenges. How to use big data effectively to improve higher education quality is an important research topic [1]. Implementation opinions on the construction of first-class undergraduate courses has been issued by the Ministry of Education. It is pointed out that: "double ten-thousand plan" of first-class undergraduate courses has been implemented. The teaching method of "double ten-thousand courses" should be advanced and interactive. The integration of modern information technology and teaching should be pushed forward. Students should be guided to actively carry out inquiry-based and individualized learning. Massive open online course, micro-lecture and flipped class are classroom teaching models based on network information technology. With the development of digital technology and information technology, they provide a new way for the construction of "double ten-thousand courses" under the background of big data. In order to create a mixed gold course which integrates online courses with classroom teaching, and to cultivate college students with solid mathematical background and professional qualities in computer science, the problems existing in the teaching of computer mathematics in local universities are analyzed, and three-dimensional flipped-blended model based on massive open online course and micro-lecture is studied.

V. Sugumaran et al. (Eds.): ICMMIA 2022, LNDECT 138, pp. 711–717, 2022.
https://doi.org/10.1007/978-3-031-05484-6_90

2   Problems of Mathematics Courses Teaching of Computer Under the Background of Big Data

With the big data era approaching, the big data of teaching and evaluation bring new challenges to the teaching mode, teaching idea, teaching method, teaching content and so on [2]. As far as the teaching mode is concerned, the teacher-centered teaching mode tends to neglect the students' subject status. It is not conducive to the cultivation of students' creativity. As far as the teaching idea is concerned, teachers need to change from relying mainly on experience to relying on both experience and big data. Only by handling and analyzing big data with experience, can teachers teach students in accordance with their aptitude. As far as the teaching method is concerned, Big data technology requires open teaching. Big data technology should be utilized to integrate traditional teaching methods with open teaching methods such as massive open online course, micro-lecture, flipped class and so on. As far as the teaching content is concerned, under the background of big data, the computer mathematics curriculum itself yields many contradictions between the class hour compression and the content deepening. The teaching content lacks the pertinence and is not closely related to computer science. As far as the teaching assessment is concerned, the examination mainly uses the theory examination paper. The specialized union is not thorough enough to realize the specialized training goal. As far as the teaching implementation is concerned, teachers are confronted with the contradictions of penetrating the content and catching up with the progress, integrating theory teaching with practice teaching, and integrating mathematical knowledge with professional background.

3   Teaching Ideas and Models of MOOC, Micro-lecture and Flipped Class

Massive open online course (or MOOC) was first proposed by Dave Cormier of Canada and Bryan Alexander of the United States in 2008 [3]. MOOC has large-scale, open, linear characteristics and operating process of clarifying goal, watching video, submitting job and completing evaluation [4, 5]. however teachers and students can hardly communicate in time. With a high registration rate but low completion rate, course design is lacking in pertinence with insufficient practice and fixed content. MOOC model can not reach the traditional teaching goal according to students' aptitude. Micro-lecture was first proposed in 2008 by David Penrose of the United States. He defines it as the actual content of a course developed by constructivist methods for the purpose of online or mobile learning [6]. The advantages of micro-lecture are short video time, small teaching capacity, fine topic selection and strong interactive application. The disadvantage of micro-lecture is that it is not suitable for practical and operational courses, and it is not conducive to students' systematic study. Micro-lecture is often combined with other models to make up its shortcomings [7]. Flipped class model or inverted classroom is a new model based on multimedia technology. Its teaching activities mainly include the following aspects: Teachers provide teaching video before class, and students watch the video and finish the task. In class teachers motivate students to discuss in groups and finally give inductive feedback [8]. According to the pyramid theory, flipped class model has realized the combination of reading, listening, seeing, speaking and doing. It is an efficient way to learn. Compared with the traditional teaching mode, flipped class model has the advantages of respecting individuality, teaching students according to their aptitude, problem-oriented, effective interaction, expanding capacity and transcending time and space. But flipped class is no substitute for traditional class. The quality of teaching video, the consciousness of students watching video and the initiative of teachers will have a direct impact on the implementation of classroom teaching mode.

4 The Feasibility Analysis of the Application of Micro-lecture, MOOC and Flipped Class Model in Mathematics Courses of Computer

It is beneficial to realize the goal of mathematics courses of computer. The mathematics curriculum objectives of computer majors in local colleges and universities can be divided into three aspects: knowledge objective, skill objective and emotion objective [9]. The knowledge objective is to understand and master the mathematics curriculum system knowledge. The skills objective is to master the basic thinking and method of modern mathematics, to exercise professional skills. The emotional objective is to improve students' mathematics core quality, to cultivate students' mathematics innovative thinking. MOOC can provide abundant course resources and platforms for the study of mathematics system knowledge. Micro-lecture has a strong pertinence. Flipped class provides the students with more space for extensive activity. MOOC, micro-lecture, and flipped class provide a convenient condition for the realization of the course objective.

It is Beneficial to improve the Ability and Quality of Teachers and Students in Local College and University. The role of teachers has changed. Teachers are changed from lecturers of theoretical knowledge to participants in teaching activities, designers of teaching content, and builders of teaching resources. These play an important role in guiding teachers to change their teaching ideas, renew their knowledge structure, improve their teaching level, strengthen their professional development and improve their information literacy. MOOC, micro-lecture and flipped class meet the needs of the differential development of college students. They can learn at their own pace or learn from other resources through online platforms. These play an important role in promoting students' ability of cooperation and exploration, self-regulation, logical reasoning and innovation.

It is Beneficial to Improve leaching Quality and Learning Efficiency. Big Data brings a lot of convenience to the use of MOOC resources, the making of micro courses, and the design of flipped courses. Through the online-offline hybrid teaching mode, teachers use big data technology to integrate the ideas, theories, development, application, culture, ideological and political elements and other information of computer mathematics courses. Thus they grasp the teaching information in time, to improve the teaching content and improve the teaching quality. Students use big data technology to learn more efficiently, to feedback accurately and to review more pertinently. They stimulate interest in study, innovate ways of the study and improve efficiency in study.

5   Construction and Practice of Mixed Teaching Mode

Mathematics core literacy is personal key ability and thinking quality which has the basic characteristics of mathematics and meets the needs of individual life-long development and social development. It includes six elements: mathematical abstraction, logical reasoning, mathematical modeling, visual imagination, mathematical operation and data analysis [10]. Mathematical creative thinking refers to the reorganization of mathematical knowledge and experience in mathematics learning under the strong sense of innovation, and the production of novel and socially valuable mathematical thinking results [11]. According to the abroad and domestic research, combining with the teaching and exploration for many years, we put forward a hybrid model. The online and offline mixed instruction model is shown in Fig. 1.

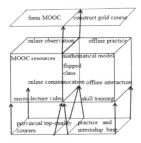

Fig. 1.   Three-dimensional mixed-modes of on-line and off-line

The above mixed model is a mixture of online teaching and offline coaching, and of online observation and offline practice, and of online learning and offline training. The mechanism of the mixed model is that the on-line flipped teaching is integrated with the off-line practical activities by taking the flipped class as the center, making use of micro-lecture scripts and relying on MOOC resources. This mode extends the flipped teaching content to the practice platform, and the practice training in turn supports the flipped teaching. Through the integration of online and offline activities, the mixed process of teaching and practical materials will be sorted out to form MOOC. Finally, MOOC is constantly improved through teaching practice, and the national first-class curriculum is gradually realized. The above mixed mode encourages students to engage in personalized learning. It facilitates the cultivation of innovative thinking in mathematics, the improvement of mathematics core literacy, and the training of computer professional skills. However, according to the different roles of teachers and students in teaching, these activities focus on different aspects. Specific activities are implemented as shown in the table below.

Probability and statistics courses cover a wide range of subjects. Any problem involving data collection, collation, analysis, visualization and interpretation is a stage where probability statistics can play its part [12]. It can be seen that probability statistics plays an important role in computer courses. And the development and design of the computer software system can turn the mathematical model of probability and statistics into a feasible algorithm and realize it. Combined with the professional background of computer, the classical probability (chapter 1 Sect. 3 of Computer class probability statistics curriculum) can be implemented according to the above model for teaching.

Analysis of the academic situation: Students have already learned the concept of classical probability and basic calculations in secondary school. The students have different development goals such as postgraduate entrance examination, civil servant entrance examination and computer enterprises. There are two types of learning methods: Major and Minor.

Setting goals: The first goal is to carry out the fundamental ask of "strengthen moral education and cultivate people". The goal of knowledge is to grasp the concepts, properties and calculations of classical probability type. The goal of skills is to develop professional skills in computer science, and to develop an awareness of and interest in the practical role of the software industry. The emotional goal is to improve the humanistic quality and patriotism by mining the ideological and political elements and carrying out the hierarchical teaching.

Integrating resources: The main resources are probability theory and Mathematics Statistics MOOC of Chinese University, the probability theory course (provincial top-quality offline open course), and probability theory micro-lecture of I-course platform.

Before-class design: The teaching team uses big data technology to create modular videos based on curriculum standards. Students learn MOOC and micro-lecture independently, and choose different modules of video for learning and self-testing. Teachers and students interact online through QQ or WeChat, message and telephone. The video mainly has four kinds of modules, such as major, revision, postgraduate examination, practice. The common parts of these four modules include five aspects: The first is to introduce Bao-lu XU (a pioneer of multivariate statistical analysis in China) to carry out patriotic education. The second is to introduce the French mathematical genius Pascal's "point number problem" and the famous French mathematician Laplace's "analytic probability theory". The main goal is to arouse the student understanding on the probability theory development history, to enhance the student's humanities accomplishment, and to stimulate student's study interest and the random thought consciousness. The third is to compare the similarities and differences of probability and frequency. Students can understand the Unity of opposites of contingency and necessity and develop their dialectical materialism thinking. The fourth is to analyze the conditions and calculation methods of classical probability type according to the classical definition of probability. Students' logical reasoning and mathematical abstraction of the mathematical core literacy is cultivated. The fifth is to enumerate the basic events by means of illustration and so on. The students' visual imagination of the mathematical core literacy is cultivated. The differences between the four modules are as follows: Re-repair module adds classical probability type of theoretical analysis. Postgraduate entrance examination module adds examination questions about classical probability type in recent years. The practice module adds the application case of the classical probability type of computer specialty.

In-class design: The implementation of classroom teaching is divided into three stages. In the first stage, discussion-typed micro-lecture is manly applied. How is the combination formula derived? Is the drawing of lots unrelated to the order? How to apply binomial distribution and hyper-geometric distribution to solve probability? How to use computational programs to optimize the construction of classical probability models? The above-mentioned problems strengthen the visual expression of concept and theory through the use of mathematical software such as drawing software and data design theory. This phase of teaching encourages students to actively participate in research and carry out random experiments. Its function was illustrated from the following aspects: Students can understand the random thoughts and grasp the meaning of random in their personal experience. Students internalize their knowledge in independent thinking and teacher-student interaction, and develop their mathematical operations and logical reasoning of mathematical core literacy. Students learn from each other in collaborative discussions, and improve their analytical and problem-solving skills. Students stimulate their motivation of learning by analyzing the lottery cases. In the second stage, demonstration-typed mini-lecture is applied to demonstrate the results. Teachers use computer technology to explain the textbook examples. Students use PPT to show the results of group discussions. This phase of teaching encourages students to improve logical thinking ability and comprehensive skills. In the third stage, feedback-typed micro-lecture is applied to carry on the summary comment. Teachers make a comprehensive assessment of students' performance before and during class, and sum up the application of the classical probability model and the random mathematics thought. Students consolidate their understanding of the content through the class practice of concrete examples.

After-class design: Students complete homework, self-assessment and mutual assessment independently. They learn the operation and use of MATLAB software through after-class extended micro-lecture, and use the software for the same-day-of-birthday problem of classical probability mathematical modeling. Such after-class design creates a space for students to develop their random thinking. Students improve their mathematical modeling of the mathematical core literacy and professional computer literacy. Teams of teachers conduct after-class tutoring and use big data to obtain student formative assessments. They gradually build their own MOOC by summarizing the flipped classroom activities, improving the teaching design, creating SPOC platform, uploading the network and so on. These have laid a solid foundation for the creation of the gold course.

Internships and practices: The activities of internships and practice are mainly carried out by online observation and communication and offline practical exercise. Students apply the classical conceptual knowledge and computer skills acquired in class to practice. Students can use MATLAB software to solve the model, and can use R-soft ware to simulate the Buffon injection experiment. Through these practices, students further consolidate the knowledge of classical probability, and improve their overall quality and professional skills.

6   Summary and Future Work

Under the background of big data, according to MOOC resources, micro-lecture script and flipped class are used to integrate the network teaching, classroom teaching and practical teaching of computer mathematics courses. The online and offline three-dimensional mixed teaching mode is constructed. This mode is a new exploration and attempt to the traditional teaching mode. It has realized the effective integration of the use of modern educational technology, the construction of mathematics courses and the cultivation of computer professional ability. This mode realizes the student-centered teaching idea and promotes the common development of teachers and students. Compared with the traditional teaching mode, students' learning enthusiasm and learning efficiency are significantly strengthened, and teachers' teaching skills are enhanced. Curriculum construction has also been vigorously promoted. One national innovation training project, one provincial quality course and one provincial teaching demonstration course have been approved. Although this model has achieved some results, the realization of this model needs the hard work of the teacher team, the active cooperation of students, parents, teachers and so on. Many problems are to be further explored and studied.

Acknowledgements. This work was supported in part by the Quality Engineering Project of Anhui Province under Grant 2018jyxm0507, in part by the Quality Engineering Project of Fuyang Normal University under Grant 2017JCJY01. And we would like to thank the reviewers for their beneficial comments and suggestions, which improves the paper.

References

1. Zhang, Y.C.: Study on Big Data to Influence the Development of Higher Education. Chongqing University (2016)
2. Fei, S.J., Ji, Y.X., Wang, L.: Teaching reform and practice of mathematical statistics teaching in information and computing science major in big data times. J. Sci. Teach. Coll. Univ. 3(10), 72–75 (2017)
3. Gad-El-Hak, M.: Educating rita: In: The Time of MOOC. Letters and Comments (2012)
4. Sun, N., Zhang, Y.H.: The operating mode and teaching interaction mechanism of MOOC. High. Educ. Dev. Eval. 31(5), 81–88 (2015)
5. Hashim, H., Ali, P.: Readiness of PSMZA lecturer on the use of Mooc platform in learning and teaching. J. Konseling dan Pendidikan 6(3), 149–159 (2018)
6. McCoy, J., Brasfield, D., Milkman, M.: The impact of macro/micro course sequencing. J. Educ. Bus. 66(4), 223–227 (1991)
7. Hu, T.S.: "Micro-lecture": new development trend of regional educational information resources. e-Educ. Res. 10, 61–65 (2011)
8. Bergmann, J., Sams, A.: Flipping the Classroom. Tech & Learning, San Bruno (2012)
9. Yang, Q.H., Lan, Y.Z., Yuan, H.Q.: Reserach on Reform and Advancement of Computer Mathematics Curriculum. Educ. Modernization 17, 51–52 (2017)
10. Lv, S.H., Wu, Z.Y.: The connotation and systems construction on mathematics core literacy. Curriculum Teach. Mater. Method 37(9), 12–17 (2017)
11. Fatah, A., Suryadi, D., Sabandar, J., Turmudi: Open-ended approach: an effort in cultivating students mathematical creative thinking ability and self-esteem in mathematics. J. Math. Educ. 7(1), 103–109 (1975)
12. Heyting, A.: Intuitionism: An Introduction, 3rd edn. North-Holland Publishing Company, North-Holland (1971)

Optimization of Food System Based on Neural Network and Grey Prediction

Zixuan Wang, Chenhao Ni, Yinan Xu, and Linzhong Zhang[✉]

Anhui Agricultural University, Hefei 230061, Anhui, China
zhanglinzhong@ahau.edu.cn

Abstract. In order to optimize the global food system, the economy and resources can develop for a long time on the basis of solving people's food security problems. This paper uses the TOPSIS comprehensive evaluation method, establishes a neural network training model and combines grey predictions to analyze the impact of changing the priority of the food system on countries with different backgrounds. The study of this issue can provide guiding opinions for the improvement of the food system, and is of positive significance to the global economic and ecological development.

Keywords: TOPSIS rating · BP neural network · Grey prediction

1 Introduction

Food security means that people can get enough safe and nutritious food to ensure normal growth and development and a positive and healthy life state [1]. However, the global food system is still not in a relatively stable state, and food insecurity in many countries and regions will cause a series of social problems [2].To achieve food security and sustainable development [3], it is necessary to optimize the food system of countries under different production systems and social backgrounds [4]. In this paper, by changing the priority levels of different influencing factors in the food system, the fairness of food distribution and the sustainability of resources and environment can be achieved as far as possible. The overall framework of this paper is shown in the Fig. 1 below.

2 Evaluation of Current Food System

In order to enable the four influencing factors of efficiency, profitability, equity and sustainability to be quantified by specific indicators, to achieve the purpose of evaluating the food system from different focuses and evaluating the relative advantages and disadvantages of the following countries' food systems. At the same time, considering the reasonable selection and availability of data, this paper uses fertilizer consumption, rural population, and mechanical efficiency as evaluation indicators for efficiency; agricultural value added is used to evaluate profitability; per capita GDP and unemployment are used as fairness evaluations Indicators: The total consumption of crude oil, raw coal, natural gas, hydropower, and energy is used as an evaluation indicator for

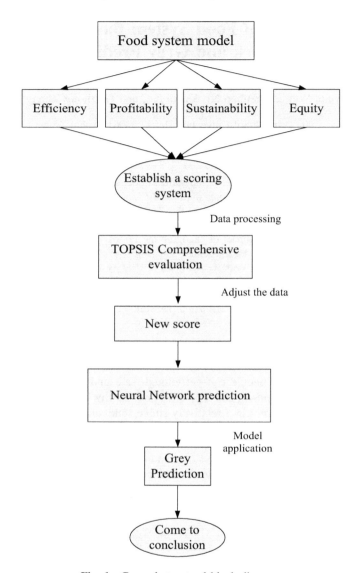

Fig. 1. General structural block diagram

environmental sustainability. Make full use of the original data and use the TOPSIS method for multi-objective decision analysis [5].

Suppose there are n objects to be evaluated, and now select m evaluation indicators to evaluate them. That is, the number of the above-mentioned countries is n, and the number of indicators selected from agricultural production, resources and environment, and the following standardized matrix can be obtained:

$$Z = \begin{bmatrix} z_{11} & z_{12} & \cdots & z_{1m} \\ z_{21} & z_{22} & \cdots & z_{2m} \\ \vdots & \vdots & \ddots & \vdots \\ z_{n1} & z_{n2} & \cdots & z_{nm} \end{bmatrix} \tag{1}$$

Defining the maximum and the minimum:

$$Z^+ = \left(Z_1^+, Z_2^+, \cdots, Z_m^+\right)$$
$$= \left(\max\{z_{11}, \cdots, z_{n1}\}, \max\{z_{12}, \cdots, z_{n2}\}, \cdots, \max\{z_{1m}, \cdots, z_{mm}\}\right) \tag{2}$$

$$Z^- = \left(Z_1^-, Z_2^-, \cdots, Z_m^-\right)$$
$$= \left(\min\{z_{11}, \cdots, z_{n1}\}, \min\{z_{12}, \cdots, z_{n2}\}, \cdots, \min\{z_{1m}, \cdots, z_{mm}\}\right) \tag{3}$$

Using MATLAB software to calculate, the results obtained after scoring by countries using the TOPSIS method are shown in Fig. 2.

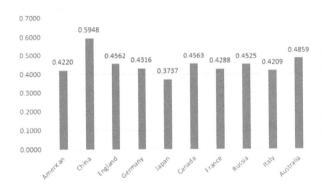

Fig. 2. First TPOSIS scoring results of countries

From the data in the figure, it is not difficult to conclude that the scores of China and Japan are quite different from those of other countries. This is caused by the different geographical environments and agricultural backgrounds of the two countries. As a large agricultural country, China has a relatively high amount of chemical fertilizers and a high proportion of the rural population, so its scores are relatively high. On the contrary, Japan is surrounded by the sea, agricultural production is not prosperous, and the demand for chemical fertilizers and agricultural equipment is low, so the results obtained are low. Differences in environmental backgrounds may make the scoring results have certain peculiarities, but overall, it does not affect the accuracy and rationality of the scoring.

Then, in order to improve the resources and environment, seek long-term sustainable development. We need to change the importance of the current global food system to different aspects, and give priority to fairness and sustainability [6]. As shown in Fig. 3, the TOPSIS score after changing the priority order, we assume that due to government intervention and policy changes, the unemployment rate and the environment are reduced to 80% of the original, and the per capita GDP is increased to 120% of the original. ten. Here, the United Kingdom and Australia are selected for index optimization. Compared with Fig. 1, it is obvious that the scores of both countries have improved, which shows that their poverty population has indeed been reduced.

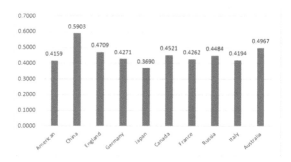

Fig. 3. TOPSIS score after changing priorities

3   Training Model Based on BP Neural Network

In this article, we want to use neural networks to establish how the proportion of poor people in the same country changes as its different indicators change. In order to predict how the proportion of poor people in different types of countries will change after fairness and sustainability are optimized, and finally get the impact that the optimized indicators can have on the global food system.

3.1   Model Establishment

BP neural network, shown in Fig. 4, is a multi-layer training network of information back propagation algorithm. Through back propagation, the weights and thresholds of the network are constantly adjusted, so that the error square sum of the entire network is continuously reduced [7].

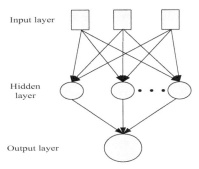

Fig. 4. Topological structure of BP neural network model

Generally speaking, the error will decrease after more training stages, but the error of the validation data set may start to increase as the network starts to overfit the training data. In the default setting, after the MSE of the validation data set is increased six times in a row, the training stops, and the best model corresponds to the smallest MSE. During the training process, the data will be divided into training set, validation set.

3.2   Model Solving

Bringing the indicators of different countries and the proportion of the corresponding poor population into the training, the results can be seen that both the mean square error and the goodness of fit have achieved good results, so we can choose this result to solve the subsequent ones. We chose to manually optimize the indicators of fairness and sustainability in different countries in the next five years. And the results for the next five years are shown in Table 1 and Table 2.

Table 1. The proportion of poor population in United States

Year	Real value/%	Predictive value/%
2012	0.96236	0.9108
2013	0.97839	0.9479
2014	0.99442	0.9520
2015	1.01045	0.9459
2016	1.02648	0.9534

Table 2. The proportion of poor population in China

Year	Real value/%	Predictive value/%
2017	0.9839	4.1
2018	0.9918	3.1
2019	0.9994	2.2
2020	1.0068	1.6
2021	1.0141	1.1

4   Forecast of the Realization Time of the Optimization Goal

Grey prediction refers to the use of GM (1,1) model to evaluate and predict the development and change rules of system behavior characteristics [8]. With this model, we can make full use of the known information of the data to obtain a discrete model that can analyze segment by segment over time, and objectively analyze a period of short-term changes in the future, so as to achieve the purpose of prediction and planning.

4.1   Modeling Establishment

After neural network training, the relationship between changes in sustainability evaluation indicators and changes in the poor population is obtained. Assuming that the implementation of external policies will affect the indicator data of fairness and sustainable development, the grey forecast is used to obtain the adjusted poverty population ratio of about five years.

a. Generate sequence
   In the grey prediction model, we set the known reference data as:

$$x^{(1)} = \left(x^{(0)}(1), x^{(0)}(2), \ldots, x^{(0)}(n)\right) \tag{4}$$

is the initial non-negative data column. Accumulate it once to get a new generated data column $x^{(1)}$, As shown in the following formula

$$x^{(1)} = \left(x^{(1)}(1), x^{(1)}(2), \ldots, x^{(1)}(n)\right) \tag{5}$$

$$x^{(1)}(m) = \sum_{i=1}^{m} x^{(0)}(i), m = 1, 2, \ldots n \tag{6}$$

b. Establish grey differential equation and corresponding whitening differential equation
   $x^{(0)}(k) + \alpha z^{(1)}(k) = b$ is regarded as the basic form of GM (1,1). The albino differential equation is as follows.

$$\frac{dx^{(1)}}{dt} + ax^{(1)} = b \tag{7}$$

where b is the control variable, α means developing coefficient, that is the $x^{(0)}$ is regarded as the dependent variable, the $z^{(1)}$ sequence is regarded as the independent variable, $z^{(1)}(k)$ means albino background value, $x^{(0)}(k)$ is the grey derivative. And the grey differential equation is introduced.

c. The corresponding time response function is obtained by solving the whitening equation as

$$\hat{x}^{(1)}(k+1) = \left(x^{(1)}(1) - \frac{b}{a}\right)e^{(-ak)} + \frac{b}{a}, k = 1, 2, \ldots, \quad n \tag{8}$$

4.2   Method Selection and Data Obtained

Taking into account the different social backgrounds of developed and developing countries, our different policy adjustments may produce different results. Therefore, to ensure that the results obtained are reasonable and accurate, we choose China and the United States as representatives of different types of countries to optimize the fairness and sustainability indicators in a good direction. In order to verify whether the optimized food system has a positive effect, this paper fitted the change trend of poverty population in China and the United States in five years. By evaluating the change in the proportion of people living in poverty,the indirect results show that the priority of improving equity and sustainability has a positive impact on the development of food system. The changes in the proportion of the poor is shown in Fig. 5.

Based on the predicted values of the poverty population data in China and the United States in the two figures, the continuous decline in China's poverty population indicates that people's basic food and clothing problems have been solved, while the sustainability and food security problems have been improved. Although the proportion of the poor in the United States is still increasing year by year, the rate of increase has declined compared to before. Therefore, bringing the importance of fairness and sustainability ahead of time has a certain positive effect on both developed and developing countries. This is also instructive for what kind of changes the global food system should make appropriately.

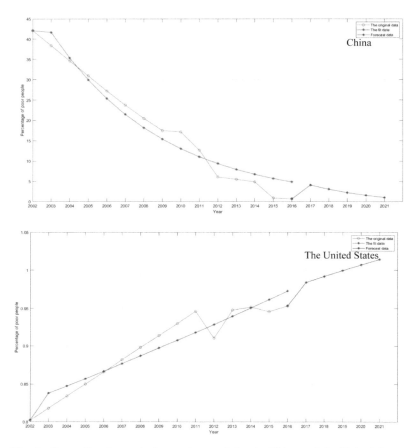

Fig. 5. Proportion forecast of poverty population in China and the United States

5   Conclusions and Future Work

In this paper, the various models established for the current situation of the global food system are progressive and deepened. Although the grain structure of some countries has certain fragility and peculiarities, in general, except for unexpected special circumstances, the model in this article can be universally applied to the evaluation of food systems in developed and developing countries. Understand, predict, make timely adjustments to different levels of demand or respond to emergencies in a timely and effective manner. On the basis of ensuring people's food, clothing and nutrition, the negative impact of agricultural production on global resources and environment can also be improved.

Acknowledgements. This work was supported by the Science Research Foundation of Anhui Provincial Colleges (KJ2018A0160). And we would like to thank the reviewers for their beneficial comments and suggestions, which improves the paper.

References

1. Campi, M., Dueñas, M., Fagiolo, G.: Specialization in food production affects global food security and food systems sustainability. World Dev. **141**, 105411 (2021)
2. Mora-Rivera, J., van Gameren, E.: The impact of remittances on food insecurity: evidence from Mexico. World Dev. **140**, 105349 (2020)
3. Nicholson, C.F., Stephens, E.C., Kopainsky, B.: Food security outcomes in agricultural systems models: case examples and priority information needs. Agric. Syst. **188**, 103030 (2021)
4. Loboguerrero, A.M., et al.: Perspective article: actions to reconfigure food systems. Global Food Secur. **26**, 100432 (2020)
5. Kumar, P.G., Meikandan, M., Sakthivadivel, D., Vigneswaran, V.S.: Selection of optimum glazing material for solar thermal applications uing TOPSIS methodology. Int. J. Ambient Energy **42**(3), 274–278 (2021)
6. Cadillo-Benalcazar, J.J., Renner, A., Giampietro, M.: A multiscale integrated analysis of the factors characterizing the sustainability of food systems in Europe. J. Environ. Manage. **271**, 110944 (2020)
7. Roy, N.B., Bhattacharya, K.: Application of Signal Processing Tools and Neural Network in Diagnosis of Power System Faults. CRC Press, Boca Raton (2021)
8. Zeng, B., Luo, C.M., Liu, S.F., Li, C.: A novel multi-variable grey forecasting model and its application in forecasting the amount of motor vehicles in Beijing. Comput. Ind. Eng. **101**, 479–489 (2016)